APPLIED DYNAMICS IN ENGINEERING

By

MICHAEL B. SPEKTOR

Industrial Press, Inc.

Industrial Press, Inc.
32 Haviland Street, Suite 3
South Norwalk, Connecticut 06854
Tel: 203-956-5593, Toll-Free: 888-528-7852
E-mail: info@industrialpress.com

Library of Congress Cataloging-in-Publication Data

Spektor, Michael B., 1933– author.
 Applied dynamics in engineering / Michael B. Spektor.
    pages cm
 Includes index.
 ISBN: 978-0-8311-3522-5 (softcover)
 Summary: "This book is a reference and guide for professionals and students of applied dynamics. It features the solutions to 96 linear differential equations of motion that describe common problems in mechanical engineering. These equations are applicable to electrical and other engineering fields. The book also includes a guiding table that enables users to easily find the descriptions of and solutions to problems of particular interest."
 1. Dynamics. I. Title.
 TA352.S64 2015
 620.1'04—dc23   2015027328

ISBN print: 978-0-8311-3522-5
ISBN ePDF: 978-0-8311-9347-8
ISBN ePUB: 978-0-8311-9348-5
ISBN eMOBI: 978-0-8311-9349-2

Sponsoring Editor: Jim Dodd
Developmental Editor: Robert Weinstein
Interior Text and Cover Designer: Janet Romano-Murray

books.industrialpress.com
ebooks.industrialpress.com

10 9 8 7 6 5 4 3 2 1

To My Family

TABLE OF CONTENTS

Chapter 7: Stiffness   177

Chapter 8: Stiffness and Friction   217

Chapter 9: Stiffness and Constant Resistance   259

Chapter 10: Stiffness, Constant Resistance, and Friction   299

Chapter 11: Damping   343

Chapter 12: Damping and Friction   375

Chapter 13: Damping and Constant Resistance   **409**

Chapter 14: Damping, Constant Resistance, and Friction   **443**

Chapter 15: Damping and Stiffness   **477**

Chapter 16: Damping, Stiffness, and Friction — 529

Chapter 17: Damping, Stiffness, and Constant Resistance — 583

Chapter 18: Damping, Stiffness, Constant Resistance, and Friction — 633

INTRODUCTION

Purposeful improvement of existing mechanical engineering systems, development of new systems, and appropriate control of their parameters all involve analytical investigations. In particular, we want to analyze the dynamics of these systems' working processes or their elements. In one of the first steps, we compose the corresponding differential equations of motion for each system or element. The analyses of solutions for these equations provide the basis for controlling the system parameters in order to achieve the required results.

The solution of a differential equation of motion represents an equation that describes the displacement of the system as a function of running time. The equation itself reflects the system's law of motion and is the main basic parameter of motion. The first and the second derivatives of the displacement with respect to time characterize respectively the system's velocity and acceleration. The displacement, the velocity, and the acceleration are the basic parameters of motion. The appropriate analysis of equations describing these parameters can reveal ways that lead to the desired improvement or development of the system.

Composing differential equations of motion that analytically describe the working processes of real-life systems often presents certain difficulties for practicing engineers, senior, and graduate students. Solving these equations can also be a very challenging process. This book is intended to make it easier to overcome these difficulties and challenges.

Based on an appropriate analysis, I concluded that the real-life common engineering problems associated with single-degree-of-freedom mechanical systems could be described by 96 differential equations of motion. In reality, the dynamics of motion for these mechanical systems is often associated with parameters that, to a certain degree, exhibit non-linear characteristics. Currently there are no methodologies for solving non-linear differential equations of motion. The existing solutions of some non-linear differential equations have an extremely limited application for mechanical engineering problems.

However, these 96 common engineering problems could be described, with a certain level of accuracy, by linear differential equations of motion. Certainly, the level of non-linearity of the parameters should be estimated in each particular case and appropriate decisions should be made during the investigation. Very often the linear differential equations of motion describe the common engineering problems in dynamics with an acceptable level of accuracy. We can compose and solve all 96 linear differential equations of motion that actually describe the possible common problems in the engineering dynamics.

This book contains all of these 96 linear differential equations of motion, their solutions, and, for the most part, the analyses of the solutions. All this is described in the corresponding 96 sections of the book. The book is organized in such a way that it is very easy to find the appropriate section containing the description of a certain problem of interest. Each section represents a stand-alone description—there is no need to look for additional references in other places of the book. The descriptions of the problems indicate the area of the possible real-life working processes of the engineering systems, emphasizing the characteristics of the loading factors applied to the systems. Explanations related to assembling the differential equation of motion and to the structure of the initial conditions of motion of the system are presented in each section. The comprehensive step-by-step methodology of solving the differential equations of motion for all possible initial conditions of motion is presented in the descriptions of all related problems. All sections contain the equations describing the displacement as functions of time. The equations for the three basic parameters of motion and the relevant analysis of these equations are presented in the majority of the sections.

In this book, the Laplace Transform Pairs represent the basis of the methodology for solving differential equations of motion.

You are not required to become familiar with the methodology of solving the differential equations of motion in order to use the solutions provided in this book. Comprehending the methodology can be achieved later when time allows. Hence, the part of each section that discusses the Laplace Transform methodology for solving the differential equation of motion may be skipped. It is possible to proceed directly to the equations describing the parameters of motion and to start their analysis toward achieving the goal of the investigation.

In order to locate the appropriate section that contains the solution of a certain engineering problem in dynamics, you must first formulate the problem in corresponding terms that are appropriate in dynamics. In other words, characteristics of the loading factors (forces or moments) applied to the known mass of the engineering system should be determined. The components of the differential equation of motion represent just loading factors. This book shows there are 96 combinations of the loading factors that could be included in the 96 differential equations of motion of common engineering systems.

The investigator should determine the loading factors that resist the motion (the resisting or reactive loading factors) and factors that cause the motion (the active or external loading factors). The loading factors represent forces in the rectilinear motion, while in the rotational motion they are moments. However, the differential equations that describe rectilinear motion are similar to those describing the rotational motion. Thus, we can discuss all issues related to solving engineering problem in dynamics using just forces or just moments. In this book, these forces are considered the loading factors.

Chapter 1 presents the comprehensive description of the characteristics of the resisting forces and the active forces. In this book, as in many other related sources, the resisting forces are placed in the left side of the differential equation of motion, while the active forces are in the right side of this equation. This placement of the loading factors provides the basis for the structure for Guiding Table 2.1 (Chapter 2), which helps readers locate the section number that corresponds to the particular engineering system. The characteristics of all existing resisting forces in dynamics are shown in the intersections of Row 2 with Columns A through E in the left side of Guiding Table 2.1.

The 16 possible combinations of resisting forces that could be applied to an engineering system are shown in the left side of Guiding Table 2.1, in Rows 3 through 18. For instance, in Row 3, just one resisting force is marked by the plus sign (+); it is the force of inertia. Therefore, this row is related to systems subjected to the force of inertia as the resisting force. In Row 13, three resisting forces are marked by the plus sign. Therefore, this row is associated with the systems subjected to the force of inertia, the damping force, and the constant resisting force as the resisting forces. In Row 18, all five resisting forces are marked by the sign plus; therefore, this row is associated with systems subjected to the force of inertia, the damping force, the stiffness force, the constant resisting force, and the friction force as the resisting forces. Identifying the row in Guiding Table 2.1 where the resisting forces are marked by the plus signs corresponding to your engineering system is the first step.

The second step consists of identifying the column related to the active force or forces applied to the system. The six columns numbered 1 through 6 on the right side of Guiding Table 2.1 are associated with these active forces. The intersection of the corresponding row and column contains the number of the section that describes the differential equations of motion and their solutions. The analyses of these solutions are based on mathematical procedures that are very familiar to graduates from engineering programs. The numeric analysis of the solutions may involve usage of appropriate computer software.

It should be mentioned that Column 1 in Guiding Table 2.1 is associated with systems in which the active force equals zero.

This book discusses the structures of the differential equations of motion and explains how to compose the equations for actual engineering systems. In addition, this book offers a straightforward, universal methodology for solving linear differential equations of motion that describe the common engineering problems in dynamics. The Laplace Transform methodology allows us to convert the differential equation of motion into an algebraic equation of motion, the processing of which involves basic algebra. Actually, there is no need to memorize the Laplace Transform fundamentals in order to use this methodology. The engineers may immediately begin to use all 96 solutions of the common engineering problems without studying this methodology. In order to use this book for getting the needed solutions, however, you should become familiar with the two first chapters, even without going into the details of the mathematics.

Chapter 1 analyzes the structure of the differential equation of motion and the components that make up the equation. This analysis allows us to identify the characteristics of all possible components that could be included in the differential equation of motion. The physical nature of these components is also explained in Chapter 1, which contains the Laplace Transform Pairs Table 1.1. This table is helpful for solving all 96 differential equations of motion associated with the common engineering problems in dynamics. For the most part, Table 1.1 represents a compilation based on numerous published sources. Using the method of decomposition, I developed several appropriate expressions that could be not found in published sources. These expressions are also included in this table.

Chapter 2 presents a few examples of composing differential equations of motion for actual engineering systems. It demonstrates the methodology of solving these equations by using the Laplace Transform Pairs. This chapter also shows the ways to analyze the basic parameters of motion. A substantial part of Chapter 2 is devoted to the applicability of the solution of the general differential equation of motion for solving differential equations of motion of particular engineering systems. This analysis concludes that the obtained solution cannot be used for solving particular problems in dynamics. Instead, this analysis indicates that it is necessary to compose the differential equations of motion and solve them for each particular engineering problem in dynamics.

Chapters 3 through 18 describe the solutions of the 96 common engineering problems in dynamics. The chapter titles reflect the resisting loading factors applied to the systems. Each chapter consists of six sections, the names of which reflect the active forces applied to the systems.

Chapter 19 shows that the engineering problems associated with two-dimensional motion can be solved using Guiding Table 2.1 and the corresponding sections. The two-dimensional motion can be described by a system of two simultaneous differential equations of rectilinear motion in two mutually perpendicular directions. Thus, the solution of a problem associated with two-dimensional motion is built on the corresponding two sections, the numbers of which can be found in Guiding Table 2.1.

It is important to emphasize that this book can also be used to solve differential equations associated with electrical engineering, electronics engineering, and other engineering fields. I hope that this book will be helpful to the engineering community for solving problems in dynamics.

Michael Spektor
October 2015

PRINCIPLES OF APPLIED DYNAMICS

Parameters of motion are important characteristics of the working process of mechanical engineering systems. The methods of applied dynamics help us analyze these parameters so that we can purposefully control these processes. One of the basic parameters — the law of motion of a mechanical system — represents the solution of a differential equation of motion, which in its turn expresses the dependence of the mechanical system's displacement as a function of time. The differential equation of motion opens the ways for us to understand the mechanical system's working process; it allows us to achieve the required level of the system's performance.

Different mechanical systems have different criteria for evaluating their performance. For instance, velocity and acceleration, as well as braking distance, are criteria of a transportation system's performance. However, an elevator's performance is characterized by its velocity and lifting capacity. Productivity, efficiency, energy consumption, and many other characteristics are also among the performance criteria for mechanical systems. Performance is generally

1

evaluated by a combination of criteria. Achievement of a mechanical system's required performance is always its improvement and development goal. Effective analysis of a system's appropriate laws of motion allows us to accomplish this goal.

Analyses of the dynamics of motion play an important role in purposefully controlling both the performance and sophistication of mechanical systems. These analyses include three steps: 1) composing a differential equation of motion of a mechanical engineering system and determining the initial conditions of motion; 2) solving this differential equation for these initial conditions; and 3) analyzing this solution to determine the basic parameters of motion, reveal the roles of the system parameters, and evaluate the influence of these parameters on performance. The description of these three steps is presented below.

1.1 Mathematical Approach to Composing the General Differential Equation of Motion

The process of motion can be characterized by acceleration, constant velocity, and deceleration (braking).

Acceleration and deceleration have identical analytical expressions and both describe the rate of velocity change as a function of time. Both are also functions of time. So, in vibratory processes, the same equation — depending on time — describes the acceleration or deceleration of the system. Actually, the term acceleration implies deceleration; therefore, the process of motion basically consists of two phases: acceleration and motion at constant velocity (uniform motion).

The dynamics of motion focuses mostly on the phase of acceleration. The analysis of this phase reveals the roles of the system parameters, the interaction between the parameters, and the mutual influence of the parameters. It also shows the ways to accomplish the desired control of the parameters. During the phase of uniform motion, the acceleration equals zero. The motion of the system is described by a linear relationship between the displacement and time. This relationship represents a formula including the displacement, the velocity, and the time; this formula allows us to determine one parameter when the other two are given.

There are no readily available formulas that describe the dependence between the displacement and the running time during the acceleration process of mechanical systems. The expression that describes the displacement as a function of the running time for the process of acceleration is considered as the law of motion of the system; it can be obtained as the solution of a differential equation of the system's motion. Therefore, the initial stage of analyzing a system's dynamics is associated with composing the differential equation of motion that reflects the real-life characteristics of the system and the circumstances of motion.

According to Newton's Second Law, the process of acceleration is caused by loading factors. The factors that cause rectilinear motion are forces, whereas rotational motion is caused by moments. The basic parameters of motion are displacement, velocity, and acceleration. All these parameters are functions of time. Displacement as a function of time is the main parameter and represents the solution of the differential equation of motion. Velocity and acceleration are respectively the first and second derivatives of the displacement. Thus, in order to obtain the analytical expression for the displacement as a function of time, we need to solve the differential equation of motion for the particular mechanical system.

From a mathematical perspective, a differential equation of motion is a second order differential equation. Consequently, its structure is predetermined by certain principles of mathematics. Usually a general second order linear differential equation has the following structure. The left side of the equation represents a sum of the following components: the second derivative of the function, the first derivative of the function, the function, the argument, and the constant value, while the right side equals zero. (A constant value can actually be considered as a function of the argument or another involved parameter to a zero power.) Very often, a differential equation is presented as having a left side and a right side populated by certain parameters that have the same structures as the components described above.

The structure of these components is similar to the structure of the parameters of motion. The acceleration represents the second derivative of the displacement as a function of time. The velocity represents the first derivative of the displacement as a function of time. The displacement represents the function of time. The time represents the argument and the constant value can be considered as a function of time or another involved parameter to the zero power. This similarity lets us apply the second order differential equations to investigate the motion of mechanical systems; the goal is to reveal the analytical expressions of the system parameters.

By the definition, a linear differential equation of motion should have constant coefficients for its components. Acceleration, velocity, displacement, time, and the constant value have different physical characteristics and are measured by different units. In order to use these parameters as a sum in any equation, they must be expressed by the same physical units. By using appropriate constant coefficients as multipliers, we can convert these components into parameters that have the same units. In mechanical engineering, these units should represent loading factors (forces or moments). Hence, the differential equation of motion should be composed of components representing forces or moments. In order to compose a differential equation of motion of an actual mechanical system, we must determine for each particular case the loading factors (forces or moments) that are applied to the system. The left and right sides of a differential equation of motion could consist of one parameter or certain sums of these loading factors.

From the mathematical point of view, there are five types of components that can be included in the left and right sides of a second order differential equation. The right side of this equation may also be equal to zero. One of these components represents the second derivative of the function (usually it is the first component), and this component should be present in each second order differential equation. The left side of a second order differential equation could consist just of the second derivative, while the right side of this equation equals zero. In general, the second order differential equation may include in each side a component or any combination of the above-mentioned five types of components, whereas the left side of the equation should contain the second derivative.

Therefore, the differential equation of motion can consist of any combination of five different loading factors in each of its sides, while one of these factors (usually in the left side of the equation) should have the structure of the second derivative. As with any second order differential equation, four of these factors are variables and the fifth has a constant value. The characteristics of these loading factors are:

1) force or moment of inertia (depends on acceleration)
2) damping force or moment (depends on velocity)
3) stiffness force or moment (depends on displacement)
4) time-dependent force or moment
5) constant force or moment

It is acceptable to represent the loading factors in two groups, namely: factors that cause the motion (accelerate the motion) and factors that resist the motion (decelerate the motion). The loading factors that accelerate the motion are external or active loading factors, whereas the loading factors that decelerate the motion represent resisting or reactive loading factors. In the processes of motion, the reactive loading factors can act in the absence of the active loading factors. However, during the action of the active loading factors, at least one reactive loading factor is always present. This factor represents the force or moment of inertia of the mechanical system.

Suppose a body is moving by inertia on a horizontal frictional surface and is surrounded by air. The body will decelerate due to the resisting forces exerted by the air and the frictional surface. In this case, there are no active forces applied to this body. The body moves due to its kinetic energy, which decreases during the deceleration and becomes equal to zero when the body stops. As a reaction to the deceleration, this body exerts a force of inertia that in this case is directed opposite to the resisting forces. Its absolute value equals the sum of the absolute values of the resisting forces. The force or moment of inertia cannot accelerate the body. Therefore, it is not an active or external force/moment; it is a reactive force/moment by nature.

Let's consider the case where an external force is applied to the same body that is moving on a frictional surface and is subjected to the same air resistance. As in the previous case, the air and the frictional surface will exert reactive forces that will resist the motion. However, in this case, the body may accelerate if the active force exceeds these two resisting forces. As a reaction to the acceleration, the body exerts a reactive force representing its force of inertia; it plays the role of a resisting force. For this case, the sum of the force of inertia and the rest of the resisting forces equals the external active force and is directed opposite to the active force. If in this case the active force equals the sum of air and friction resisting forces, the body will move with a constant velocity and the force of inertia will be equal to zero.

Based on all these considerations, it is acceptable to include in the left side of the differential equation of motion all resisting or reactive loading factors, and in the right side all external or active loading factors. In order to describe the methodology of composing differential equations of motion for actual engineering problems, we will first assemble the general differential equation of motion that includes all possible loading factors. We will analyze each component of this equation from the point of view of real-life engineering problems in dynamics. This analysis allows us to clarify the sequence of steps needed to compose the differential equation of motion for actual systems. There is no difference in the structure of differential equations of rectilinear or rotational motion. The steps of composing the differential equations for both types of motion are the same and the components for both cases are similar. Thus, further descriptions in this book are based on forces that are associated with rectilinear motion, keeping in mind that all related methodologies and principles are applicable to rotational motion.

The vast majority of actual common problems associated with the dynamics of mechanical systems can be described with a certain level of approximation by linear differential equations of motion. This book focuses just on linear differential equations of motion of single degree-of-freedom mechanical systems. Even if one component of a differential equation of motion is non-linear, the differential equation is non-linear. As it is known, at the present time, there are no methodologies for solving non-linear differential equations of motion. The existing catalogs containing solutions for certain non-linear differential equations have a very limited applicability for engineering problems in dynamics.

The general differential equation of motion for a hypothetical mechanical system moving horizontally and consisting of all possible loading factors has the following shape:

$$m\frac{d^2x}{dt^2} + C\frac{dx}{dt} + Kx + A_0\sin(\omega_0 t + \lambda_0) + B\frac{t}{\tau} + P + F$$

$$= \alpha m\frac{d^2x}{dt^2} + C_1\frac{dx}{dt} + K_1 x + A\sin(\omega_1 t + \lambda) + Q(\rho + \frac{\mu t}{\tau}) + R \qquad \textbf{(1.1.1)}$$

where m is the mass of the mechanical system, x is the system's displacement that represents a function of time, and t is the running time (the argument). The rest of the parameter notations are explained below.

The initial conditions of motion for a general case are:

$$\text{for} \quad t = 0 \quad x = s_o; \quad \frac{dx}{dt} = v_0 \qquad \textbf{(1.1.2)}$$

where s_0 and v_0 are the initial displacement and initial velocity respectively.

As stated above, there are five types of loading factors that can be included in the left and right sides of a differential equation of motion. However, the left side of equation (1.1.1) has seven components, whereas the right side has six components. But we can see that the fourth and fifth components in the left and right sides of equation (1.1.1) are different versions of functions of time. Therefore, each of them represents the loading factor that depends on the same parameter and should be considered as belonging to the same type of factors. The sixth and seventh components in the left side of the equation are constant forces and also should be considered as one loading factor. So, there are actually just five types of loading factors in each side of equation (1.1.1). The analysis of these loading factors is presented below.

1.2   Analysis of the Resisting Forces

All resisting forces are reactive forces. The force of inertia represents the reaction of a rigid body (the mechanical system) to its acceleration/deceleration, whereas the rest of the resisting forces represent the reaction of the surrounding media to the interaction with this rigid body. The surrounding media implies fluid media (air, water, oil, etc.) and solid state media including the soil and engineering materials such as metals, polymers, wood, concrete, and other.

The interaction between the rigid body and the media represents all kind of deformations of the media, including forging, penetration, cutting, distortion, etc.

The left side of the differential equation of motion comprises one or a sum of resisting forces. The analysis of these forces follows the order in which they are positioned in equation (1.1.1).

1) The *force of inertia*, which depends on the acceleration of the mechanical system, equals the product of multiplying the mass m of the system by its acceleration $\frac{d^2x}{dt^2}$. Linear differential equations have constant coefficients. Therefore, the mass represents a constant coefficient at the acceleration. The force of inertia is present in each differential equation of motion. If $\frac{d^2x}{dt^2} = 0$, then the system is in uniform motion and the displacement of the system equals the product of multiplying the constant velocity of the system by the running time.

In the vast majority of actual engineering problems, the mass of the mechanical system represents a constant value. Consequently, the force of inertia is a linear function of the acceleration. However, in transportation systems powered by internal combustion engines, the masses of these systems decrease as a result of fuel consumption. The masses of movable systems can change due to snow, rain, evaporation, etc. If the mass is a variable value, the differential equation becomes non-linear and it is not considered in this book.

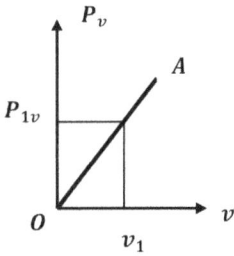
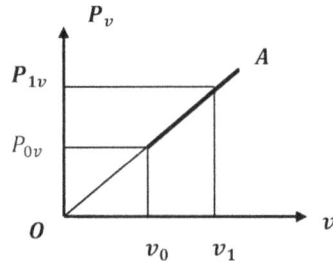

1.1a 1.1b

Figures 1.1a and 1.1b   Graphs of force-velocity relationships

2) The *damping force* depends on the velocity of the mechanical system. The damping force equals the product of multiplying the constant damping coefficient C by the velocity of the system. Damping forces are exerted by fluid media (liquids and gases) as a reaction of their interaction with a movable body. The graphs in Figure 1.1 illustrate the relationships between the damping force P_v and the velocity.

The straight line OA in Figures 1.1a and 1.1b represents the relationship between the damping force and the velocity. Figure 1.1a shows a case when the initial velocity equals zero, whereas in Figure 1.1b, the initial velocity equals v_0. According to these graphs, the damping coefficient C can be determined from the following expression:

$$C = \frac{P_{1v}}{v_1} \tag{1.2.1}$$

Because the motion of bodies occurs in fluid environments, the damping forces are always present. Damping forces are also exerted by special hydraulic links that are widely used in mechanical engineering systems. These links represent various shock absorbers intended to soften the action of vibratory and impact loading. In many actual cases, damping forces play an insignificant role in comparison to other forces and may be ignored. There are no readily available formulas for determining the values of damping coefficients that depend on the viscosity of the fluid, its temperature, geometric characteristics of the movable systems, and other factors. In each particular case, it is necessary to obtain related experimental data in order to determine the value of the damping coefficient.

Usually the experimental data should be processed graphically to express the relationship between the damping force and the velocity of the system, as shown in Figure 1.1. If the graph is linear, the damping coefficient is a constant value and the damping force represents a linear function of the velocity. If this graph is shaped as a curved line, the damping coefficient is changing its values from point to point on the graph. In this case, the damping force is a non-linear function of the velocity, and the differential equation of motion is non-linear. In many practical cases, the non-linearity is insignificant and the damping force may be considered as a linear function of the velocity. In case of strong non-linearity, you may want to apply the methodology of the piece-wise linear approximation presented in the author's book, *Solving Engineering Problems in Dynamics*, published by Industrial Press.

A dashpot represents the hydraulic link mentioned above. Its schematic image is used in physical models of movable systems to indicate the presence of a damping force that is exerted by the fluid media or shock absorber. In these cases, the dashpot symbolizes just the resisting damping force; it does not represent a physical part of the mechanical system, and, consequently, does not have any mass. Figure 1.2 shows a model of a system subjected to a damping force that is represented by a dashpot. The rigid body (1) moves along the x-axis and is rigidly connected to the piston (2) that is moving inside of the cylinder (3) which is securely attached to the non-movable support (4). The cylinder is filled with a fluid, the flow of which is restricted by calibrated orifices. The liquid flowing through these orifices exerts a resisting damping force applied to the rigid body. The characteristic of the dashpot is the damping coefficient C.

In some situations, the mechanical systems may be subjected to the action of two sources of damping resistance. For instance, a ship experiences a combined resistance caused by air and water. In this case, the air and water damping forces are acting in parallel.

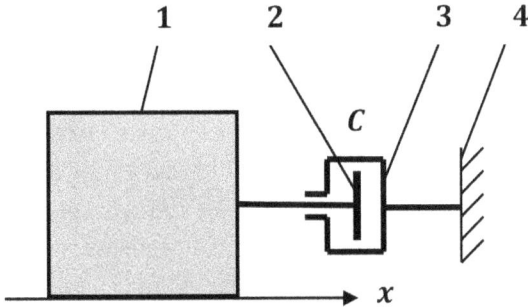

Figure 1.2 Model of a system with a dashpot

In some mechanical systems, the shock absorbers may act sequentially, although sometimes they simultaneously act in parallel and sequentially. If a mechanical system is subjected to the action of several damping forces, the resultant damping coefficient C equals the sum of each particular damping coefficient, regardless if the dashpots act in parallel or sequentially:

$$C = \sum_{1}^{n} C_i \qquad (1.2.2)$$

where C_i is a particular damping coefficient, n is the number of dashpots, and:

$$i = 1, 2, 3, ..., n$$

Some engineering materials, such as certain polymers, exhibit viscous properties during their interaction with rigid bodies. These materials belong to viscolastic and viscoelastoplastic media that exert damping resisting forces as a reaction to their deformation.

Differential equation of motion do not always contain damping forces because in some cases these forces are insignificant.

3) The *stiffness force* depends on the displacement and equals the product of multiplying the stiffness coefficient K by the system's displacement x. The reaction of an elastoplastic or viscoelastoplastic medium to its deformation represents a resisting force

that is often considered as the frontal resistance force applied to the forehead of the deformer. The vast majority of engineering materials are elastoplastic. Normally the first phase of the deformation of an engineering material exhibits mainly features of elasticity, whereas the next phase is associated with the plasticity features of the material. The interaction of the system with the media results in different types of deformation of the media such as penetration, cutting, etc. The measure of displacement of the system (deformer) is considered equivalent to measure of deformation of the media.

For many engineering materials, the deformation force on the phase of elasticity is characterized as a linear function of the deformation. Actually, this deformation force is the stiffness force, and the deformation, as stated above, represents the displacement of the system. The deformation force on the phase of plastic deformations usually has a constant value. However, some elastoplastic and viscoelastoplastic media are characterized by a phase of strain-strengthening during their plastic deformation process.

Figure 1.3 shows the force-displacement relationship on the phases of elastic deformation, plastic deformation with strain-strengthening, and plastic deformation. The broken line *OABC* illustrates the dependence of the force P_s on the displacement s of a system interacting with an elastoplastic or viscoelastoplastic medium that is characterized by strain-strengthening. By its nature, this force represents the frontal resistance force of these media to their deformation (penetration, cutting, etc.). The straight line *OA* is the graph reflecting the relationship between the force and displacement on the phase of elastic deformation. According to Figure 1.3, the stiffness coefficient *K* for the phase of elastic deformation is calculated from the following equation:

$$K = \frac{P_{1s}}{s_1} \qquad (1.2.3)$$

The straight line *AB* characterizes the relationship between the stiffness force and displacement on the phase of plastic

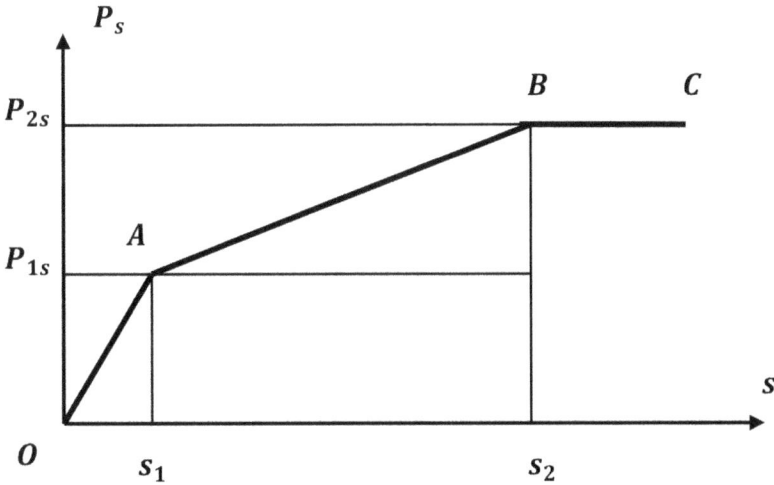

Figure 1.3 Graph of force-displacement relationship

deformations with strain-strengthening. Based on Figure 1.3, the stiffness coefficient K_{12} for this phase is determined from the following expression:

$$K_{12} = \frac{P_{2s} - P_{1s}}{s_2 - s_1} \tag{1.2.4}$$

The straight line BC in Figure 1.3 represents the graph that reflects the relationship between the force and displacement on the phase of plastic deformations. As this graph shows, the resisting force on this phase has a constant value.

During their deformation, some engineering materials exert resisting forces that are proportional to the velocity of the deformation. These materials possess the properties of viscosity and the above-mentioned resisting forces actually represent damping forces. The materials that possess viscosity properties represent the viscoelastic and viscoelastoplastic media. These media exert damping forces during all phases of their deformation. Thus, these media simultaneously exert damping forces and stiffness forces during the phase of elastic deformation. During the phase of plastic deformation with strain-strengthening, the viscoelastoplastic media exert damping forces, stiffness forces, and constant resisting forces at the same time. Finally, these media exert damping forces and constant resisting forces during the phase of plastic deformation. Depending on the shape of the deformer or the rigid body interacting with the viscoelastic, elastoplastic, or viscoelastoplastic media, these media can exert on each phase of deformation a friction force in addition to those resisting forces already mentioned. Usually this friction force has a constant value.

Specific elastic links that represent springs are used in some mechanical systems and very often work together with dashpots. There are formulas to calculate the stiffness coefficients (spring constants) of springs. Some elements of mechanical engineering systems can be considered as specific elastic links. All kinds of structural elements (such as beams, brackets, booms, and frames) and all kinds of machine elements (including shafts, axes, etc.) may represent specific elastic links. However, in order to determine the stiffness coefficient of a specific elastic link or above-mentioned media subjected to deformation, it is necessary to obtain appropriate experimental data processed in graphs that reflect the relationship between the force and the deformation (displacement). These graphs are similar to the graph shown in Figure 1.3.

If the relationship between the stiffness force and displacement is linear, the stiffness coefficient has a constant value and the differential equation of motion is linear. If there is a strong nonlinearity in the dependence between the stiffness force and the displacement, use the same approach described above for the damping force. In the physical models, the presence of a stiffness force is expressed by a schematic image of a spring, as shown in Figure 1.4. In this figure, the body (1) moves along the x axis and is subjected to the action of a resisting stiffness force exerted by the spring (2) having a stiffness coefficient K. The spring is connected to the body and to a non-movable support (3).

In reality, a mechanical system can consist of several springs that may act in parallel or sequentially, or in a combination of both. The resultant stiffness coefficient K_p of springs acting in parallel can be calculated based on the following formula:

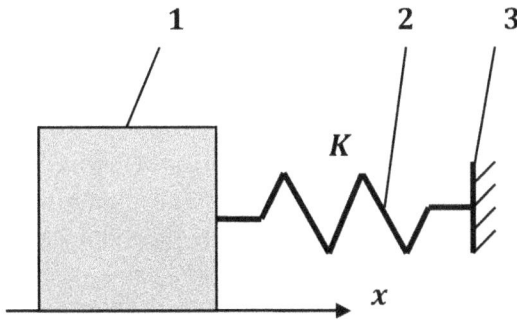

Figure 1.4 Model of a system with a spring

$$K_p = \sum_1^n K_i \qquad (1.2.5)$$

where K_i is the stiffness coefficient of a particular spring, n is the number of springs, and:

$$i = 1, 2, 3, \ldots, n$$

In cases when the springs act sequentially, the resultant stiffness coefficient K_s can be calculated using the following expression:

$$\frac{1}{K_s} = \sum_1^n \frac{1}{K_i} \qquad (1.2.6)$$

where the notations are similar to the previous case. In case of a combined action of springs acting in parallel and sequentially, the resultant stiffness coefficient K is determined from the following equation:

$$K = K_p + K_s \qquad (1.2.7)$$

Resisting stiffness forces appear in certain pneumatic systems during the process of air compression.

Stiffness forces are not present in each differential equation of motion.

4) The *resisting force representing a harmonic function depending on time* is the product of multiplying the amplitude of the

exciting force A_0 by the harmonic function having a frequency of vibration ω_0 while λ_0 is an angle that determines the magnitude and the direction of the exciting force at the moment when $t = 0$. Mathematically it is justifiable to have this force in the left side of the differential equation. However, in real engineering systems, it is very difficult to imagine the presence of a resisting (or reactive) harmonic force. Therefore, this force should not be included in the left side of the differential equation of motion.

5) The *time-dependent force* is a resisting force that also represents a function of time and equals the product of multiplying a certain constant force B by a linear function of time t, as t increases from zero to a limited interval of time τ that the time-dependent force is acting. This force is relevant from a mathematical perspective. However, in real life, it is extremely problematic to come up with this type of a time-dependent resisting (or reactive) force. For this reason, it should be not included in the left side of the differential equation of motion.

6) The *constant resisting force P* usually represents the reaction of elastoplastic or viscoelastoplastic media to the deformation; it is the frontal resistance force. In cases of upward motion, this force can also be associated with the weight of the system. In cases of interaction with some elastoplastic or viscoelastoplastic media, the constant resisting forces appear on the phase of plastic deformation (as well as plastic deformation with strain-strengthening). As seen in Figure 1.3, during the phase of plastic deformation with strain-strengthening, the constant resisting force equals P_{1s}. However, during the phase of plastic deformation (without strain-strengthening), this force equals P_{2s}. Figure 1.3 also indicates that, during the phase of plastic deformation with strain-strengthening, the frontal resistance force consists of a superposition of two forces that appear simultaneously: the constant resisting force P_{1s} and the stiffness force having the stiffness coefficient calculated from equation (1.2.4).

Certain considerations should be emphasized regarding the behavior of constant resisting forces applied to vibratory systems that interact with elastoplastic or viscoelastoplastic media. In this regard, we should consider two possible situations. In the first, the initial velocity of the system equals zero and the motion of the system is caused by a harmonic force. When a vibratory system is moving horizontally and deforms the elastoplastic or viscoelastoplastic medium, the constant resisting forces appear as a reaction of the medium to its deformation. The deformation occurs during one half of the vibratory cycle that represents the loading stage of the medium.

During the second half of the cycle, the system moves in the opposite direction. It distances itself from the medium, resulting in disappearance of the constant resisting forces. The displacement in the opposite direction represents the unloading stage of the medium. The loading stage is characterized by the presence of constant resisting forces. However, during the unloading stage, these forces are absent. Therefore, the same differential equation of motion cannot describe both stages. The differential equation of motion can be valid for just one half (usually the first half) of the cycle of vibrations when the system is deforming the medium. The second half of this cycle can be described by another differential equation of motion for which the constant resisting forces equal zero. The initial displacement for this equation represents the system's displacement at the end of the first half of the cycle, which coincides with the loading stage.

In the second situation, the system possesses a positive velocity or the system, in addition to the harmonic force, is subjected to corresponding active forces such as a constant active force or a time-dependent force. The system continues to deform the medium during the second half of the cycle and continues to overcome the constant resisting force. Therefore, the differential equation of motion is valid during the complete cycle.

Let's consider a case when the motion of a vibratory system occurs in the vertical direction without interacting with an elastoplastic or viscoelastoplastic medium. In this case, the force of gravity of the system is directed downward and plays the role of a constant resisting force during the upward motion. During the system's downward motion, the force of gravity does not change its direction and is directed downward. In this case, the differential equation of motion describes the complete cycle of vibrations.

Based on all these considerations, this book presents just one differential equation of motion describing the motion during the complete cycle of vibrations. The analysis of this equation's solution provides the sufficient information about the parameters of motion of the system.

7) The *dry friction force F* is a constant resisting force that acts on the interface of sliding rigid bodies as a reaction to their relative motion. Dry friction forces also act on the lateral surface of a penetrator or cutter that deforms or cuts an elastoplastic or viscoelastoplastic medium. In this case, the friction force represents the lateral resistance forces and, depending on the shape of the deformer, appears simultaneously with other resisting forces. In the case of elastoplastic media, the friction force can act together with a) the stiffness force during the phase of elastic deformations; b) the constant resisting force during the phase of plastic deformations; or c) the superposition of the constant resisting force and stiffness force during the phase of plastic deformation with stain-strengthening. In the case of viscoelastoplastic media, the damping forces will be also acting together with all the resisting forces, as in the cases a), b), and c) of the elastoplastic media.

The reason why the friction force F is considered separately from the constant resisting force P can be found in the specific behavior of dry friction forces. These forces are always directed opposite to the direction of the system's velocity. In a vibratory process, the friction force instantaneously changes its direction when the system comes to a stop and starts to move into the opposite direction.

Similar to the constant resisting forces, we should consider the same two situations discussed above. In the first situation, the system does not possess any initial velocity and is subjected to the action of a harmonic force only. Here, the differential equation of motion is valid just to the end of the first half of the cycle of vibrations when the velocity becomes equal to zero; at this moment, the velocity and the friction force are instantaneously changing their directions. The differential equation of motion that describes the displacement during the first half of the cycle cannot by itself account for the change of the direction of the friction force. Consequently, it becomes not applicable for the description of the second half of the cycle of vibrations.

In order to describe the displacement of the system during the second half of the cycle of vibrations, another differential equation of motion should be composed reflecting the change of the direction of the friction force. In comparison with the behavior of the constant resisting force — in the case of deformation of an elastoplastic or viscoelastoplastic medium — the friction force does not disappear during the second half of the cycle. It just changes its direction. For the second differential equation of motion, the initial displacement equals the displacement of the system at the end of the first half of the period of vibrations.

The second situation occurs when the system possesses an initial velocity or additional active forces applied to the system. During the second half of the vibratory cycle, the system continues to move into the same direction as in the first half of the cycle; the friction force does not change its direction. Therefore, for this situation, the differential equation of motion is valid during the complete cycle of the vibrations.

All of these considerations allow this book to present the investigations of vibratory systems subjected to the action of a friction force based on just one differential equation of motion describing the motion during the complete cycle of vibrations. The analysis of this equation's solution provides the needed information regarding the parameters of the system's motion.

There are a few more details characterizing the role and behavior of the constant resisting force and friction force with regard to vibratory systems; however, there is no need for this text to discuss these details.

In concluding this section, note that for actual engineering systems, the left side of the differential equation of motion can consist minimally of one resisting force (the force of inertia) and at most of five resisting forces (the force of inertia, the damping force, the stiffness force, the constant resistance force, and the dry friction force).

1.3 Analysis of the Active Forces

The right side of equation (1.1.1) includes components that represent active (external) forces. In the majority of common engineering problems, the characteristics of these forces are defined during formulation of the problem. In most cases, they have constant values or can be considered as linear functions. The analysis of the active forces is presented in the order they are placed in the right side of the equation.

1) The *force of inertia* is included in the right side of the differential equation of motion based strictly on mathematical considerations. The force of inertia is multiplied by a dimensionless coefficient α, where $\alpha < 1$. If $\alpha = 1$, the forces of inertia in both sides of the equation will be canceled out. As a result, there will be no second order differential equation of motion. If $\alpha > 1$, the force of inertia will become an active force — which is impossible. The force of inertia is a reactive force by nature; as such, it should not be included in the right side of the equation.

However, in wheeled transportation systems, the force of inertia causes the redistribution of the system's weight between the rear and front axles. In a two-wheel drive system, this weight redistribution results in the change of the normal pressure of the front and rear wheels on ground. In turn, that change results in the change of the active force that accelerates the system. This redistribution is revealed when determining the reactions in the front and rear wheels by solving the equations of statics.

As a result of this distribution, a fraction of the force of inertia appears in the expression for the active force applied to the system. Thus, a fraction of the force of inertia that is a part of the active force is included in the right side of the differential equation of motion. This case is described in the author's book, *Solving Engineering Problems in Dynamics*. It is difficult to imagine other cases where a fraction of the force inertia represents an active force. In general, as mentioned above, the force of inertia is a reactive force; it should not be included in the right side of the differential equation of motion.

2) The *active force depending on the velocity*. These forces appear in some pneumatically operating systems during the process of accelerating pistons under the action of compressed air. In these systems, the active force that depends on the velocity decreases its value with the increase of the velocity of the piston. Therefore, the structure of the expression that reflects the characteristics of this force may have the following shape:

$$R_v = R_{0v} - C_1 \frac{dx}{dt} \qquad (1.3.1)$$

where R_v is the decreasing active force depending on the velocity, R_{0v} is the pressure force acting on the beginning of the acceleration process, and C_1 is a coefficient that is similar to the damping coefficient of the compressed air for the particular machine. As seen from equation (1.3.1), the second term in the right side of this equation is negative. This term has the structure of a negative damping force. When being transferred from the right side of the differential equation of motion to the left side, this term will play the role of a resisting damping force. It is difficult to imagine a real-life problem where the active force increases with the increase of the velocity. All considerations regarding the ways of determining the characteristics of the damping coefficients for the active forces are the same as for the resisting damping forces considered above for similar components in the left side of equation (1.1.1). In actual engineering problems, the active forces depending on velocity do not appear often.

3) The *active force that depends on the displacement of the system*. For example, the active force could cause the motion of a

piston in an engine of internal combustion or the motion of a projectile in the barrel of a firearm. The active forces in these mechanical systems are actually decreasing with the increase of the displacement. The structure of the expression that reflects the behavior of these forces is similar to the structure of the active force according to equation (1.3.1) and may have the following shape:

$$R_s = R_{0s} - K_1 x \qquad (1.3.2)$$

where R_s is the decreasing active force depending on the displacement, R_{0s} is the initial active force, and K_1 is the spring constant that is similar to the stiffness coefficient. The second term in the right side of equation (1.3.2) is negative. After transferring this term from the right side of the differential equation of motion to the left side, this term will play the role of a resisting stiffness force. However, when the acceleration is caused by an attracting magnetic force, the active force increases with the increase of the displacement.

In order to determine the characteristics of the spring constant (or stiffness coefficient) for the active forces that depend on the displacement, we first need the appropriate experimental data. This data should undergo the same analysis presented above for resisting stiffness forces. In real-life engineering problems, these types of active forces seldom appear.

4) The *active force representing a harmonic function of time or a harmonic force*. The harmonic force equals the product of multiplying the amplitude of the exciting force A by a harmonic function having ω_1 as the frequency of the vibrations. Also, λ is the angle associated with the initial value and direction of the exciting force at the moment when $t = 0$, and $0 \leq \lambda \leq 2\pi$. The harmonic function has the following structure:

$$\sin(\omega_1 t + \lambda) = \sin \omega_1 t \cos \lambda + \cos \omega_1 t \sin \lambda$$

Hence, both the sine and the cosine are included in equation (1.1.1) and represent a more general approach to accounting the harmonic function of time. In certain cases, it may be accepted that $\lambda = 0$.

The harmonic forces are used in numerous vibratory systems.

5) The *active time-dependent force* represents the product of multiplying a certain constant force Q by a linear function of time t. The time-dependent force acts during a predetermined interval time τ, with $t \leq \tau$ while ρ and μ are dimensionless coefficients having the following values: $\rho \geq 0$ and $\mu \neq 0$. For example, consider the acceleration process of a transportation system in which the active force gradually increases during a certain interval of time ($\mu > 0$). However, when the coefficient μ is negative, the time-dependent force is decreasing. This may happen in certain deceleration (braking) processes. Because the time-dependent force is acting during a limited, predetermined interval of time, it is reasonable to assume that these forces are used during the initial steps of working processes.

6) The *constant active force* R has a wide application in many mechanical systems.

Because the right side of a differential equation of motion could be equal to zero, we should consider one more case that is characterized by the absence of active forces. In real-life conditions, the deceleration (braking) processes of certain mechanical systems occur in the absence of active forces. In these cases, the motion of these systems is determined by their initial conditions of motion. If the system has an initial velocity, the system's motion is caused by its kinetic energy. If there are no resisting forces, then according to Newton's first law, the system will be in uniform motion in the horizontal direction. In the presence of the resisting forces, the system will decelerate and come to a stop. If a vibratory system possesses an initial displacement, the system's motion is caused by the potential energy. In this case, the system becomes subjected to the action of forces that are exerted by a deformed spring or elastic medium that releases the potential energy of its deformation. In cases when resisting forces are absent, the system will perform free vibrations in the horizontal direction; when resisting forces are present, these vibrations will be damped.

1.4 The General Differential Equation of Motion of Engineering Systems

The analysis of the components of equation (1.1.1) allows us to present the general differential equation of motion, made up of five components in each side:

$$m\frac{d^2x}{dt^2} + C\frac{dx}{dt} + Kx + P + F$$

$$= R + C_1\frac{dx}{dt} + K_1x + A\sin(\omega_1 t + \lambda) + Q(\rho + \frac{\mu t}{\tau}) \qquad (1.4.1)$$

The initial conditions of motion for this equation are given by expression (1.1.2).

In order to compose the differential equation of motion, it is first necessary to formulate the problem in the most accurate way, identifying the resisting forces, the active forces, and the initial conditions of motion. The initial conditions of motion characterize the state of the system at the beginning of the process of motion — in other words, when time equals zero. The state of the system's motion is characterized by two parameters of motion: by the displacement and by the velocity. The solution of the differential equation of motion depends on the initial conditions of motion. The same differential equation of motion will have different solutions for various initial conditions of motion.

The structure of the differential equation of rotational motion is similar to the structure of equation (1.4.1). Thus, the general differential equation of motion for a system that is rotating around its horizontal axis and is subjected to all possible resisting and active moments reads:

$$J\frac{d^2\theta}{dt^2} + C_2\frac{d\theta}{dt} + K_2\theta + M_P + M_F$$

$$= M_R + C_3\frac{d\theta}{dt} + K_3\theta + M_A\sin(\omega_2 t + \lambda) + M_Q\left(\rho + \frac{\mu t}{\tau}\right) \qquad (1.4.2)$$

where J is the moment of inertia of the system, θ is the angular displacement, C_2 and K_2 are respectively the damping and stiffness coefficients of the corresponding resisting moments, M_p is a constant resisting moment, M_F is a constant frictional resisting moment, M_R is a constant active moment, C_3 and K_3 are respectively the damping and stiffness coefficients of the corresponding active moments, M_A is the amplitude of a harmonic active moment, ω_2 is the frequency of the harmonic function, λ is the angle characterizing the initial value and direction of the exciting moment, M_Q is the constant value of an active moment at the beginning of the motion, ρ and μ are constant dimensionless coefficients, and $\tau > 0$, $\rho \geq 0$; $\mu \neq 0$, while $t \leq \tau$.

The initial conditions of motion for the equation (1.4.2) are:

$$\text{for} \quad t = 0, \quad \theta = \theta_0; \quad \frac{d\theta}{dt} = \Omega_0 \qquad (1.4.3)$$

where θ_0 and Ω_0 are the initial angular displacement and the initial angular velocity respectively.

1.5 Methodology of Solving Differential Equations of Motion

All linear differential equations of motion can be solved using the methodology based on the Laplace Transform. You do not need to memorize the fundamentals of Laplace Transform in order to use this methodology, as will become apparent. The methodology is universal and straightforward. All mathematical procedures associated with solving differential equations of motion represent conventional algebraic actions. The methodology consists of three steps, which can be summarized as 1) converting, 2) solving, and 3) inverting.

1) In the first step, we convert the differential equation of motion with the initial conditions of motion from the time domain into the Laplace domain. In the Laplace domain, the independent variable is a complex argument. This conversion from the time domain into the Laplace domain is accomplished by the help of Laplace Transform pairs that represent matching expressions in time and Laplace domains. Tables of matching pairs can be found in numerous publications.

Table 1.1 lists a set of Laplace Transform pairs that is basically sufficient for solving the vast majority of the differential equations of motion of common actual mechanical engineering systems. This table consists of two columns. The expressions in the time domain are presented in the left column, and the matching expressions in the Laplace domain are presented in the right column. The matching expressions have the same ordinal numbers. To convert the differential equation of motion from the time domain into the Laplace domain, identify each time domain component of the equation in the left column and replace it with the matching expression from the right column. As a result of this conversion, the differential equation of motion in the time domain will be replaced with an algebraic equation of motion in the Laplace domain.

2)　The second step is associated with solving the Laplace domain algebraic equation of motion obtained in the first step. Some of the components of this algebraic equation contain the Laplace domain displacement function as multipliers. Using regular algebraic procedures, we should factor this function in order to obtain an algebraic equation that has in its left side this function of the displacement in the Laplace domain, whereas its right side has an expression consisting of one or a sum of rational proper fractions.

3)　In the third step, we use Laplace Transform pairs to invert algebraic equation obtained in the second step from the Laplace domain into the time domain. The inversion of this equation represents the time domain solution of the differential equation of motion with the initial conditions of motion. During this step, the table is used in the reverse order. Each component of the algebraic equation in Laplace domain should be identified in the right column of the table, then replaced with the matching expression from the left column that is in time domain. Sometimes the existing tables of Laplace Transform pairs do not have all representations of the components in the right side of the algebraic equation in Laplace domain. The components in the right side of the equation are proper rational fractions. If these fractions are not represented in Table 1.1, they probably will not be found in other publications.

In such cases, apply the method of decomposition to these fractions in order to resolve them into simpler fractions that are represented in the tables. However, I have already identified a certain amount of fractions that are not represented in published tables of Laplace Transform pairs and, applying the method of decomposition, resolved them into simpler fractions. The results of these decompositions are included in Table 1.1. For the vast majority of differential equations of motion associated with solving common real-life engineering problems, Table 1.1 should comprise all the expressions that could be needed to invert the equations from the Laplace domain into equations in the time domain.

Examples that demonstrate the application of this three-step methodology for solving differential equations of motion are presented below.

1.6   The Table of Laplace Transform Pairs

The table of Laplace Transform pairs (Table 1.1) consists of two columns. The left column contains expressions in the time domain, whereas the right column consists of the matching expressions in the Laplace domain. Each matching pair has the same ordinal number. Let's consider a few examples of matching pairs:

$$1)\ \ x(t) \leftrightarrow x(l)\ \text{ or }\ x \leftrightarrow x(l)$$

where $x(t)$ or x is the displacement of the system as a function of time (in time domain), while $x(l)$ is the displacement as a function of the complex argument l (in Laplace domain). Table 1.1 uses the letter x to represent the displacement in any direction.

$$2)\ \ \theta(t) \leftrightarrow \theta(l)\ \text{ or }\ \theta \leftrightarrow \theta(l)$$

where $\theta(t)$ or θ is the angular displacement of the system as a function of time (in time domain), while $\theta(l)$ is the angular displacement as a function of the complex argument l (in Laplace domain).

$$3)\ \ constant \leftrightarrow constant$$

where a constant value is the same in both domains.

$$4)\ \ t \leftrightarrow \frac{1}{l}$$

where t is the running time — the argument — (in time domain), whereas l is the complex argument (in Laplace domain).

For the convenience of performing the first step of this methodology for converting differential equations of motion from the time domain into the Laplace domain, the components are placed at the beginning of the table according to the sequence they are presented in the equations. The rest of the Laplace Transform pairs are placed in the table in certain groups depending on the structures of the numerators and denominators of the fractions in the right column and on the powers of the complex argument l.

The use of Table 1.1 when solving differential equations of motion is demonstrated in Chapter 2.

Table 1.1 Laplace Transform Pairs		
Time domain functions	Laplace domain functions	
1. $x(t)$ or x, or $\theta(t)$, or θ	1. $x(l)$ or $\theta(l)$	
2. t	2. $\dfrac{1}{l}$	
3. $\dfrac{d^2x}{dt^2}$ or $\dfrac{d^2\theta}{dt^2}$	3. $l^2x(l)-lv_0-l^2s_0$ or $l^2\theta(l)-l\Omega_0-l^2\theta_0$ where s_0 and θ_0 are respectively the initial displacement and initial angular displacement, and v_0 and Ω_0 are respectively the initial velocity and initial angular velocity.	
4. $\dfrac{dx}{dt}$ or $\dfrac{d\theta}{dt}$	4. $lx(l)-ls_0$ or $l\theta(l)-l\theta_0$ where s_0 and θ_0 are respectively the initial displacement and initial angular displacement.	

5.	*constant*	5.	*constant*
6.	$\sin \omega t$	6.	$\dfrac{\omega l}{l^2 + \omega^2}$
7.	$\cos \omega t$	7.	$\dfrac{l^2}{l^2 + \omega^2}$
8.	$\dfrac{t^n}{n!}$	8.	$\dfrac{1}{l^n}$ Where n is a positive integer.
9.	$\dfrac{e^{nt} - 1}{n}$	9.	$\dfrac{1}{l-n}$ where n here and in the following expressions in the table is a positive constant value.
10.	$\dfrac{1 - e^{-nt}}{n}$	10.	$\dfrac{1}{l+n}$
11.	e^{nt}	11.	$\dfrac{l}{l-n}$
12.	e^{-nt}	12.	$\dfrac{l}{l+n}$
13.	$\dfrac{1}{n}[t + \dfrac{1}{n}(e^{-nt} - 1)]$	13.	$\dfrac{1}{l(l+n)}$
14.	$\dfrac{1}{\omega^2}(1 - \cos \omega t)$	14.	$\dfrac{1}{l^2 + \omega^2}$ where ω here and in the following expressions in the table is a positive constant value.
15.	$\dfrac{1}{n^2}[t - \dfrac{2}{n} + e^{-nt}\left(\dfrac{2}{n} + t\right)]$	15.	$\dfrac{1}{l(l+n)^2}$

16.	$\dfrac{1}{\omega^2}(t-\dfrac{1}{\omega}\sin\omega t)$	16.	$\dfrac{1}{l(l^2+\omega^2)}$
17.	$\dfrac{1}{\omega}\sinh\omega t$	17.	$\dfrac{l}{l^2-\omega^2}$
18.	$\cosh\omega t$	18.	$\dfrac{l^2}{l^2-\omega^2}$
19.	$\dfrac{n(1-\cos\omega t)}{\omega^2(\omega^2+n^2)}-\dfrac{\sin\omega t}{\omega(\omega^2+n^2)}+\dfrac{1-e^{-nt}}{n(\omega^2+n^2)}$	19.	$\dfrac{1}{(l+n)(l^2+\omega^2)}$
20.	$\dfrac{1-\cos\omega t}{\omega^2+n^2}+\dfrac{n\sin\omega t}{\omega(\omega^2+n^2)}-\dfrac{1-e^{-nt}}{\omega^2+n^2}$	20.	$\dfrac{l}{(l+n)(l^2+\omega^2)}$
21.	$\dfrac{1}{\omega^2}(\cosh\omega t-1)$	21.	$\dfrac{1}{l^2-\omega^2}$
22.	$\dfrac{1}{\omega^2+n^2}[1-e^{-nt}(\cos\omega t+\dfrac{n}{\omega}\sin\omega t)]$	22.	$\dfrac{1}{(l+n)^2+\omega^2}$
23.	$\dfrac{1}{n^2-\omega^2}[1-e^{-nt}(\cosh\omega t+\dfrac{n}{\omega}\sinh\omega t)]$	23.	$\dfrac{1}{(l+n)^2-\omega^2}$
24.	$\dfrac{1}{\omega}e^{-nt}\sin\omega t$	24.	$\dfrac{l}{(l+n)^2+\omega^2}$
25.	$\dfrac{1}{\omega}e^{-nt}\sinh\omega t$	25.	$\dfrac{l}{(l+n)^2-\omega^2}$
26.	$e^{-nt}\cos\omega t$	26.	$\dfrac{l(l+n)}{(l+n)^2+\omega^2}$
27.	$e^{-nt}(\cos\omega t-\dfrac{n}{\omega}\sin\omega t)$	27.	$\dfrac{l^2}{(l+n)^2+\omega^2}$
28.	$e^{-nt}(\omega\cosh\omega t-\dfrac{n}{\omega}\sinh\omega t)$	28.	$\dfrac{l^2}{(l+n)^2-\omega^2}$
29.	$\dfrac{1}{\omega^2}(\dfrac{\sinh\omega t}{\omega}-t)$	29.	$\dfrac{1}{l(l^2-\omega^2)}$

30.	$e^{-nt}\cosh\omega t$	30.	$\dfrac{l(l+n)}{(l+n)^2-\omega^2}$
31.	$\dfrac{1}{4n}\left[t^2-\dfrac{1}{n}t-\dfrac{1}{2n^2}\left(e^{-2nt}-1\right)\right]$	31.	$\dfrac{1}{l^2(l+2n)}$
32.	$\dfrac{t^2}{n^2+\omega^2}-\dfrac{2nt}{\left(n^2+\omega^2\right)^2}+\dfrac{3n^2-\omega^2}{\left(n^2+\omega^2\right)^3}$ $-(1-e^{-nt}\cos\omega t)-\dfrac{n\left(n^2-3\omega^2\right)}{\omega\left(n^2+\omega^2\right)^3}e^{-nt}\sin\omega t$	32.	$\dfrac{1}{l^2[(l+n)^2+\omega^2]}$
33.	$\dfrac{\omega\sin\varphi t-\varphi\sin\omega t}{\varphi\omega(\omega^2-\varphi^2)}$	33.	$\dfrac{l}{\left(l^2+\omega^2\right)(l^2+\varphi^2)}$
34.	$\dfrac{\cos\varphi t-\cos\omega t}{\omega^2-\varphi^2}$	34.	$\dfrac{l^2}{\left(l^2+\omega^2\right)(l^2+\varphi^2)}$
35.	$e^{-nt}\cosh\omega t$	35.	$\dfrac{l(l+n)}{(l+n)^2-\omega^2}$
36.	$\dfrac{1}{n^2}[1-e^{-nt}(1+nt)]$	36.	$\dfrac{1}{(l+n)^2}$
37.	te^{-nt}	37.	$\dfrac{l}{(l+n)^2}$
38.	$(1-nt)e^{-nt}$	38.	$\dfrac{l^2}{(l+n)^2}$
39.	$\dfrac{1}{\omega(\omega^2+n^2)^2}\{(\omega^2+n^2)\omega t-2n\omega$ $-e^{-nt}[(\omega^2-n^2)\sin\omega t-2n\omega\cos\omega t]\}$	39.	$\dfrac{1}{l[(l+n)^2+\omega^2]}$
40.	$\dfrac{1}{\omega(n^2-\omega^2)^2}[(n^2-\omega^2)(\omega t+e^{-nt}\sinh\omega t)$ $-2n\omega(1-\cosh\omega t)]$	40.	$\dfrac{1}{l[(l+n)^2-\omega^2]}$
41.	$\dfrac{t(1+e^{-nt})}{\omega^2+n^2}-\dfrac{2(1-e^{-nt})}{n(\omega^2+n^2)}$	41.	$\dfrac{1}{l(l+n)^2}$

42.	$\dfrac{1}{\left(\varepsilon^2-\delta^2\right)^2+4n^2\varepsilon^2}\{2n[e^{-nt}(\cos\omega t$ $+\dfrac{n}{\omega}\sin\omega t)-\cos\varepsilon t]-\left(\varepsilon^2-\delta^2\right)$ $\times(\dfrac{1}{\varepsilon}\sin\varepsilon t-\dfrac{1}{\omega}e^{-nt}\sin\omega t)\}\quad \delta^2=n^2+\omega^2$	42.	$\dfrac{l}{\left(l^2+\varepsilon^2\right)[(l+n)^2+\omega^2]}$
43.	$\dfrac{1}{\left(\varepsilon^2-\delta^2\right)^2+4n^2\varepsilon^2}\{(\varepsilon^2-\delta^2)$ $\times[e^{-nt}(\cos\omega t+\dfrac{n}{\omega}\sin\omega t)-\cos\varepsilon t]$ $+2n\varepsilon(\sin\varepsilon t-\dfrac{\varepsilon}{\omega}e^{-nt}\sin\omega t)\}\quad \delta^2=n^2+\omega^2$	43.	$\dfrac{l^2}{\left(l^2+\varepsilon^2\right)[(l+n)^2+\omega^2]}$
44.	$\dfrac{1}{\omega^2n^2}t-\dfrac{1}{n^2-\omega^2}(\dfrac{1}{\omega^3}\sin\omega t-\dfrac{1}{n^3}\sin nt)$	44.	$\dfrac{1}{l\left(l^2+\omega^2\right)(l^2+n^2)}$
45.	$\dfrac{1}{\omega^4}(\cos\omega t-1)-\dfrac{1}{\omega^2}t^2$	45.	$\dfrac{1}{l^2(l^2+\omega^2)}$
46.	$\dfrac{1}{4n^2\varepsilon^2+(\varepsilon^2-n^2)^2}\{2n[e^{nt}(1+nt)-\cos\varepsilon t]$ $-\left(\varepsilon^2-n^2\right)(\dfrac{1}{\varepsilon}\sin\varepsilon t-te^{-nt})\}$	46.	$\dfrac{l}{\left(l^2+\varepsilon^2\right)(l+n)^2}$
47.	$\dfrac{1}{4n^2\varepsilon^2+\left(\varepsilon^2-n^2\right)^2}\{2n\varepsilon(\sin\varepsilon t-\varepsilon te^{-nt})$ $-\left(\varepsilon^2-n^2\right)[e^{-nt}(1+nt)-\cos\varepsilon t]\}$	47.	$\dfrac{l^2}{\left(l^2+\varepsilon^2\right)(l+n)^2}$
48.	$\dfrac{1}{4n^2\varepsilon^2+\left(\varepsilon^2+\omega^2-n^2\right)^2}\{e^{-nt}[2n\cosh\omega t$ $+\dfrac{1}{\omega}(\varepsilon^2+\omega^2+n^2)\sinh\omega t]-2n\cos\varepsilon t$ $-\dfrac{1}{\varepsilon}(\varepsilon^2+\omega^2-n^2)\sin\varepsilon t\}$	48.	$\dfrac{l}{\left(l^2+\varepsilon^2\right)[(l+n)^2-\omega^2]}$
49.	$\dfrac{(\varepsilon^2+\omega^2-n^2)}{4n^2\varepsilon^2+(\varepsilon^2+\omega^2-n^2)^2}\{\dfrac{\varepsilon^2}{\omega^2}(1-\cos\varepsilon t)$ $-\dfrac{\varepsilon^2}{n^2-\omega^2}[1-e^{-nt}(\cosh\omega t+\dfrac{n}{\omega}\sinh\omega t)]$ $+\dfrac{2n\varepsilon}{\varepsilon^2+\omega^2-n^2}(\sin\varepsilon t-\dfrac{\varepsilon}{\omega}e^{-nt}\sinh\omega t)\}$	49.	$\dfrac{l^2}{\left(l^2+\varepsilon^2\right)[(l+n)^2-\omega^2]}$

2

COMMON ENGINEERING PROBLEMS IN DYNAMICS

The concept of using the solution of the general differential equation of motion to obtain solutions for particular problems is of significant interest. This concept can potentially provide an effective methodology for solving common engineering problems in dynamics. But does this concept achieve this goal? In analytical investigations, it is acceptable to develop a general solution for a certain complex problem that includes several specific and interrelated problems with the goal of solving the specific problems.

The solution of a particular differential equation of motion contains fewer loading factors than the solution of the general equation. By eliminating from the solution of the general differential equation those loading factors that are not present in the particular equation, it may be possible to obtain the solution of the particular equation.

This chapter presents the detailed analysis of the structure of the general differential equation of motion, the solution of this equation based on the Laplace Transform methodology, and considerations regarding the applicability of this solution for solving common engineering problems.

It will be helpful to consider first a few examples of composing the differential equations of motion for actual engineering systems, solving these equations using the step-by-step Laplace Transform methodology (includes Table 1.1, the Laplace Transform Pairs), and analyzing these solutions.

2.1 Composing Differential Equations of Motion, Their Solutions, and Analyses

In dynamics of engineering systems, the problems are associated with acceleration/deceleration processes. This section considers examples that illustrate the investigation of these processes.

2.1.1 Deceleration of a System Moving on the Ground

Suppose we want to investigate the braking (deceleration) process of a mechanical system moving on the ground in the horizontal direction in order to determine the braking distance and time, and also the maximum values of the deceleration and the force applied to the system. The values of the braking distance and time represent the important criteria of the braking mechanism's performance. The maximum value of the force is needed to carry out the stress analysis of the system. The maximum value of the deceleration for public transportation systems should be verified with the norms of deceleration for public health and safety.

The investigation has three stages: a) composing the differential equation of motion and determining the initial conditions of motion, b) solving this equation with these initial conditions, and c) analyzing the solution.

2.1.1a Composing the Differential Equation of Motion

During the braking process, the source of energy is disengaged and the system's motion occurs due to its kinetic energy. No active forces are applied to the system. Hence, the right side of the differential equation of motion equals zero. The system is subjected to the action of reactive resisting forces exerted by the air as the damping force, and by the ground as the dry friction force. Figure 2.1 shows the model of a system subjected to a damping force and a friction force.

Usually the model of the system allows us to see the problem in its entirety. Because the force of inertia is present in each differential equation of motion, there is no need to show this force in the corresponding models. In Figure 2.1, m is the mass of the system, C is the damping coefficient of the damping force represented by the dashpot, and F is the friction force. It is assumed that the damping coefficient has a constant value. The resisting forces should be included in the left side of the equation. Thus, the left side of the differential equation of motion for the braking process represents a sum of three forces: the force of inertia, the damping force, and the friction force.

Based on the considerations mentioned, including equation 1.4.1 and Figure 2.1, we can compose the following differential equation of motion:

$$m\frac{d^2x}{dt^2} + C\frac{dx}{dt} + F = 0 \qquad\qquad (2.1.1)$$

The initial conditions represent the state of the motion at the beginning of the process. In general, the position of the system can be determined by indicating a certain distance s_0 from the

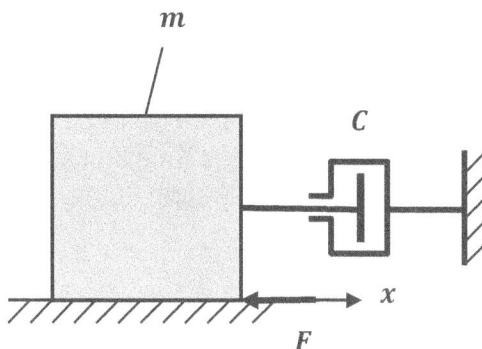

Figure 2.1 Model of a system subjected to a friction force and a damping force

origin — or the initial displacement can be equal zero. Because the system possesses a certain amount of kinetic energy, consequently, the system possesses a certain initial velocity v_0. These considerations allow us to formulate the following initial conditions of motion for a general case:

$$\text{for} \quad t = 0 \quad x = s_0; \quad \frac{dx}{dt} = v_0 \qquad (2.1.2)$$

2.1.1b Solving the Differential Equation of Motion

1) The first step consists of converting differential equation of motion (2.1.1) with the initial conditions of motion (2.1.2) from the time domain into the Laplace domain based on Table 1.1. Dividing equation (2.1.1) by m, we obtain:

$$\frac{d^2x}{dt^2} + 2n\frac{dx}{dt} + f = 0 \qquad (2.1.3)$$

where:

$$2n = \frac{C}{m} \qquad (2.1.4)$$

while n is the damping factor, and:

$$f = \frac{F}{m} \qquad (2.1.5)$$

Using Laplace Transform Pairs 3, 4, and 5 from Table 1.1, we convert differential equation of motion (2.1.3) with the initial conditions of motion (2.1.2) from the time domain into the Laplace domain. As a result of conversion, we obtain the algebraic equation of motion in Laplace domain:

$$l^2x(l) - lv_0 - l^2s_0 + 2nlx(l) - 2nls_0 + f = 0 \qquad (2.1.6)$$

2) During the second step, we need to solve equation (2.1.6) for the displacement in Laplace domain $x(l)$. After performing the

appropriate algebraic procedures with equation (2.1.6), we obtain the solution for this equation:

$$x(l) = \frac{v_0 + 2ns_0}{l + 2n} + \frac{ls_0}{l + 2n} - \frac{f}{l(l + 2n)} \qquad (2.1.7)$$

3) The third step deals with the inversion of equation (2.1.7) from the Laplace domain into the time domain. Using pairs 1, 10, 12, and 13 from Table 1.1, we perform the inversion and obtain the solution of differential equation of motion (2.1.1) with the initial conditions of motion (2.1.2):

$$x = \frac{v_0 + 2ns_0}{2n}(1 - e^{-2nt}) + s_0 e^{-2nt} - \frac{f}{2n}[t + \frac{1}{2n}(e^{-2nt} - 1)]$$

Applying conventional algebraic procedures to this equation, we transform it to the following shape:

$$x = \frac{1}{2n}[v_0 + 2ns_0 + \frac{f}{2n} - ft - (v_0 + \frac{f}{2n})e^{-2nt}] \qquad (2.1.8)$$

2.1.1c Analyzing the Solution of the Differential Equation of Motion

Our goal is to determine the braking distance and time, and also the maximum values of the acceleration and force applied to the system. At the end of the braking process, the velocity equals zero. Taking the first derivative of equation (2.1.8), we determine the velocity:

$$\frac{dx}{dt} = \left(v_0 + \frac{f}{2n}\right)e^{-2nt} - \frac{f}{2n} \qquad (2.1.9)$$

Equating the left side of equation (2.1.9) to zero, we can determine the braking time T:

$$0 = \left(v_0 + \frac{f}{2n}\right)e^{-2nT} - \frac{f}{2n}$$

This equation can be transformed into the following shape:

$$e^{-2nT} = \frac{f}{2nv_0 + f} \tag{2.1.10}$$

Solving equation (2.1.10) for the braking time T, we have:

$$T = \frac{1}{2n} \ln \frac{2nv_0 + f}{f} \tag{2.1.11}$$

Substituting equations (2.1.10) and (2.1.11) into equation (2.1.8), we calculate the displacement s of the system:

$$s = \frac{1}{2n} \left(v_0 + 2ns_0 - \frac{f}{2n} \ln \frac{2nv_0 + f}{f} \right) \tag{2.1.12}$$

The initial displacement should not be included in the braking distance. Thus, eliminating the initial displacement s_0 from equation (2.1.12), we determine the braking distance s_{br}:

$$s_{br} = \frac{1}{2n} \left(v_0 - \frac{f}{2n} \ln \frac{2nv_0 + f}{f} \right) \tag{2.1.13}$$

The second derivative of equation (2.1.8) represents the deceleration of the system:

$$\frac{d^2x}{dt^2} = -(2nv_0 + f)e^{-2nt} \tag{2.1.14}$$

The maximum value of the deceleration occurs when the velocity equals zero. Thus, combining equation (2.1.10) with equation (2.1.14), we determine the maximum value of the deceleration a_{max}:

$$a_{max} = -f \tag{2.1.15}$$

However, as can be seen from equation (2.1.14), at the beginning of the braking process when $t = 0$ the minimum value of the deceleration a_{min} equals:

$$a_{min} = -(2nv_0 + f) \tag{2.1.16}$$

Comparing equations (2.1.15) and (2.1.16), it becomes clear that the maximum absolute value a_{abs} of the deceleration occurs at the beginning of the braking process. This value should be considered further. Thus, based on equation (2.1.16), we calculate the maximum absolute value of the deceleration:

$$a_{abs} = |\,2nv_0 + f\,|$$

Substituting notations (2.1.4) and (2.1.5) into this equation, we obtain:

$$a_{abs} = \frac{Cv_0 + F}{m} \qquad (2.1.17)$$

According to equation (2.1.17), the value of the deceleration should be in compliance with the norms of public health and safety.

Multiplying equation (2.1.17) by m, we determine the maximum force R_{max} applied to the system:

$$R_{max} = Cv_0 + F \qquad (2.1.18)$$

The appropriate stress analysis of the system should be based on the value of the force according to equation (2.1.18).

Thus, all three stages of the investigation of the braking process of the system are carried out.

2.1.2　Acceleration of a Water Vessel

In this example, the goal is to investigate the acceleration process of a water vessel subjected to the action of a constant active force. We want to determine the vessel's maximum achievable velocity, the maximum value of the acceleration, the maximum force applied to the vessel, and the power of the source of energy. Assume the combined action of the air and water resisting forces results in a damping force that has a constant damping coefficient.

The maximum velocity is one of the criteria of the vessel's performance. The maximum value of the force applied to the vessel during its acceleration plays the main role in the corresponding stress calculations of the vessel. The maximum value of the acceleration should be in compliance with the norms of public health and safety.

The investigation consists of three stages: a) composing the differential equation of motion and determining the initial conditions of motion, b) solving this equation with these initial conditions, and c) analyzing the solution.

2.1.2a Composing the Differential Equation of Motion

The vessel is accelerated by a constant active force; it is also subjected to the action of the force of inertia and a resisting damping force, which is characterized by a constant damping coefficient. Figure 2.2 presents the model of a system subjected to an active constant force and a resisting damping force. In this model, m is the mass of the vessel, R is the active force, and C is the damping coefficient associated with the dashpot. The left side of the differential equation of motion includes the sum of the force of inertia and damping force, while the right side contains the active force. Based on these considerations, equation (1.4.1), and Figure 2.2, we compose the following differential equation of motion:

$$m\frac{d^2x}{dt^2} + C\frac{dx}{dt} = R \qquad (2.1.19)$$

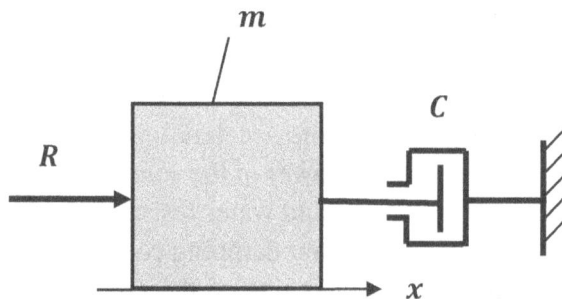

Figure 2.2 Model of a system subjected to a constant active force and a damping force

The initial conditions of motion in this case are:

$$\text{for} \quad t = 0 \quad x = 0; \quad \frac{dx}{dt} = 0 \qquad \textbf{(2.1.20)}$$

2.1.2b Solving the Differential Equation of Motion

1) We start by converting differential equation of motion (2.1.19) with the initial conditions of motion (2.1.20) from the time domain into the Laplace domain, based on Table 1.1. Dividing equation (2.1.19) by m, and using notation (2.1.4), we may write:

$$\frac{d^2x}{dt^2} + 2n\frac{dx}{dt} = r \qquad \textbf{(2.1.21)}$$

where:

$$r = \frac{R}{m} \qquad \textbf{(2.1.22)}$$

Using Laplace Transform Pairs 3, 4, and 5 from Table 1.1, we convert differential equation of motion (2.1.21) with the initial conditions of motion (2.1.20) from the time domain into the corresponding algebraic equation of motion in the Laplace domain:

$$l^2 x(l) - lv_0 + 2nlx(l) = r \qquad \textbf{(2.1.23)}$$

2) Now we need to solve equation (2.1.23) for the displacement $x(1)$ in the Laplace domain. Applying basic algebra, we obtain the solution:

$$x(l) = \frac{r}{l(l+2n)} \qquad \textbf{(2.1.24)}$$

3) Using pairs 1 and 13 from Table 1.1, we invert equation (2.1.24) from the Laplace domain into the time domain and obtain the solution of differential equation of motion (2.1.19) with the initial conditions of motion (2.1.20):

$$x = \frac{r}{2n}[t + \frac{1}{2n}(e^{-2nt} - 1)] \qquad \textbf{(2.1.25)}$$

2.1.2c Analyzing the Solution of the Differential Equation of Motion

Taking the first and second derivatives of equation (2.1.25), we determine the velocity and acceleration respectively:

$$\frac{dx}{dt} = \frac{r}{2n}(1 - e^{-2n}) \qquad (2.1.26)$$

$$\frac{d^2x}{dt^2} = re^{-2nt} \qquad (2.1.27)$$

The velocity reaches its maximum value when the acceleration becomes equal to zero. Actually, equation (2.1.27) approaches zero when the time tends to infinity. Therefore, we may write:

$$e^{-2nT} \to 0 \qquad (2.1.28)$$

where T is the time duration of the acceleration process.

Substituting expression (2.1.28) into equation (2.1.26), we obtain:

$$v_{max} \to \frac{r}{2n} \qquad (2.1.29)$$

where v_{max} is the maximum value of the velocity. Substituting notations (2.1.4) and (2.1.22) into expression (2.1.29), we have:

$$v_{max} \to \frac{R}{C} \qquad (2.1.30)$$

Because equation (2.1.27) is a decreasing function, it becomes clear that the maximum value of the acceleration occurs at the beginning of the accelerating process. Thus, equating time t to zero in equation (2.1.27), we determine the maximum value of the acceleration a_{max}:

$$a_{max} = r$$

Substituting notations (2.1.4) and (2.1.22) into this equation, we obtain:

$$a_{max} = \frac{R}{m} \qquad (2.1.31)$$

The value of the acceleration according to equation (2.1.31) should be in compliance with the norms of public health and safety.

In general, the maximum force applied to the system in any given time equals the sum of the force of inertia and the resisting forces for this time.

Multiplying the mass of the vessel by the acceleration, according to equation (2.1.31), and accounting that when $t = 0$ the damping force equals zero, we determine the maximum force applied to the vessel R_{max}:

$$R_{max} = R \qquad (2.1.32)$$

However, when the velocity tends to its maximum value, the acceleration and the force of inertia approach zero. At this moment of time, the maximum value of the force applied to the system equals the resisting force that represents the product of multiplying expression (2.1.30) by C. As a result, we obtain the same value of the maximum force, as determined according to equation (2.1.32). The corresponding stress calculations of the vessel should be based on the force according to equation (2.1.32).

The power N of the system represents the product of multiplying the maximum force according to equation (2.1.32) by the maximum velocity according to expression (2.1.30), and we obtain:

$$N = \frac{R^2}{C} \qquad (2.1.33)$$

Thus, the investigation of the acceleration process of the vessel is completed.

2.2 Analysis of the Structure of the General Differential Equation of Motion

In order to compose the differential equation of motion of a certain mechanical engineering system, it is necessary to get the information regarding all possible loading factors that characterize the problem. Actually, all loading factors that are relevant to common engineering problems are described above. The general differential equation of motion (1.4.1) of a hypothetical engineering system was already composed and introduced in the previous chapter. The general initial conditions of motion imply that, at the beginning of the process of motion, the system's displacement and velocity have specific values. Assume that composing and solving the general differential equation of motion with general initial conditions of motion may become the universal source for obtaining solutions for real-life common engineering problems in dynamics. Also assume that by eliminating from the general solution those components that are not present in specific problems, it will become possible to obtain the solutions for these problems.

Rearranging the order of the components in the right side of equation (1.4.1), we may write:

$$m\frac{d^2x}{dt^2} + C\frac{dx}{dt} + Kx + P + F$$

$$= R + A\sin(\omega_1 t + \lambda) + Q\left(\rho + \frac{\mu t}{\tau}\right) + C_1\frac{dx}{dt} + K_1 x \qquad (2.2.1)$$

The general initial conditions are:

$$\text{for} \quad t = 0 \quad x = s_0; \quad \frac{dx}{dt} = v_0 \qquad (2.2.2)$$

Combining the similar components from both sides of the equation, we have:

$$m\frac{d^2x}{dt^2} + C\frac{dx}{dt} - C_1\frac{dx}{dt} + Kx - K_1 x + P + F$$

$$= R + A\sin(\omega_1 t + \lambda) + Q\left(\rho + \frac{\mu t}{\tau}\right) \qquad (2.2.3)$$

Denoting:

$$C_0 = C - C_1 \tag{2.2.4}$$

and:

$$K_0 = K - K_1 \tag{2.2.5}$$

we rewrite equation (2.2.3) as follows:

$$m\frac{d^2x}{dt^2} + C_0\frac{dx}{dt} + K_0x + P + F$$

$$= R + A\sin(\omega_1 t + \lambda) + Q\left(\rho + \frac{\mu t}{\tau}\right) \tag{2.2.6}$$

The damping force associated with the damping coefficient C_0 is always positive because the active force, depending on the velocity, is usually decreasing, and represents a negative value in the right side of the differential equation of motion. In the vast majority of cases, the stiffness force associated with the stiffness coefficient K_0 is also positive. However, we have to keep in mind that in a few rare cases this force could become negative, or equal to zero. We will continue solving equation (2.2.6) assuming that the stiffness force associated with K_0 is positive. However, during the analysis of the solution of equation (2.2.6), we will address the situation where the force related to K_0 is negative.

Thus, equation (2.2.6) represents the general differential equation of motion that includes all possible loading factors that characterize real-life common engineering problems in dynamics.

2.3 Solution of the General Differential Equation of Motion

The analysis of the solution of general differential equation of motion (2.2.6) with the general conditions of motion (2.2.2) will let us verify the applicability of the above-mentioned concept for solving specific common engineering problems. Dividing equation (2.2.6) by m, we obtain:

$$\frac{d^2x}{dt^2} + 2n_0\frac{dx}{dt} + \omega_0^2 x + p + f = r + a\sin(\omega_1 t + \lambda) + q\left(\rho + \frac{\mu t}{\tau}\right) \tag{2.2.7}$$

where:

$$2n_0 = \frac{C_0}{m} \tag{2.2.8}$$

$$\omega_0^2 = \frac{K_0}{m} \tag{2.2.9}$$

$$p = \frac{P}{m} \tag{2.2.10}$$

$$f = \frac{F}{m} \tag{2.2.11}$$

$$r = \frac{R}{m} \tag{2.2.12}$$

$$a = \frac{A}{m} \tag{2.2.13}$$

$$q = \frac{Q}{m} \tag{2.2.14}$$

Transforming the right side of equation (2.2.7), we have:

$$\frac{d^2x}{dt^2} + 2n_0\frac{dx}{dt} + \omega_0^2 x + p + f$$

$$= r + q\rho + \frac{q\mu t}{\tau} + a\sin\omega_1 t\cos\lambda + a\cos\omega_1 t\sin\lambda \tag{2.2.15}$$

We continue the solution assuming that ω_0^2 is positive.

Using Laplace Transform Pairs 3, 4, 1, 5, 2, 6, and 7 from Table 1.1, we convert the differential equation of motion (2.2.7) from the time domain into the Laplace domain, and obtain the resulting algebraic equation of motion in Laplace domain:

$$l^2 x(l) - lv_0 - l^2 s_0 + 2n_0 lx(l) - 2n_0 ls_0 + \omega_0^2 x(l) + p + f$$

$$= r + q\rho + \frac{q\mu}{\tau l} + \frac{a\omega_1 l\cos\lambda}{l^2 + \omega_1^2} + \frac{al^2\sin\lambda}{l^2 + \omega_1^2} \tag{2.2.16}$$

Applying to equation (2.2.16) the appropriate algebraic procedures, we solve this equation for the displacement in Laplace domain $x(1)$:

$$x(l) = \frac{r + qp - p - f}{l^2 + 2ln_0 + \omega_0^2} + \frac{l(v_0 + 2n_0 s_0)}{l^2 + 2ln_0 + \omega_0^2} + \frac{l^2 s_0}{l^2 + 2ln_0 + \omega_0^2}$$

$$+ \frac{\mu q}{\tau l(l^2 + 2ln_0 + \omega_0^2)} + \frac{la\omega_1 \cos \lambda}{(l^2 + \omega_1^2)(l^2 + 2ln_0 + \omega_0^2)}$$

$$+ \frac{l^2 a \sin \lambda}{(l^2 + \omega_1^2)(l^2 + 2ln_0 + \omega_0^2)} \qquad (2.2.17)$$

In order to invert equation (2.2.17) into the time domain using Table 1.1, we first transform the denominators of the fractions to the shape that is accepted in similar expressions in this table. Applying basic algebra to the denominators, we may write:

$$l^2 + 2ln_0 + \omega_0^2 = l^2 + 2ln_0 + \omega_0^2 + n_0^2 - n_0^2 = (l + n_0)^2 + \omega^2 \quad (2.2.18)$$

where:

$$\omega^2 = \omega_0^2 - n_0^2 \qquad (2.2.19)$$

It is very important to emphasize that ω^2 could be positive, negative, or equal zero. This depends on the value of the coefficient K_0. We assume for now that $\omega^2 > 0$; however, the two other options will be discussed below. Transforming the denominators in equation (2.2.17) according to expression (2.2.18), we write:

$$x(l) = \frac{r + qp - p - f}{(l + n_0)^2 + \omega^2} + \frac{l(v_0 + 2n_0 s_0)}{(l + n_0)^2 + \omega^2} + \frac{l^2 s_0}{(l + n_0)^2 + \omega^2}$$

$$+ \frac{\mu q}{\tau l[(l + n_0)^2 + \omega^2]} + \frac{la\omega_1 \cos \lambda}{(l^2 + \omega_1^2)[(l + n_0)^2 + \omega^2]}$$

$$+ \frac{l^2 a \sin \lambda}{(l^2 + \omega_1^2)[(l + n_0)^2 + \omega^2]} \qquad (2.2.20)$$

Using the pairs 1, 22, 24, 27, 39, 42, and 43 from Table 1.1, we invert equation (2.2.20) from the Laplace domain into the time domain, and obtain the solution of differential equation of motion (2.2.7) with the initial conditions of motion (2.2.2):

$$x = \frac{r + \rho q - p - f}{\omega_0^2}[1 - e^{-n_0 t}(\cos \omega t + \frac{n_0}{\omega}\sin \omega t)]$$

$$+ \frac{v_0 + 2n_0 s_0}{\omega}e^{-n_0 t}\sin \omega t + s_0 e^{-n_0 t}(\cos \omega t - \frac{n_0}{\omega}\sin \omega t)$$

$$+ \frac{\mu q}{\tau \omega(\omega^2 + n^2)^2}\{(\omega^2 + n_0^2)\omega t - 2n_0\omega - e^{-n_0 t}[(\omega^2 - n_0^2)$$

$$\times \sin \omega t - 2n_0\omega \cos \omega t]\}$$

$$+ \frac{a\omega_1 \cos \lambda}{(\omega_1^2 - \omega_0^2)^2 + 4n_0^2\omega_1^2}\{2n_0[e^{-n_0 t}(\cos \omega t + \frac{n_0}{\omega}\sin \omega t) - \cos \omega_1 t]$$

$$- (\omega_1^2 - \omega_0^2)(\frac{1}{\omega_1}\sin \omega_1 t - \frac{1}{\omega}e^{-n_0 t}\sin \omega t)\}$$

$$+ \frac{a \sin \lambda}{(\omega_1^2 - \omega_0^2)^2 + 4n_0^2\omega_1^2}\{(\omega_1^2 - \omega_0^2)[e^{-n_0 t}(\cos \omega t - \frac{n_0}{\omega}\sin \omega t) - \cos \omega_1 t]$$

$$+ 2n_0\omega_1(\sin \omega_1 t - \frac{\omega_1}{\omega}e^{-n_0 t}\sin \omega t)\} \tag{2.2.21}$$

Now let us consider the case when $\omega^2 = 0$. The attempt to substitute zero instead of ω into equation (2.2.21) leads to the necessity of dividing some of this equation's components by zero. This action is not permissible in mathematics and, by itself, destroys the opportunity of using solution (2.2.21) to obtain solutions of differential equations of motion for all other engineering systems.

The case when $\omega^2 < 0$ occurs when the force associated with the coefficient K_0 is negative; it can be seen from equation (2.2.20) that the denominators in all the fractions will contain the negative ω^2. In this case, the inversion into the time domain requires a different set of Laplace Transform Pairs in comparison with the case when ω^2 is positive. Consequently, the solution would be different from the solution according to equation (2.2.21). This difference is one more reason why the solution of the general differential equation of motion is not applicable for solving all other common problems in dynamics.

However, in spite of this, equation (2.2.21) could be used for obtaining the solutions of differential equations of motion for a limited number of certain engineering systems. It is not obvious in which cases the solution of the general differential equation of motion is applicable. In addition, equation (2.2.21) is both cumbersome and transcendental. Its derivatives (which are required for the analysis of motion) are even more so, and in general the analysis of this equation is extremely complicated. It would be incomparably easier to investigate the common engineering problems in dynamics by composing the appropriate differential equations of motion, solving them, and analyzing the solutions.

Thus, the above-mentioned concept that is based on utilizing equation (2.2.21) for obtaining solutions for all other common engineering problems in dynamics is not applicable.

2.4 Identifying the Common Engineering Problems in Dynamics

In equation (2.2.6), it is justifiable to eliminate the subscripts for the damping and stiffness coefficients. The general differential equation of motion then has the following shape:

$$m\frac{d^2x}{dt^2} + C\frac{dx}{dt} + Kx + P + F = R + A\sin(\omega_1 t + \lambda) + Q\left(\rho + \frac{\mu t}{\tau}\right)$$

$$(2.2.22)$$

Analyzing this equation, we can determine the total number of possible linear differential equations of rectilinear motion that could be used to describe real-life common engineering problems in dynamics. The left side consists of five resisting forces. The force of inertia should be present in the left side of each equation, and the structure of the simplest left side of the equation is characterized by the force of inertia only. The number of all possible combinations that can be obtained from the other four resisting forces is calculated from the following formula:

$$N = 2^n \tag{2.2.23}$$

where N is the number of combinations and n is the number of resisting forces (without the force of inertia); thus, $n = 4$. Consequently, $N = 2^4 = 16$. In order to represent the structures of each of the 16 combinations of the resisting forces, we create a truth table with four variable parameters. The truth table contains 16 rows. Each of these 16 rows represents a corresponding combination of resisting forces. According to formula (2.2.23) and the truth table, one of these 16 combinations does not contain any parameters (or any resisting forces). Usually the first or the last row of the truth table is the row without parameters (the empty row). By adding to the truth table a column consisting of forces of inertia, we can populate the empty row with the force of inertia. The modified truth table becomes applicable for determining the structures of the left sides of all possible differential equations of motion for common actual mechanical systems.

In order to calculate the possible number of differential equations of motion associated with actual engineering systems performing rectilinear motion, we need to determine the number of possible combinations of active forces that could be included in the right side of the differential equation of motion. Analyzing the components in the right side of equation (2.2.22), and accounting that R and Q are constant forces of the same nature, it becomes clear there are six combinations of active forces, including the case when the active force equals zero. These combinations are presented below:

1) zero
2) the constant active force R
3) the harmonic force $A\sin(\omega_1 t + \lambda)$
4) the time-dependent force $Q\left(\rho + \dfrac{\mu t}{\tau}\right)$
5) the sum of the constant active force and the harmonic force $R + A\sin(\omega_1 t + \lambda)$

6) the sum of the harmonic force and the time-dependent force
$Asin(\omega_1 t + \lambda) + Q\left(\rho + \frac{\mu t}{\tau}\right)$

Multiplying the 16 possible combinations of resisting forces by 6 possible combinations of active forces, we determine there are 96 possible combinations of linear differential equations of motion describing common mechanical engineering problems. The structure of the simplest differential equation of motion consists of two components. The left side of the equation includes the force of inertia, whereas the right side has one active force, or the active force equals zero. An example of the simplest differential equation of motion follows:

$$m\frac{d^2t}{dt^2} = 0 \qquad\qquad (2.1.24)$$

For this example, the initial conditions of motion are:

$$\text{for} \quad t = 0 \quad x = 0; \quad \frac{dx}{dt} = v_0 \qquad\qquad (2.1.25)$$

In this case, the system is moving due to its kinetic energy.

It is possible to solve all 96 differential equations of motion. Presenting these solutions in the book could be beneficial for advancing the application of analytical investigations in the area of engineering dynamics. However, practicing engineers should have a straightforward and simple way to find the required solution for their particular engineering problem without searching the entire book. This way is explained below.

2.5 The Structure and Use of the Guiding Table

Engineering dynamics covers 96 common real-life problems. The appropriate descriptions of these problems are presented in corresponding sections of the book. These sections are associated with combinations of the resisting and active forces applied to each particular engineering system. The structure of the discussion in each section is similar, as follows:

1) The resisting and the active forces applied to the engineering system. These forces compose the differential equation of motion and identify the possible practical application of the system.

2) The differential equation of motion of the system and the initial conditions of motion.

3) The three-step Laplace Transform methodology of solving the differential equation of motion with the initial conditions of motion. The methodology consists of the following steps:

   a) Converting the differential equation of motion with the initial conditions of motion from the time domain into the Laplace domain using the Laplace Transform Pairs presented in Table 1.1, and obtaining the corresponding algebraic equation of motion in the Laplace domain.

   b) Solving the obtained algebraic equation of motion for the Laplace domain displacement $x(l)$ as a function of the complex argument l.

   c) Inverting this solution from the Laplace domain into the time domain using the Table 1.1, and getting the solution of the differential equation of motion with its initial conditions of motion in the time domain.

4) The equations of the laws of motion that represent the solutions of the differential equation of motion for all possible initial conditions of motion.

The analytical descriptions of the three basic parameters of motion (displacement, velocity, and acceleration) are given in each section of Chapters 3 through 14 and also in Chapter 19. Chapters 15 through 18 present the solutions of the differential equations of certain engineering systems. These solutions are transcendental and cumbersome; their derivatives are even more so. The analyses of the solutions, representing the laws of motion, could be performed based on computational symbolic and numeric methods. Thus, Chapters 15 through 18 present only the solutions of the differential equations of motion (the laws of motion).

In each section, the analysis of the solution begins with determining the first and second derivatives of the displacement as a function of time. Based on the obtained equations of the displacement, velocity, and acceleration of the system, further analysis determines the appropriate criteria of the system performance and other related engineering requirements and aspects of the problem.

Each section contains a complete description of the problem's solution, including the explanation of its parameters and notations. This organization eliminates the need to look for explanations earlier in the book. All sections use identical notation for the resisting and active forces; there is no need to repeat the meanings of these notations in each section.

For the reader's convenience, here are the notations used for these forces:

1) $m \frac{d^2x}{dt^2}$ is the force of inertia, m is the mass of the mechanical system, x is the displacement or the position coordinate of the system and is a function of the running time t, which is the argument.

2) $C \frac{dx}{dt}$ is the resisting damping force, C is the damping coefficient.

3) Kx is the resisting stiffness force, K is the stiffness coefficient.

4) P is the constant resisting force.

5) F is the constant friction force.

6) R is the constant active force.

7) $A \sin(\omega_1 t + \lambda)$ is the harmonic force, where A is the exciting force amplitude and ω_1 is the frequency of the harmonic function, while λ is the angle associated with the initial value and direction of the exciting force at the moment when $t = 0$, and $0 \leq \lambda \leq 2\pi$.

8) $Q(\rho + \frac{\mu t}{\tau})$ is the time-dependent force, where Q is a constant force, ρ and μ are dimensionless coefficients, $\rho \geq 0$ and $\mu \neq 0$, and τ is the time that the process can last, while $t \leq \tau$.

The numbers of the sections describing all 96 common engineering problems correspond to the order in Guiding Table 2.1, the structure of which is based on the modified truth table described earlier. In this table, Row 1 indicates the ordinal order of the parameters.

Guiding Table 2.1

	Resisting Forces					Active Forces					
1	**A**	**B**	**C**	**D**	**E**	**1**	**2**	**3**	**4**	**5**	**6**
2	$m\frac{d^2x}{dt^2}$	$C\frac{dx}{dt}$	kx	P	F	0	R	$A\sin(\omega_1 t+\lambda)$	$Q(\rho+\frac{\mu t}{\tau})$	$R+A\sin(\omega_1 t+\lambda)$	$A\sin(\omega_1 t+\lambda)+Q(\rho+\frac{\mu t}{\tau})$
3	+					3.1	3.2	3.3	3.4	3.5	3.6
4	+				+	4.1	4.2	4.3	4.4	4.5	4.6
5	+			+		5.1	5.2	5.3	5.4	5.5	5.6
6	+			+	+	6.1	6.2	6.3	6.4	6.5	6.6
7	+		+			7.1	7.2	7.3	7.4	7.5	7.6
8	+		+		+	8.1	8.2	8.3	8.4	8.5	8.6
9	+		+	+		9.1	9.2	9.3	9.4	9.5	9.6
10	+		+	+	+	10.1	10.2	10.3	10.4	10.5	10.6
11	+	+				11.1	11.2	11.3	11.4	11.5	11.6
12	+	+			+	12.1	12.2	12.3	12.4	12.5	12.6
13	+	+		+		13.1	13.2	13.3	13.4	13.5	13.6
14	+	+		+	+	14.1	14.2	14.3	14.4	14.5	14.6
15	+	+	+			15.1	15.2	15.3	15.4	15.5	15.6
16	+	+	+		+	16.1	16.2	16.3	16.4	16.5	16.6
17	+	+	+	+		17.1	17.2	17.3	17.4	17.5	17.6
18	+	+	+	+	+	18.1	18.2	18.3	18.4	18.5	18.6

Numbers of Sections

Row 2 contains the notations of the resisting and active forces in their ordinal way.

Guiding Table 2.1 has two sides. The left side presents the resisting forces and includes Columns A through E. In this side of the table, Row 2 indicates the notations of the resisting forces, namely: $m \frac{d^2x}{dt^2}$ (the force of inertia), $C \frac{dx}{dt}$ (the damping force), Kx (the stiffness force), P (the constant resisting force), and F (the friction force). Rows 3 through 18 indicate the resisting forces applied to specific engineering systems. In these rows, the cells marked by the plus sign (+) indicate the presence of these resisting forces for particular problems.

These resisting forces constitute the left side of the differential equation of motion for this particular system. These forces also indicate the row that lists the numbers of the sections describing the corresponding problems. For instance, if a certain engineering system is subjected to the action of the sum of the force of inertia, the damping force, and the stiffness force, Row 15 contains the numbers of the corresponding sections, and the left sides of the differential equations of motion in these sections are made up of the resisting forces mentioned above.

The right side of Guiding Table 2.1 summarizes the active forces and the section numbers for their corresponding problems. Columns 1 through 6 list the notations for the active forces, including: 0 (zero), R (the constant active force), $A \sin(\omega_1 t + \lambda)$ (the harmonic force), $Q(\rho + \frac{\mu t}{\tau})$ (the time-dependent force), $R + A \sin(\omega_1 t + \lambda)$ (the sum of the constant active force and the harmonic force), and $A \sin(\omega_1 t + \lambda) + Q(\rho + \frac{\mu t}{\tau})$ (the sum of the harmonic force and time-dependent force).

The remaining cells represent the combinations of resisting and active forces. The number within each cell indicates the section in the text that describes a particular engineering problem that corresponds to that combination of forces. Suppose we have a problem characterized by the resisting forces listed in Row 15. That problem also has a constant active force R applied to the system, shown in Column 2. The intersection of Row 15 and Column 2 indicates that the description of this problem is presented in section 15.2 of this text.

Consider a different problem where we determine that the resisting forces consist of the force of inertia, the damping force, and
the friction force, whereas the active force is the harmonic force.
This combination goes with Row 12 and Column 3 and we find their
intersection; the solution of this problem is presented in section 12.3.

As indicated by Guiding Table 2.1, there are 16 combinations of
resisting forces and, hence, 16 chapters that describe the solutions of
the 96 problems (Chapters 3 through 18). Each of these chapters has
six sections, based on the active forces associated with those problems.

Let's consider a few more examples. Chapter 3 covers problems that are characterized by the force of inertia as the only resisting force. In that chapter, the left side of each differential equation
of motion consists only of the force of inertia. Chapter 4 describes
problems that are characterized by the sum of the force of inertia
and the friction force. Therefore, the left sides of all the differential
equations of motion in Chapter 4 include the sum of the force of
inertia and the friction force. Similarly, all the rows in the left side
of Guiding Table 2.1 reflect the structures of the left sides of their
corresponding differential equations of motion.

The right sides of all differential equations of motion associated with Column 1 are identical and equal to zero. Similarly for
Columns 2 through 6, the right sides of the related equations consist of an active force or (sum of forces) associated with the forces
identified at the top of the respective columns. Each cell reflects the
resisting and active forces applied to the system; the combination
of forces determines the structure of the corresponding differential
equations of motion. For instance, section 8.4 describes a system
subjected to the action of the force of inertia, the stiffness force,
and the friction force as the resisting forces; and the time-dependent
force as the active force. Therefore, the left side of the differential
equation of motion described in section 8.4 represents the sum of
the force of inertia, the stiffness force, and the friction force; the
right side consists of the time-dependant force.

The same differential equation of motion has various solutions — depending on the initial conditions of motion. Therefore,

the description of each particular problem encompasses the solutions for all possible initial conditions of motion, which are:

1) the general initial conditions of motion:

$$\text{for} \quad t = 0 \quad x = s_0; \quad \frac{dx}{dt} = v_0$$

where s_0 and v_0 are the initial displacement and initial velocity respectively

2) the initial displacement equals zero:

$$\text{for} \quad t = 0 \quad x = 0; \quad \frac{dx}{dt} = v_0$$

3) the initial velocity equals zero:

$$\text{for} \quad t = 0 \quad x = s_0; \quad \frac{dx}{dt} = 0$$

4) both the initial displacement and velocity equal zero:

$$\text{for} \quad t = 0 \quad x = 0; \quad \frac{dx}{dt} = 0$$

This book includes one additional chapter that deals with applied engineering dynamics. Chapter 19 describes the actual engineering systems performing two-dimensional motion, which occurs in vertical and horizontal directions. The analytical investigations of such problems are based on a system of two simultaneous differential equations of motion. These equations describe the rectilinear motion of the system in each of the two directions. Guiding Table 2.1 is completely applicable to two-dimensional problems. Engineers should determine the resisting and active forces that are applied to the system in each of these two directions. According to these two sets of forces, it is necessary to find in Guiding Table 2.1 the numbers of the two sections that describe the motion in these directions. More details for solving problems in two-dimensional motion are explained in Chapter 19.

FORCE OF INERTIA

This chapter describes engineering problems that are characterized by the force of inertia as the only resisting force. In Row 3 of Guiding Table 2.1 the cell related to the force of inertia indicates the only resisting force marked by the plus sign. The intersections of this row with Columns 1 through 6 indicate the six sections included in this chapter. Each section discusses different active forces — the section titles reflect which active forces.

Throughout this chapter, the left sides of the differential equations of motion are identical and represent the force of inertia. The right sides of these equations reflect the characteristics of the active forces applied to the systems.

3.1 Active Force Equals Zero

According to Guiding Table 2.1, this section describes an engineering problem associated with the action of the force of inertia (Row 3) in the absence of an active force (Column 1). (Refer to section 1.3 for a review of the motion of a system in the absence

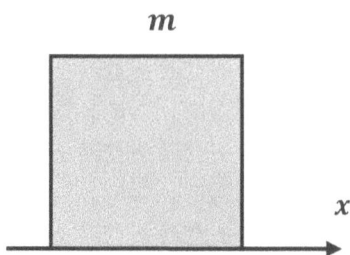

Figure 3.1 Model of a system moving by inertia

of active forces.) From a practical point of view, this problem has a limited application.

    This system moves in the horizontal direction. It is required to determine the parameters of motion. Figure 3.1 shows the model of a system moving by inertia. The above-mentioned considerations and the model in Figure 3.1 allow us to compose the differential equation of motion for the engineering system related to the current problem. The left side of the differential equation of motion for this problem includes the force of inertia while the right side equals zero. Thus, the differential equation of motion of this system reads:

$$m\frac{d^2x}{dt^2} = 0 \tag{3.1.1}$$

    Differential equation of motion (3.1.1) has different solutions for various initial conditions of motion. These solutions and their analyses follow.

3.1.1 General Initial Conditions

    The general initial conditions of motion are:

$$\text{for} \quad t = 0 \quad x = s_0; \quad \frac{dx}{dt} = v_0 \tag{3.1.2}$$

where s_0 and v_0 are the initial displacement an initial velocity respectively.

    Dividing equation (3.1.1) by m, we have:

$$\frac{d^2x}{dt^2} = 0 \tag{3.1.3}$$

Applying Laplace Transform Pairs 3 and 5 from Table 1.1, we convert differential equation of motion (3.1.3) with the initial conditions of motion (3.1.2) from the time domain into the Laplace domain and obtain the corresponding algebraic equation of motion in the Laplace domain:

$$l^2 x(l) - l v_0 - l^2 s_0 = 0 \qquad (3.1.4)$$

Solving equation (3.1.4) for the displacement $x(l)$ in the Laplace domain, we write:

$$x(l) = s_0 + \frac{v_0}{l} \qquad (3.1.5)$$

Using pairs 1, 5, and 2 from Table 1.1, we invert equation (3.1.5) from the Laplace domain into the time domain and obtain the solution of differential equation of motion (3.1.1) with the initial conditions of motion (3.1.2):

$$x = s_0 + v_0 t \qquad (3.1.6)$$

Taking the first derivative of equation (3.1.6), we determine the velocity of the system:

$$\frac{dx}{dt} = v_0 \qquad (3.1.7)$$

Equation (3.1.6) allows us to determine the displacement of the system for a given time, calculating the time needed to move a certain distance.

3.1.2 Initial Displacement Equals Zero

The initial conditions of motion are:

$$\text{for} \quad t = 0 \quad x = 0; \quad \frac{dx}{dt} = v_0 \qquad (3.1.8)$$

The solution of differential equation of motion (3.1.1) with the initial conditions of motion (3.1.8) reads:

$$x = v_0 t \qquad (3.1.9)$$

The velocity of the system is the same as in the previous case.

3.2 Constant Force *R*

This section describes a system subjected to the action of the force of inertia (Row 3 of Guiding Table 2.1) and the constant active force (Column 2). The current problem could reflect the acceleration process of a system moving in the horizontal direction while the resisting forces are insignificant. It is necessary to determine the basic parameters of motion and the maximum values of the acceleration and the force applied to the system.

The model of a system subjected to a constant active force is shown in Figure 3.2.

Based on the above-mentioned considerations and Figure 3.2, we can compose the left and right sides of the differential equation of motion for this system. The left side consists of the force of inertia, while the right side includes a constant active force. Hence, the differential equation of motion of the system reads:

$$m\frac{d^2x}{dt^2} = R \tag{3.2.1}$$

Differential equation of motion (3.2.1) has different solutions for various initial conditions of motion. These solutions and their analyses follow.

3.2.1 General Initial Conditions

The general initial conditions of motion are:

$$\text{for} \quad t = 0 \quad x = s_0; \quad \frac{dx}{dt} = v_0 \tag{3.2.2}$$

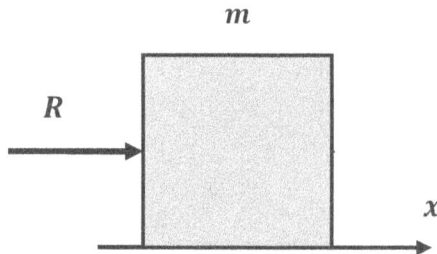

Figure 3.2 Model of a system subjected to a constant active force

where s_0 and v_0 are the initial displacement and initial velocity respectively. Dividing equation (3.2.1) by m, we have:

$$\frac{d^2x}{dt^2} = r \tag{3.2.3}$$

where:

$$r = \frac{R}{m} \tag{3.2.4}$$

Using Laplace Transform Pairs 3 and 5 from Table 1.1, we convert differential equation of motion (3.2.3) with the initial conditions of motion (3.2.2) from the time domain into the Laplace domain and obtain the corresponding algebraic equation of motion in the Laplace domain:

$$l^2 x(l) - lv_0 - l^2 s_0 = r \tag{3.2.5}$$

Solving equation (3.2.5) for the displacement $x(l)$ in the Laplace domain, we may write:

$$x(l) = s_0 + \frac{v_0}{l} + \frac{r}{l^2} \tag{3.2.6}$$

In order to invert equation (3.2.6) from the Laplace domain into the time domain, we use pairs $1, 5, 2,$ and 8 from Table 1.1 and obtain the solution of differential equation of motion (3.2.1) with the initial conditions of motion (3.2.2):

$$x = s_0 + v_0 t + \frac{rt^2}{2} \tag{3.2.7}$$

Taking the first derivative of equation (3.2.7), we determine the velocity of the system:

$$\frac{dx}{dt} = v_0 + rt \tag{3.2.8}$$

The acceleration of the system represents the second derivative of equation (3.2.7):

$$\frac{d^2x}{dt^2} = r \qquad (3.2.9)$$

Equation (3.2.9) shows that the acceleration a has a constant value and it is equal:

$$a = r \qquad (3.2.10)$$

Substituting into equation (3.2.10) notation (3.2.4), we write:

$$a = \frac{R}{m} \qquad (3.2.11)$$

If this case is associated with a public transportation system, it should be verified that the acceleration according to equation (3.2.11) complies with the norms of public safety and health. The maximum force R_{max} applied to the system in this case equals the force of inertia and represents the product of multiplying the mass m of the system by its acceleration according to equation (3.2.11). Thus, we have:

$$R_{max} = R \qquad (3.2.12)$$

The stress analysis of the system should be based on the value of the force according to equation (3.2.12).

Equations (3.2.7) and (3.2.8) allow calculating the distance and the velocity for a given time respectively.

3.2.2 Initial Displacement Equals Zero

The initial conditions of motion are:

$$\text{for} \quad t = 0 \quad x = 0; \quad \frac{dx}{dt} = v_0 \qquad (3.2.13)$$

Solving differential equation of motion (3.2.1) with the initial conditions of motion (3.2.13) we have:

$$x = v_0 t + \frac{rt^2}{2} \qquad (3.2.14)$$

The velocity, the acceleration, and the maximum value of the force are determined according to equations (3.2.8), (3.2.11), and (3.2.12) respectively.

3.2.3 Initial Velocity Equals Zero
The initial conditions of motion are:

$$\text{for} \quad t = 0 \quad x = s_0; \quad \frac{dx}{dt} = 0 \qquad (3.2.15)$$

The solution of differential equation of motion (3.2.1) with the initial conditions of motion (3.2.15) reads:

$$x = s_0 + \frac{rt^2}{2} \qquad (3.2.16)$$

Taking the first derivative of equation (3.2.16), we determine the velocity of the system:

$$\frac{dx}{dt} = rt \qquad (3.2.17)$$

The acceleration and the maximum value of the force are calculated according to equations (3.2.11) and (3.2.12) respectively.

3.2.4 Both the Initial Displacement and Velocity Equal Zero
The initial conditions of motion are:

$$\text{for} \quad t = 0 \quad x = 0; \quad \frac{dx}{dt} = 0 \qquad (3.2.18)$$

Solving differential equation of motion (3.2.1) with the initial conditions of motion (3.2.18), we obtain:

$$x = \frac{rt^2}{2} \qquad (3.2.19)$$

The velocity, the acceleration, and the force are determined by equations (3.2.17), (3.2.11), and (3.2.12) respectively.

3.3 Harmonic Force $A \sin(\omega_1 t + \lambda)$

The intersection of Row 3 with Column 3 in Guiding Table 2.1 indicates the number of this section, which describes a system characterized by the action of the force of inertia and the harmonic force.

Harmonic forces are used in vibratory systems that have a variety of applications. The system is moving in the horizontal direction. We want to determine the basic parameters of motion, and the characteristics of the forces applied to the system.

The model of a system subjected to the action of a harmonic force is presented in Figure 3.3. Accounting for the considerations mentioned above and Figure 3.3, we assemble the left and right sides of the differential equation of motion. The left side consists of the force of inertia, while the right side includes the harmonic force. Hence, the differential equation of motion reads:

$$m \frac{d^2 x}{dt^2} = A \sin(\omega_1 t + \lambda) \tag{3.3.1}$$

The differential equation of motion (3.3.1) has different solutions for various initial conditions of motion. These solutions and their analyses follow.

3.3.1 General Initial Conditions

The general initial conditions of motion are:

$$\text{for} \quad t = 0 \quad x = s_0; \quad \frac{dx}{dt} = v_0 \tag{3.3.2}$$

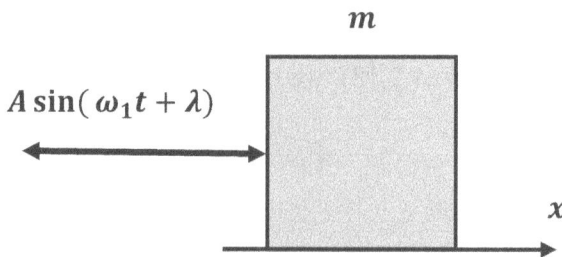

Figure 3.3 Model of a system subjected to a harmonic force

where s_0 and v_0 are the initial displacement and initial velocity respectively.

Transforming the sinusoidal function in equation (3.3.1), we may write:

$$m\frac{d^2x}{dt^2} = A\sin\omega_1 t\cos\lambda + A\cos\omega_1 t\sin\lambda \qquad (3.3.3)$$

Dividing equation (3.3.3) by m, we obtain:

$$\frac{d^2x}{dt^2} = a\sin\omega_1 t\cos\lambda + a\cos\omega_1 t\sin\lambda \qquad (3.3.4)$$

where:

$$a = \frac{A}{m} \qquad (3.3.5)$$

Using Laplace Transform Pairs 3, 6, and 7 from Table 1.1, we convert differential equation of motion (3.3.4) with the initial conditions of motion (3.3.2) from the time domain into the Laplace domain. As a result, we obtain the corresponding algebraic equation of motion in the Laplace domain:

$$l^2 x(l) - l v_0 - l^2 s_0 = \frac{a\omega_1 l}{l^2 + \omega_1^2}\cos\lambda + \frac{a l^2}{l^2 + \omega_1^2}\sin\lambda \qquad (3.3.6)$$

The solution of equation (3.3.6) for the Laplace domain displacement $x(l)$ reads:

$$x(l) = s_0 + \frac{v_0}{l} + \frac{a\omega_1}{l(l^2 + \omega_1^2)}\cos\lambda + \frac{a}{l^2 + \omega_1^2}\sin\lambda \qquad (3.3.7)$$

Based on pairs 1, 5, 2, 16, and 14 from Table 1.1, we invert equation (3.3.7) from the Laplace domain into the time domain and obtain the solution of differential equation of motion (3.3.1) with the initial conditions of motion (3.3.2):

$$x = s_0 + v_0 t + \frac{a}{\omega_1}(t - \frac{1}{\omega_1}\sin\omega_1 t)\cos\lambda + \frac{a}{\omega_1^2}(1 - \cos\omega_1 t)\sin\lambda \qquad (3.3.8)$$

Applying the appropriate algebraic procedures to equation (3.3.8), we obtain:

$$x = s_0 + v_0 t + \frac{at}{\omega_1}\cos\lambda + \frac{a}{\omega_1^2}\sin\lambda - \frac{a}{\omega_1^2}(\sin\omega_1 t \cos\lambda + \cos\omega_1 t \sin\lambda)$$

This equation could be further transformed into the following shape:

$$x = s_0 + v_0 t + \frac{at}{\omega_1}\cos\lambda + \frac{a}{\omega_1^2}\sin\lambda - \frac{a}{\omega_1^2}\sin(\omega_1 t + \lambda) \quad \textbf{(3.3.9)}$$

Taking the first derivative of equation (3.3.9), we determine the velocity of the system:

$$\frac{dx}{dt} = v_0 + \frac{a}{\omega_1}\cos\lambda - \frac{a}{\omega_1}\cos(\omega_1 t + \lambda) \qquad \textbf{(3.3.10)}$$

The second derivative of equation (3.3.9) represents the acceleration of the system:

$$\frac{d^2 x}{dt^2} = a\,\sin(\omega_1 t + \lambda) \qquad \textbf{(3.3.11)}$$

Equation (3.3.10) shows that the system reaches the extreme values of acceleration a_{ext} when:

$$\sin(\omega_1 t + \lambda) = \pm 1 \qquad \textbf{(3.3.12)}$$

Combining equations (3.3.12) and (3.3.11) and substituting notation (3.3.5), we have:

$$a_{ext} = \pm\frac{A}{m} \qquad \textbf{(3.3.13)}$$

Multiplying equation (3.3.13) by m, we obtain the extreme values of the force R_{ext} applied to the system:

$$R_{ext} = \pm A \qquad \textbf{(3.3.14)}$$

Hence, according to equation (3.3.14), the system is subjected to the action of a completely reversed loading cycle characterized by the following maximum R_{max} and minimum R_{min} values of the forces:

$$R_{max} = A \qquad (3.3.15)$$

and:

$$R_{min} = -A \qquad (3.3.16)$$

The stress analysis in this case should include appropriate fatigue calculations based on the forces according to equations (3.3.15) and (3.3.16).

3.3.2 Initial Displacement Equals Zero

The initial conditions of motion are:

$$\text{for} \quad t = 0 \quad x = 0; \quad \frac{dx}{dt} = v_0 \qquad (3.3.17)$$

Solving differential equation of motion (3.3.1) with the initial conditions of motion (3.3.17), we obtain:

$$x = v_0 t + \frac{at}{\omega_1} \cos \lambda + \frac{a}{\omega_1^2} \sin \lambda - \frac{a}{\omega_1^2} \sin(\omega_1 t + \lambda) \qquad (3.3.18)$$

Velocity, acceleration, and the forces applied to the system are calculated according to equations (3.3.10), (3.3.11), (3.3.15), and (3.3.16) respectively.

3.3.3 Initial Velocity Equals Zero

The initial conditions of motion are:

$$\text{for} \quad t = 0 \quad x = s_0; \quad \frac{dx}{dt} = 0 \qquad (3.3.19)$$

The solution of differential equation (3.3.1) with the initial conditions of motion (3.3.19) reads:

$$x = s_0 + \frac{at}{\omega_1} \cos \lambda + \frac{a}{\omega_1^2} \sin \lambda - \frac{a}{\omega_1^2} \sin(\omega_1 t + \lambda) \qquad (3.3.20)$$

Taking the first derivative of equation (3.3.19), we determine the velocity:

$$\frac{dx}{dt} = \frac{a}{\omega_1}\cos\lambda - \frac{a}{\omega_1}\cos(\omega_1 t + \lambda) \qquad (3.3.21)$$

The equations for the acceleration and the forces applied to the system are the same as for the previous cases.

3.3.4 Both the Initial Displacement and Velocity Equal Zero
The initial conditions of motion are:

$$\text{for} \quad t = 0 \quad x = 0; \quad \frac{dx}{dt} = 0 \qquad (3.3.22)$$

Solving differential equation of motion (3.3.1) with the initial conditions of motion (3.3.22), we obtain:

$$x = \frac{at}{\omega_1}\cos\lambda + \frac{a}{\omega_1^2}\sin\lambda - \frac{a}{\omega_1^2}\sin(\omega_1 t + \lambda) \qquad (3.3.23)$$

Equations (3.3.21) and (3.3.11) describe respectively the velocity and acceleration for this case. The forces applied to the system are determined by equations (3.3.15) and (3.3.16).

3.4 Time-Dependent Force $Q\left(\rho + \frac{\mu t}{\tau}\right)$
Guiding Table 2.1 shows that this section represents the intersection of Row 3 and Column 4. Consequently, the current problem is characterized by the force of inertia as a resisting force and the time-dependent force as the active force.

This problem could be related to the analysis of the acceleration of a system subjected to a time-dependent force that is acting a predetermined interval of time. The system is moving in the horizontal direction. It is necessary to determine the values of the displacement, the velocity, the acceleration, and the force applied to the system at the end of the predetermined time, as well as the power of the energy source.

Figure 3.4 shows the model of a system subjected to the time-dependent force. We can now compose the differential equation of

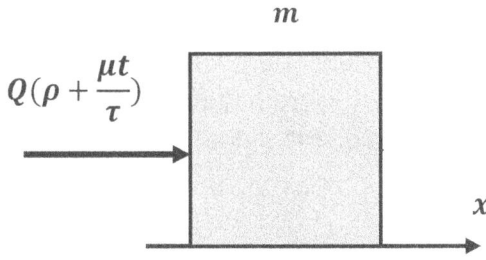

Figure 3.4 Model of a system subjected to a time-dependent force

motion that includes in the left side the force of inertia, and in the right side the time-dependent force. Thus, the differential equation of motion for this case reads:

$$m\frac{d^2x}{dt^2} = Q(\rho + \frac{\mu t}{\tau})$$ (3.4.1)

Differential equation of motion (3.4.1) has different solutions for various initial conditions of motion. These solutions and their analyses are presented below.

3.4.1 General Initial Conditions

The general initial conditions of motion are:

$$\text{for} \quad t = 0 \quad x = s_0; \quad \frac{dx}{dt} = v_0$$ (3.4.2)

where s_0 and v_0 are the initial displacement and initial velocity respectively.

Dividing equation (3.4.1) by m, we have:

$$\frac{d^2x}{dt^2} = q(\rho + \frac{\mu t}{\tau})$$ (3.4.3)

where:

$$q = \frac{Q}{m}$$ (3.4.4)

Based on Laplace Transform Pairs 3, 5, and 2 from Table 1.1, we convert differential equation of motion (3.4.3) with the initial conditions of motion (3.4.2) from the time domain into the Laplace domain, and obtain the corresponding algebraic equation of motion in the Laplace domain:

$$l^2 x(l) - l v_0 - l^2 s_0 = q\rho + \frac{q\mu}{\tau l} \qquad (3.4.5)$$

Solving equation (3.4.5) for the Laplace domain displacement $x(l)$, we have:

$$x(l) = s_0 + \frac{v_0}{l} + \frac{q\rho}{l^2} + \frac{q\mu}{\tau l^3} \qquad (3.4.6)$$

Applying pairs 1, 5, 2, and 8 from Table 1.1 to equation (3.4.6), we invert this equation from the Laplace domain into the time domain and obtain the solution of differential equation of motion (3.4.1) with the initial conditions of motion (3.4.2):

$$x = s_0 + v_0 t + \frac{q\rho}{2} t^2 + \frac{q\mu}{6\tau} t^3 \qquad (3.4.7)$$

Taking the first derivative of equation (3.4.7), we determine the velocity of the system:

$$\frac{dx}{dt} = v_0 + q\rho t + \frac{q\mu}{2\tau} t^2 \qquad (3.4.8)$$

The second derivative of equation (3.4.7) describes the acceleration of the system:

$$\frac{d^2 x}{dt^2} = q\rho + \frac{q\mu}{\tau} t \qquad (3.4.9)$$

The acceleration process lasts the interval of time that equals τ. Therefore, substituting time τ into equations (3.4.7), (3.4.8), and (3.4.9) and recalling notation (3.4.4), we determine respectively the values of the displacement s, the velocity v, and the acceleration a at the end of the time τ.

$$s = s_0 + v_0\tau + \frac{Q}{m}\tau^2(\frac{\rho}{2} + \frac{\mu}{6}) \qquad (3.4.10)$$

$$v = v_0 + \frac{Q}{m}\tau(\rho + \frac{\mu}{2}) \qquad (3.4.11)$$

$$a = \frac{Q}{m}(\rho + \mu) \qquad (3.4.12)$$

For public transportation systems, the acceleration according to equation (3.4.12) should comply with the norms of public health and safety.

For this case, the force applied to the system equals the force of inertia. Multiplying the mass m by the acceleration according to equation (3.4.12), we obtain the force R_0 that is applied to the system:

$$R_0 = Q(\rho + \mu) \qquad (3.4.13)$$

The stress analysis should be based on the force according to equation (3.4.13). The power N of the energy source equals the product of multiplying this force by the velocity according to equation (3.4.11):

$$N = Q(\rho + \mu)[v_0 + \frac{Q}{m}\tau\left(\rho + \frac{\mu}{2}\right)]$$

3.4.2 Initial Displacement Equals Zero

The initial conditions of motion are:

$$\text{for} \quad t = 0 \quad x = 0; \quad \frac{dx}{dt} = v_0 \qquad (3.4.14)$$

The solution of differential equation of motion (3.4.1) with the initial conditions (3.4.14) reads:

$$x = v_0 t + \frac{q\rho}{2}t^2 + \frac{q\mu}{6\tau}t^3 \qquad (3.4.15)$$

Substituting the interval of time τ into equation (3.4.15), we determine the value of the displacement for this case:

$$s = v_0\tau + q\tau^2\left(\frac{\rho}{2} + \frac{\mu}{6}\right) \qquad \textbf{(3.4.16)}$$

The rest of the parameters are the same as for the previous case.

3.4.3 Initial Velocity Equals Zero

The initial conditions of motion are:

$$\text{for} \quad t = 0 \quad x = s_0; \quad \frac{dx}{dt} = 0 \qquad \textbf{(3.4.17)}$$

Solving differential equation of motion (3.4.1) with the initial conditions (3.4.17), we obtain:

$$x = s_0 + \frac{q\rho}{2}t^2 + \frac{q\mu}{6\tau}t^3 \qquad \textbf{(3.4.18)}$$

Taking the first derivative of equation (3.4.18), we determine the velocity of the system:

$$\frac{dx}{dt} = q\rho t + \frac{q\mu}{2\tau}t^2 \qquad \textbf{(3.4.19)}$$

Substituting the time interval τ into equations (3.4.18) and (3.4.19), we determine the values of the displacement and velocity respectively:

$$s = s_0 + q\tau^2\left(\frac{\rho}{2} + \frac{\mu}{6}\right) \qquad \textbf{(3.4.20)}$$

$$v = q\tau\left(\rho + \frac{\mu}{2}\right) \qquad \textbf{(3.4.21)}$$

The values of the acceleration and the force applied to the system can be determined from equations (3.4.12) and (3.4.13) respectively. For this case, the power N of the system equals the product of multiplying the force according to equation (3.4.13) by the velocity according to equation (3.4.21):

$$N = \frac{Q^2\tau}{m}(\rho+\mu)(\rho+\frac{\mu}{2}) \qquad (3.4.22)$$

3.4.4 Both the Initial Displacement and Velocity Equal Zero

The initial conditions of motion are:

$$\text{for} \quad t = 0 \quad x = 0; \quad \frac{dx}{dt} = 0 \qquad (3.4.23)$$

The solution of differential equation of motion (3.4.1) with the initial conditions of motion (3.4.23) reads:

$$x = \frac{q\rho}{2}t^2 + \frac{q\mu}{6\tau}t^3 \qquad (3.4.24)$$

Substituting into equation (3.4.24) the time interval τ, we determine the displacement for this case:

$$s = q\tau^2(\frac{\rho}{2}+\frac{\mu}{6}) \qquad (3.4.25)$$

The values of the velocity, acceleration, applied force, and power can be determined according to the following equations respectively: (3.4.21), (3.4.12), (3.4.13), and (3.4.22).

3.5 Constant Active Force R and Harmonic Force $A \sin(\omega_1 t + \lambda)$

This section, which represents the intersection of Row 3 and Column 5 of Guiding Table 2.1, describes a problem associated with the force of inertia, the constant active force, and the harmonic force. The current problem could be related to the analysis of motion of a vibratory system subjected to the action of a constant active force.

The system is moving in the horizontal direction. We want to determine the basic parameters of motion and the characteristics of the forces applied to the system. Figure 3.5 presents the model

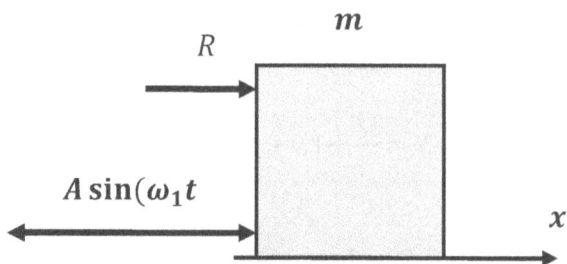

Figure 3.5 Model of a system subjected to a constant active force and a harmonic force

of a system subjected to the action of a constant active force and a harmonic force.

Based on the considerations above and Figure 3.5, it is possible to describe the left and right sides of the differential equation of motion of the system. The left side consists of the force of inertia, while the right side includes the sum of the constant active force and the harmonic force. Hence, the differential equation of motion reads:

$$m\frac{d^2x}{dt^2} = R + A\sin(\omega_1 t + \lambda) \tag{3.5.1}$$

Differential equation of motion (3.5.1) has different solutions for various initial conditions of motion. These solutions and their analyses follow.

3.5.1 General Initial Conditions

The general initial conditions of motion are:

$$\text{for} \quad t = 0 \quad x = s_0; \quad \frac{dx}{dt} = v_0 \tag{3.5.2}$$

where s_0 and v_0 are the initial displacement and initial velocity respectively.

Transforming the sinusoidal function in equation (3.5.1) and dividing the latter by m, we may write:

$$\frac{d^2x}{dt^2} = r + a\sin\omega_1 t \cos\lambda + a\cos\omega_1 t \sin\lambda \tag{3.5.3}$$

where

$$r = \frac{R}{m} \qquad (3.5.4)$$

$$a = \frac{A}{m} \qquad (3.5.5)$$

Using Laplace Transform Pairs 3, 5, 6, and 7 from Table 1.1, we convert differential equation of motion (3.5.3) with the initial conditions of motion (3.5.2) from the time domain into the Laplace domain. As a result, we obtain the corresponding algebraic equation of motion in the Laplace domain:

$$l^2 x(l) - l v_0 - l^2 s_0 = r + \frac{a \omega_1 l}{l^2 + \omega_1^2} \cos \lambda + \frac{a l^2}{l^2 + \omega_1^2} \sin \lambda \qquad (3.5.6)$$

Solving equation (3.5.6) for the Laplace domain displacement, $x(l)$ we have:

$$x(l) = s_0 + \frac{v_0}{l} + \frac{r}{l^2} + \frac{a \omega_1}{l(l^2 + \omega_1^2)} \cos \lambda + \frac{a}{l^2 + \omega_1^2} \sin \lambda \qquad (3.5.7)$$

Based on pairs 1, 5, 2, 8, 16, and 14 from Table 1.1, we invert equation (3.5.7) from the Laplace domain into the time domain and obtain the solution of differential equation of motion (3.5.1) with the initial conditions of motion (3.5.2):

$$x = s_0 + v_0 t + \frac{r}{2} t^2 + \frac{a}{\omega_1} (t - \frac{1}{\omega_1} \sin \omega_1 t) \cos \lambda + \frac{a}{\omega_1^2} (1 - \cos \omega_1 t) \sin \lambda$$

$$(3.5.8)$$

Applying basic algebra to equation (3.5.8), we write:

$$x = s_0 + v_0 t + \frac{r}{2} t^2 + \frac{at}{\omega_1} \cos \lambda$$

$$+ \frac{a}{\omega_1^2} \sin \lambda - \frac{a}{\omega_1^2} (\sin \omega_1 t \cos \lambda + \cos \omega_1 t \sin \lambda)$$

Transforming this equation further, we have:

$$x = s_0 + v_0 t + \frac{r}{2} t^2 + \frac{at}{\omega_1} \cos \lambda + \frac{a}{\omega_1^2} \sin \lambda - \frac{a}{\omega_1^2} \sin(\omega_1 t + \lambda) \qquad \textbf{(3.5.9)}$$

Taking the first derivative of equation (3.5.9), we determine the velocity of the system:

$$\frac{dx}{dt} = v_0 + rt + \frac{a}{\omega_1} \cos \lambda - \frac{a}{\omega_1} \cos(\omega_1 t + \lambda) \qquad \textbf{(3.5.10)}$$

The second derivative of equation (3.5.9) represents the acceleration of the system:

$$\frac{d^2 x}{dt^2} = r + a \sin(\omega_1 t + \lambda) \qquad \textbf{(3.5.11)}$$

It can be seen from equation (3.5.11) that the acceleration reaches its extreme values at the following condition:

$$\sin(\omega_1 t + \lambda) = \pm 1 \qquad \textbf{(3.5.12)}$$

Combining equation (3.5.11) with equation (3.5.12), we determine the extreme values of the acceleration a_{ext}:

$$a_{ext} = r \pm a \qquad \textbf{(3.5.13)}$$

Substituting notations (3.5.4) and (3.5.5) into equation (3.5.13), we have:

$$a_{ext} = \frac{R \pm A}{m} \qquad \textbf{(3.5.14)}$$

Multiplying equation (3.5.14) by m, we obtain the extreme values of force R_{ext} applied to the system:

$$R_{ext} = R \pm A \qquad \textbf{(3.5.15)}$$

According to equation (3.5.15), the system is subjected to the action of a random loading cycle for which the maximum force R_{max} equals:

$$R_{max} = R + A$$

while the minimum force R_{min} equals:

$$R_{min} = R - A$$

The stress analysis of the system for this case should be based on the appropriate fatigue calculations.

3.5.2 Initial Displacement Equals Zero

The initial conditions of motion are:

$$\text{for} \quad t = 0 \quad x = 0; \quad \frac{dx}{dt} = v_0 \qquad (3.5.16)$$

The solution of differential equation of motion (3.5.1) with the initial conditions of motion (3.5.16) reads:

$$x = v_0 t + \frac{r}{2}t^2 + \frac{at}{\omega_1}\cos\lambda + \frac{a}{\omega_1^2}\sin\lambda - \frac{a}{\omega_1^2}\sin(\omega_1 t + \lambda) \quad (3.5.17)$$

The rest of the parameters are the same as in the previous case.

3.5.3 Initial Velocity Equals Zero

The initial conditions of motion are:

$$\text{for} \quad t = 0 \quad x = s_0; \quad \frac{dx}{dt} = 0 \qquad (3.5.18)$$

Solving differential equation of motion (3.5.1) with the initial conditions of motion (3.5.18), we obtain:

$$x = s_0 + \frac{r}{2}t^2 + \frac{at}{\omega_1}\cos\lambda + \frac{a}{\omega_1^2}\sin\lambda - \frac{a}{\omega_1^2}\sin(\omega_1 t + \lambda) \quad (3.5.19)$$

Taking the first derivative of equation (3.5.19), we determine the velocity of the system:

$$\frac{dx}{dt} = rt + \frac{a}{\omega_1}\cos\lambda - \frac{a}{\omega_1}\cos(\omega_1 t + \lambda) \qquad (3.5.20)$$

The rest of the parameters can be determined from equations (3.5.11), (3.5.14) and (3.5.15).

3.5.4 Both the Initial Displacement and Velocity Equal Zero

The initial conditions of motion are:

$$\text{for} \quad t = 0 \quad x = 0; \quad \frac{dx}{dt} = 0 \qquad (3.5.21)$$

The solution of differential equation of motion (3.5.1) with the initial conditions of motion (3.5.21) reads:

$$x = \frac{r}{2}t^2 + \frac{at}{\omega_1}\cos\lambda + \frac{a}{\omega_1^2}\sin\lambda - \frac{a}{\omega_1^2}\sin(\omega_1 t + \lambda) \qquad (3.5.22)$$

The rest of the parameters are the same as for the previous case.

3.6 Harmonic Force $A\sin(\omega_1 t + \lambda)$ and Time-Dependent Force $Q\left(\rho + \frac{\mu t}{\tau}\right)$

This section represents the intersection of Row 3 and Column 6 in Guiding Table 2.1. The current problem is characterized by the force of inertia as the resisting force and the sum of a harmonic force and a time-dependent force as the active forces. Certain vibratory systems on the beginning of their working process may be subjected to the action of a time-dependent force that acts a limited interval of time.

The system is moving in the horizontal direction. We want to determine the basic parameters of motion and the characteristics of the forces applied to the system. Figure 3.6 shows the model of a system subjected to the action of the harmonic force and a time-dependent force.

According to the considerations above and Figure 3.6, it is possible to describe the structure of the differential equation of motion. The left side includes the force of inertia, while the right side consists of the harmonic force and the time-dependent force. Thus, the differential equation of motion reads:

$$m\frac{d^2x}{dt^2} = A\sin(\omega_1 t + \lambda) + Q(\rho + \frac{\mu t}{\tau}) \qquad (3.6.1)$$

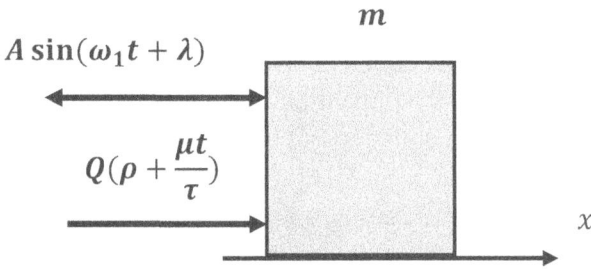

Figure 3.6 Model of a system subjected to a harmonic force and a time-dependent force

Differential equation of motion (3.6.1) has different solutions for various initial conditions of motion. These solutions and their analyses follow.

3.6.1 General Initial Conditions

The general initial conditions of motion are:

$$\text{for} \quad t = 0 \quad x = s_0; \quad \frac{dx}{dt} = v_0 \qquad (3.6.2)$$

where s_0 and v_0 are the initial displacement and initial velocity respectively.

Transforming equation (3.6.1), we may write:

$$m\frac{d^2x}{dt^2} = A\sin\omega_1 t\cos\lambda + A\cos\omega_1 t\sin\lambda + Q\rho + \frac{Q\mu}{\tau}t \qquad (3.6.3)$$

Dividing equation (3.6.3) by m, we have:

$$\frac{d^2x}{dt^2} = a\sin\omega_1 t\cos\lambda + a\cos\omega_1 t\sin\lambda + q\rho + \frac{q\mu}{\tau}t \qquad (3.6.4)$$

where:

$$a = \frac{A}{m} \qquad (3.6.5)$$

$$q = \frac{Q}{m} \qquad (3.6.6)$$

Using Laplace Transform Pairs $3, 6, 7, 5$, and 2 from Table 1.1, we convert differential equation of motion $(3.6.4)$ with the initial conditions of motion $(3.6.2)$ from the time domain into the Laplace domain. The resulting algebraic equation of motion in Laplace domain reads:

$$l^2 x(l) - l v_0 - l^2 s_0 = \frac{a \omega_1 l}{l^2 + \omega_1^2} \cos \lambda + \frac{a l^2}{l^2 + \omega_1^2} \sin \lambda + q \rho + \frac{q \mu}{\tau l} \quad (3.6.7)$$

Solving equation $(3.6.7)$ for the displacement $x(l)$ in the Laplace domain, we have:

$$x(l) = s_0 + \frac{v_0}{l} + \frac{a \omega_1}{l(l^2 + \omega_1^2)} \cos \lambda + \frac{a}{l^2 + \omega_1^2} \sin \lambda + \frac{q \rho}{l^2} + \frac{q \mu}{\tau l^3} \quad (3.6.8)$$

Applying pairs $1, 5, 2, 16, 14$, and 8 from Table 1.1, we invert equation $(3.6.8)$ from the Laplace domain into the time domain, and obtain the solution of differential equation of motion $(3.6.1)$ with the initial conditions of motion $(3.6.2)$:

$$x = s_0 + v_0 t + \frac{a}{\omega_1} \left(t - \frac{1}{\omega_1} \sin \omega_1 t \right) \cos \lambda$$

$$+ \frac{a}{\omega_1^2} (1 - \cos \omega_1 t) \sin \lambda + \frac{q \rho t^2}{2} + \frac{q \mu t^3}{6 \tau} \quad (3.6.9)$$

Applying the appropriate algebra to equation $(3.6.9)$, we write:

$$x = s_0 + \left(v_0 + \frac{a}{\omega_1} \cos \lambda \right) t + \frac{q \rho}{2} t^2 + \frac{q \mu t^3}{6 \tau} + \frac{a}{\omega_1^2} \sin \lambda$$

$$- \frac{a}{\omega_1^2} (\sin \omega_1 t \cos \lambda + \cos \omega_1 t \sin \lambda) \quad (3.6.10)$$

Based on further transformations of equation $(3.6.10)$, we have:

$$x = s_0 + \left(v_0 + \frac{a}{\omega_1} \cos \lambda \right) t + \frac{q \rho}{2} t^2 + \frac{q \mu t^3}{6 \tau} + \frac{a}{\omega_1^2} \sin \lambda - \frac{a}{\omega_1^2} \sin(\omega_1 t + \lambda)$$

$$(3.6.11)$$

Taking the first derivative of equation (3.6.11), we determine the velocity of the system:

$$\frac{dx}{dt} = v_0 + \frac{a}{\omega_1}\cos\lambda + q\rho t + \frac{q\mu t^2}{2\tau} - \frac{a}{\omega_1}\cos(\omega_1 t + \lambda) \quad (3.6.12)$$

The second derivative of equation (3.6.11) characterizes the system's acceleration:

$$\frac{d^2 x}{dt^2} = q\rho + \frac{q\mu t}{\tau} + a\sin(\omega_1 t + \lambda) \quad (3.6.13)$$

This process of motion lasts the interval of time τ. Substituting τ into equation (3.6.13), we obtain the value of the acceleration a_τ at the end of the time interval τ.

$$a_\tau = q\rho + q\mu + a\sin(\omega_1 \tau + \lambda) \quad (3.6.14)$$

The acceleration reaches its extreme values a_{ext} at the following conditions:

$$\sin(\omega_1 \tau + \lambda) = \pm 1 \quad (3.6.15)$$

Combining equations (3.6.14) and (3.6.15), we determine the extreme values of the acceleration:

$$a_{ext} = q\rho + q\mu \pm a \quad (3.6.16)$$

Substituting notations (3.6.5) and (3.6.6) into equation (3.6.16), we have:

$$a_{ext} = \frac{Q}{m}(\rho + \mu) \pm \frac{A}{m} \quad (3.6.17)$$

Multiplying equation (3.6.17) by m, we determine the extreme values of the forces R_{ext} that are applied to the system:

$$R_{ext} = Q(\rho + \mu) \pm A \quad (3.6.18)$$

As seen from equation (3.6.18), the system is subjected to a random loading cycle for which the maximum force R_{max} and the minimum force R_{min} respectively equal:

$$R_{max} = Q(\rho + \mu) + A$$

$$R_{min} = Q(\rho + \mu) - A$$

The stress analysis for this case should be based on appropriate fatigue calculations.

3.6.2 Initial Displacement Equals Zero
The initial conditions of motion are:

$$\text{for} \quad t = 0 \quad x = 0; \quad \frac{dx}{dt} = v_0 \qquad \textbf{(3.6.19)}$$

The solution of differential equation of motion (3.6.1) with the initial conditions of motion (3.6.19) reads:

$$x = \left(v_0 + \frac{a}{\omega_1}\cos \lambda\right)t + \frac{q\rho}{2}t^2 + \frac{q\mu t^3}{6\tau} + \frac{a}{\omega_1^2}\sin \lambda - \frac{a}{\omega_1^2}\sin(\omega_1 t + \lambda)$$

$$\textbf{(3.6.20)}$$

The rest of the parameters are the same as for the previous case.

3.6.3 Initial Velocity Equals Zero
The initial conditions of motion are:

$$\text{for} \quad t = 0 \quad x = s_0; \quad \frac{dx}{dt} = 0 \qquad \textbf{(3.6.21)}$$

Solving differential equation of motion (3.6.1) with the initial conditions of motion (3.6.21), we obtain:

$$x = s_0 + \frac{at}{\omega_1}\cos \lambda + \frac{q\rho}{2}t^2 + \frac{q\mu t^3}{6\tau} + \frac{a}{\omega_1^2}\sin \lambda - \frac{a}{\omega_1^2}\sin(\omega_1 t + \lambda)$$

$$\textbf{(3.6.22)}$$

Taking the first derivative of equation (3.6.22), we determine the velocity for this case:

$$\frac{dx}{dt} = \frac{a}{\omega_1}\cos\lambda + q\rho t + \frac{q\mu t^2}{2\tau} - \frac{a}{\omega_1}\cos(\omega_1 t + \lambda) \quad \textbf{(3.6.23)}$$

The rest of the parameters are the same as for the previous case.

3.6.4 Both the Initial Displacement and Velocity Equal Zero

The initial conditions of motion are:

$$\text{for} \quad t = 0 \quad x = 0; \quad \frac{dx}{dt} = 0 \quad \textbf{(3.6.24)}$$

The solution of differential equation of motion (3.6.1) with the initial conditions of motion (3.6.24) reads:

$$x = \frac{at}{\omega_1}\cos\lambda + \frac{q\rho}{2}t^2 + \frac{q\mu t^3}{6\tau} + \frac{a}{\omega_1^2}\sin\lambda - \frac{a}{\omega_1^2}\sin(\omega_1 t + \lambda) \quad \textbf{(3.6.25)}$$

The rest of the parameters are the same as for the previous case.

4

FRICTION

This chapter describes engineering systems that are subjected to both the force of inertia and the friction force as the resisting forces. The section numbers for this chapter are shown in Row 4 of Guiding Table 2.1, which has plus signs for both of these resisting forces. The intersections of this row with Columns 1 through 6 indicate the six sections in this chapter. The section names reflect the active forces applied to the systems. The left sides of the differential equations in all sections of this chapter are identical and consist of the force of inertia and friction force. The right sides of these equations in the sections differ from each other by the active forces applied to the systems.

4.1 Active Force Equals Zero

According to Guiding Table 2.1 this section describes engineering systems experiencing the action of the force of inertia and the friction force (Row 4) in the absence of active forces (Column 1). Key factors related to the motion of a system in the absence of active forces were discussed earlier in section 1.3.

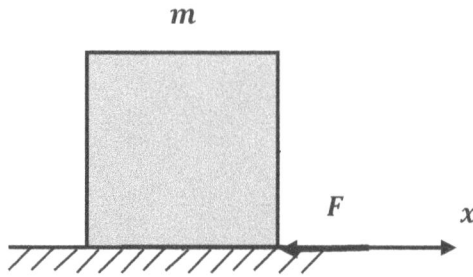

Figure 4.1 Model of a system subjected to a friction force

The current problem represents a deceleration (braking) process of a mechanical system. In real-life conditions, the movable systems are also subjected to air resistance, which in some cases could be insignificant.

The system is moving in the horizontal direction. We want to determine the basic parameters of motion, the braking distance, and the value of the deceleration, as well as the force applied to the system. Figure 4.1 shows a system subjected to the action of a friction force.

The above-mentioned considerations and the model in Figure 4.1 let us compose the differential equation of motion of the system. The left side includes the force of inertia and the friction force, while the right side of the equation equals zero. Thus, the differential equation of motion reads:

$$m\frac{d^2x}{dt^2} + F = 0 \qquad (4.1.1)$$

Differential equation of motion (4.1.1) has different solutions for various initial conditions of motion. These solutions and their analyses are presented below.

4.1.1 General Initial Conditions
The general initial conditions of motion are:

$$\text{for} \quad t = 0 \quad x = s_0; \quad \frac{dx}{dt} = v_0 \qquad (4.1.2)$$

where s_0 and v_0 are the initial displacement and initial velocity respectively.

Dividing equation (4.1.1) by m, we have:

$$\frac{d^2x}{dt^2} + f = 0 \qquad (4.1.3)$$

where:

$$f = \frac{F}{m} \qquad (4.1.4)$$

Using Laplace Transform Pairs 3 and 5 from Table 1.1, we convert differential equation of motion (4.1.3) with the initial conditions of motion (4.1.2) from the time domain into the Laplace domain and obtain the corresponding algebraic equation of motion in the Laplace domain:

$$l^2 x(l) - l v_0 - l^2 s_0 + f = 0 \qquad (4.1.5)$$

Solving equation (4.1.5) for the displacement $x(l)$ in the Laplace domain we have:

$$x(l) = s_0 + \frac{v_0}{l} - \frac{f}{l^2} \qquad (4.1.6)$$

Based on pairs 1, 5, 2, and 8 from Table 1.1, we invert equation (4.1.6) from the Laplace domain into the time domain and obtain the solution of differential equation of motion (4.1.1) with the initial conditions of motion (4.1.2):

$$x = s_0 + v_0 t - \frac{f}{2}t^2 \qquad (4.1.7)$$

Taking the first derivative of the equation (4.1.7), we determine the velocity of the system:

$$\frac{dx}{dt} = v_0 - ft \qquad (4.1.8)$$

The second derivative of equation (4.1.7) represents the acceleration (in this case the deceleration) of the system:

$$\frac{d^2x}{dt^2} = -f \qquad \textbf{(4.1.9)}$$

Equation (4.1.9) shows that the deceleration is a constant value. Substituting notation (4.1.4) into equation (4.1.9), we determine the value of the acceleration/deceleration a_0:

$$a_0 = -\frac{F}{m} \qquad \textbf{(4.1.10)}$$

Thus, the absolute value of the acceleration is:

$$a_0 = |\frac{F}{m}| \qquad \textbf{(4.1.11)}$$

For public transportation systems, the value of the acceleration according to equation (4.1.11) should comply with the norms of public health and safety.

Multiplying the mass m by the absolute value of the acceleration according to equation (4.1.11), we determine the value of the force R_0 applied to the system:

$$R_0 = |F| \qquad \textbf{(4.1.12)}$$

The stress calculations of the system should be performed based on the value of the force according to equation (4.1.12).

At the end of the braking process, the velocity of the system equals zero. Equating the left side of equation (4.1.8) to zero and performing the needed algebraic procedures, we determine the time T that the braking process lasts:

$$T = \frac{v_0}{f} \qquad \textbf{(4.1.13)}$$

Combining equations (4.1.7) and (4.1.13), we calculate the displacement s of the system:

$$s = s_0 + \frac{v_0^2}{2f} \qquad \textbf{(4.1.14)}$$

The braking distance does not include the initial displacement. Excluding the initial displacement from equation (4.1.14), we determine the braking distance s_{br}:

$$s_{br} = \frac{v_0^2}{2f} \qquad (4.1.15)$$

4.1.2 Initial Displacement Equals Zero
The initial conditions of motion are:

$$\text{for} \quad t = 0 \quad x = 0; \quad \frac{dx}{dt} = v_0 \qquad (4.1.16)$$

The solution of differential equation of motion (4.1.1) with the initial conditions of motion (4.1.16) reads:

$$x = v_0 t - \frac{f}{2} t^2 \qquad (4.1.17)$$

The rest of the parameters and their analysis are the same as for the previous case.

4.2 Constant Force R
This section describes the mechanical systems subjected to the action of the force of inertia and friction force as the resisting forces (Row 4 of Guiding Table 2.1) and the constant active force (Column 2).

The system is moving in the horizontal direction. We want to determine the basic parameters of motion, the value of the acceleration, and the force applied to the system. The model of a mechanical system subjected to the action of a constant active force and a friction force is shown in Figure 4.2.

Using the considerations presented above and the model in Figure 4.2, we can compose the differential equation of motion of this system. The left side consists of the force of inertia and the friction force, while the right side includes the constant active force. Therefore, the differential equation of motion reads:

$$m \frac{d^2 x}{dt^2} + F = R \qquad (4.2.1)$$

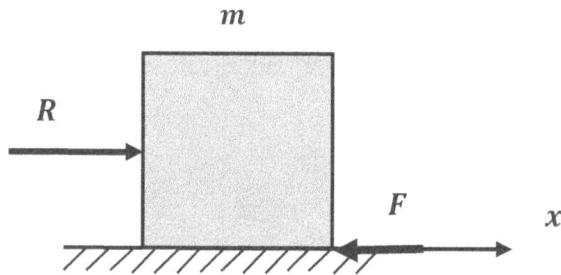

Figure 4.2 Model of a system subjected to a constant active force and a friction force

Differential equation of motion (4.2.1) has different solutions for various initial conditions of motion. These solutions and their analyses are presented below.

4.2.1 General Initial Conditions

The general initial conditions of motion are:

$$\text{for} \quad t = 0 \quad x = s_0; \quad \frac{dx}{dt} = v_0 \qquad \text{(4.2.2)}$$

where s_0 and v_0 are the initial displacement and initial velocity respectively.

Dividing equation (4.2.1) by m, we write:

$$\frac{d^2x}{dt^2} + f = r \qquad \text{(4.2.3)}$$

where:

$$f = \frac{F}{m} \qquad \text{(4.2.4)}$$

$$r = \frac{R}{m} \qquad \text{(4.2.5)}$$

Using Laplace Transform Pairs 3 and 5 from Table 1.1, we convert differential equation of motion (4.2.3) with the initial conditions of motion (4.2.2) from the time domain into the Laplace domain and obtain the corresponding algebraic equation of motion in Laplace domain:

$$l^2 x(l) - lv_0 - l^2 s_0 + f = r \qquad (4.2.6)$$

The solution of equation (4.2.6) for the Laplace domain displacement $x(l)$ reads:

$$x(l) = s_0 + \frac{v_0}{l} + \frac{r-f}{l^2} \qquad (4.2.7)$$

Based on pairs 1, 5, 2, and 8 from Table 1.1, we invert equation (4.2.7) from the Laplace domain into the time domain and obtain the solution of the differential equation of motion (4.2.1) with the initial conditions of motion (4.2.2):

$$x = s_0 + v_0 t + \frac{r-f}{2} t^2 \qquad (4.2.8)$$

The first derivative of equation (4.2.8) represents the velocity of the system:

$$\frac{dx}{dt} = v_0 + (r - f)t \qquad (4.2.9)$$

Taking the second derivative of equation (4.2.8), we determine the acceleration:

$$\frac{d^2 x}{dt^2} = r - f \qquad (4.2.10)$$

Equation (4.2.10) shows that the acceleration for this case represents a constant value. Substituting notations (4.2.4) and (4.2.5) into equation (4.2.10), we determine the value of the acceleration a_0:

$$a_0 = \frac{R - F}{m} \qquad (4.2.11)$$

For public transportation systems, the acceleration according to equation (4.2.11) should comply with the norms of public health and safety.

In general, the force applied to the system represents the sum of the force of inertia and the resisting forces. Adding the friction force F to the product of multiplying equation (4.2.11) by the mass m, we calculate the force R_0 applied to the system:

$$R_0 = R \qquad (4.2.12)$$

The stress calculations of the system should be based on the force according to equation (4.2.12).

4.2.2 Initial Displacement Equals Zero

The initial conditions of motion are:

$$\text{for} \quad t = 0 \quad x = 0; \quad \frac{dx}{dt} = v_0 \qquad (4.2.13)$$

The solution of differential equation of motion (4.2.1) with the initial conditions of motion (4.2.13) reads:

$$x = v_0 t + \frac{r - f}{2} t^2 \qquad (4.2.14)$$

The rest of the analysis is the same as in the previous case.

4.2.3 Initial Velocity Equals Zero

The initial conditions of motion are:

$$\text{for} \quad t = 0 \quad x = s_0; \quad \frac{dx}{dt} = 0 \qquad (4.2.15)$$

Solving differential equation of motion (4.2.1) with the initial conditions of motion (4.2.15), we have:

$$x = s_0 + \frac{r - f}{2} t^2 \qquad (4.2.16)$$

Taking the first derivative of equation (4.2.16), we determine the velocity of the system:

$$\frac{dx}{dt} = (r - f) t \qquad (4.2.17)$$

The rest of the analysis is the same as in the previous cases.

4.2.4 Both the Initial Displacement and Velocity Equal Zero

The initial conditions of motion are:

$$\text{for} \quad t = 0 \quad x = 0; \quad \frac{dx}{dt} = 0 \qquad \textbf{(4.2.18)}$$

The solution of differential equation of motion (4.2.1) with the initial conditions of motion (4.2.18) reads:

$$x = \frac{r - f}{2} t^2 \qquad \textbf{(4.2.19)}$$

The velocity and the acceleration can be determined from equations (4.2.17) and (4.2.10) respectively. The analysis of the parameters is the same as in the previous cases.

4.3 Harmonic Force $A \sin(\omega_1 t + \lambda)$

This section describes problems that are characterized by the force of inertia and friction force as the resisting forces (Row 4 of Guiding Table 2.1), and the action of a harmonic force (Column 3). In general, a harmonic force applied to a mechanical system causes a vibratory motion of the system.

The system is moving in the horizontal direction. We want to determine the basic parameters of motion and the characteristics of the forces applied to the system. Figure 4.3 shows the model of a system subjected to the action of a harmonic force and a friction force.

According to the considerations above and the model in Figure 4.3, we can compose the differential equation of motion. The left side includes the force of inertia and the friction force, while the right side consists of the harmonic force. Therefore, the differential equation of motion reads:

$$m \frac{d^2 x}{dt^2} + F = A \sin(\omega_1 t + \lambda) \qquad \textbf{(4.3.1)}$$

Keep in mind that equation (4.3.1) is valid just to the moment when the velocity of the system becomes equal to zero and changes its

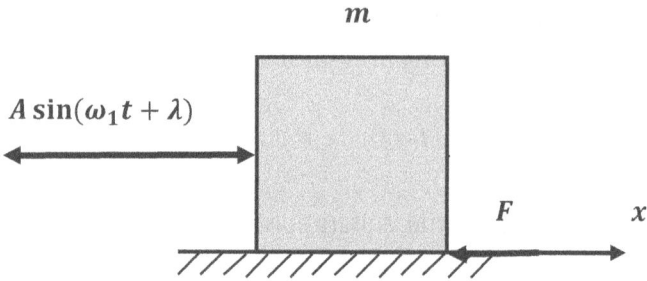

Figure 4.3 Model of a system subjected to a harmonic force and a friction force

direction, causing the friction force to change its direction as well. Considerations related to the behavior of a friction force applied to a vibratory system are presented in section 1.2.

The differential equation of motion (4.3.1) has different solutions for various initial conditions of motion. These solutions and their analyses are presented below.

4.3.1 General Initial Conditions

The general initial conditions of motion read:

$$\text{for} \quad t = 0 \quad x = s_0; \quad \frac{dx}{dt} = v_0 \qquad (4.3.2)$$

where s_0 and v_0 are the initial displacement and initial velocity respectively.

Transforming the sinusoidal function in equation (4.3.1) and dividing the latter by m, we have:

$$\frac{d^2x}{dt^2} + f = a\sin\omega_1 t\cos\lambda + a\cos\omega_1 t\sin\lambda \qquad (4.3.3)$$

where:

$$f = \frac{F}{m} \qquad (4.3.4)$$

$$a = \frac{A}{m} \qquad (4.3.5)$$

Using Laplace Transform Pairs 3, 5, 6, and 7 from Table 1.1, we convert differential equation of motion (4.3.3) with the initial conditions of motion (4.3.2) from the time domain into the Laplace domain and obtain the corresponding algebraic equation of motion in the Laplace domain:

$$l^2 x(l) - lv_0 - l^2 s_0 + f = a\cos\lambda \frac{\omega_1 l}{l^2 + \omega_1^2} + a\sin\lambda \frac{l^2}{l^2 + \omega_1^2} \quad (4.3.6)$$

Solving equation (4.3.6) for the displacement $x(l)$ in the Laplace domain, we have:

$$x(l) = s_0 + \frac{v_0}{l} - \frac{f}{l^2} + a\cos\lambda \frac{\omega_1}{l(l^2 + \omega_1^2)} + a\sin\lambda \frac{1}{l^2 + \omega_1^2} \quad (4.3.7)$$

Based on pairs 1, 5, 2, 8, 16, and 14 from Table 1.1, we invert equation (4.3.7) from Laplace domain into the time domain and obtain the solution of differential equation of motion (4.3.1) with the initial conditions of motion (4.3.2):

$$x = s_0 + v_0 t - \frac{f}{2}t^2 + \frac{a\cos\lambda}{\omega_1}\left(t - \frac{1}{\omega_1}\sin\omega_1 t\right) + \frac{a\sin\lambda}{\omega_1^2}(1 - \cos\omega_1 t)$$

$$(4.3.8)$$

Performing some conventional transformations with equation (4.3.8), we write:

$$x = s_0 + v_0 t - \frac{f}{2}t^2 + \frac{at}{\omega_1}\cos\lambda + \frac{a}{\omega_1^2}\sin\lambda - \frac{a}{\omega_1^2}\sin(\omega_1 t + \lambda) \quad (4.3.9)$$

Taking the first derivative of equation (4.3.9), we determine the velocity of the system:

$$\frac{dx}{dt} = v_0 - ft + \frac{a}{\omega_1}[\cos\lambda - \cos(\omega_1 t + \lambda)] \quad (4.3.10)$$

The second derivative of equation (4.3.9) allows calculating the acceleration:

$$\frac{d^2 x}{dt^2} = -f + a\sin(\omega_1 t + \lambda) \quad (4.3.11)$$

Due to the initial velocity, the system may perform a complete vibratory cycle during the forward motion. Therefore, the extreme values of the acceleration according to equation (4.3.11) occur at the following conditions:

$$\sin(\omega_1 t + \lambda) = \pm 1 \qquad (4.3.12)$$

Hence, combining equations (4.3.11) and (4.3.12), we determine the extreme values of the acceleration a_{ext}:

$$a_{ext} = -f \pm a \qquad (4.3.13)$$

Substituting notations (4.3.4) and (4.3.5) into equation (4.3.13), we write:

$$a_{ext} = \frac{-F \pm A}{m} \qquad (4.3.14)$$

Adding friction force F to the product of multiplying equation (4.3.14) by the mass m, we determine the extreme values R_{ext} of the forces applied to the system:

$$R_{ext} = \pm A \qquad (4.3.15)$$

According to equation (4.3.15), the system is subjected to the action of a completely reversed loading cycle having the maximum value of the force $R_{max} = A$ and the minimum force $R_{min} = -A$. The stress analysis should include fatigue considerations.

4.3.2 Initial Displacement Equals Zero
The initial conditions of motion are:

$$\text{for} \quad t = 0 \quad x = 0; \quad \frac{dx}{dt} = v_0 \qquad (4.3.16)$$

The solution of differential equation of motion (4.3.1) with the initial conditions of motion (4.3.16) reads:

$$x = v_0 t - \frac{f}{2} t^2 + \frac{at}{\omega_1} \cos \lambda + \frac{a}{\omega_1^2} \sin \lambda - \frac{a}{\omega_1^2} \sin(\omega_1 t + \lambda) \qquad (4.3.17)$$

The velocity, acceleration, and applied forces are the same as for the previous case.

4.3.3   Initial Velocity Equals Zero

The initial conditions of motion are:

$$\text{for} \quad t = 0 \quad x = s_0; \quad \frac{dx}{dt} = 0 \qquad (4.3.18)$$

Solving differential equation of motion (4.3.1) with the initial conditions of motion (4.3.18), we obtain:

$$x = s_0 - \frac{f}{2}t^2 + \frac{at}{\omega_1}\cos\lambda + \frac{a}{\omega_1^2}\sin\lambda - \frac{a}{\omega_1^2}\sin(\omega_1 t + \lambda) \qquad (4.3.19)$$

Taking the first derivative of equation (4.3.19), we determine the velocity of the system:

$$\frac{dx}{dt} = -ft + \frac{a}{\omega_1}[\cos\lambda - \cos(\omega_1 t + \lambda)] \qquad (4.3.20)$$

The acceleration can be calculated from equation (4.3.11). The values of the applied forces are the same as in the previous cases.

4.3.4   Both the Initial Displacement and Velocity Equal Zero

The initial conditions of motion are:

$$\text{for} \quad t = 0 \quad x = 0; \quad \frac{dx}{dt} = 0 \qquad (4.3.21)$$

In this case, the motion begins when the exciting force exceeds the friction force. The solution of differential equation of motion (4.3.1) with the initial conditions of motion (4.3.21) reads:

$$x = -\frac{f}{2}t^2 + \frac{at}{\omega_1}\cos\lambda + \frac{a}{\omega_1^2}\sin\lambda - \frac{a}{\omega_1^2}\sin(\omega_1 t + \lambda) \quad (4.3.22)$$

The rest of the parameters and their analysis are the same as in the previous case.

4.4 Time-Dependent Force $Q\left(\rho+\frac{\mu t}{\tau}\right)$

According to Guiding Table 2.1 this section describes problems characterized by the force of inertia and the friction force as the resisting forces (Row 4) and the time-dependent force as the active force (Column 4). Some mechanical systems, at the beginning of their working processes, may utilize the time-dependent force that is acting during a certain interval of time. For example, we may consider a transportation system in which the active force is increasing during a predetermined interval of time. By the end of this time, the system reaches a certain velocity.

The system is moving in the horizontal direction. We want to determine the values of the displacement, the velocity, the acceleration, the force applied to the system, and the power of the energy source. All these values should be calculated for the end of time interval τ. The model of a system subjected to the action of a time-dependent force and a friction force is shown in Figure 4.4.

Based on the considerations above and the model in Figure 4.4, we can compose the differential equation of motion for the current system. The left side includes the force of inertia and the friction force, while the right side consists of the time-dependent force. Thus, the differential equation of motion reads:

$$m\frac{d^2x}{dt^2}+F=Q(\rho+\frac{\mu t}{\tau}) \qquad (4.4.1)$$

Equation (4.4.1) describes the motion of a system in which $Q\rho > F$.

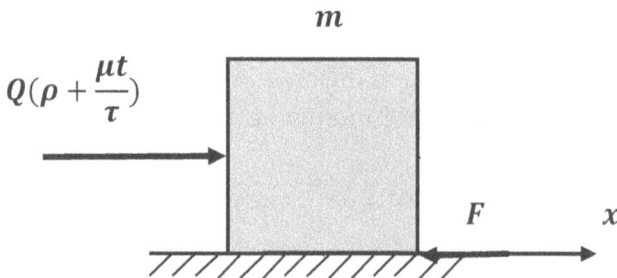

Figure 4.4 Model of a system subjected to a time-dependent force and a friction force

Differential equation of motion (4.4.1) has different solutions for various initial conditions of motion. These solutions and their analyses are presented below.

4.4.1 General Initial Conditions

The general initial conditions of motion are:

$$\text{for} \quad t = 0 \quad x = s_0; \quad \frac{dx}{dt} = v_0 \qquad (4.4.2)$$

where s_0 and v_0 are the initial displacement and initial velocity respectively.

Dividing equation (4.4.1) by m, we obtain:

$$\frac{d^2x}{dt^2} + f = q(\rho + \frac{\mu t}{\tau}) \qquad (4.4.3)$$

where:

$$f = \frac{F}{m} \qquad (4.4.4)$$

$$q = \frac{Q}{m} \qquad (4.4.5)$$

Based on Laplace Transform Pairs 3, 5, and 2 from Table 1.1, we convert differential equation of motion (4.4.3) with the initial conditions of motion (4.4.2) from the time domain into the Laplace domain and obtain the corresponding algebraic equation of motion in the Laplace domain:

$$l^2 x(l) - l v_0 - l^2 s_0 + f = q(\rho + \frac{\mu}{\tau l}) \qquad (4.4.6)$$

Solving equation (4.4.6) for the Laplace domain displacement $x(l)$, we have:

$$x(l) = s_0 + \frac{v_0}{l} + \frac{q\rho - f}{l^2} + \frac{q\mu}{\tau l^3} \qquad (4.4.7)$$

Using pairs 1, 5, 2, and 8 from Table 1.1, we invert equation (4.4.7) from the Laplace domain into the time domain and obtain the solution of differential equation of motion (4.4.1) with the initial conditions of motion (4.4.2):

$$x = s_0 + v_0 t + \frac{1}{2}(q\rho - f)t^2 + \frac{q\mu}{6\tau}t^3 \qquad (4.4.8)$$

Taking the first derivative of equation (4.4.8), we determine the velocity of the system:

$$\frac{dx}{dt} = v_0 + (q\rho - f)t + \frac{q\mu}{2\tau}t^2 \qquad (4.4.9)$$

The second derivative of equation (4.4.8) represents the acceleration of the system:

$$\frac{d^2x}{dt^2} = q\rho - f + \frac{q\mu}{\tau}t \qquad (4.4.10)$$

In the current problem, differential equation (4.4.1) describes the motion that is lasting an interval of time τ. Equating the running time t in equations (4.4.8), (4.4.9), and (4.4.10) to the interval of time τ, we obtain respectively the values of the displacement s, the velocity v, and the acceleration a at the end of time τ.

$$s = s_0 + v_0\tau + \frac{3q\rho - 3f + q\mu}{6}\tau^2 \qquad (4.4.11)$$

$$v = v_0 + \frac{2q\rho - 2f + q\mu}{2}\tau \qquad (4.4.12)$$

$$a = q(\rho + \mu) - f \qquad (4.4.13)$$

The value of the force R_0 applied to the system at the end of time τ is equal to the sum of the force of inertia and the friction force. Multiplying equation (4.4.13) by m, we determine the force of inertia. Adding the friction force F to the force of inertia, and substituting notations (4.4.4) and (4.4.5), we determine the force applied to the system:

$$R_0 = Q(\rho + \mu) \qquad (4.4.14)$$

The stress calculations of the system should be based on the force according to equation (4.4.14).

Multiplying the force according to equation (4.4.14) by the velocity according to equation (4.4.12), we determine the power N of the energy source:

$$N = Q(\rho + \mu)[v_0 + \frac{Q(2\rho + \mu) - 2F}{2m}\tau]$$

4.4.2 Initial Displacement Equals Zero

The initial conditions of motion are:

$$\text{for} \quad t = 0 \quad x = 0; \quad \frac{dx}{dt} = v_0 \qquad (4.4.15)$$

The solution of differential equation of motion (4.4.1) with the initial conditions of motion (4.4.15) reads:

$$x = v_0 t + \frac{1}{2}(q\rho - f)t^2 + \frac{q\mu}{6\tau}t^3 \qquad (4.4.16)$$

Using equation (4.4.16), we calculate the displacement at the end of time τ.

$$s = v_0\tau + \frac{3q\rho - 3f + q\mu}{6}\tau^2 \qquad (4.4.17)$$

The rest of the parameters and their analysis are the same as for the previous case.

4.4.3 Initial Velocity Equals Zero

The initial conditions of motion are:

$$\text{for} \quad t = 0 \quad x = s_0; \quad \frac{dx}{dt} = 0 \qquad (4.4.18)$$

Solving differential equation of motion (4.4.1) with the initial conditions of motion (4.4.18), we have:

$$x = s_0 + \frac{1}{2}(q\rho - f)t^2 + \frac{q\mu}{6\tau}t^3 \qquad (4.4.19)$$

Taking the first derivative of equation (4.4.19), we determine the velocity of the system:

$$\frac{dx}{dt} = (q\rho - f)t + \frac{q\mu}{2\tau}t^2 \qquad (4.4.20)$$

Substituting the time τ into equations (4.4.19) and (4.4.20), we calculate the displacement and the velocity respectively:

$$s = s_0 + \frac{3q\rho - 3f + q\mu}{6}\tau^2 \qquad (4.4.21)$$

$$v = \frac{2q\rho - 2f + q\mu}{2}\tau \qquad (4.4.22)$$

Multiplying the force according to equation (4.4.14) by the velocity according to equation (4.4.22), we determine the power:

$$N = Q(\rho + \mu)\frac{Q(2\rho + \mu) - 2F}{2m}\tau$$

The rest of the parameters and their analysis is the same as in the previous case.

4.4.4 Both the Initial Displacement and Velocity Equal Zero

The initial conditions of motion are:

$$\text{for} \quad t = 0 \quad x = 0; \quad \frac{dx}{dt} = 0 \qquad (4.4.23)$$

The solution of differential equation of motion (4.4.1) with the initial conditions of motion (4.4.23) reads:

$$x = \frac{1}{2}(q\rho - f)t^2 + \frac{q\mu}{6\tau}t^3 \qquad (4.4.24)$$

According to equation (4.4.24), the displacement at the end of time τ equals:

$$s = \frac{3q\rho - 3f + q\mu}{6}\tau^2 \qquad (4.4.25)$$

The rest of the parameters and their analysis are the same as in the previous case.

4.5 Constant Force R and Harmonic Force $A\sin(\omega_1 t + \lambda)$

This section describes problems that have the force of inertia and the friction force as the resisting forces (Row 4 in Guiding Table 2.1) and both the constant active force and the harmonic force (Column 5). The current problem could be related to the working process of certain vibratory systems. Usually the constant active force exceeds the resisting forces and there are no interruptions in the vibratory motion of the system.

The system is moving in the horizontal direction. We want to determine the basic parameters of motion and the characteristics of the forces applied to the system. The model of a system subjected to the action of a constant active force, a harmonic force, and a friction force is shown in Figure 4.5.

Based on these considerations and the model in Figure 4.5, we can describe the structure of the differential equation of motion of the system. The left side consists of the force of inertia and the friction force, while the right side of this equation includes the constant active force and the harmonic force. Hence, the differential equation of motion reads:

$$m\frac{d^2x}{dt^2} + F = R + A\sin(\omega_1 t + \lambda) \qquad (4.5.1)$$

The considerations related to the behavior of the friction force applied to a vibratory system are discussed in section 1.2. Differential equation of motion (4.5.1) has different solutions for various initial conditions of motion. These solutions and their analyses are presented below.

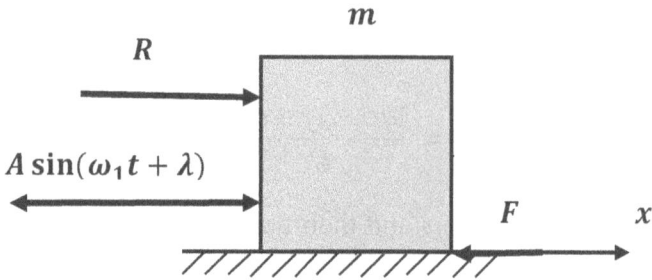

Figure 4.5 Model of a system subjected to a constant active force, a harmonic fore, and a friction force

4.5.1 General Initial Conditions

The general initial conditions of motion are:

$$\text{for} \quad t = 0 \quad x = s_0; \quad \frac{dx}{dt} = v_0 \qquad (4.5.2)$$

where s_0 and v_0 are the initial displacement and initial velocity respectively.

Transforming the sinusoidal function in equation (4.5.1) and dividing this equation by m, we have:

$$\frac{d^2x}{dt^2} + f = r + a \sin \omega_1 t \cos \lambda + a \cos \omega_1 t \sin \lambda \qquad (4.5.3)$$

where:

$$f = \frac{F}{m} \qquad (4.5.4)$$

$$r = \frac{R}{m} \qquad (4.5.5)$$

$$a = \frac{A}{m} \qquad (4.5.6)$$

Using Laplace Transform Pairs 3, 5, 6, and 7 from Table 1.1, we convert differential equation of motion (4.5.3) with the initial conditions of motion (4.5.2) from the time domain into the Laplace

domain. The resulting algebraic equation of motion in the Laplace domain reads:

$$l^2 x(l) - lv_0 - l^2 s_0 + f = r + \frac{a\omega_1 l}{l^2 + \omega_1^2} \cos \lambda + \frac{al^2}{l^2 + \omega_1^2} \sin \lambda \quad \textbf{(4.5.7)}$$

Solving equation (4.5.7) for the Laplace domain displacement $x(l)$, we have:

$$x(l) = s_0 + \frac{v_0}{l} + \frac{r - f}{l^2} + \frac{a\omega_1}{l(l^2 + \omega_1^2)} \cos \lambda + \frac{a}{l^2 + \omega_1^2} \sin \lambda \quad \textbf{(4.5.8)}$$

Based on pairs 1, 5, 2, 8, 16, and 14 from Table 1.1, we invert equation (4.5.8) from the Laplace domain into the time domain and obtain the solution of differential equation of motion (4.5.1) with the initial conditions of motion (4.5.2):

$$x = s_0 + v_0 t + \frac{r - f}{2} t^2 + \frac{a \cos \lambda}{\omega_1} (t - \frac{1}{\omega_1} \sin \omega_1 t) + \frac{a \sin \lambda}{\omega_1^2} (1 - \cos \omega_1 t)$$

$$\textbf{(4.5.9)}$$

Performing algebraic procedures with equation (4.5.9), we have:

$$x = s_0 + v_0 t + \frac{r - f}{2} t^2 + \frac{at}{\omega_1} \cos \lambda + \frac{a}{\omega_1^2} \sin \lambda$$

$$- \frac{a}{\omega_1^2} (\sin \omega_1 t \cos \lambda + \cos \omega_1 t \sin \lambda)$$

Further transformation of this equation reads:

$$x = s_0 + v_0 t + \frac{r - f}{2} t^2 + \frac{at}{\omega_1} \cos \lambda + \frac{a}{\omega_1^2} \sin \lambda - \frac{a}{\omega_1^2} \sin(\omega_1 t + \lambda)$$

$$\textbf{(4.5.10)}$$

The first derivative of equation (4.5.10) represents the velocity of the system:

$$\frac{dx}{dt} = v_0 + (r - f)t + \frac{a}{\omega_1} \cos \lambda - \frac{a}{\omega_1} \cos(\omega_1 t + \lambda) \quad \textbf{(4.5.11)}$$

Taking the second derivative of equation (4.5.10), we determine the acceleration:

$$\frac{d^2x}{dt^2} = r - f + a\,\sin(\omega_1 t + \lambda) \qquad (4.5.12)$$

The presence of the constant active force lets us assume that the extreme values of the acceleration according to equation (4.5.12) occur at the following conditions:

$$\sin(\omega_1 t + \lambda) = \pm 1 \qquad (4.5.13)$$

Combining equation (4.5.12) with equation (4.5.13), we determine the extreme values of the acceleration a_{ext}:

$$a_{ext} = r - f \pm a \qquad (4.5.14)$$

Substituting notations (4.5.4), (4.5.5), and (4.5.6) into equation (4.5.14), we may write:

$$a_{ext} = \frac{R - F + A}{m} \qquad (4.5.15)$$

Adding the friction force F to the product of multiplying equation (4.5.15) by m, we obtain the extreme values of the force R_{ext}:

$$R_{ext} = R \pm A \qquad (4.5.16)$$

According to equation (4.5.16), the system is subjected to the action of a random loading cycle for which the maximum force $R_{max} = R + A$, while the minimum force $R_{min} = R - A$. Therefore, the stress calculations for this case should be based on appropriate fatigue considerations.

4.5.2 Initial Displacement Equals Zero

The initial conditions of motion are:

$$\text{for} \quad t = 0 \quad x = 0; \quad \frac{dx}{dt} = v_0 \qquad (4.5.17)$$

The solution of differential equation of motion (4.5.1) with the initial conditions of motion (4.5.17) reads:

$$x = v_0 t + \frac{r-f}{2}t^2 + \frac{at}{\omega_1}\cos\lambda + \frac{a}{\omega_1^2}\sin\lambda - \frac{a}{\omega_1^2}\sin(\omega_1 t + \lambda) \quad \textbf{(4.5.18)}$$

The rest of the parameters are the same as in the previous case.

4.5.3 Initial Velocity Equals Zero
The initial conditions of motion are:

$$\text{for} \quad t = 0 \quad x = s_0; \quad \frac{dx}{dt} = 0 \qquad \textbf{(4.5.19)}$$

Solving differential equation of motion (4.5.1) with the initial conditions of motion (4.5.19), we have:

$$x = s_0 + \frac{r-f}{2}t^2 + \frac{at}{\omega_1}\cos\lambda + \frac{a}{\omega_1^2}\sin\lambda - \frac{a}{\omega_1^2}\sin(\omega_1 t + \lambda) \quad \textbf{(4.5.20)}$$

The first derivative of equation (4.5.20) represents the velocity of the system:

$$\frac{dx}{dt} = (r-f)t + \frac{a}{\omega_1}\cos\lambda - \frac{a}{\omega_1}\cos(\omega_1 t + \lambda) \qquad \textbf{(4.5.21)}$$

The rest of the parameters can be determined from equations (4.5.12), (4.5.15) and (4.5.16).

4.5.4 Both the Initial Displacement and Velocity Equal Zero
The initial conditions of motion are:

$$\text{for} \quad t = 0 \quad x = 0; \quad \frac{dx}{dt} = 0 \qquad \textbf{(4.5.22)}$$

The solution of differential equation of motion (4.5.1) with the initial conditions of motion (4.5.22) reads:

$$x = \frac{r-f}{2}t^2 + \frac{at}{\omega_1}\cos\lambda + \frac{a}{\omega_1^2}\sin\lambda - \frac{a}{\omega_1^2}\sin(\omega_1 t + \lambda) \qquad \textbf{(4.5.23)}$$

The rest of the parameters are the same as for the previous case.

4.6 Harmonic Force $A\sin(\omega_1 t + \lambda)$ and Time-Dependent Force $Q\left(\rho + \frac{\mu t}{\tau}\right)$

This section describes problems characterized by the force of inertia and the friction force as the resisting forces (Row 4 in Guiding Table 2.1) and the harmonic force and the time-dependent force (Column 6) as the active forces. Sometimes in the beginning of certain working processes, vibratory systems can be subjected to the action of a time-dependent force that acts during a limited interval of time. In these kinds of systems, the initial value of the active force $Q\rho$ exceeds the value of the friction force, resulting in the system's vibratory motion without a delay.

The system is moving in the horizontal direction. We want to determine the basic parameters of motion and the characteristics of the forces applied to the system. Figure 4.6 represents the model of a system subjected to the action of a harmonic force, a time-dependent force, and a friction force. Based on the considerations above and the model in Figure 4.6, we can compose the differential equation of motion. The left side includes the force of inertia and the friction force, while the right side consists of the harmonic force and the time-dependent force. The differential equation of motion reads:

$$m\frac{d^2x}{dt^2} + F = A\sin(\omega_1 t + \lambda) + Q(\rho + \frac{\mu t}{\tau}) \qquad \textbf{(4.6.1)}$$

The considerations related to the behavior of the friction force applied to a vibratory system are discussed in section 1.2.

Transforming equation (4.6.1), we may write:

$$m\frac{d^2x}{dt^2} + F = A\sin\omega_1 t\cos\lambda + A\cos\omega_1 t\sin\lambda + Q\rho + \frac{Q\mu}{\tau}t \qquad \textbf{(4.6.2)}$$

Differential equation of motion (4.6.1) has different solutions for various initial conditions of motion. These solutions and their analyses are presented below.

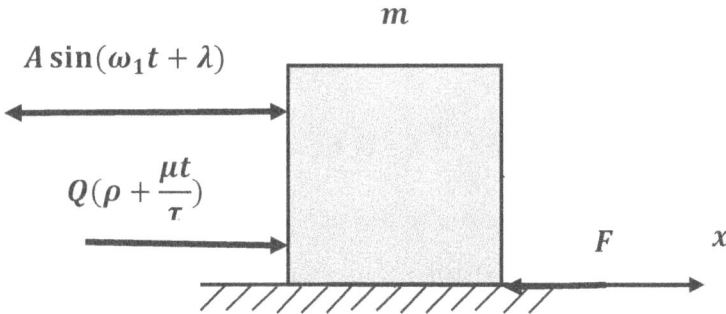

Figure 4.6 Model of a system subjected to a harmonic force, a time-dependent force, and a friction force

4.6.1 General Initial Conditions

The general initial conditions of motion are:

$$\text{for} \quad t = 0 \quad x = s_0; \quad \frac{dx}{dt} = v_0 \qquad (4.6.3)$$

where s_0 and v_0 are the initial displacement and initial velocity respectively.

Dividing equation (4.6.2) by m, we may write:

$$\frac{d^2x}{dt^2} + f = a\sin\omega_1 t\cos\lambda + a\cos\omega_1 t\sin\lambda + q\rho + \frac{q\mu}{\tau}t \quad (4.6.4)$$

where:

$$f = \frac{F}{m} \qquad (4.6.5)$$

$$a = \frac{A}{m} \qquad (4.6.6)$$

$$q = \frac{Q}{m} \qquad (4.6.7)$$

Using Laplace Transform Pairs 3, 5, 6, 7, and 2 from Table 1.1, we convert differential equation of motion (4.6.4) with the initial conditions of motion (4.6.3) from the time domain into the Laplace

domain. The resulting algebraic equation of motion in the Laplace domain reads:

$$x(l) - lv_0 - l^2 s_0 + f = \frac{a\omega_1 l}{l^2 + \omega_1^2} \cos\lambda + \frac{al^2}{l^2 + \omega_1^2} \sin\lambda + q\rho + \frac{q\mu}{\tau l}$$

(4.6.8)

Solving equation (4.6.8) for the displacement in Laplace domain $x(l)$, we obtain:

$$x(l) = s_0 + \frac{v_0}{l} + \frac{a\omega_1}{l(l^2 + \omega_1^2)} \cos\lambda + \frac{a}{l^2 + \omega_1^2} \sin\lambda + \frac{q\rho - f}{l^2} + \frac{q\mu}{\tau l^3}$$

(4.6.9)

Based on pairs 1, 5, 2, 16, 14, and 8 from Table 1.1, we invert equation (4.6.8) into the time domain. The inversion represents the solution of differential equation of motion (4.6.1) with the initial conditions of motion (4.6.2):

$$x = s_0 + v_0 t + \frac{a\cos\lambda}{\omega_1}(t - \frac{1}{\omega_1}\sin\omega_1 t) + \frac{a\sin\lambda}{\omega_1^2}(1 - \cos\omega_1 t)$$
$$+ \frac{(q\rho - f)t^2}{2} + \frac{q\mu t^3}{6\tau}$$

(4.6.10)

Applying appropriate algebraic procedures to equation (4.6.10), we have:

$$x = s_0 + (v_0 + \frac{a}{\omega_1}\cos\lambda)t + \frac{(q\rho - f)}{2}t^2 + \frac{q\mu t^3}{6\tau} + \frac{a}{\omega_1^2}\sin\lambda$$
$$- \frac{a}{\omega_1^2}(\sin\omega_1 t\cos\lambda + \cos\omega_1 t\sin\lambda)$$

(4.6.11)

Upon further transformations of equation (4.6.11), we write:

$$x = s_0 + (v_0 + \frac{a}{\omega_1}\cos\lambda)t + \frac{(q\rho - f)}{2}t^2 + \frac{q\mu t^3}{6\tau}$$
$$+ \frac{a}{\omega_1^2}\sin\lambda - \frac{a}{\omega_1^2}\sin(\omega_1 t + \lambda)$$

(4.6.12)

Taking the first derivative of equation (4.6.12), we determine the velocity of the system:

$$\frac{dx}{dt} = v_0 + \frac{a}{\omega_1}\cos\lambda + (q\rho - f)t + \frac{q\mu t^2}{2\tau} - \frac{a}{\omega_1}\cos(\omega_1 t + \lambda) \quad \textbf{(4.6.13)}$$

The second derivative of equation (4.6.12) characterizes the system's acceleration:

$$\frac{d^2x}{dt^2} = q\rho - f + \frac{q\mu t}{\tau} + a\sin(\omega_1 t + \lambda) \qquad \textbf{(4.6.14)}$$

Substituting the time interval τ into equation (4.6.14), we calculate the value of the acceleration a_τ at the end of this time interval:

$$a_\tau = q\rho - f + q\mu + a\sin(\omega_1\tau + \lambda) \qquad \textbf{(4.6.15)}$$

Because the system possesses an initial velocity, the system may perform a complete vibratory cycle before the velocity becomes equal to zero. Therefore, the extreme values of the acceleration according to equation (4.6.15) occur at the following conditions:

$$\sin(\omega_1\tau + \lambda) = \pm 1 \qquad \textbf{(4.6.16)}$$

Combining equations (4.6.15) and (4.6.16), we determine the extreme values of the acceleration a_{ext}:

$$a_{ext} = q\rho - f + q\mu \pm a \qquad \textbf{(4.6.17)}$$

Substituting notations (4.6.5), (4.6.6), and (4.6.7) into equation (4.6.7), we have:

$$a_{ext} = (\rho + \mu) - \frac{F}{m} \pm \frac{A}{m} \qquad \textbf{(4.6.18)}$$

Adding the friction force F to the product of multiplying equation (4.6.18) by m, we determine the extreme values of the forces R_{ext} applied to the system:

$$R_{ext} = Q(\rho + \mu) \pm A \qquad \textbf{(4.6.19)}$$

According to equation (4.6.19), the system is subjected to a random loading cycle having $R_{max} = Q(\rho + \mu) + A$ and $R_{min} = Q(\rho + \mu) - A$. The stress calculations should include fatigue considerations.

4.6.2 Initial Displacement Equals Zero

The initial conditions of motion are:

$$\text{for} \quad t = 0 \quad x = 0; \quad \frac{dx}{dt} = v_0 \tag{4.6.20}$$

The solution of differential equation of motion (4.6.1) with the initial conditions of motion (4.6.20) reads:

$$x = (v_0 + \frac{a}{\omega_1}\cos\lambda)t + \frac{q\rho - f}{2}t^2 + \frac{q\mu t^3}{6\tau} + \frac{a}{\omega_1^2}\sin\lambda - \frac{a}{\omega_1^2}\sin(\omega_1 t + \lambda) \tag{4.6.21}$$

The rest of the parameters are the same as for the previous case.

4.6.3 Initial Velocity Equals Zero

The initial conditions of motion are:

$$\text{for} \quad t = 0 \quad x = s_0; \quad \frac{dx}{dt} = 0 \tag{4.6.22}$$

Solving differential equation of motion (4.6.1) with the initial conditions of motion (4.6.22), we obtain:

$$x = s_0 + \frac{at}{\omega_1}\cos\lambda + \frac{q\rho - f}{2}t^2 + \frac{q\mu t^3}{6\tau} + \frac{a}{\omega_1^2}\sin\lambda - \frac{a}{\omega_1^2}\sin(\omega_1 t + \lambda) \tag{4.6.23}$$

Taking the first derivative of equation (4.6.23), we determine the velocity of the system:

$$\frac{dx}{dt} = \frac{a}{\omega_1}\cos\lambda + (q\rho - f)t + \frac{q\mu t^2}{2\tau} - \frac{a}{\omega_1}\cos(\omega_1 t + \lambda) \tag{4.6.24}$$

The rest of the parameters are the same as for the previous case.

4.6.4 Both the Initial Displacement and Velocity Equal Zero

The initial conditions of motion are:

$$\text{for} \quad t = 0 \quad x = 0; \quad \frac{dx}{dt} = 0 \qquad (4.6.25)$$

The solution of differential equation of motion (4.6.1) with the initial conditions of motion (4.6.25) reads:

$$x = \frac{at}{\omega_1}\cos\lambda + \frac{q\rho - f}{2}t^2 + \frac{q\mu t^3}{6\tau} + \frac{a}{\omega_1^2}\sin\lambda - \frac{a}{\omega_1^2}\sin(\omega_1 t + \lambda)$$

$$(4.6.26)$$

The rest of the parameters are the same as for the previous case.

5

CONSTANT RESISTANCE

This chapter focuses on engineering problems characterized by the force of inertia and the constant resisting force as the resisting forces. In Row 5 of Guiding Table 2.1, these two forces are marked by the plus sign. The intersections of this row with Columns 1 through 6 indicate the sections of this chapter with respect to different active forces.

The engineering systems described in this chapter could be intended for interaction with an elastoplastic or viscoelastoplastic medium that exerts a constant resisting force during its plastic deformation. More information related to the deformation of different types of media is presented in section 1.2. In case of an upward motion, the weight of the system represents a constant resisting force.

Throughout this chapter, the left sides of the differential equations of motion in each section are identical and made up by the force of inertia and the constant resisting force. The right sides of these equations consist of the active forces applied to the system. The section titles reflect the active forces involved in the corresponding problems.

5.1 Active Force Equals Zero

This section describes an engineering system subjected to the action of the force of inertia and a constant resisting force (Row 5 in Guiding Table 2.1) as the resisting forces in the absence of an active force (Column 1). For example, this system could interact with an elastoplastic or viscoelastoplastic medium that exerts a constant resisting force as a reaction to its plastic deformation. More information regarding the deformation of these media is presented in section 1.2. The considerations related to the motion of a system in the absence of active forces are discussed in section 1.3.

The system is moving in the horizontal direction. We want to determine the basic parameters of motion, the value of displacement of the system, and the maximum value of the force applied to the system. Figure 5.1 shows the model of a system subjected to the action of a constant resisting force.

Based on the considerations above and the model in Figure 5.1, we can compose the differential equation of motion of the system. The left side includes the force of inertia and the constant resisting force while the right side equals zero. Thus, the differential equation of motion reads:

$$m\frac{d^2x}{dt^2} + P = 0 \qquad (5.1.1)$$

Differential equation of motion (5.1.1) has different solutions for various initial conditions of motion. These solutions and their analyses are presented below.

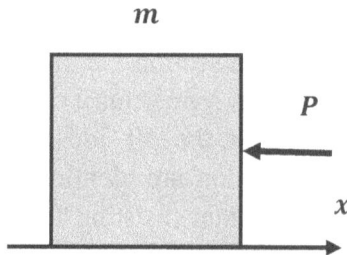

Figure 5.1 Model of a system subjected to a constant resisting force

5.1.1 General Initial Conditions

The general initial conditions of motion are:

$$\text{for} \quad t = 0 \quad x = s_0; \quad \frac{dx}{dt} = v_0 \qquad \text{(5.1.2)}$$

where s_0 and v_0 are the initial displacement and initial velocity respectively. Dividing equation (5.1.1) by m, we have:

$$\frac{d^2 x}{dt^2} + p = 0 \qquad \text{(5.1.3)}$$

where:

$$p = \frac{P}{m} \qquad \text{(5.1.4)}$$

Using Laplace Transform Pairs 3 and 5 from Table 1.1, we convert differential equation of motion (5.1.3) with the initial conditions of motion (5.1.2) from the time domain into the Laplace domain and obtain the resulting algebraic equation of motion in Laplace domain:

$$l^2 x(l) - l v_0 - l^2 s_0 + p = 0 \qquad \text{(5.1.5)}$$

Solving equation (5.1.5) for the displacement in the Laplace domain $x(l)$, we may write:

$$x(l) = s_0 + \frac{v_0}{l} - \frac{p}{l^2} \qquad \text{(5.1.6)}$$

Based on pairs 1, 5, 2, and 8 from Table 1.1, we invert equation (5.1.6) from the Laplace domain into the time domain and obtain the solution of differential equation of motion (5.1.1) with the initial conditions of motion (5.1.2):

$$x = s_0 + v_0 t - \frac{p}{2} t^2 \qquad \text{(5.1.7)}$$

The first derivative of equation (5.1.7) represents the velocity of the system:

$$\frac{dx}{dt} = v_0 - pt \qquad (5.1.8)$$

Taking the second derivative of equation (5.1.7), we determine the acceleration (in this case it is the deceleration) of the system:

$$\frac{d^2x}{dt^2} = -p \qquad (5.1.9)$$

Equation (5.1.9) shows that the deceleration in this case is a constant value. Substituting notation (5.1.4) into equation (5.1.9), we determine the value of the deceleration a_0:

$$a_0 = -\frac{P}{m} \qquad (5.1.10)$$

Thus, the absolute value of the deceleration is:

$$a_0 = |\frac{P}{m}| \qquad (5.1.11)$$

We calculate the absolute value of the force R_0 applied to the system by multiplying the mass m by the absolute value of the deceleration according to equation (5.1.11):

$$R_0 = |P| \qquad (5.1.12)$$

The stress analysis of the system should be based on the force according to equation (5.1.12). At the end of the process, the velocity of the system equals zero. Equating the left side of equation (5.1.8) to zero and performing the needed algebraic procedures, we determine the time T that the process lasts:

$$T = \frac{v_0}{p} \qquad (5.1.13)$$

Combining equations (5.1.7) and (5.1.13), we determine the value of the displacement s of the system:

$$s = s_0 + \frac{v_0^2}{2p} \qquad (5.1.14)$$

5.1.2 Initial Displacement Equals Zero

The initial conditions of motion are:

$$\text{for} \quad t = 0 \quad x = 0; \quad \frac{dx}{dt} = v_0 \qquad (5.1.15)$$

Solving differential equation of motion (5.1.1) with the initial conditions of motion (5.1.15), we have:

$$x = v_0 t - \frac{p}{2} t^2 \qquad (5.1.16)$$

The rest of the parameters and their analysis are the same as for the previous case.

5.2 Constant Force R

According to Guiding Table 2.1, this section considers problems characterized by the force of inertia and a constant resisting force as the resisting forces (Row 5) along with a constant active force (Column 2). The working process of systems related to this problem may be associated with the phase of plastic deformation of an elastoplastic or viscoelastoplastic medium that exerts a constant resisting force as a reaction to its plastic deformation (more considerations related to the deformation of these types of media are discussed in section 1.2).

The system is moving in the horizontal direction. We want to determine the basic parameters of motion and the maximum value of the force applied to the system. The model of a mechanical system subjected to the action of a constant active and a constant resisting force is shown in Figure 5.2.

The considerations above and the model in Figure 5.2 let us compose the differential equation of motion, the left side of which includes the force of inertia and the constant resisting force, while

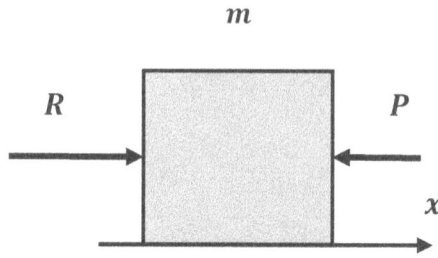

Figure 5.2 Model of a system subjected to a constant active force and a constant resisting force

the right side consists of the constant active force. Thus, the differential equation of motion reads:

$$m\frac{d^2x}{dt^2} + P = R \tag{5.2.1}$$

Differential equation of motion (5.2.1) has different solutions for various initial conditions of motion. These solutions and their analyses follow.

5.2.1 General Initial Conditions

The general initial conditions of motion are:

$$\text{for} \quad t = 0 \quad x = s_0; \quad \frac{dx}{dt} = v_0 \tag{5.2.2}$$

where s_0 and v_0 are the initial displacement and initial velocity respectively.

Dividing equation (5.2.1) by m, we obtain:

$$\frac{d^2x}{dt^2} + p = r \tag{5.2.3}$$

where:

$$p = \frac{P}{m} \tag{5.2.4}$$

$$r = \frac{R}{m} \tag{5.2.5}$$

Using the Laplace Transform Pairs 3 and 5 from Table 1.1, we convert differential equation of motion (5.2.3) with the initial conditions of motion (5.2.2) from the time domain into the Laplace domain. The resulting algebraic equation of motion in the Laplace domain reads:

$$l^2x(l) - lv_0 - l^2s_0 + p = r \qquad (5.2.6)$$

Solving equation (5.2.6) for the Laplace domain displacement $x(l)$ we have:

$$x(l) = s_0 + \frac{v_0}{l} + \frac{r-p}{l^2} \qquad (5.2.7)$$

Based on pairs 1, 5, 2, and 8 from Table 1.1, we invert equation (5.2.7) from the Laplace domain into the time domain and obtain the solution of differential equation of motion (5.2.1) with the initial conditions of motion (5.2.2):

$$x = s_0 + v_0 t + \frac{r-p}{2} t^2 \qquad (5.2.8)$$

Taking the first derivative of equation (5.2.8), we determine the velocity of the system:

$$\frac{dx}{dt} = v_0 + (r-p)t \qquad (5.2.9)$$

The second derivative of equation (5.2.8) represents the acceleration of the system:

$$\frac{d^2x}{dt^2} = r - p \qquad (5.2.10)$$

Equation (5.2.10) shows that the acceleration for this case represents a constant value. Substituting notations (5.2.4) and (5.2.5) into equation (5.2.10), we determine the value of the acceleration a_0:

$$a_0 = \frac{R-P}{m} \qquad (5.2.11)$$

In general, the maximum force applied to the system is equal to the sum of the force of inertia and the resisting forces. Adding the constant resisting force P to the product of multiplying equation (5.2.11) by mass m, we determine the maximum force applied to the system R_{max}:

$$R_{max} = R$$

The stress analysis of the system should be based on this value of the force R_{max}.

5.2.2 Initial Displacement Equals Zero
The initial conditions of motion are:

$$\text{for} \quad t = 0 \quad x = 0; \quad \frac{dx}{dt} = v_0 \qquad (5.2.12)$$

Solving differential equation of motion (5.2.1) with the initial conditions of motion (5.2.12), we write:

$$x = v_0 t + \frac{r - p}{2} t^2 \qquad (5.2.13)$$

The rest of the analysis is the same as in the previous case.

5.2.3 Initial Velocity Equals Zero
The initial conditions are:

$$\text{for} \quad t = 0 \quad x = s_0; \quad \frac{dx}{dt} = 0 \qquad (5.2.14)$$

The solution of differential equation of motion (5.2.1) with the initial conditions of motion (5.2.14) reads:

$$x = s_0 + \frac{r - p}{2} t^2 \qquad (5.2.15)$$

Taking the first derivative of equation (5.2.15), we determine the velocity of the system:

$$\frac{dx}{dt} = (r - p)t \qquad (5.2.16)$$

The rest of the parameters and their analysis is the same as in the previous cases.

5.2.4 Both the Initial Displacement and Velocity Equal Zero

The initial conditions of motion are:

$$\text{for} \quad t = 0 \quad x = 0; \quad \frac{dx}{dt} = 0 \qquad (5.2.17)$$

Solving differential equation of motion (5.2.1) with the initial conditions of motion (5.2.17), we obtain:

$$x = \frac{r-p}{2} t^2 \qquad (5.2.18)$$

The velocity and the acceleration can be determined according to equations (5.2.16) and (5.2.10) respectively. The rest of the analysis of the parameters is the same as in the previous cases.

5.3 Harmonic Force $A \sin(\omega_1 t + \lambda)$

This section describes a system subjected to the action of the force of inertia and the constant resisting force as the resisting forces (Row 5 in Guiding Table 2.1), and the harmonic force as the active force (Column 3). The current problem could be related to the working processes of some vibratory systems that interact with elastoplastic or viscoelastoplastic media that exert constant resisting forces as the reaction of these media to their plastic deformation. The considerations related to the deformation of these types of media as well as to the behavior of a constant resisting force applied to a vibratory system are discussed in section 1.2.

The system is moving in the horizontal direction. We want to determine the basic parameters of motion and the characteristics of the forces applied to the system. Figure 5.3 shows the model of a system subjected to the action of a harmonic force and a constant resisting force.

Based on the above considerations and the model in Figure 5.3, we can compose the differential equation of motion. The left side includes the force of inertia and the constant resisting force, while

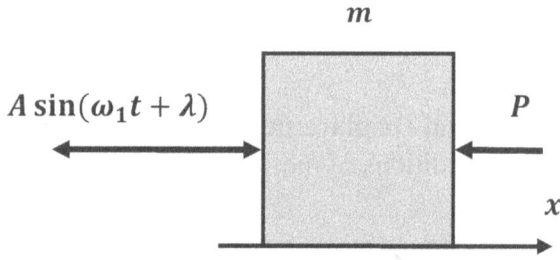

$$m$$

$$A\sin(\omega_1 t + \lambda) \qquad\qquad P$$

$$x$$

Figure 5.3 Model of a system subjected to a harmonic force and a constant resisting force

the right side consists of the harmonic force. Therefore, the differential equation of motion reads:

$$m\frac{d^2x}{dt^2} + P = A\sin(\omega_1 t + \lambda) \qquad (5.3.1)$$

Differential equation of motion (5.3.1) has different solutions for various initial conditions of motion. These solutions and their analyses are presented below.

5.3.1 General Initial Conditions

The general initial conditions of motion are:

$$\text{for} \quad t = 0 \quad x = s_0; \quad \frac{dx}{dt} = v_0 \qquad (5.3.2)$$

where s_0 and v_0 are the initial displacement and initial velocity respectively.

Transforming the sinusoidal function in equation (5.3.1) and dividing this equation by m, we have:

$$\frac{d^2x}{dt^2} + p = a\sin\omega_1 t \cos\lambda + a\cos\omega_1 t \sin\lambda \qquad (5.3.3)$$

where:

$$p = \frac{P}{m} \qquad (5.3.4)$$

$$a = \frac{A}{m} \qquad (5.3.5)$$

Using Laplace Transform Pairs 3, 5, 6, and 7 from Table 1.1, we convert differential equation of motion (5.3.3) with the initial conditions of motion (5.3.2) from the time domain into the Laplace domain. The resulting algebraic equation of motion in Laplace domain reads:

$$l^2 x(l) - l v_0 - l^2 s_0 + p = \frac{\omega_1 l}{l^2 + \omega_1^2} a \cos \lambda + \frac{l^2}{l^2 + \omega_1^2} a \sin \lambda \qquad (5.3.6)$$

Solving equation (5.3.6) for the displacement $x(l)$ in the Laplace domain, we have:

$$x(l) = s_0 + \frac{v_0}{l} - \frac{p}{l^2} + \frac{\omega_1}{l(l^2 + \omega_1^2)} a \cos \lambda + \frac{l^2}{l^2 + \omega_1^2} a \sin \lambda \qquad (5.3.7)$$

Based on pairs 1, 5, 2, 8, 16, and 14 from Table 1.1, we invert equation (5.3.7) from the Laplace domain into the time domain. The inversion represents the solution of differential equation of motion (5.3.1) with the initial conditions of motion (5.3.2):

$$x = s_0 + v_0 t - \frac{p}{2} t^2 + \frac{a \cos \lambda}{\omega_1} \left(t - \frac{1}{\omega_1} \sin \omega_1 t \right) + \frac{a \sin \lambda}{\omega_1^2} (1 - \cos \omega_1 t)$$

$$(5.3.8)$$

Performing some conventional transformations with equation (4.3.8), we write:

$$x = s_0 + v_0 t - \frac{p}{2} t^2 + \frac{at}{\omega_1} \cos \lambda + \frac{a}{\omega_1^2} \sin \lambda - \frac{a}{\omega_1^2} \sin(\omega_1 t + \lambda) \qquad (5.3.9)$$

Taking the first derivative of equation (5.3.9), we determine the velocity of the system:

$$\frac{dx}{dt} = v_0 - pt + \frac{a}{\omega_1} [\cos \lambda - \cos(\omega_1 t + \lambda)] \qquad (5.3.10)$$

The second derivative of equation (5.3.9) represents the acceleration:

$$\frac{d^2x}{dt^2} = -p + a\sin(\omega_1 t + \lambda) \qquad (5.3.11)$$

Due to the initial velocity, the system could perform a complete vibratory cycle while moving forward. This lets us consider that the system reaches the extreme values of the acceleration at the following conditions:

$$\sin\left(\omega_1 t + \lambda\right) = \pm 1 \qquad (5.3.12)$$

Combining equations (5.3.11) and (5.3.12), we calculate the extreme values of the acceleration a_{ext}:

$$a_{ext} = -p \pm a \qquad (5.3.13)$$

Substituting notations (5.3.4) and (5.3.5) into equation (5.3.13), we have:

$$a_{ext} = \frac{-P \pm A}{m} \qquad (5.3.14)$$

Adding the constant resisting force P to the product of multiplying equation (5.3.14) by m, we determine the extreme values of the forces of R_{ext} that are applied to the system:

$$R_{ext} = \pm A \qquad (5.3.15)$$

Equation (5.3.15) shows that the system is subjected to the action of a completely reversed loading cycle. The stress analysis should be based on fatigue calculations where the maximum force $R_{max} = A$ while the minimum force $R_{min} = -A$.

5.3.2 Initial Displacement Equals Zero

The initial conditions of motion are:

$$\text{for} \quad t = 0 \quad x = 0; \quad \frac{dx}{dt} = v_0 \qquad (5.3.16)$$

Solving differential equation of motion (5.3.1) with the initial conditions of motion (5.3.16), we obtain:

$$x = v_0 t - \frac{p}{2}t^2 + \frac{at}{\omega_1}\cos\lambda + \frac{a}{\omega_1^2}\sin\lambda - \frac{a}{\omega_1^2}\sin(\omega_1 t + \lambda) \quad \textbf{(5.3.17)}$$

The rest of the parameters are the same as for the previous case.

5.3.3 Initial Velocity Equals Zero
The initial conditions of motion are:

$$\text{for} \quad t = 0 \quad x = s_0; \quad \frac{dx}{dt} = 0 \quad \textbf{(5.3.18)}$$

The solution of differential equation of motion (5.3.1) with the initial conditions of motion (5.3.17) reads:

$$x = s_0 - \frac{p}{2}t^2 + \frac{at}{\omega_1}\cos\lambda + \frac{a}{\omega_1^2}\sin\lambda - \frac{a}{\omega_1^2}\sin(\omega_1 t + \lambda) \quad \textbf{(5.3.19)}$$

The first derivative of equation (5.3.18) represents the velocity:

$$\frac{dx}{dt} = -pt + \frac{a}{\omega_1}[\cos\lambda - \cos(\omega_1 t + \lambda)] \quad \textbf{(5.3.20)}$$

The acceleration and the forces could be calculated using equation (5.3.11). As discussed in section 1.2, when the initial velocity equals zero, the current differential equation could be valid just for the first half of the vibratory cycle. However, it is reasonable to perform the stress calculations based on equation (5.3.15).

5.3.4 Both the Initial Displacement and Velocity Equal Zero
The initial conditions of motion are:

$$\text{for} \quad t = 0 \quad x = 0; \quad \frac{dx}{dt} = 0 \quad \textbf{(5.3.21)}$$

Solving differential equation of motion (5.3.1) with the initial conditions of motion (5.2.21), we have:

$$x = -\frac{p}{2}t^2 + \frac{at}{\omega_1}\cos\lambda + \frac{a}{\omega_1^2}\sin\lambda - \frac{a}{\omega_1^2}\sin(\omega_1 t + \lambda) \quad (5.3.22)$$

The rest of the parameters and their analysis is the same as in the previous case.

5.4 Time-Dependent Force $Q\left(\rho + \frac{\mu t}{\tau}\right)$

The problems described in this section are characterized by the sum of the force of inertia and the constant resisting force as the resisting forces (Row 5 of Guiding Table 2.1) and the time-dependent force as the active force (Column 4).

The current problem could be related to some working processes of engineering systems involved in the interaction with an elastoplastic or viscoelastoplastic medium that exerts a constant resisting force as the reaction of its plastic deformation (more information regarding the deformation of these types of media is presented in section 1.2).

During these processes, the active force gradually increases from its minimum value that exceeds the value of the constant resisting force. The increasing of the active force lasts a predetermined interval of time. By the end of this interval, the system reaches a certain velocity.

The system is moving in the horizontal direction. We want to determine the values of the displacement, the velocity, the acceleration, the force applied to the system, and the power of the energy source. The values of these parameters should be calculated for the end of the interval of time τ.

The model of the system subjected to the action of a time-dependent force and a constant resisting force is shown in Figure 5.4. Based on the considerations above and the model in Figure 5.4, we can describe the structure of the differential equation of motion for this case. The left side includes the force of inertia and the constant resisting force, while the right side consists of the time-dependent force. Thus, the differential equation of motion reads:

$$m\frac{d^2x}{dt^2} + P = Q(\rho + \frac{\mu t}{\tau}) \quad (5.4.1)$$

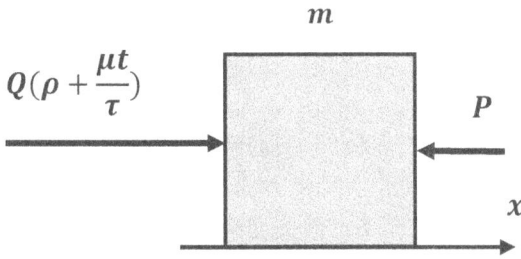

Figure 5.4 Model of a system subjected to a time-dependent force and a constant resisting force

Note that in equation (5.4.1) $Q\rho > P$. Differential equation of motion (5.4.1) has different solutions for various initial conditions of motion. These solutions and their analyses are presented below.

5.4.1 General Initial Conditions

The general initial conditions of motion read:

$$\text{for} \quad t = 0 \quad x = s_0; \quad \frac{dx}{dt} = v_0 \qquad (5.4.2)$$

where s_0 and v_0 are the initial displacement and initial velocity respectively.

Dividing equation (5.4.1) by m, we have:

$$\frac{d^2x}{dt^2} + p = q(\rho + \frac{\mu t}{\tau}) \qquad (5.4.3)$$

where:

$$p = \frac{P}{m} \qquad (5.4.4)$$

$$q = \frac{Q}{m} \qquad (5.4.5)$$

Using Laplace Transform Pairs 3, 5, and 2 from Table 1.1, we convert differential equation of motion (5.4.3) with the initial conditions of motion (5.4.2) from the time domain into the Laplace

domain. The resulting algebraic equation of motion in the Laplace domain reads:

$$l^2 x(l) - l v_0 - l^2 s_0 + p = q(\rho + \frac{\mu}{\tau l}) \tag{5.4.6}$$

Solving equation (5.4.6) for the Laplace domain displacement $x(l)$, we write:

$$x(l) = s_0 + \frac{v_0}{l} + \frac{q\rho - p}{l^2} + \frac{q\mu}{\tau l^3} \tag{5.4.7}$$

Based on pairs 1, 5, 2, and 8 from Table 1.1, we invert equation (5.4.7) from the Laplace domain into the time domain and obtain the solution of differential equation of motion (5.4.1) with the initial conditions of motion (5.4.2):

$$x = s_0 + v_0 t + \frac{1}{2}(q\rho - p)t^2 + \frac{q\mu}{6\tau}t^3 \tag{5.4.8}$$

The first derivative of equation (5.4.8) represents the velocity of the system:

$$\frac{dx}{dt} = v_0 + (q\rho - p)t + \frac{q\mu}{2\tau}t^2 \tag{5.4.9}$$

Taking the second derivative of equation (5.4.8), we determine the acceleration of the system:

$$\frac{d^2 x}{dt^2} = q\rho - p + \frac{q\mu}{\tau}t \tag{5.4.10}$$

Differential equation (5.4.1) describes the motion that lasts an interval of time τ. Thus, replacing the running time t in equations (5.4.8), (5.4.9), and (5.4.10) by the interval of time τ, we obtain the values of the displacement s, the velocity v, and the acceleration a at the end of time τ respectively:

$$s = s_0 + v_0 \tau + \frac{3q\rho - 3p + q\mu}{6}\tau^2 \tag{5.4.11}$$

$$v = v_0 + \frac{2q\rho - 2p + q\mu}{2}\tau \qquad (5.4.12)$$

$$a = q(\rho + \mu) - p \qquad (5.4.13)$$

The value of the force R_0 applied to the system at the end of time τ is equal to the sum of the force of inertia and the constant resisting force. Adding the constant resisting force P to the product of multiplying equation (5.4.13) by m and substituting notations (5.4.4) and (5.4.5), we have:

$$R_0 = Q(\rho + \mu) \qquad (5.4.14)$$

The stress calculations of the system should be based on the force according to equation (5.4.14). The product of multiplying the force according to equation (5.4.14) by the velocity according to equation (5.4.12) represents the power N of the energy source:

$$N = Q(\rho + \mu)[v_0 + \frac{Q(2\rho + \mu) - 2P}{2m}\tau]$$

5.4.2 Initial Displacement Equals Zero
The initial conditions of motion are:

$$\text{for} \quad t = 0 \quad x = 0; \quad \frac{dx}{dt} = v_0 \qquad (5.4.15)$$

The solution of differential equation of motion (5.4.1) with the initial conditions of motion (5.4.15) reads:

$$x = v_0 t + \frac{1}{2}(q\rho - p)t^2 + \frac{q\mu}{6\tau}t^3 \qquad (5.4.16)$$

Replacing the running time t in equation (5.4.16) by time τ, we determine the displacement at the end of this time:

$$s = v_0\tau + \frac{3q\rho - 3p + q\mu}{6}\tau^2 \qquad (5.4.17)$$

The rest of the parameters and their analysis are the same as for the previous case.

5.4.3 Initial Velocity Equals Zero

The initial conditions of motion are:

$$\text{for} \quad t = 0 \quad x = s_0; \quad \frac{dx}{dt} = 0 \qquad (5.4.18)$$

The solution of differential equation of motion (5.4.1) with the initial condition of motion (5.4.18) reads:

$$x = s_0 + \frac{1}{2}(q\rho - p)t^2 + \frac{q\mu}{6\tau}t^3 \qquad (5.4.19)$$

The first derivative of equation (5.4.19) represents the velocity of the system:

$$\frac{dx}{dt} = (q\rho - p)t + \frac{q\mu}{2\tau}t^2 \qquad (5.4.20)$$

Substituting the time τ into equations (5.4.19) and (5.4.20), we calculate the values of the displacement and the velocity at the end of this time interval respectively:

$$s = s_0 + \frac{3q\rho - 3p + q\mu}{6}\tau^2 \qquad (5.4.21)$$

$$v = \frac{2q\rho - 2p + q\mu}{2}\tau \qquad (5.4.22)$$

Multiplying the force according to equation (5.4.14) by the velocity according to equation (5.4.22), we determine the power:

$$N = Q(\rho + \mu)\frac{Q(2\rho + \mu) - 2P}{2m}\tau$$

The rest of the parameters and their analysis are the same as in the previous case.

5.4.4 Both the Initial Displacement and Velocity Equal Zero

The initial conditions of motion are:

$$\text{for} \quad t = 0 \quad x = 0; \quad \frac{dx}{dt} = 0 \qquad (5.4.23)$$

The solution of differential equation of motion (5.4.1) with the initial conditions of motion (5.4.23) reads:

$$x = \frac{1}{2}\left(q\rho - p\right)t^2 + \frac{q\mu}{6\tau}t^3 \qquad (5.4.24)$$

Substituting the time τ into equation (5.4.24), we determine the displacement at the end of this time interval:

$$s = \frac{3q\rho - 3p + q\mu}{6}\tau^2 \qquad (5.4.25)$$

The rest of the parameters and their analysis are the same as in the previous case.

5.5   Constant Force R and Harmonic Force $A\sin(\omega_1 t + \lambda)$

According to Guiding Table 2.1, the problems in this section are characterized by the action of the force of inertia and the constant resisting force as the resisting forces (Row 5), and by the action of an active constant force and a harmonic force (Column 5).

The current problem could be associated with a vibratory system intended for interaction with an elastoplastic or viscoelastoplastic medium that exerts a constant resisting force as a reaction to its plastic deformation. Usually the constant active force exceeds the constant resisting force; as a result, there are no interruptions in the vibratory motion of the system. The considerations related to the deformation of the above-mentioned media and to the behavior of a constant resisting force applied to a vibratory system are discussed in section 1.2.

The system is moving in the horizontal direction. We want to determine the basic parameters of motion and the characteristics of the forces applied to the system. Figure 5.5 shows the model of a system subjected to the action of a constant active force, a harmonic force, and a constant resisting force.

Based on the considerations above and the model in Figure 5.5, we can describe the structure of the differential equation of motion of the system. The left side consists of the force of inertia and the constant resisting force, while the right side includes the constant

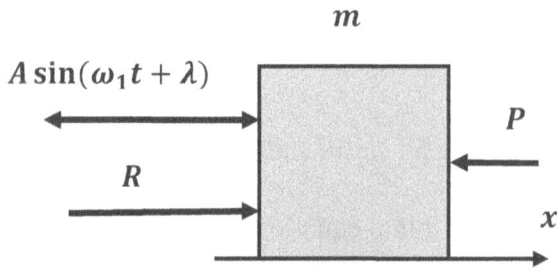

Figure 5.5 Model of a system subjected to a harmonic force, a constant active force, and a constant resisting force

active force and the harmonic force. Hence, the differential equation of motion reads:

$$m\frac{d^2x}{dt^2} + P = R + A\sin(\omega_1 t + \lambda) \qquad (5.5.1)$$

Differential equation of motion (5.5.1) has different solutions for various initial conditions of motion. These solutions and their analyses are presented below.

5.5.1 General Initial Conditions

The general initial conditions of motion are:

$$\text{for} \quad t = 0 \quad x = s_0; \quad \frac{dx}{dt} = v_0 \qquad (5.5.2)$$

where s_0 and v_0 are the initial displacement and initial velocity respectively.

Transforming the sinusoidal function in equation (5.5.1) and dividing this equation by m, we have:

$$\frac{d^2x}{dt^2} + p = r + a\sin\omega_1 t\cos\lambda + a\cos\omega_1 t\sin\lambda \qquad (5.5.3)$$

where:

$$p = \frac{P}{m} \qquad (5.5.4)$$

$$r = \frac{R}{m} \tag{5.5.5}$$

$$a = \frac{A}{m} \tag{5.5.6}$$

Based on Laplace Transform Pairs 3, 5, 6, and 7 from Table 1.1, we convert differential equation of motion (5.5.3) with its initial conditions of motion (5.5.2) from the time domain into the Laplace domain. The resulting algebraic equation of motion in the Laplace domain reads:

$$l^2 x(l) - lv_0 - l^2 s_0 + p = r + \frac{a\omega_1 l}{l^2 + \omega_1^2} \cos\lambda + \frac{al^2}{l^2 + \omega_1^2} \sin\lambda \tag{5.5.7}$$

Solving equation (5.5.7) for the Laplace domain displacement $x(l)$, we have:

$$x(l) = s_0 + \frac{v_0}{l} + \frac{r - p}{l^2} + \frac{a\omega_1}{l(l^2 + \omega_1^2)} \cos\lambda + \frac{a}{l^2 + \omega_1^2} \sin\lambda \tag{5.5.8}$$

Using pairs 1, 5, 2, 8, 16, and 14 from Table 1.1, we invert equation (5.5.8) from the Laplace domain into the time domain and obtain the solution of the differential equation of motion (5.5.1) with the initial conditions of motion (5.5.2):

$$x = s_0 + v_0 t + \frac{r - p}{2} t^2 + \frac{a\cos\lambda}{\omega_1} \left(t - \frac{1}{\omega_1}\sin\omega_1 t\right) + \frac{a\sin\lambda}{\omega_1^2}(1 - \cos\omega_1 t) \tag{5.5.9}$$

Applying algebraic procedures to equation (5.5.9), we have:

$$x = s_0 + v_0 t + \frac{r - p}{2} t^2 + \frac{at}{\omega_1}\cos\lambda$$

$$+ \frac{a}{\omega_1^2}\sin\lambda - \frac{a}{\omega_1^2}(\sin\omega_1 t \cos\lambda + \cos\omega_1 t \sin\lambda)$$

Further transformation of this equation reads:

$$x = s_0 + v_0 t + \frac{r - p}{2} t^2 + \frac{at}{\omega_1}\cos\lambda + \frac{a}{\omega_1^2}\sin\lambda - \frac{a}{\omega_1^2}\sin(\omega_1 t + \lambda) \tag{5.5.10}$$

Taking the first derivative of equation (5.5.10), we determine the velocity of the system:

$$\frac{dx}{dt} = v_0 + (r - p)t + \frac{a}{\omega_1}\cos\lambda - \frac{a}{\omega_1}\cos(\omega_1 t + \lambda) \quad \textbf{(5.5.11)}$$

The second derivative of equation (5.5.10) represents the acceleration:

$$\frac{d^2 x}{dt^2} = r - p + a\sin(\omega_1 t + \lambda) \quad \textbf{(5.5.12)}$$

Equation (5.5.12) shows that the acceleration reaches its extreme values when:

$$\sin(\omega_1 t + \lambda) = \pm 1 \quad \textbf{(5.5.13)}$$

Combining equations (5.5.12) and (5.5.13), we determine the extreme values of the acceleration a_{ext}:

$$a_{ext} = r - p \pm a \quad \textbf{(5.5.14)}$$

Substituting notations (5.5.4), (5.5.5), and (5.5.6) into equation (5.5.14), we may write:

$$a_{ext} = \frac{R - P \pm A}{m} \quad \textbf{(5.5.15)}$$

Adding the constant resisting force P to the product of multiplying equation (5.5.15) by m, we obtain the extreme values of the force R_{ext} applied to the system:

$$R_{ext} = R \pm A \quad \textbf{(5.5.16)}$$

Equation (5.5.16) shows that the system is subjected to a random loading cycle for which the maximum force $R_{max} = R + A$, while the minimum force $R_{min} = R - A$.

The stress analysis for this case should be based on appropriate fatigue considerations.

5.5.2 Initial Displacement Equals Zero

The initial conditions of motion are:

$$\text{for} \quad t = 0 \quad x = 0; \quad \frac{dx}{dt} = v_0 \qquad (5.5.17)$$

The solution of differential equation of motion (5.5.1) with the initial conditions of motion (5.5.17) reads:

$$x = v_0 t + \frac{r-p}{2} t^2 + \frac{at}{\omega_1} \cos \lambda + \frac{a}{\omega_1^2} \sin \lambda - \frac{a}{\omega_1^2} \sin(\omega_1 t + \lambda) \quad (5.5.18)$$

The rest of the parameters are the same as in the previous case.

5.5.3 Initial Velocity Equals Zero

The initial conditions of motion are:

$$\text{for} \quad t = 0 \quad x = s_0; \quad \frac{dx}{dt} = 0 \qquad (5.5.19)$$

Solving differential equation of motion (5.5.1) with the initial conditions of motion (5.5.19), we obtain:

$$x = s_0 + \frac{r-p}{2} t^2 + \frac{at}{\omega_1} \cos \lambda + \frac{a}{\omega_1^2} \sin \lambda - \frac{a}{\omega_1^2} \sin(\omega_1 t + \lambda) \quad (5.5.20)$$

Taking the first derivative of equation (5.5.20), we determine the velocity of the system:

$$\frac{dx}{dt} = (r-p)t + \frac{a}{\omega_1} \cos \lambda - \frac{a}{\omega_1} \cos(\omega_1 t + \lambda) \qquad (5.5.21)$$

The rest of the parameters can be determined from equations (5.5.12), (5.5.15), and (5.5.16).

5.5.4 Both the Initial Displacement and Velocity Equal Zero

The initial conditions of motion are:

$$\text{for} \quad t = 0 \quad x = 0; \quad \frac{dx}{dt} = 0 \qquad (5.5.22)$$

The solution of differential equation of motion (5.5.1) with the initial conditions of motion (5.5.22) reads:

$$x = \frac{r-p}{2}t^2 + \frac{at}{\omega_1}\cos\lambda + \frac{a}{\omega_1^2}\sin\lambda - \frac{a}{\omega_1^2}\sin(\omega_1 t + \lambda) \qquad (5.5.23)$$

The rest of the parameters are the same as for the previous case.

5.6 Harmonic Force $A\sin(\omega_1 t + \lambda)$ and Time-Dependent Force $Q\left(\rho + \frac{\mu t}{\tau}\right)$

This section describes systems subjected to the action of the force of inertia and the constant resisting force as the resisting forces (Row 5 in Guiding Table 2.1), and to the harmonic force and time-dependent force as the active forces (Column 6). The current problem could be associated with a vibratory system intended for interaction with an elastoplastic or viscoelastoplastic medium that as a reaction to its plastic deformation exerts a constant resisting force (more information related to the deformation of these media is presented in section 1.2). During the initial stage of the working process, the system could be subjected along with the harmonic force to a time-dependent force that acts a limited interval of time. In systems of this kind, the initial value of the active force $Q\rho$ exceeds the value of the constant resisting force.

The system is moving in the horizontal direction. We want to determine the basic parameters of motion and the characteristics of the forces applied to the system. Figure 5.6 shows the model of a system subjected to the action of a harmonic force, a time-dependent force, and a constant resisting force.

The considerations above and the model in Figure 5.6 let us compose the differential equation of motion of the system. The left side includes the force of inertia and the constant resisting force,

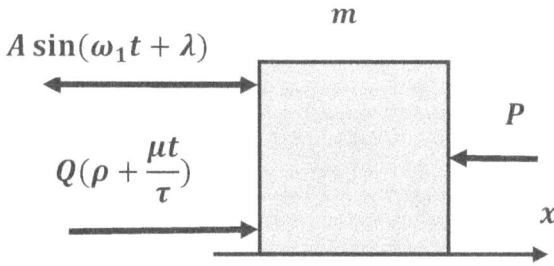

Figure 5.6 Model of a system subjected to a harmonic force, a time-dependent force, and a constant resisting force

while the right side consists of the sum of the harmonic force and the time-dependent force. Thus, the differential equation of motion reads:

$$m\frac{d^2x}{dt^2} + P = A\sin(\omega_1 t + \lambda) + Q\left(\rho + \frac{\mu t}{\tau}\right) \qquad (5.6.1)$$

Transforming equation (5.6.1), we write

$$m\frac{d^2x}{dt^2} + P = A\sin\omega_1 t\cos\lambda + A\cos\omega_1 t\sin\lambda + Q\rho + \frac{Q\mu}{\tau}t \qquad (5.6.2)$$

Differential equation of motion (5.6.1) has different solutions for various initial conditions of motion. These solutions and their analyses are presented below.

5.6.1 General Initial Conditions

The general initial conditions of motion are:

$$\text{for} \quad t = 0 \quad x = s_0; \quad \frac{dx}{dt} = v_0 \qquad (5.6.3)$$

where s_0 and v_0 are the initial displacement and initial velocity respectively.

Dividing equation (5.6.2) by m, we have:

$$\frac{d^2x}{dt^2} + p = a\sin\omega_1 t\cos\lambda + a\cos\omega_1 t\sin\lambda + q\rho + \frac{q\mu}{\tau}t \qquad (5.6.4)$$

where:

$$p = \frac{P}{m} \tag{5.6.5}$$

$$a = \frac{A}{m} \tag{5.6.6}$$

$$q = \frac{Q}{m} \tag{5.6.7}$$

Applying Laplace Transform Pairs 3, 5, 6, 7, and 2 from Table 1.1, we convert differential equation of motion (5.6.4) with the initial conditions of motion (5.6.3) from the time domain into the Laplace domain. The resulting algebraic equation of motion in the Laplace domain reads:

$$x(l)l^2 - lv_0 - l^2 s_0 + p = \frac{a\omega_1 l}{l^2 + \omega_1^2} \cos \lambda + \frac{al^2}{l^2 + \omega_1^2} \sin \lambda + qp + \frac{q\mu}{\tau l}$$

$$(5.6.8)$$

Solving equation (5.6.8) for the displacement in Laplace domain $x(l)$, we have:

$$x(l) = s_0 + \frac{v_0}{l} + \frac{a\omega_1}{l(l^2 + \omega_1^2)} \cos \lambda + \frac{a}{l^2 + \omega_1^2} \sin \lambda + \frac{qp - p}{l^2} + \frac{q\mu}{\tau l^3}$$

$$(5.6.9)$$

Using pairs 1, 5, 2, 16, 14, and 8 from Table 1.1, we invert equation (5.6.9) from the Laplace domain into the time domain and obtain the solution of differential equation of motion (5.6.1) with the initial conditions of motion (5.6.3):

$$x = s_0 + v_0 t + \frac{a \cos \lambda}{\omega_1}\left(t - \frac{1}{\omega_1}\sin \omega_1 t\right) + \frac{a \sin \lambda}{\omega_1^2}(1 - \cos \omega_1 t)$$

$$+ \frac{(qp - p)t^2}{2} + \frac{q\mu t^3}{6\tau} \tag{5.6.10}$$

Applying basic algebra to equation (5.6.10), we have:

$$x = s_0 + (v_0 + \frac{a}{\omega_1}\cos\lambda)t + \frac{(q\rho - p)}{2}t^2 + \frac{q\mu t^3}{6\tau} + \frac{a}{\omega_1^2}\sin\lambda$$

$$-\frac{a}{\omega_1^2}(\sin\omega_1 t \cos\lambda + \cos\omega_1 t \sin\lambda) \qquad \textbf{(5.6.11)}$$

Upon further transformations of equation (5.6.11), we may write:

$$x = s_0 + (v_0 + \frac{a}{\omega_1}\cos\lambda)t + \frac{(q\rho - p)}{2}t^2 + \frac{q\mu t^3}{6\tau}$$

$$+ \frac{a}{\omega_1^2}\sin\lambda - \frac{a}{\omega_1^2}\sin(\omega_1 t + \lambda) \qquad \textbf{(5.6.12)}$$

The first derivative of equation (5.6.12) represents the velocity of the system:

$$\frac{dx}{dt} = v_0 + \frac{a}{\omega_1}\cos\lambda + (q\rho - p)t + \frac{q\mu t^2}{2\tau} - \frac{a}{\omega_1}\cos(\omega_1 t + \lambda) \quad \textbf{(5.6.13)}$$

Taking the second derivative of equation (5.6.12), we determine the acceleration:

$$\frac{d^2 x}{dt^2} = q\rho - p + \frac{q\mu t}{\tau} + a\sin(\omega_1 t + \lambda) \qquad \textbf{(5.6.14)}$$

This process of motion lasts during the time interval τ. Substituting this time interval into equation (5.6.14), we calculate the value of the acceleration a_τ:

$$a_\tau = q\rho - p + q\mu + a\sin(\omega_1 \tau + \lambda) \qquad \textbf{(5.6.15)}$$

The acceleration reaches its extreme values a_{ext} at the following condition:

$$\sin(\omega_1 \tau + \lambda) = \pm 1 \qquad \textbf{(5.6.16)}$$

Combining equations (5.6.15) and (5.6.16), we determine the extreme values of the acceleration a_{ext}:

$$a_{ext} = q\rho - p + q\mu \pm a \qquad (5.6.17)$$

Substituting notations (5.6.5), (5.6.6), and (5.6.7) into equation (5.6.17), we have:

$$a_{ext} = \frac{Q}{m}(\rho + \mu) - \frac{P}{m} \pm \frac{A}{m} \qquad (5.6.18)$$

Adding the constant resisting force P to the product of multiplying equation (5.6.18) by m, we determine the extreme values of the force R_{ext} applied to the system:

$$R_{ext} = Q(\rho + \mu) \pm A \qquad (5.6.19)$$

Equation (5.6.19) shows that the system is subjected to the random loading cycle having the maximum force $R_{max} = Q(\rho + \mu) + A$ and the minimum force $R_{min} = Q(\rho + \mu) - A$. The stress analysis for this case should be based on appropriate fatigue calculations.

5.6.2 Initial Displacement Equals Zero

The initial conditions of motion are:

$$\text{for} \quad t = 0 \quad x = 0; \quad \frac{dx}{dt} = v_0 \qquad (5.6.20)$$

The solution of differential equation of motion (5.6.1) with the initial conditions of motion (5.6.20) reads:

$$x = \left(v_0 + \frac{a}{\omega_1}\cos\lambda\right)t + \frac{q\rho - p}{2}t^2 + \frac{q\mu t^3}{6\tau} + \frac{a}{\omega_1^2}\sin\lambda - \frac{a}{\omega_1^2}\sin(\omega_1 t + \lambda) \qquad (5.6.21)$$

The rest of the parameters are the same as for the previous case.

5.6.3 Initial Velocity Equals Zero

The initial conditions of motion are:

$$\text{for} \quad t = 0 \quad x = s_0; \quad \frac{dx}{dt} = 0 \qquad (5.6.22)$$

Solving differential equation of motion (5.6.1) with the initial conditions of motion (5.6.22), we have:

$$x = s_0 + \frac{at}{\omega_1}\cos\lambda + \frac{q\rho - p}{2}t^2 + \frac{q\mu t^3}{6\tau} + \frac{a}{\omega_1^2}\sin\lambda - \frac{a}{\omega_1^2}\sin(\omega_1 t + \lambda)$$

$$(5.6.23)$$

Taking the first derivative of equation (5.6.23), we determine the velocity of the system:

$$\frac{dx}{dt} = \frac{a}{\omega_1}\cos\lambda + (q\rho - p)t + \frac{q\mu t^2}{2\tau} - \frac{a}{\omega_1}\cos(\omega_1 t + \lambda) \qquad (5.6.24)$$

The rest of the parameters are the same as for the previous case.

5.6.4 Both the Initial Displacement and Velocity Equal Zero
The initial conditions of motion are:

$$\text{for} \quad t = 0 \quad x = 0; \quad \frac{dx}{dt} = 0 \qquad (5.6.25)$$

The solution of differential equation of motion (5.6.2) with the initial conditions of motion (5.6.25) reads:

$$x = \frac{at}{\omega_1}\cos\lambda + \frac{q\rho - p}{2}t^2 + \frac{q\mu t^3}{6\tau} + \frac{a}{\omega_1^2}\sin\lambda - \frac{a}{\omega_1^2}\sin(\omega_1 t + \lambda)$$

$$(5.6.26)$$

The rest of the parameters are the same as for the previous case.

6

CONSTANT RESISTANCE AND FRICTION

This chapter considers engineering systems subjected to the force of inertia, the constant resisting force, and the friction force as the resisting forces. These forces are marked with the plus signs in Row 6 of Guiding Table 2.1. The intersections of this row with Columns 1 through 6 indicate the active forces applied to the systems, and described in the individual sections of this chapter.

The left sides of the differential equation of motion in each section of this chapter are identical; they consist of the force of inertia, the constant resisting force, and the friction force. The right sides of these equations consist of the active forces applied to the systems. The section titles reflect the characteristics of the active forces associated with the problems.

The problems in this chapter include the working processes of engineering systems that could interact with elastoplastic or visco-elastoplastic media. During the phase of plastic deformation, these media can exert constant resisting and friction forces. Considerations related to the deformation of these types of media are presented in section 1.2.

6.1 Active Force Equals Zero

The combination of forces in this section is indicated by the intersection of Row 6 and Column 1 in Guiding Table 2.1. The problems are characterized by the action of the force of inertia, the constant resisting force, and the friction force as the resisting forces in the absence of active forces. The basic principles of a system's motion in the absence of active forces are discussed in section 1.3. The current problem could be related to the working processes of systems that interact with an elastoplastic or viscoelastoplastic medium that exerts a constant resisting force and a friction force as a reaction to its plastic deformations (see section 1.2 for related information).

The system is moving in the horizontal direction. We want to determine the basic parameters of motion, the value of the displacement of the system at the end of the process, and the maximum value of the force applied to the system. Figure 6.1 shows the model of a system subjected to the action of a constant resisting force and a friction force.

Based on the considerations above and the model in Figure 6.1, we can compose the differential equation of motion of the system.

The left side of this equation includes the force of inertia, the constant resisting force, and the friction force, while the right side of

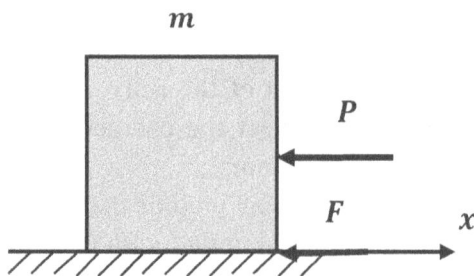

Figure 6.1 Model of a system subjected to a constant resisting force and a friction force

the equation equals zero. Thus, the differential equation of motion reads:

$$m\frac{d^2x}{dt^2} + P + F = 0 \qquad (6.1.1)$$

Differential equation of motion (6.1.1) has different solutions for various initial conditions of motion. These solutions and their analyses are presented below.

6.1.1 General Initial Conditions
The general initial conditions of motion are:

$$\text{for} \quad t = 0 \quad x = s_0; \quad \frac{dx}{dt} = v_0 \qquad (6.1.2)$$

where s_0 and v_0 are the initial displacement and initial velocity respectively.

Dividing equation (6.1.1) by m, we have:

$$\frac{d^2x}{dt^2} + p + f = 0 \qquad (6.1.3)$$

where:

$$p = \frac{P}{m} \qquad (6.1.4)$$

$$f = \frac{F}{m} \qquad (6.1.5)$$

Using Laplace Transform Pairs 3 and 5 from Table 1.1, we convert differential equation of motion (6.1.3) with the initial conditions of motion (6.1.2) from the time domain into the Laplace domain. The resulting algebraic equation of motion in the Laplace domain reads:

$$l^2x(l) - lv_0 - l^2s_0 + p + f = 0 \qquad (6.1.6)$$

Solving equation (6.1.6) for the displacement $x(l)$ in the Laplace domain, we may write:

$$x(l) = s_0 + \frac{v_0}{l} - \frac{p+f}{l^2} \qquad \textbf{(6.1.7)}$$

Based on pairs 1, 5, 2, and 8 from Table 1.1, we invert equation (6.1.7) from the Laplace domain into the time domain and obtain the solution of differential equation of motion (6.1.1) with the initial conditions of motion (6.1.2):

$$x = s_0 + v_0 t - \frac{p+f}{2} t^2 \qquad \textbf{(6.1.8)}$$

Taking the first derivative of equation (6.1.8), we determine the velocity of the system:

$$\frac{dx}{dt} = v_0 - (p+f)t \qquad \textbf{(6.1.9)}$$

The second derivative of equation (6.1.8) represents the acceleration (in this case the deceleration) of the system:

$$\frac{d^2 x}{dt^2} = -p - f \qquad \textbf{(6.1.10)}$$

Equation (6.1.10) shows that in this case the deceleration has a constant value. Substituting notations (6.1.4) and (6.1.5) into equation (6.1.10) and denoting the constant value of the deceleration by a_0, we have:

$$a_0 = -\frac{P+F}{m} \qquad \textbf{(6.1.11)}$$

Hence, the absolute value of the deceleration is:

$$a_0 = \left| \frac{P+F}{m} \right| \qquad \textbf{(6.1.12)}$$

We calculate the absolute value of the force R_0 applied to the system by multiplying the mass m by the absolute value of the deceleration according to equation (6.1.12):

$$R_0 = |P + F|$$ (6.1.13)

The stress analysis of the system should be based on the force according to equation (6.1.13).

Equating the left side of equation (6.1.9) to zero, we calculate the duration of time T that the process lasts:

$$T = \frac{v_0}{p + f}$$ (6.1.14)

Combining equations (6.1.8) and (6.1.14), we determine the displacement s of the system:

$$s = s_0 + \frac{v_0^2}{2(p + f)}$$ (6.1.15)

6.1.2 Initial Displacement Equals Zero

The initial conditions of motion are:

$$\text{for} \quad t = 0 \quad x = 0; \quad \frac{dx}{dt} = v_0$$ (6.1.16)

The solution of differential equation of motion (6.1.1) with the initial conditions of motion (6.1.16) reads:

$$x = v_0 t - \frac{p + f}{2} t^2$$ (6.1.17)

The rest of the parameters and their analysis are the same as for the previous case.

6.2 Constant Force R

This section describes problems associated with the force of inertia, the constant resisting force, and the friction force as the resisting forces (Row 6 in Guiding Table 2.1) along with the constant active force (Column 2). These problems are related to the engineering systems interacting with elastoplastic or viscoelastoplastic media that exert constant resisting and a friction force as a reaction to their plastic deformation (more related information is presented in section 1.2).

The system is moving in the horizontal direction. We want to determine the basic parameters of motion and the maximum value of the force applied to the system. The model of a mechanical system subjected to the action of the constant active force, the constant resisting force, and the friction force is shown in Figure 6.2.

Based on the considerations above and on the model in Figure 6.2, we can assemble the system's differential equation of motion. The left side consists of the sum of the force of inertia, the constant resisting force, and the friction force as the resisting forces, while the right side includes the constant active force. Thus the differential equation of motion reads:

$$m\frac{d^2x}{dt^2} + P + F = R \qquad (6.2.1)$$

Differential equation of motion (6.2.1) has different solutions for various initial conditions of motion. These solutions and their analyses are presented below.

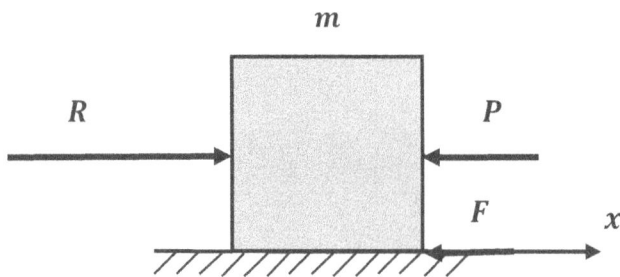

Figure 6.2 Model of a system subjected to a constant active force, a constant resisting force, and a friction force

6.2.1. General Initial Conditions

The general initial conditions have the following shape:

$$\text{for} \quad t = 0 \quad x = s_0; \quad \frac{dx}{dt} = v_0 \qquad \textbf{(6.2.2)}$$

where s_0 and v_0 are the initial displacement and initial velocity respectively.

Dividing equation (6.2.1) by m, we obtain:

$$\frac{d^2x}{dt^2} + p + f = r \qquad \textbf{(6.2.3)}$$

where:

$$p = \frac{P}{m} \qquad \textbf{(6.2.4)}$$

$$f = \frac{F}{m} \qquad \textbf{(6.2.5)}$$

$$r = \frac{R}{m} \qquad \textbf{(6.2.6)}$$

Applying Laplace Transform Pairs 3 and 5 from Table 1.1, we convert differential equation of motion (6.2.3) with the initial conditions of motion (6.2.2) from the time domain into the Laplace domain. The resulting algebraic equation of motion in the Laplace domain reads:

$$l^2 x(l) - l v_0 - l^2 s_0 + p + f = r \qquad \textbf{(6.2.7)}$$

Solving equation (6.2.7) for the Laplace domain displacement $x(l)$, we have:

$$x(l) = s_0 + \frac{v_0}{l} + \frac{r - p - f}{l^2} \qquad \textbf{(6.2.8)}$$

Based on pairs 1, 5, 2, and 8 from Table 1.1, we invert equation (6.2.8) from the Laplace domain into the time domain, and obtain

the solution of differential equation of motion (6.2.1) with the initial conditions of motion (6.2.2):

$$x = s_0 + v_0 t + \frac{r - p - f}{2} t^2 \qquad (6.2.9)$$

The first derivative of equation (6.2.9) represents the velocity of the system:

$$\frac{dx}{dt} = v_0 + (r - p - f)t \qquad (6.2.10)$$

Taking the second derivative of equation (6.2.9), we determine the acceleration of the system:

$$\frac{d^2 x}{dt^2} = r - p - f \qquad (6.2.11)$$

Equation (6.2.11) shows that the acceleration for this case represents a constant value.

Substituting notations (6.2.4), (6.2.5), and (6.2.6) into equation (6.2.11), we determine the value of the acceleration a_0:

$$a_0 = \frac{R - P - F}{m} \qquad (6.2.12)$$

In general, the maximum force applied to the system is equal to the sum of the force of inertia and the resisting forces. Adding the sum of the constant resisting force P and friction force F to the product of multiplying equation (6.2.12) by the mass m, we obtain the maximum force applied to the system R_{max}:

$$R_{max} = R \qquad (6.2.13)$$

The stress analysis of the system should be based on the force according to equation (6.2.13).

6.2.2 Initial Displacement Equals Zero

The initial conditions of motion are:

$$\text{for} \quad t = 0 \quad x = 0; \quad \frac{dx}{dt} = v_0 \qquad (6.2.14)$$

Solving differential equation of motion (6.2.1) with the initial conditions of motion (6.2.14), we have:

$$x = v_0 t + \frac{r - p - f}{2} t^2 \qquad (6.2.15)$$

The rest of the analysis is the same as in the previous case.

6.2.3 Initial Velocity Equals Zero

The initial conditions of motion are:

$$\text{for} \quad t = 0 \quad x = s_0; \quad \frac{dx}{dt} = 0 \qquad (6.2.16)$$

The solution of differential equation of motion (6.2.1) with the initial conditions of motion (6.2.16) reads:

$$x = s_0 + \frac{r - p - f}{2} t^2 \qquad (6.2.17)$$

The first derivative of equation (6.2.16) represents the velocity for this case:

$$\frac{dx}{dt} = (r - p - f) t \qquad (6.2.18)$$

The rest of the analysis is the same as in the previous cases.

6.2.4 Both the Initial Displacement and Velocity Equal Zero

The initial conditions of motion are:

$$\text{for} \quad t = 0 \quad x = 0; \quad \frac{dx}{dt} = 0 \qquad (6.2.19)$$

Solving differential equation of motion (6.2.1) with the initial conditions of motion (6.2.19), we have:

$$x = \frac{r - p - f}{2} t^2 \qquad (6.2.20)$$

The velocity and the acceleration can be determined according to equations (6.2.18) and (6.2.11) respectively. The analysis of the parameters is the same as in the previous cases.

6.3 Harmonic Force $A \sin(\omega_1 t + \lambda)$

This section discusses engineering systems characterized by the action of the force of inertia, constant resisting force, and friction force as the resisting forces (Row 6 in Guiding Table 2.1), and harmonic force as the active force (Column 3). The current problem could be related to the working processes of vibratory systems interacting with elastoplastic or viscoelastoplastic media that exert constant resisting and friction forces as a reaction to their plastic deformation (more considerations related to the deformation of the media and to the behavior of the constant resisting and friction forces applied to a vibratory system are presented in section 1.2).

The system is moving in the horizontal direction. We want to determine the basic parameters of motion and the characteristics of the forces applied to the system. Figure 6.3 shows the model of a system subjected to the action of a harmonic force, a constant resisting force, and a friction force.

Based on the considerations above and the model in Figure 6.3, we can compose the left and right sides of the differential equation of motion. Thus, the left side consists of the force of inertia, the constant resisting force, and the friction force. The right side consists

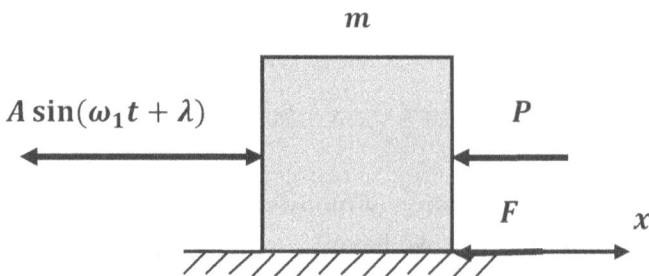

Figure 6.3 Model of a system subjected to a harmonic force, a constant resisting force, and a friction force

of the harmonic force. Therefore, the differential equation of motion reads:

$$m\frac{d^2x}{dt^2} + P + F = A\sin(\omega_1 t + \lambda) \qquad (6.3.1)$$

Differential equation of motion (6.3.1) has different solutions for various initial conditions of motion. These solutions and their analyses are presented below.

6.3.1 General Initial Conditions

The general initial conditions of motion are:

$$\text{for} \quad t = 0 \quad x = s_0; \quad \frac{dx}{dt} = v_0 \qquad (6.3.2)$$

where s_0 and v_0 are the initial displacement and initial velocity respectively.

Transforming the sinusoidal function in equation (6.3.1) and dividing this equation by m, we have:

$$\frac{d^2x}{dt^2} + p + f = a\sin\omega_1 t\cos\lambda + a\cos\omega_1 t\sin\lambda \qquad (6.3.3)$$

where:

$$p = \frac{P}{m} \qquad (6.3.4)$$

$$f = \frac{F}{m} \qquad (6.3.5)$$

$$a = \frac{A}{m} \qquad (6.3.6)$$

Applying Laplace Transform Pairs 3, 5, 6, and 7 from Table 1.1, we convert differential equation of motion (6.3.3) from the time domain into the Laplace domain. The resulting algebraic equation of motion in the Laplace domain reads:

$$l^2 x(l) - lv_0 - l^2 s_0 + p + f = a\cos\lambda\frac{\omega_1 l}{l^2 + \omega_1^2} + a\sin\lambda\frac{l^2}{l^2 + \omega_1^2} \qquad (6.3.7)$$

Solving equation (6.3.7) for the displacement $x(l)$ in the Laplace domain, we have:

$$x(l) = s_0 + \frac{v_0}{l} - \frac{p+f}{l^2} + \frac{a\omega_1 \cos \lambda}{l(l^2 + \omega_1^2)} + \frac{a \sin \lambda}{l^2 + \omega_1^2} \qquad (6.3.8)$$

Using pairs $1, 5, 2, 8, 16$, and 14 from Table 1.1, we invert equation (6.3.8) from the Laplace domain into the time domain, and obtain the solution of differential equation of motion (6.3.1) with the initial conditions of motion (6.3.2):

$$x = s_0 + v_0 t - \frac{p+f}{2} t^2 + \frac{a \cos \lambda}{\omega_1}\left(t - \frac{1}{\omega_1} \sin \omega_1 t\right) + \frac{a \sin \lambda}{\omega_1^2}(1 - \cos \omega_1 t)$$

$$(6.3.9)$$

Applying some basic algebra to equation (6.3.9), we have

$$x = s_0 + v_0 t - \frac{p+f}{2} t^2 + \frac{at}{\omega_1} \cos \lambda + \frac{a}{\omega_1^2} \sin \lambda - \frac{a}{\omega_1^2} \sin(\omega_1 t + \lambda)$$

$$(6.3.10)$$

Taking the first derivative of equation (6.3.10), we determine the velocity of the system:

$$\frac{dx}{dt} = v_0 - (p+f)t + \frac{a}{\omega_1}[\cos \lambda - \cos(\omega_1 t + \lambda)] \qquad (6.3.11)$$

The second derivative of equation (6.3.10) represents the acceleration:

$$\frac{d^2 x}{dt^2} = -p - f + a \sin(\omega_1 t + \lambda) \qquad (6.3.12)$$

Due to the initial velocity of the system, it can be assumed that during the first cycle of vibrations the system continues to move in the positive direction. Therefore, according to equation (6.3.12), the extreme values of the acceleration a_{ext} occur at the following conditions:

$$\sin(\omega_1 t + \lambda) = \pm 1 \qquad (6.3.13)$$

Combining equations (6.3.12) and (6.3.13), we calculate the extreme values of the acceleration:

$$a_{ext} = -p - f \pm a \qquad (6.3.14)$$

Substituting notations (6.3.4), (6.3.5), and (6.3.6) into equation (6.3.14), we have:

$$a_{ext} = \frac{-P - F \pm A}{m} \qquad (6.3.15)$$

Adding the constant resisting force P and the friction force F to the product of multiplying equation (6.3.15) by m, we determine the extreme values of the forces R_{ext} that are applied to the system:

$$R_{ext} = \pm A \qquad (6.3.16)$$

The system is subjected to the action of a completely reversed loading cycle. The stress analysis should be based on fatigue calculations where the maximum force $R_{max} = A$, while the minimum force $R_{min} = -A$.

6.3.2 Initial Displacement Equals Zero

The initial conditions of motion are:

$$\text{for} \quad t = 0 \quad x = 0; \quad \frac{dx}{dt} = v_0 \qquad (6.3.17)$$

The solution of differential equation of motion (6.3.1) with the initial conditions of motion (6.3.17) reads:

$$x = v_0 t - \frac{p+f}{2} t^2 + \frac{at}{\omega_1} \cos \lambda + \frac{a}{\omega_1^2} \sin \lambda - \frac{a}{\omega_1^2} \sin(\omega_1 t + \lambda) \qquad (6.3.18)$$

The velocity, the acceleration, and the forces are the same as for the previous case.

6.3.3 Initial Velocity Equals Zero

$$\text{For} \quad t = 0 \quad x = s_0; \quad \frac{dx}{dt} = 0 \qquad (6.3.19)$$

Solving differential equation of motion (6.3.1) with the initial conditions of motion (6.3.19), we obtain:

$$x = s_0 - \frac{p+f}{2}t^2 + \frac{at}{\omega_1}\cos\lambda + \frac{a}{\omega_1^2}\sin\lambda - \frac{a}{\omega_1^2}\sin(\omega_1 t + \lambda) \quad \textbf{(6.3.20)}$$

Taking the first derivative of equation (6.3.20), we determine the velocity:

$$\frac{dx}{dt} = -(p+f)t + \frac{a}{\omega_1}[\cos\lambda - \cos(\omega_1 t + \lambda)] \quad \textbf{(6.3.21)}$$

The acceleration and the force can be calculated from equations (6.3.12) and (6.1.3.16) respectively.

6.3.4 Both the Initial Displacement and Velocity Equal Zero

The initial conditions of motion are:

$$\text{for} \quad t = 0 \quad x = 0; \quad \frac{dx}{dt} = 0 \quad \textbf{(6.3.22)}$$

The solution of differential equation of motion (6.3.1) with the initial conditions of motion (6.2.22) reads:

$$x = -\frac{p+f}{2}t^2 + \frac{at}{\omega_1}\cos\lambda + \frac{a}{\omega_1^2}\sin\lambda - \frac{a}{\omega_1^2}\sin(\omega_1 t + \lambda) \quad \textbf{(6.3.23)}$$

The rest of the parameters and their analysis is the same as in the previous case.

6.4 Time-Dependent Force $Q\left(\rho + \frac{\mu t}{\tau}\right)$

The engineering problems described in this section are associated with systems subjected to the action of the force of inertia, the constant resisting force, and the friction force as the resisting forces (Row 6 in Guiding Table 2.1) and to the time-dependent force as the active force (Column 4). The current problems could be related to the interaction of mechanical systems with an elastoplastic or viscoelastoplastic medium that exerts a constant resisting force and a friction force as the reaction to plastic deformation (more related considerations are

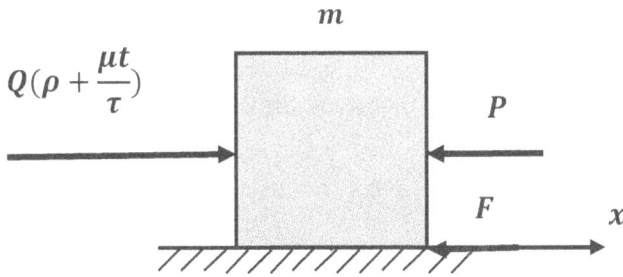

Figure 6.4 Model of a system subjected to a time-dependent force, a constant resisting force, and a friction force

discussed in section 1.2). In some mechanical systems at the beginning of the working process, the active force is increasing from a level that exceeds the resisting force, in other words $Q\rho > P + F$.

The system is moving in the horizontal direction. We want to determine the basic parameters of motion, the values of the displacement, the velocity, the acceleration, the forces applied to the system at the end of time τ, and the power of the energy source. The model of a system subjected to a time-dependent force, a constant resisting force, and a friction force is shown in Figure 6.4.

Accounting for the considerations above and the model in Figure 6.4, we can describe the structure of the system's differential equation of motion. The left side of the equation consists of the force of inertia, the constant resisting force, and the friction force, while the right side consists of the time-dependent force. Thus, the differential equation of motion reads:

$$m\frac{d^2x}{dt^2} + P + F = Q(\rho + \frac{\mu t}{\tau}) \qquad (6.4.1)$$

The differential equation of motion (6.4.1) has different solutions for various initial conditions of motion. These solutions and their analyses are presented below.

6.4.1 General Initial Conditions

The general initial conditions of motion are:

$$\text{for} \quad t = 0 \quad x = s_0; \quad \frac{dx}{dt} = v_0 \qquad (6.4.2)$$

where s_0 and v_0 are the initial displacement and initial velocity respectively.

Dividing equation (6.4.1) by m, we have:

$$\frac{d^2x}{dt^2} + p + f = q(\rho + \frac{\mu t}{\tau}) \qquad (6.4.3)$$

where:

$$p = \frac{P}{m} \qquad (6.4.4)$$

$$f = \frac{F}{m} \qquad (6.4.5)$$

$$q = \frac{Q}{m} \qquad (6.4.6)$$

Applying Laplace Transform Pairs 3, 5, and 2 from Table 1.1, we convert differential equation of motion (6.4.3) with the initial conditions of motion (6.4.2) from the time domain into the Laplace domain. The resulting algebraic equation of motion in the Laplace domain reads:

$$l^2 x(l) - l v_0 - l^2 s_0 + p + f = q(\rho + \frac{\mu}{\tau l}) \qquad (6.4.7)$$

Solving equation (6.4.7) for the Laplace domain displacement $x(l)$, we write:

$$x(l) = s_0 + \frac{v_0}{l} + \frac{q\rho - p - f}{l^2} + \frac{q\mu}{\tau l^3} \qquad (6.4.8)$$

Based on pairs 1, 5, 2, and 8 from Table 1.1, we invert equation (6.4.8) from the Laplace domain into the time domain, and obtain the solution of differential equation of motion (6.4.1) with the initial conditions of motion (6.4.2):

$$x = s_0 + v_0 t + \frac{1}{2}(q\rho - p - f)t^2 + \frac{q\mu}{6\tau}t^3 \qquad (6.4.9)$$

The first derivative of equation (5.4.8) represents the velocity of the system:

$$\frac{dx}{dt} = v_0 + (q\rho - p - f)t + \frac{q\mu}{2\tau}t^2 \qquad (6.4.10)$$

Taking the second derivative of equation (6.4.9), we determine the acceleration of the system:

$$\frac{d^2x}{dt^2} = q\rho - p - f + \frac{q\mu}{\tau}t \qquad (6.4.11)$$

Differential equation (6.4.1) describes the motion that is lasting an interval of time τ. Thus, in equations (6.4.9), (6.4.10), and (6.4.11), replacing the running time t by the interval of time τ, we obtain respectively the values of the displacement s, the velocity v, and the acceleration a at the end of this interval of time:

$$s = s_0 + v_0\tau + \frac{3q\rho - 3(p+f) + q\mu}{6}\tau^2 \qquad (6.4.12)$$

$$v = v_0 + \frac{2q\rho - 2(p+f) + q\mu}{2}\tau \qquad (6.4.13)$$

$$a = q(\rho + \mu) - p - f \qquad (6.4.14)$$

The value of the force R_0 applied to the system at the end of time τ is equal to the sum of the force of inertia, the constant resisting force, and friction force. Adding the resisting forces P and F to the product of multiplying equation (6.4.14) by m, and substituting notations (6.4.4), (6.4.5), and (6.4.6), we determine the force applied to the system:

$$R_0 = Q(\rho + \mu) \qquad (6.4.15)$$

The stress calculations of the system should be based on the force according to equation (6.4.15).

Substituting notations (6.4.4), (6.4.5), and (6.4.6) into the equation of velocity (6.4.13) and multiplying the latter by the force according to equation (6.4.15), we obtain the energy source's power N:

$$N = Q(\rho + \mu)[v_0 + \frac{2Q\rho - 2P - 2F + Q\mu}{2m}\tau]$$

6.4.2 Initial Displacement Equals Zero

The initial conditions of motion are:

$$\text{for} \quad t = 0 \quad x = 0; \quad \frac{dx}{dt} = v_0 \qquad (6.4.16)$$

The solution of differential equation of motion (6.4.1) with the initial conditions of motion (6.4.16) reads:

$$x = v_0 t + \frac{1}{2}(q\rho - p - f)t^2 + \frac{q\mu}{6\tau}t^3 \qquad (6.4.17)$$

Replacing the running time t by time τ in equation (6.4.17), we determine the displacement at the end of this time:

$$s = s_0 + \frac{3q\rho - 3(p + f) + q\mu}{6}\tau^2 \qquad (6.4.18)$$

The rest of the parameters and their analysis are the same as for the previous case.

6.4.3 Initial Velocity Equals Zero

The initial conditions of motion are:

$$\text{for} \quad t = 0 \quad x = s_0; \quad \frac{dx}{dt} = 0 \qquad (6.4.19)$$

Solving differential equation of motion (6.4.1) with the initial condition of motion (6.4.19), we obtain:

$$x = s_0 + \frac{1}{2}(q\rho - p - f)t^2 + \frac{q\mu}{6\tau}t^3 \qquad (6.4.20)$$

The first derivative of equation (6.4.20) represents the velocity of the system:

$$\frac{dx}{dt} = (q\rho - p - f)t + \frac{q\mu}{2\tau}t^2 \tag{6.4.21}$$

Replacing the running time t by the time τ in equations (6.4.20) and (6.4.21), we calculate the values of the displacement and the velocity respectively:

$$s = s_0 + \frac{3q\rho - 3(p+f) + q\mu}{6}\tau^2 \tag{6.4.22}$$

$$v = \frac{2q\rho - 2(p+f) + q\mu}{2}\tau \tag{6.4.23}$$

The product of multiplying the force according to equation (6.4.15) by the velocity according to equation (6.4.23) represents the power of the system:

$$N = Q(\rho + \mu)\frac{Q(2\rho + \mu) - 2(P + F)}{2m}\tau$$

The rest of the parameters and their analysis are the same as in the previous case.

6.4.4 Both the Initial Displacement and Velocity Equal Zero

In this case the initial conditions are:

$$\text{for} \quad t = 0 \quad x = 0; \quad \frac{dx}{dt} = 0 \tag{6.4.24}$$

The solution of differential equation of motion (6.4.1) with the initial conditions of motion (6.4.24) reads:

$$x = \frac{1}{2}(q\rho - p - f)t^2 + \frac{q\mu}{6\tau}t^3 \tag{6.4.25}$$

Substituting the time τ into equation (6.4.25), we determine the displacement:

$$s = \frac{3q\rho - 3(p+f) + q\mu}{6}\tau^2 \qquad (6.4.26)$$

The rest of the parameters and their analysis are the same as in the previous case.

6.5 Constant Force R and Harmonic Force $A\sin(\omega_1 t + \lambda)$

This section describes engineering systems characterized by the action of the force of inertia, the constant resisting force, and the friction force as the resisting forces (Row 6 in Guiding Table 2.1), and the constant active force and harmonic force (Column 5). The current problem could be related to vibratory systems intended for interaction with elastoplastic or viscoelastoplastic media. The working processes of these systems are sometimes associated with the combined action of a constant active force and a harmonic force. Both elastoplastic and viscoelastoplastic media exert simultaneously a constant resisting force and a friction force as a reaction to their plastic deformation. (More related information regarding the media deformation and the behavior of a constant resisting force and a friction force applied to a vibratory system can be found in section 1.2).

Figure 6.5 shows the model of a system subjected to a constant active force, a harmonic force, a constant resisting force, and a friction force. The considerations above and the model in Figure 6.5

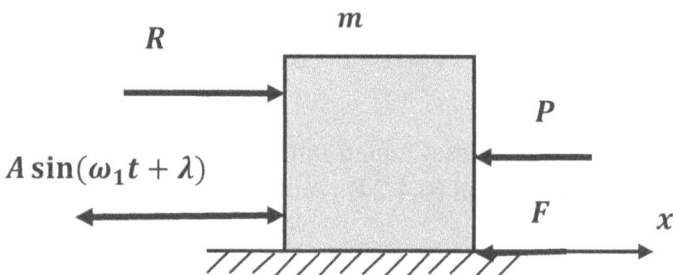

Figure 6.5 Model of a system subjected to a constant active force, a harmonic force, a constant resisting force, and a friction force

allow us to describe the structure of the system's differential equation of motion.

The left side of the equation consists of the sum of the force of inertia, the constant resisting force, and the friction force. The right side includes the constant active force and the harmonic force. Therefore, the differential equation of motion reads:

$$m\frac{d^2x}{dt^2} + P + F = R + A\sin(\omega_1 t + \lambda) \qquad (6.5.1)$$

Differential equation of motion (6.4.1) has different solutions for various initial conditions of motion. These solutions and their analyses are presented below.

6.5.1 General Initial Conditions

The general initial conditions of motion are:

$$\text{for} \quad t = 0 \quad x = s_0; \quad \frac{dx}{dt} = v_0 \qquad (6.5.2)$$

where s_0 and v_0 are the initial displacement and initial velocity respectively.

Transforming the sinusoidal function in equation (6.5.1) and dividing the latter by m, we have:

$$\frac{d^2x}{dt^2} + p + f = r + a\sin\omega_1 t\cos\lambda + a\cos\omega_1 t\sin\lambda \qquad (6.5.3)$$

where:

$$p = \frac{P}{m} \qquad (6.5.4)$$

$$f = \frac{F}{m} \qquad (6.5.5)$$

$$r = \frac{R}{m} \qquad (6.5.6)$$

$$a = \frac{A}{m} \qquad (6.5.7)$$

Using Laplace Transform Pairs 3, 5, 6, and 7 from Table 1.1, we convert differential equation of motion (6.5.3) with its initial conditions of motion (6.5.2) from the time domain into the Laplace domain. The resulting algebraic equation of motion in the Laplace domain reads:

$$l^2 x(l) - l v_0 - l^2 s_0 + p + f = r + \frac{a \omega_1 l}{l^2 + \omega_1^2} \cos \lambda + \frac{a l^2}{l^2 + \omega_1^2} \sin \lambda \quad \textbf{(6.5.8)}$$

Solving equation (6.5.8) for the Laplace domain displacement $x(l)$, we have:

$$x(l) = s_0 + \frac{v_0}{l} + \frac{r - p - f}{l^2} + \frac{a \omega_1}{l(l^2 + \omega_1^2)} \cos \lambda + \frac{a}{l^2 + \omega_1^2} \sin \lambda \quad \textbf{(6.5.9)}$$

Based on pairs 1, 5, 2, 8, 16, and 14 from Table 1.1, we invert equation (6.5.9) from the Laplace domain into the time domain and obtain the solution of differential equation of motion (6.5.1) with the initial conditions of motion (6.5.2):

$$x = s_0 + v_0 t + \frac{r - p - f}{2} t^2 + \frac{a \cos \lambda}{\omega_1} \left(t - \frac{1}{\omega_1} \sin \omega_1 t \right)$$

$$+ \frac{a \sin \lambda}{\omega_1^2} (1 - \cos \omega_1 t) \qquad \textbf{(6.5.10)}$$

Performing algebraic operations with equation (6.5.10), we have:

$$x = s_0 + v_0 t + \frac{r - p - f}{2} t^2 + \frac{at}{\omega_1} \cos \lambda$$

$$+ \frac{a}{\omega_1^2} \sin \lambda - \frac{a}{\omega_1^2} (\sin \omega_1 t \cos \lambda + \cos \omega_1 t \sin \lambda)$$

Further transformation of this equation reads:

$$x = s_0 + v_0 t + \frac{r - p - f}{2} t^2 + \frac{at}{\omega_1} \cos \lambda + \frac{a}{\omega_1^2} \sin \lambda - \frac{a}{\omega_1^2} \sin(\omega_1 t + \lambda)$$

$$\textbf{(6.5.11)}$$

The first derivative of equation (6.5.11) represents the velocity of the system:

$$\frac{dx}{dt} = v_0 + (r - p - f)t + \frac{a}{\omega_1}\cos\lambda - \frac{a}{\omega_1}\cos(\omega_1 t + \lambda) \qquad \textbf{(6.5.12)}$$

Taking the second derivative of equation (6.5.11), we determine the acceleration:

$$\frac{d^2 x}{dt^2} = r - p - f + a\sin(\omega_1 t + \lambda) \qquad \textbf{(6.5.13)}$$

Equation (6.5.13) shows that the acceleration reaches its extreme values when:

$$\sin(\omega_1 t + \lambda) = \pm 1 \qquad \textbf{(6.5.14)}$$

Combining equation (6.5.13) with equation (6.5.14), we determine the extreme values of the acceleration a_{ext}:

$$a_{ext} = r - p - f \pm a \qquad \textbf{(6.5.15)}$$

Substituting notations (6.5.4), (6.5.5), (6.5.6), and (6.5.7) into equation (6.5.15), we may write:

$$a_{ext} = \frac{R - P - F \pm A}{m} \qquad \textbf{(6.5.16)}$$

Adding the constant resisting force P and the friction force F to the product of multiplying equation (6.5.16) by m, we determine the extreme values of the force R_{ext} applied to the system:

$$R_{ext} = R \pm A \qquad \textbf{(6.5.17)}$$

Equation (6.5.17) shows that the system is subjected to a random loading cycle having the maximum force $R_{max} = R + A$ and the minimum force $R_{min} = R - A$. The stress analysis for this case should be based on appropriate fatigue considerations.

6.5.2 Initial Displacement Equals Zero

The initial conditions of motion are:

$$\text{for} \quad t = 0 \quad x = 0; \quad \frac{dx}{dt} = v_0 \qquad \text{(6.5.18)}$$

Solving differential equation of motion (6.5.1) with the initial conditions of motion (6.5.18), we have:

$$x = v_0 t + \frac{r - p - f}{2} t^2 + \frac{at}{\omega_1} \cos \lambda + \frac{a}{\omega_1^2} \sin \lambda - \frac{a}{\omega_1^2} \sin(\omega_1 t + \lambda)$$

$$\text{(6.5.19)}$$

The rest of the parameters are the same as in the previous case.

6.5.3 Initial Velocity Equals Zero

The initial conditions of motion are:

$$\text{for} \quad t = 0 \quad x = s_0; \quad \frac{dx}{dt} = 0 \qquad \text{(6.5.20)}$$

The solution of differential equation of motion (6.5.1) with the initial conditions of motion (6.5.20) reads:

$$x = s_0 + \frac{r - p - f}{2} t^2 + \frac{at}{\omega_1} \cos \lambda + \frac{a}{\omega_1^2} \sin \lambda - \frac{a}{\omega_1^2} \sin(\omega_1 t + \lambda) \quad \text{(6.5.21)}$$

The first derivative of equation (6.5.21) represents the velocity of the system:

$$\frac{dx}{dt} = (r - p - f)t + \frac{a}{\omega_1} \cos \lambda - \frac{a}{\omega_1} \cos(\omega_1 t + \lambda) \quad \text{(6.5.22)}$$

The rest of the parameters can be determined from equations (6.5.13), (6.5.16), and (6.5.17).

6.5.4 Both the Initial Displacement and Velocity Equal Zero

The initial conditions of motion are:

$$\text{for} \quad t = 0 \quad x = 0; \quad \frac{dx}{dt} = 0 \qquad \text{(6.5.23)}$$

Solving differential equation of motion (6.5.1) with the initial conditions of motion (6.5.23), we obtain:

$$x = \frac{r-p-f}{2}t^2 + \frac{at}{\omega_1}\cos\lambda + \frac{a}{\omega_1^2}\sin\lambda - \frac{a}{\omega_1^2}\sin(\omega_1 t + \lambda) \qquad (6.5.24)$$

The rest of the parameters are the same as for the previous case.

6.6 Harmonic Force $A\sin(\omega_1 t + \lambda)$ and Time-Dependant Force $Q\left(\rho + \frac{\mu t}{\tau}\right)$

According to Guiding Table 2.1, this section describes engineering systems characterized by the action of the force of inertia, the constant resisting force, and the friction force as the resisting forces (Row 6), and the harmonic force and a time-dependent force as active forces (Column 6). The current problem could be associated with a vibratory system intended for the interaction with elastoplastic or viscoelastoplastic media. (Additional information related to the deformation of these media and considerations regarding the behavior of a constant resisting force and a friction force applied to a vibratory system are discussed in section 1.2.)

This problem could reflect the initial phase of the working process of a vibratory system subjected to a time-dependent force that acts a limited interval of time. An elastoplastic or viscoelastoplastic medium can simultaneously exert a constant resisting force and a friction force as a reaction to its plastic deformation. In these kinds of systems, the initial value of the active force exceeds the value of the resultant of the resisting forces $(Q\rho > P + F)$.

The system is moving in the horizontal direction. We want to determine the basic parameters of motion and the characteristics of the forces applied to the system.

Figure 6.6 represents the model of a system subjected to the action of a harmonic force, a time-dependent force, a constant resisting force, and a friction force. Based on the considerations above and the model in Figure 6.6, we can compose the differential equation of motion. Thus, the left side of this equation consists of the force of inertia, the constant resisting force, and the friction force, while the

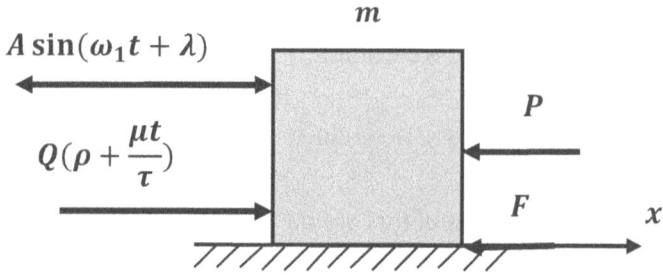

Figure 6.6 Model of a system subjected to a harmonic force, a time-dependent force, a constant resisting force, and a friction force

right side includes the harmonic force and the time-dependent force. Therefore, the differential equation of motion reads:

$$m\frac{d^2x}{dt^2} + P + F = A\sin(\omega_1 t + \lambda) + Q(\rho + \frac{\mu t}{\tau}) \qquad (6.6.1)$$

Differential equation of motion (6.6.1) has different solutions for various initial conditions of motion. These solutions and their analyses are presented below.

6.6.1 General Initial Conditions

The general initial conditions of motion are:

$$\text{for} \quad t = 0 \quad x = s_0; \quad \frac{dx}{dt} = v_0 \qquad (6.6.2)$$

where s_0 and v_0 are the initial displacement and initial velocity respectively.

Transforming equation (6.6.1), we may write:

$$m\frac{d^2x}{dt^2} + P + F = A\sin\omega_1 t\cos\lambda + A\cos\omega_1 t\sin\lambda + Q\rho + \frac{Q\mu}{\tau}t$$

$$(6.6.3)$$

Dividing equation (6.6.3) by m, we have:

$$\frac{d^2x}{dt^2} + p + f = a\sin\omega_1 t\cos\lambda + a\cos\omega_1 t\sin\lambda + q\rho + \frac{q\mu}{\tau}t \qquad (6.6.4)$$

where:

$$p = \frac{P}{m} \tag{6.6.5}$$

$$f = \frac{F}{m} \tag{6.6.6}$$

$$a = \frac{A}{m} \tag{6.6.7}$$

$$q = \frac{Q}{m} \tag{6.6.8}$$

Using Laplace Transform Pairs $3, 5, 6, 7, 5$, and 2 from Table 1.1, we convert differential equation of motion $(6.6.4)$ with the initial conditions of motion $(6.6.2)$ from the time domain into the Laplace domain. The resulting algebraic equation of motion in the Laplace domain reads:

$$x(l) - lv_0 - l^2 s_0 + p + f = \frac{a\omega_1 l}{l^2 + \omega_1^2} \cos \lambda + \frac{al^2}{l^2 + \omega_1^2} \sin \lambda + qp + \frac{q\mu}{\tau l}$$
$$\tag{6.6.9}$$

Solving equation $(6.6.9)$ for the displacement in Laplace domain $x(l)$, we have:

$$x(l) = s_0 + \frac{v_0}{l} + \frac{a\omega_1}{l(l^2 + \omega_1^2)} \cos \lambda + \frac{a}{l^2 + \omega_1^2} \sin \lambda + \frac{qp - p - f}{l^2} + \frac{q\mu}{\tau l^3}$$
$$\tag{6.6.10}$$

Based on pairs $1, 5, 2, 16, 14$, and 8 from Table 1.1, we invert equation $(6.6.10)$ from the Laplace domain into the time domain and obtain the solution of differential equation of motion $(6.6.1)$ with the initial conditions of motion $(6.6.2)$:

$$x = s_0 + v_0 t + \frac{a \cos \lambda}{\omega_1}\left(t - \frac{1}{\omega_1}\sin \omega_1 t\right) + \frac{a \sin \lambda}{\omega_1^2}(1 - \cos \omega_1 t)$$
$$+ \frac{(qp - p - f)t^2}{2} + \frac{q\mu t^3}{6\tau} \tag{6.6.11}$$

Applying basic algebraic procedures to equation (6.6.11), we may write:

$$x = s_0 + \left(v_0 + \frac{a}{\omega_1}\cos\lambda\right)t + \frac{(q\rho - p - f)}{2}t^2 + \frac{q\mu t^3}{6\tau} + \frac{a}{\omega_1^2}\sin\lambda$$

$$-\frac{a}{\omega_1^2}(\sin\omega_1 t\cos\lambda + \cos\omega_1 t\sin\lambda) \tag{6.6.12}$$

After further transformations of equation (6.6.12), we have:

$$x = s_0 + \left(v_0 + \frac{a}{\omega_1}\cos\lambda\right)t + \frac{(q\rho - p - f)}{2}t^2 + \frac{q\mu t^3}{6\tau}$$

$$+\frac{a}{\omega_1^2}\sin\lambda - \frac{a}{\omega_1^2}\sin(\omega_1 t + \lambda) \tag{6.6.13}$$

Taking the first derivative of equation (6.6.13), we determine the velocity of the system:

$$\frac{dx}{dt} = v_0 + \frac{a}{\omega_1}\cos\lambda + (q\rho - p - f)t + \frac{q\mu t^2}{2\tau} - \frac{a}{\omega_1}\cos(\omega_1 t + \lambda)$$

$$\tag{6.6.14}$$

The second derivative of equation (6.6.13) represents the system's acceleration:

$$\frac{d^2x}{dt^2} = q\rho - p - f + \frac{q\mu t}{\tau} + a\sin(\omega_1 t + \lambda) \tag{6.6.15}$$

Because the motion lasts during the time interval τ, we may substitute this time into equation (6.6.15) and calculate the value of the acceleration a_τ at the end of this time interval:

$$a_\tau = q\rho - p - f + q\mu + a\sin(\omega_1 \tau + \lambda) \tag{6.6.16}$$

The acceleration reaches its extreme values a_{ext} when:

$$\sin(\omega_1 \tau + \lambda) = \pm 1 \tag{6.6.17}$$

Combining equations (6.6.16) and (6.6.17), we determine the extreme values of the acceleration a_{ext}:

$$a_{ext} = q\rho - p - f + q\mu \pm a \qquad \text{(6.6.18)}$$

Substituting notations (6.6.5), (6.6.6), (6.6.7), and (6.6.8) into equation (6.6.18), we write:

$$a_{ext} = \frac{Q}{m}(\rho + \mu) - \frac{P + F}{m} \pm \frac{A}{m} \qquad \text{(6.6.19)}$$

Adding the constant resisting force P and the friction force F to the product of multiplying equation (6.6.19) by m, we determine the extreme values of the forces R_{ext} that are applied to the system:

$$R_{ext} = Q(\rho + \mu) \pm A \qquad \text{(6.6.20)}$$

Equation (6.6.20) shows that the system is subjected to a random loading cycle for which the maximum force $R_{max} = Q(\rho + \mu) + A$, while the minimum force $R_{min} = Q(\rho + \mu) - A$.

The stress analysis for this case should be based on appropriate fatigue calculations.

6.6.2 Initial Displacement Equals Zero

The initial conditions of motion are:

$$\text{for} \quad t = 0 \quad x = 0; \quad \frac{dx}{dt} = v_0 \qquad \text{(6.6.21)}$$

Solving differential equation of motion (6.6.1) with the initial conditions of motion (6.6.21), we have:

$$x = t(v_0 + \frac{a}{\omega_1} \cos \lambda) + \frac{q\rho - p - f}{2} t^2 + \frac{q\mu t^3}{6\tau}$$

$$+ \frac{a}{\omega_1^2} \sin \lambda - \frac{a}{\omega_1^2} \sin(\omega_1 t + \lambda) \qquad \text{(6.6.22)}$$

The rest of the parameters are the same as for the previous case.

6.6.3 Initial Velocity Equals Zero

The initial conditions of motion are:

$$\text{for} \quad t = 0 \quad x = s_0; \quad \frac{dx}{dt} = 0 \tag{6.6.23}$$

The solution of differential equation of motion (6.6.1) with the initial conditions of motion (6.6.23) reads:

$$x = s_0 + \frac{at}{\omega_1}\cos\lambda + \frac{q\rho - p - f}{2}t^2 + \frac{q\mu t^3}{6\tau} + \frac{a}{\omega_1^2}\sin\lambda - \frac{a}{\omega_1^2}\sin(\omega_1 t + \lambda) \tag{6.6.24}$$

Taking the first derivative of equation (6.6.24), we determine the velocity of the system:

$$\frac{dx}{dt} = \frac{a}{\omega_1}\cos\lambda + (q\rho - p - f) - t + \frac{q\mu t^2}{2\tau} - \frac{a}{\omega_1}\cos(\omega_1 t + \lambda) \tag{6.6.25}$$

The rest of the parameters are the same as for the previous case.

6.6.4 Both the Initial Displacement and Velocity Equal Zero

The initial conditions of motion are:

$$\text{for} \quad t = 0 \quad x = 0; \quad \frac{dx}{dt} = 0 \tag{6.6.26}$$

Solving differential equation of motion (6.6.1) with the initial conditions of motion (6.6.26), we have:

$$x = \frac{at}{\omega_1}\cos\lambda + \frac{q\rho - p - f}{2}t^2 + \frac{q\mu t^3}{6\tau} + \frac{a}{\omega_1^2}\sin\lambda - \frac{a}{\omega_1^2}\sin(\omega_1 t + \lambda) \tag{6.6.27}$$

The rest of the parameters are the same as for the previous case.

7

STIFFNESS

This chapter covers engineering systems characterized by the force of inertia and the stiffness force as resisting forces. These forces are marked by the plus sign in Row 7 of Guiding Table 2.1. The intersection of this row with Columns 1 through 6 indicates the active forces applied to these systems, and discussed respectively in this chapter's six sections. Throughout this chapter, the left sides of the differential equations of motion are identical, consisting of the force of inertia and the stiffness force. The right sides of these differential equations vary by the active forces applied to the systems.

The engineering problems in this chapter could be associated with mechanical systems intended to interact with elastoplastic or viscoelastoplastic media or with specific elastic links.

7.1 Active Force Equals Zero

According to Guiding Table 2.1, the engineering problems in this section are characterized by the force of inertia and the stiffness force as the resisting forces (Row 7), but with no active force present (Column 1). (Considerations related to the motion of a system in the absence of active forces are discussed in section 1.3.)

The current problem could represent a system performing free vibration or a system interacting with a specific elastic link. It also could be related to a system involved in the phase of elastic deformation of an elastoplastic or viscoelastoplastic medium that exerts a stiffness force as a reaction to the deformation (more related information is presented in section 1.2).

The system is moving in the horizontal direction. We want to determine the basic parameters of motion, their extreme values, and the characteristics of the forces applied to the elastic link. The model of a system subjected to the action of a stiffness force is shown in Figure 7.1.

Based on the considerations above and the model in Figure 7.1, we can compose the differential equation of motion of the system. The left side consists of the force of inertia and the stiffness force, while the right side equals zero. Thus, the differential equation of motion reads:

$$m\frac{d^2x}{dt^2} + Kx = 0 \qquad (7.1.1)$$

Differential equation of motion (7.1.1) has different solutions for various initial conditions of motion. These solutions and their analyses are presented below.

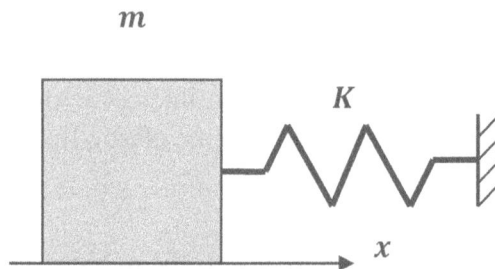

Figure 7.1 Model of a system subjected to a stiffness force

7.1.1 General Initial Conditions

The general initial conditions of motion are:

$$\text{for}\quad t=0\quad x=s_0;\quad \frac{dx}{dt}=v_0 \tag{7.1.2}$$

where s_0 and v_0 are the initial displacement and initial velocity respectively.

Dividing equation (7.1.1) by m, we have:

$$\frac{d^2x}{dt^2}+\omega^2 x=0 \tag{7.1.3}$$

where ω is the natural frequency of the vibratory system and:

$$\omega^2=\frac{K}{m} \tag{7.1.4}$$

Using Laplace Transform Pairs 3, 1, and 5 from Table 1.1, we convert differential equation of motion (7.1.3) with the initial conditions of motion (7.1.2) from the time domain into the Laplace domain, and obtain the corresponding algebraic equation of motion in the Laplace domain:

$$l^2 x(l)-lv_0-l^2 s_0+\omega^2 x(l)=0 \tag{7.1.5}$$

Solving equation (7.1.5) for the displacement $x(l)$ in the Laplace domain, we have:

$$x(l)=\frac{lv_0}{l^2+\omega^2}+\frac{l^2 s_0}{l^2+\omega^2} \tag{7.1.6}$$

Based on pairs 1, 5, 6, and 7 from Table 1.1, we invert equation (7.1.6) from the Laplace domain into the time domain, and obtain the solution of differential equation of motion (7.1.1) with the initial conditions of motion (7.1.2):

$$x=\frac{v_0}{\omega}\sin\omega t+s_0\cos\omega t \tag{7.1.7}$$

Taking the first derivative of equation (7.1.7), we determine the velocity of the system:

$$\frac{dx}{dt} = v_0 \cos \omega t - s_0 \omega \sin \omega t \qquad (7.1.8)$$

The second derivative of equation (7.1.7) represents the acceleration:

$$\frac{d^2 x}{dt^2} = -v_0 \omega \sin \omega t - s_0 \omega^2 \cos \omega t \qquad (7.1.9)$$

In order to simplify the analysis of equations (7.1.7), (7.1.8), and (7.1.9), we apply some conventional procedures to the coefficients at the trigonometric functions in equation (7.1.7). Based on these procedures, we may write:

$$\frac{s_0^2 \omega^2}{s_0^2 \omega^2 + v_0^2} + \frac{v_0^2}{s_0^2 \omega^2 + v_0^2} = 1 = (\sin \alpha)^2 + (\cos \alpha)^2 \quad (7.1.10)$$

where:

$$\sin \alpha = \frac{s_0 \omega}{\sqrt{s_0^2 \omega^2 + v_0^2}} \qquad (7.1.11)$$

$$\cos \alpha = \frac{v_0}{\sqrt{s_0^2 \omega^2 + v_0^2}} \qquad (7.1.12)$$

Substituting equations (7.1.11) and (7.1.12) into equation (7.1.7), we obtain:

$$x = \frac{\sqrt{s_0^2 \omega^2 + v_0^2}}{\omega} (\sin \omega t \cos \alpha + \cos \omega t \sin \alpha) \qquad (7.1.13)$$

and finally we have:

$$x = \frac{\sqrt{s_0^2 \omega^2 + v_0^2}}{\omega} \sin(\omega t + \alpha) \qquad (7.1.14)$$

The first and second derivatives of equation (7.1.14) represent the velocity and the acceleration respectively:

$$\frac{dx}{dt} = \sqrt{s_0^2\omega^2 + v_0^2}\,\cos(\omega t + \alpha) \qquad (7.1.15)$$

$$\frac{d^2x}{dt^2} = -\omega\,\sqrt{s_0^2\omega^2 + v_0^2}\,\sin(\omega t + \alpha) \qquad (7.1.16)$$

Taking for equations (7.1.14) and (7.1.16) that:

$$\sin(\omega t + \alpha) = \pm 1 \qquad (7.1.17)$$

we determine the extreme values of the displacement s_{ext} and acceleration a_{ext} respectively:

$$s_{ext} = \pm\frac{\sqrt{s_0^2\omega^2 + v_0^2}}{\omega} \qquad (7.1.18)$$

$$a_{ext} = \pm(-\omega\,\sqrt{s_0^2\omega^2 + v_0^2}\,) \qquad (7.1.19)$$

Similarly, taking for equation (7.1.14) that:

$$\cos(\omega t + \alpha) = \pm 1 \qquad (7.1.20)$$

we determine the extreme values of the velocity v_{ext}:

$$v_{ext} = \pm\sqrt{s_0^2\omega^2 + v_0^2} \qquad (7.1.21)$$

Multiplying equation (7.1.18) by K and using notation (7.1.4), we determine the extreme values of the forces R_{ext} applied to the elastic link:

$$R_{ext} = \pm\sqrt{K^2 s_0^2 + mKv_0^2} \qquad (7.1.22)$$

Equation (7.1.22) shows that the elastic link is subjected to a completely reversed loading cycle where the maximum force R_{max} and minimum force R_{min} respectively are:

$$R_{max} = \sqrt{K^2 s_0^2 + mKv_0^2} \qquad (7.1.23)$$

$$R_{min} = -\sqrt{K^2 s_0^2 + mKv_0^2} \qquad (7.1.24)$$

Equations (7.1.23) and (7.1.24) indicate that the stress calculations of the elastic link should be based on fatigue considerations.

7.1.2 Initial Displacement Equals Zero

The initial conditions of motion are:

$$\text{for} \quad t = 0 \quad x = 0; \quad \frac{dx}{dt} = v_0 \tag{7.1.25}$$

The solution of differential equation of motion (7.1.1) with the initial conditions of motion (7.1.25) reads:

$$x = \frac{v_0}{\omega} \sin \omega t \tag{7.1.26}$$

The first derivative of equation (7.1.26) represents the velocity:

$$\frac{dx}{dt} = v_0 \cos \omega t \tag{7.1.27}$$

Taking the second derivative of equation (7.1.26), we determine the acceleration:

$$\frac{d^2 x}{dt^2} = -v_0 \omega \sin \omega t \tag{7.1.28}$$

The extreme values of the displacement and acceleration occur when:

$$\sin \omega t = \pm 1 \tag{7.1.29}$$

Combining equations (7.1.26) and (7.1.28) with equation (7.1.29), we determine the extreme values of the displacement s_{ext} and acceleration a_{ext} respectively:

$$s_{ext} = \pm \frac{v_0}{\omega} \tag{7.1.30}$$

$$a_{ext} = \pm (-v_0 \omega) \tag{7.1.31}$$

Similarly, taking for equation (7.1.27) that:

$$\cos \omega t = \pm 1 \qquad (7.1.32)$$

we determine the extreme values of the velocity V_{ext}:

$$v_{ext} = \pm v_0 \qquad (7.1.33)$$

Multiplying equation (7.1.30) by K and substituting notation (7.1.4), we determine the forces applied to the elastic link: $R_{max} = v_0 \sqrt{mK}$ and $R_{min} = -v_0 \sqrt{mK}$.

The elastic link is subjected to a completely reversed loading cycle; therefore, the stress calculations should be based on fatigue considerations.

7.1.3 Initial Velocity Equals Zero

Because the model in Figure 7.1 implies that the system moves in the positive direction at the beginning of the motion, we should take the initial displacement with the negative sign. Therefore, the initial conditions of motion are:

$$\text{for} \quad t = 0 \quad x = -s_0; \quad \frac{dx}{dt} = 0 \qquad (7.1.34)$$

Solving differential equation of motion (7.1.1) with the initial conditions of motion (7.1.34), we obtain:

$$x = -s_0 \cos \omega t \qquad (7.1.35)$$

The first derivative of equation (7.1.35) represents the velocity:

$$\frac{dx}{dt} = \omega \sin \omega t \qquad (7.1.36)$$

Taking the second derivative of equation (7.1.35), we determine the acceleration:

$$\frac{d^2 x}{dt^2} = s_0 \omega^2 \cos \omega t \qquad (7.1.37)$$

The extreme values of the displacement and acceleration occur when:

$$\cos \omega t = \pm 1 \tag{7.1.38}$$

Combining equations (7.1.35) and (7.1.37) with equation (7.1.38), we determine the extreme values of the displacement s_{ext} and acceleration a_{ext} respectively:

$$s_{ext} = \pm s_0 \tag{7.1.39}$$

$$a_{ext} = \pm(-s_0 \omega^2) \tag{7.1.40}$$

Similarly, taking for equation (7.1.36) that:

$$\sin \omega t = \pm 1 \tag{7.1.41}$$

we determine the extreme values of the velocity V_{ext}:

$$v_{ext} = \pm s_0 \omega \tag{7.1.42}$$

Multiplying equation (7.1.39) by K, we determine the forces applied to the elastic link:

$$R_{max} = K s_0 \text{ and } R_{min} = -K s_0$$

The elastic link is subjected to a completely reversed loading cycle; therefore, the stress calculations should be based on fatigue considerations.

7.2 Constant Force R

This section describes engineering systems experiencing the force of inertia and the stiffness force as the resisting forces (Row 7). Guiding Table 2.1 indicates the current problem is also characterized by the constant active force (Column 2). Numerous mechanical systems interact with elastoplastic or viscoelastoplastic media and with specific elastic links, while being subjected to the action of a constant active force. These media exert a stiffness force as a reaction to the elastic deformation (more related considerations are presented in section 1.2).

The system is moving in the horizontal direction. The model of a system subjected to the action of a constant active force and a stiffness force is shown in Figure 7.2.

According to the considerations above and the model in Figure 7.2, we can compose the left and right sides of this system's differential equation of motion. The left side of this equation consists of the force of inertia and the stiffness force, while the right side includes a constant active force. Hence, the differential equation of motion reads:

$$m\frac{d^2x}{dt^2} + Kx = R \qquad (7.2.1)$$

As it is explained in section 1.3, in some situations a force that depends on the displacement could play the role of an active force. For instance, in case of magnetic interaction, the attraction force P_s can be described by the following equation:

$$P_s = R + Kx \qquad (7.2.2)$$

where $R > 0$. In this case, the differential equation of motion reads:

$$m\frac{d^2x}{dt^2} - Kx = R \qquad (7.2.3)$$

For the cases according to equations (7.2.1) and (7.2.3), we want to determine the basic parameters of motion, while for the case according to equation (7.2.1), we also want to determine the extreme values of these basic parameters, and the forces applied to the elastic

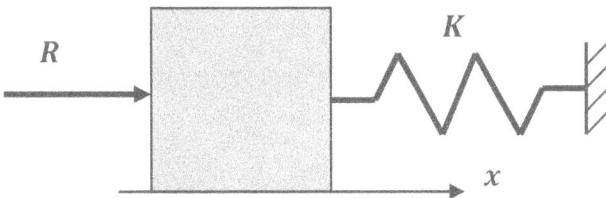

Figure 7.2 Model of a system subjected to a constant active force and a stiffness force

link. The differential equations of motion (7.2.1) and (7.2.3) have different solutions for various initial conditions of motion. These solutions and their analyses are presented below.

7.2.1 General Initial Conditions

The general initial conditions of motion are:

$$\text{for} \quad t = 0 \quad x = s_0; \quad \frac{dx}{dt} = v_0 \tag{7.2.4}$$

where s_0 and v_0 are the initial displacement and initial velocity respectively.

a) Equation (7.2.1)

Dividing this equation by m, we obtain:

$$\frac{d^2x}{dt^2} + \omega^2 x = r \tag{7.2.5}$$

where ω is the natural frequency of the system and:

$$\omega^2 = \frac{K}{m} \tag{7.2.6}$$

$$r = \frac{R}{m} \tag{7.2.7}$$

Using Laplace Transform Pairs 3, 1, and 5 from Table 1.1, we convert differential equation of motion (7.2.5) with the initial conditions of motion (7.2.4) from the time domain into the Laplace domain, and obtain the corresponding algebraic equation of motion in the Laplace domain:

$$l^2 x(l) - lv_0 - l^2 s_0 + \omega^2 x(l) = r \tag{7.2.8}$$

Solving equation (7.2.6) for the displacement $x(l)$ in Laplace domain, we have:

$$x(l) = \frac{lv_0}{l^2 + \omega^2} + \frac{l^2 s_0}{l^2 + \omega^2} + \frac{r}{l^2 + \omega^2} \tag{7.2.9}$$

Based on pairs 1, 5, 6, 7, and 14 from Table 1.1, we invert equation (7.2.9) from the Laplace domain into the time domain. The inversion represents the solution of differential equation of motion (7.2.1) with the initial conditions of motion (7.2.4):

$$x = \frac{v_0}{\omega}\sin\omega t + s_0\cos\omega t + \frac{r}{\omega^2}(1-\cos\omega t) \qquad (7.2.10)$$

In order to simplify the analysis of equation (7.2.10), we rewrite it in the following shape:

$$x = \frac{r}{\omega^2} + \frac{1}{\omega^2}[v_0\omega\sin\omega t - (r - s_0\omega^2)\cos\omega t] \qquad (7.2.11)$$

Using the coefficients at the trigonometric functions in equation (7.2.21), we compose the following expression:

$$\frac{(r - s_0\omega^2)^2}{(r - s_0\omega^2)^2 + v_0^2\omega^2} + \frac{v_0^2\omega^2}{(r - s_0\omega^2)^2 + v_0^2\omega^2} = 1 = (\sin\beta)^2 + (\cos\beta)^2$$

$$(7.2.12)$$

According to expression (7.2.12), we denote:

$$\sin\beta = -\frac{(r - s_0\omega^2)^2}{(r - s_0\omega^2)^2 + v_0^2\omega^2} \qquad (7.2.13)$$

$$\cos\beta = \frac{v_0^2\omega^2}{\left(r - s_0\omega^2\right)^2 + v_0^2\omega^2} \qquad (7.2.14)$$

Combining equations (7.2.13) and (7.2.14) with equation (7.2.11), we have:

$$x = \frac{1}{\omega^2}[r + \sqrt{\left(r - s_0\omega^2\right)^2 + v_0^2\omega^2}\,\sin(\omega t - \beta)] \qquad (7.2.15)$$

Taking the first derivative of equation (7.2.15), we determine the velocity of the system:

$$\frac{dx}{dt} = \frac{1}{\omega}\sqrt{\left(r - s_0\omega^2\right)^2 + v_0^2\omega^2}\,\cos(\omega t - \beta) \qquad (7.2.16)$$

The second derivative of equation (7.2.15) represents the acceleration:

$$\frac{d^2x}{dt^2} = -\sqrt{\left(r - s_0\omega^2\right)^2 + v_0^2\omega^2}\;\sin(\omega t - \beta) \qquad (7.2.17)$$

The velocity becomes equal to zero when $\cos(\omega t - \beta) = 0$ and, consequently, we may write:

$$\sin(\omega t - \beta) = \pm 1 \qquad (7.2.18)$$

Combining equations (7.2.15) and (7.2.17) with equation (7.2.18), we determine the extreme values of the displacement s_{ext} and acceleration a_{ext} respectively:

$$s_{ext} = \frac{1}{\omega^2}[r \pm \sqrt{\left(r - s_0\omega^2\right)^2 + v_0^2\omega^2}] \qquad (7.2.19)$$

$$a_{ext} = \pm[-\sqrt{\left(r - s_0\omega^2\right)^2 + v_0^2\omega^2}] \qquad (7.2.20)$$

Taking for equation (7.2.18) that $\cos(\omega t - \beta) = \pm 1$, we determine the extreme values of the velocity v_{ext}:

$$v_{ext} = \pm\frac{1}{\omega}\sqrt{\left(r - s_0\omega^2\right)^2 + v_0^2\omega^2} \qquad (7.2.21)$$

Multiplying equation (7.2.19) by K and using notations (7.2.6) and (7.2.6), we determine the extreme values of the forces R_{ext} applied to the elastic link:

$$R_{ext} = R \pm \sqrt{(R - Ks_0)^2 + mKv_0^2} \qquad (7.2.22)$$

According to equation (7.2.22), the elastic link is subjected to a random loading cycle having the maximum force:

$$R_{max} = R + \sqrt{(R - Ks_0)^2 + mKv_0^2}$$

and the minimum force:

$$R_{min} = R - \sqrt{(R - Ks_0)^2 + mKv_0^2}$$

The stress calculations should be based on fatigue considerations.

b) Equation (7.2.3)

Dividing this equation by m, we may write:

$$\frac{d^2x}{dt^2} - \omega^2 x = r \qquad (7.2.23)$$

Converting differential equation of motion (7.2.23) with the initial conditions of motion (7.2.4) from the time domain into the Laplace domain, we obtain the corresponding algebraic equation of motion in the Laplace domain:

$$l^2 x(l) - lv_0 - l^2 s_0 - \omega^2 x(l) = r \qquad (7.2.24)$$

Solving equation (7.2.24) for the displacement $x(l)$ in Laplace domain, we obtain:

$$x(l) = \frac{lv_0}{l^2 - \omega^2} + \frac{l^2 s_0}{l^2 - \omega^2} + \frac{r}{l^2 - \omega^2} \qquad (7.2.25)$$

Using pairs 1, 17, 18, and 21 from Table 1.1, we invert equation (7.2.25) from the Laplace domain into the time domain, and obtain the solution of differential equation of motion (7.2.3) with the initial conditions of motion (7.2.4):

$$x = \frac{v_0}{\omega} \sinh \omega t + s_0 \cosh \omega t + \frac{r}{\omega^2} (\cosh \omega t - 1) \qquad (7.2.26)$$

Equation (7.2.26) describes a system in rectilinear translational motion.

Taking the first and second derivatives of equation (7.2.26), we determine respectively the velocity and the acceleration:

$$\frac{dx}{dt} = v_0 \cosh \omega t + s_0 \omega \sinh \omega t + \frac{r}{\omega} \sinh \omega t \qquad (7.2.27)$$

$$\frac{d^2x}{dt^2} = v_0 \omega \sinh \omega t + s_0 \cosh \omega t + r \cosh \omega t \qquad (7.2.28)$$

7.2.2 Initial Displacement Equals Zero
The initial conditions of motion are:

$$\text{for} \quad t=0 \quad x=0; \quad \frac{dx}{dt}=v_0 \qquad (7.2.29)$$

a) Equation (7.2.1)
The solution of differential equation of motion (7.2.1) with the initial conditions of motion (7.2.29) reads:

$$x = \frac{r}{\omega^2}(1-\cos\omega t) + \frac{v_0}{\omega}\sin\omega t \qquad (7.2.30)$$

Applying similar procedures to equation (7.2.30) as in the previous case, we may write:

$$x = \frac{1}{\omega^2}[r + \sqrt{r^2 + v_0^2\omega^2}\,\sin(\omega t - \beta_1)] \qquad (7.2.31)$$

where:

$$\sin\beta_1 = -\frac{r}{\sqrt{r^2 + v_0^2\omega^2}} \qquad (7.2.32)$$

$$\cos\beta_1 = \frac{v_0\omega}{\sqrt{r^2 + v_0^2\omega^2}} \qquad (7.2.33)$$

Taking the first and second derivatives of equation (7.2.31), we determine the velocity and acceleration respectively:

$$\frac{dx}{dt} = \frac{1}{\omega}\sqrt{r^2 + v_0^2\omega^2}\,\cos(\omega t - \beta_1) \qquad (7.2.34)$$

$$\frac{d^2x}{dt^2} = -\sqrt{r^2 + v_0^2\omega^2}\,\sin(\omega t - \beta_1) \qquad (7.2.35)$$

At the moment of time when the velocity becomes equal to zero, the displacement and the acceleration obtain their extreme values. At this moment of time, we have:

$$\sin(\omega t + \beta_1) = \pm 1 \qquad (7.2.36)$$

Combining equation (7.2.36) with equations (7.2.31) and (7.2.35), we calculate the extreme values of the displacement s_{ext} and the acceleration a_{ext} respectively:

$$s_{ext} = \frac{1}{\omega^2}(r \pm \sqrt{r^2 + v_0^2 \omega^2})$$ (7.2.37)

$$a_{ext} = \pm(-\sqrt{r^2 + v_0^2 \omega^2})$$ (7.2.38)

Taking for equation (7.2.34) that $\cos(\omega t + \beta_1) = \pm 1$, we obtain the extreme values of the velocity v_{ext}:

$$v_{ext} = \pm \frac{1}{\omega}\sqrt{r^2 + v_0^2 \omega^2}$$

Multiplying equation (7.2.37) by K and substituting notations (7.2.6) and (7.2.7), we calculate the forces applied to the elastic link:

$$R_{max} = R + \sqrt{R^2 + mK v_0^2}$$

and

$$R_{min} = R - \sqrt{R^2 + mK v_0^2} \,.$$

The stress calculations should be based on fatigue considerations with respect to a random loading cycle.

b) **Equation (7.2.3)**

The solution of differential equation of motion (7.2.3) with the initial conditions of motion (7.2.29) reads:

$$x = \frac{v_0}{\omega}\sinh \omega t + \frac{r}{\omega^2}(\cosh \omega t - 1)$$ (7.2.39)

Equation (7.2.39) shows that the system is in rectilinear translational motion. Taking the first and second derivatives of equation (7.2.39), we determine the velocity and acceleration respectively:

$$\frac{dx}{dt} = v_0 \cosh \omega t + \frac{r}{\omega}\sinh \omega t$$

$$\frac{d^2 x}{dt^2} = v_0 \omega \sinh \omega t + r \cosh \omega t$$

7.2.3 Initial Velocity Equals Zero

The initial conditions of motion are:

$$\text{for} \quad t = 0 \quad x = s_0; \quad \frac{dx}{dt} = 0 \qquad (7.2.40)$$

a) Equation (7.2.1)

Solving differential equation of motion (7.2.1) with the initial conditions of motion (7.2.40), we obtain:

$$x = \frac{1}{\omega^2}[r - (r - s_0\omega^2)\cos\omega t] \qquad (7.2.41)$$

Taking the first and second derivatives of equation (7.2.41), we determine the velocity and acceleration respectively:

$$\frac{dx}{dt} = \frac{1}{\omega}(r - s_0\omega^2)\sin\omega t \qquad (7.2.42)$$

$$\frac{d^2x}{dt^2} = (r - s_0\omega^2)\cos\omega t \qquad (7.2.43)$$

Taking for equations (7.2.41) and (7.2.43) that $\cos\omega t = \pm 1$, we can determine the extreme values of the displacement s_{ext} and the acceleration a_{ext} respectively:

$$s_{ext} = \frac{1}{\omega^2}[r \pm (r - s_0\omega^2)] \qquad (7.2.44)$$

$$a_{ext} = \pm(r - s_0\omega^2) \qquad (7.2.45)$$

Accepting for equation (7.2.42) that $\sin\omega t = \pm 1$, we calculate the extreme values of the velocity v_{ext}:

$$v_{ext} = \pm\frac{1}{\omega}\left(r - s_0\omega^2\right) \qquad (7.2.46)$$

Multiplying equation (7.2.44) by K and substituting the notations (7.2.6) and (7.2.7), we determine the extreme values of the forces R_{ext} applied to the elastic link:

$$R_{ext} = R \pm (R - Ks_0)$$

Therefore, the link is subjected to a random loading cycle having $R_{max} = 2R - Ks_0$ and $R_{min} = Ks_0$.

The stress calculations should be performed for a random loading cycle including fatigue considerations.

b) Equation (7.2.3)

Solving differential equation of motion (7.2.3) with the initial conditions of motion (7.2.40), we obtain:

$$x = s_0 \cosh \omega t + \frac{r}{\omega^2}(\cosh \omega t - 1) \qquad (7.2.47)$$

According to equation (7.2.47), the system is performing rectilinear translational motion. Taking the first and second derivatives of equation (7.2.47), we determine the velocity and acceleration respectively:

$$\frac{dx}{dt} = s_0 \omega \sinh \omega t + \frac{r}{\omega} \sinh \omega t$$

$$\frac{d^2 x}{dt^2} = s_0 \cosh \omega t + r \cosh \omega t$$

7.2.4 Both the Initial Displacement and Velocity Equal Zero

The initial conditions of motion are:

$$\text{for} \quad t = 0 \quad x = 0; \quad \frac{dx}{dt} = 0 \qquad (7.2.48)$$

a) Equation (7.2.1)

The solution of differential equation of motion (7.2.1) with the initial conditions of motion (7.2.48) reads:

$$x = \frac{r}{\omega^2}(1 - \cos \omega t) \qquad (7.2.49)$$

Taking the first derivative of equation (7.2.49), we determine the velocity:

$$\frac{dx}{dt} = \frac{r}{\omega} \sin \omega t \qquad (7.2.50)$$

The second derivative of equation (7.2.49) represents the acceleration:

$$\frac{d^2x}{dt^2} = r\cos\omega t \qquad (7.2.51)$$

Because the velocity becomes equal to zero at the moment of time when $\sin\omega t = 0$, the extreme values of the displacement and acceleration occur when:

$$\cos\omega t = \pm 1 \qquad (7.2.52)$$

Combining equations (7.2.49) and (7.2.51) with equation (7.2.52), we can determine the extreme values of the displacement s_{ext} and the acceleration a_{ext} respectively:

$$s_{ext} = \frac{r}{\omega^2}[1 - (\pm 1)] \qquad (7.2.53)$$

$$a_{ext} = \pm r \qquad (7.2.54)$$

Multiplying equation (7.4.53) by K and using notations (7.2.6) and (7.2.7), we determine the maximum force R_{max} and minimum force R_{min} applied to the elastic link: $R_{max} = 2R$ while $R_{min} = 0$. The stress analysis should be performed for a repeated loading cycle including fatigue calculations.

b) **Equation (7.2.3)**

The solution of differential equation of motion (7.2.3) with the initial conditions of motion (7.2.48) reads:

$$x = \frac{r}{\omega^2}(\cosh\omega t - 1) \qquad (7.2.55)$$

Equation (7.2.55) shows that the system is in rectilinear translational motion. Taking the first and second derivatives of equation (7.2.55), we determine the velocity and acceleration respectively:

$$\frac{dx}{dt} = \frac{r}{\omega}\sinh\omega t$$

$$\frac{d^2x}{dt^2} = r\cosh\omega t$$

7.3 Harmonic Force $A\sin(\omega_1 t + \lambda)$

This section describes engineering systems characterized by the action of the force of inertia and the stiffness force as the resisting forces (Row 7 in Guiding Table 2.1) and the harmonic force as the active force (Column 3). The current problem could be associated with engineering systems comprising elastic links and subjected to the action of harmonic forces. These systems have a variety of applications. This problem could also be related to a vibratory system interacting with an elastoplastic or viscoelastoplastic medium. During the phase of elastic deformation, these types of media exert a stiffness force as a reaction to the deformation (more related considerations are discussed in section 1.2).

The system is moving in the horizontal direction. We want to determine the basic parameters of motion. The model of a system subjected to the action of a harmonic force and a stiffness force is shown in Figure 7.3.

Based on the considerations above and the model in Figure 7.3, we can compose the left and right sides of the differential equation of motion. Thus, the left side consists of the force of inertia and the stiffness force, while the right side includes the harmonic force. Therefore, we can write:

$$m\frac{d^2x}{dt^2} + Kx = A\sin(\omega_1 t + \lambda) \qquad (7.3.1)$$

Differential equation of motion (7.3.1) has different solutions for various initial conditions of motion. These solutions and their analyses are presented below.

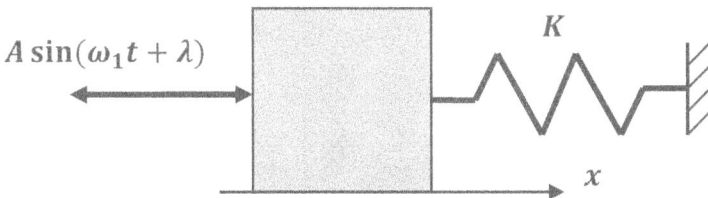

Figure 7.3 Model of a system subjected to a harmonic force and a stiffness force

7.3.1 General Initial Conditions

The general initial conditions of motion are:

$$\text{for} \quad t = 0 \quad x = s_0; \quad \frac{dx}{dt} = v_0 \tag{7.3.2}$$

where s_0 and v_0 are the initial displacement and initial velocity respectively.

Transforming the sinusoidal function in equation (7.3.1) and dividing the latter by m, we have:

$$\frac{d^2x}{dt^2} + \omega^2 x = a \sin \omega_1 t \cos \lambda + a \cos \omega_1 t \sin \lambda \tag{7.3.3}$$

where ω is the natural frequency of the system and:

$$\omega^2 = \frac{K}{m} \tag{7.3.4}$$

$$a = \frac{A}{m} \tag{7.3.5}$$

Using Laplace Transform Pairs 3, 5, 1, 6, and 7 from Table 1.1, we convert differential equation of motion (7.3.3) with the initial conditions of motion (7.3.2) from the time domain into the Laplace domain, and obtain the resulting algebraic equation of motion in the Laplace domain:

$$l^2 x(l) - l v_0 - l^2 s_0 + \omega^2 x(l) = \frac{a \omega_1 l}{l^2 + \omega_1^2} \cos \lambda + \frac{a l^2}{l^2 + \omega_1^2} \sin \lambda \tag{7.3.6}$$

Applying basic algebra to equation (7.3.6), we may write:

$$x(l)\left(l^2 + \omega^2\right) = l v_0 + l^2 s_0 + \frac{a \omega_1 l}{l^2 + \omega_1^2} \cos \lambda + \frac{a l^2}{l^2 + \omega_1^2} \sin \lambda \tag{7.3.7}$$

Solving equation (7.3.7) for the Laplace domain displacement $x(l)$, we have:

$$x(l) = \frac{l v_0}{l^2 + \omega^2} + \frac{l^2 s_0}{l^2 + \omega^2} + \frac{l a \omega_1}{(l^2 + \omega^2)(l^2 + \omega_1^2)} \cos \lambda$$

$$+ \frac{l^2 a}{(l^2 + \omega^2)(l^2 + \omega_1^2)} \sin \lambda \tag{7.3.8}$$

Based on pairs 1, 6, 7, 33, and 34 from Table 1.1, we invert equation (7.3.8) from the Laplace domain into the time domain, and obtain the solution of differential equation of motion (7.3.1) with the initial conditions of motion (7.3.2):

$$x = \frac{v_0}{\omega} \sin \omega t + s_0 \cos \omega t + \frac{a(\omega \sin \omega_1 t - \omega_1 \sin \omega t) \cos \lambda}{\omega(\omega^2 - \omega_1^2)}$$

$$+ \frac{a(\cos \omega_1 t - \cos \omega t) \sin \lambda}{\omega^2 - \omega_1^2} \tag{7.3.9}$$

Taking the first derivative of equation (7.3.9), we determine the velocity of the system:

$$\frac{dx}{dt} = v_0 \cos \omega t - \omega s_0 \sin \omega t + \frac{a\omega_1(\cos \omega_1 t - \cos \omega t) \cos \lambda}{\omega^2 - \omega_1^2}$$

$$- \frac{a(\omega_1 \sin \omega_1 t - \omega \sin \omega t) \sin \lambda}{\omega^2 - \omega_1^2} \tag{7.3.10}$$

The second derivative of equation (7.3.9) represents the acceleration of the system:

$$\frac{d^2 x}{dt^2} = -v_0 \omega \sin \omega t - \omega^2 s_0 \cos \omega t - \frac{a\omega_1(\omega_1 \sin \omega_1 t - \omega \sin \omega t) \cos \lambda}{\omega^2 - \omega_1^2}$$

$$- \frac{a(\omega_1^2 \cos \omega_1 t - \omega^2 \cos \omega t) \sin \lambda}{\omega^2 - \omega_1^2} \tag{7.3.11}$$

7.3.2 Initial Displacement Equals Zero

The initial conditions of motion are:

$$\text{for} \quad t = 0 \quad x = 0; \quad \frac{dx}{dt} = v_0 \tag{7.3.12}$$

The solution of differential equation of motion (7.3.1) with the initial conditions of motion (7.3.12) reads:

$$x = \frac{v_0}{\omega} \sin \omega t + \frac{a(\omega \sin \omega_1 t - \omega_1 \sin \omega t) \cos \lambda}{\omega(\omega^2 - \omega_1^2)}$$

$$+ \frac{a(\cos \omega_1 t - \cos \omega t) \sin \lambda}{\omega^2 - \omega_1^2} \tag{7.3.13}$$

The first derivative of equation (7.3.13) represents the velocity:

$$\frac{dx}{dt} = v_0 \cos \omega t + \frac{a\omega_1 (\cos \omega_1 t - \cos \omega t) \cos \lambda}{\omega^2 - \omega_1^2}$$

$$- \frac{a(\omega_1 \sin \omega_1 t - \omega \sin \omega t) \sin \lambda}{\omega^2 - \omega_1^2} \qquad (7.3.14)$$

Taking the second derivative of equation (7.3.13), we determine the acceleration:

$$\frac{d^2 x}{dt^2} = -\omega v_0 \sin \omega t - \frac{a\omega_1 (\omega_1 \sin \omega_1 t - \omega \sin \omega t) \cos \lambda}{\omega^2 - \omega_1^2}$$

$$- \frac{a(\omega_1^2 \cos \omega_1 t - \omega^2 \cos \omega t) \sin \lambda}{\omega^2 - \omega_1^2} \qquad (7.3.15)$$

7.3.3 Initial Velocity Equals Zero

The initial conditions of motion are:

$$\text{for} \quad t = 0 \quad x = s_0; \quad \frac{dx}{dt} = 0 \qquad (7.3.16)$$

Solving differential equation of motion (7.3.1) with the initial conditions of motion (7.3.16), we obtain:

$$x = s_0 \cos \omega t + \frac{a(\omega \sin \omega_1 t - \omega_1 \sin \omega t) \cos \lambda}{\omega(\omega^2 - \omega_1^2)}$$

$$+ \frac{a(\cos \omega_1 t - \cos \omega t) \sin \lambda}{\omega^2 - \omega_1^2} \qquad (7.3.17)$$

Taking the first derivative of equation (7.3.17), we determine the velocity:

$$\frac{dx}{dt} = -\omega s_0 \sin \omega t + \frac{a\omega_1 (\cos \omega_1 t - \cos \omega t) \cos \lambda}{\omega^2 - \omega_1^2}$$

$$- \frac{a(\omega_1 \sin \omega_1 t - \omega \sin \omega t) \sin \lambda}{\omega^2 - \omega_1^2} \qquad (7.3.18)$$

The second derivative of equation (7.3.17) represents the acceleration:

$$\frac{d^2x}{dt^2} = -\omega^2 s_0 \cos\omega t - \frac{a\omega_1(\omega_1 \sin\omega_1 t - \omega\sin\omega t)\cos\lambda}{\omega^2 - \omega_1^2}$$

$$-\frac{a(\omega_1^2 \cos\omega_1 t - \omega^2 \cos\omega t)\sin\lambda}{\omega^2 - \omega_1^2} \qquad (7.3.19)$$

7.3.4 Both the Initial Displacement and Velocity Equal Zero

The initial conditions of motion are:

$$\text{for} \quad t = 0 \quad x = 0; \quad \frac{dx}{dt} = 0 \qquad (7.3.20)$$

The solution of differential equation of motion (7.3.1) with the initial conditions of motion (7.3.20) reads:

$$x = \frac{a(\omega \sin\omega_1 t - \omega_1 \sin\omega t)\cos\lambda}{\omega(\omega^2 - \omega_1^2)}$$

$$+ \frac{a(\cos\omega_1 t - \cos\omega t)\sin\lambda}{\omega^2 - \omega_1^2} \qquad (7.3.21)$$

Taking the first derivative of equation (7.3.21), we determine the velocity:

$$\frac{dx}{dt} = \frac{a\omega_1(\cos\omega_1 t - \cos\omega t)\cos\lambda}{\omega^2 - \omega_1^2}$$

$$- \frac{a(\omega_1 \sin\omega_1 t - \omega\sin\omega t)\sin\lambda}{\omega^2 - \omega_1^2} \qquad (7.3.22)$$

The second derivative of equation (7.3.21) represents the acceleration:

$$\frac{d^2x}{dt^2} = -\frac{a\omega_1(\omega_1 \sin\omega_1 t - \omega\sin\omega t)\cos\lambda}{\omega^2 - \omega_1^2}$$

$$- \frac{a(\omega_1^2 \cos\omega_1 t - \omega^2 \cos\omega t)\sin\lambda}{\omega^2 - \omega_1^2} \qquad (7.3.23)$$

7.4 Time-Dependent Force $Q\left(\rho + \frac{\mu t}{\tau}\right)$

This section focuses on engineering systems characterized by the action of the force of inertia and the stiffness force as the resisting forces (Row 7 in Guiding Table 2.1) and the time-dependent force as the active force (Column 4). The current problems could be related to systems interacting with a specific elastic link or with an elasto-plastic or viscoelastoplastic medium. Sometimes at the beginning of the working processes, the system is subjected to a time-dependent force that acts for a predetermined interval of time. The media mentioned above exert stiffness forces during the phase of elastic deformation (more related considerations are discussed in section 1.2).

The system is moving in the horizontal direction. We want to determine the basic parameters of motion, their values at the end of the predetermined interval of time, and the characteristics of the forces applied to the elastic link. The model of the system subjected to the action of a time-dependent force and a stiffness force is shown in Figure 7.4.

Based on the considerations above and the model in Figure 7.4, it is possible to assemble the left and right sides of the differential equation of motion. The left side consists of the force of inertia and the stiffness force, while the right side includes the time-dependent force. Thus, the differential equation of motion reads:

$$m\frac{d^2x}{dt^2} + Kx = Q(\rho + \frac{\mu t}{\tau}) \qquad (7.4.1)$$

Differential equation of motion (7.4.1) has different solutions for various initial conditions of motion. These solutions and their analyses are presented below.

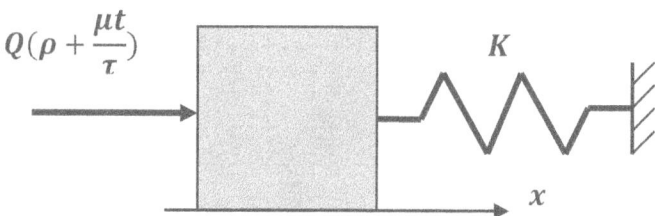

Figure 7.4 Model of system subjected to a time-dependent force and a stiffness force

7.4.1 General Initial Conditions

The general initial conditions of motion are:

$$\text{for} \quad t = 0 \quad x = s_0; \quad \frac{dx}{dt} = v_0 \qquad (7.4.2)$$

where s_0 and v_0 are the initial displacement and initial velocity respectively.

Dividing equation (7.4.1) by m, we have:

$$\frac{d^2x}{dt^2} + \omega^2 x = q(\rho + \frac{\mu t}{\tau}) \qquad (7.4.3)$$

where ω is the natural frequency of the system and:

$$\omega^2 = \frac{K}{m} \qquad (7.4.4)$$

$$q = \frac{Q}{m} \qquad (7.4.5)$$

Using Laplace Transform Pairs 3, 5, 1, and 2 from Table 1.1, we convert differential equation of motion (7.4.3) with the initial conditions of motion (7.4.2) from the time domain into the Laplace domain. The resulting algebraic equation of motion in the Laplace domain reads:

$$l^2 x(l) - lv_0 - l^2 s_0 + \omega^2 x(l) = q\rho + \frac{q\mu}{\tau l} \qquad (7.4.6)$$

Solving equation (7.4.6) for the Laplace domain displacement $x(l)$, we have:

$$x(l) = \frac{lv_0}{l^2 + \omega^2} + \frac{l^2 s_0}{l^2 + \omega^2} + \frac{q\rho}{l^2 + \omega^2} + \frac{q\mu}{\tau l(l^2 + \omega^2)} \qquad (7.4.7)$$

Based on pairs 1, 6, 7, 14, and 16 from Table 1.1, we invert equation (7.4.7) from the Laplace domain into the time domain and obtain the solution of differential equation of motion (7.4.1) with the initial conditions of motion (7.4.2):

$$x = \frac{v_0}{\omega}\sin\omega t + s_0\cos\omega t + \frac{q\rho}{\omega^2}(1 - \cos\omega t) + \frac{q\mu}{\tau\omega^2}(t - \frac{1}{\omega}\sin\omega t) \qquad (7.4.8)$$

Applying algebraic procedures to equation (7.4.8), we may write:

$$x = \frac{q\rho}{\omega^2} + \frac{q\mu t}{\tau\omega^2} + (s_0 - \frac{q\rho}{\omega^2})\cos\omega t + (\frac{v_0}{\omega} - \frac{q\mu}{\tau\omega^3})\sin\omega t \quad \textbf{(7.4.9)}$$

Taking the first derivative of equation (7.4.9), we determine the velocity:

$$\frac{dx}{dt} = \frac{q\mu}{\tau\omega^2} - (s_0\omega - \frac{q\rho}{\omega})\sin\omega t + (v_0 - \frac{q\mu}{\tau\omega^2})\cos\omega t \quad \textbf{(7.4.10)}$$

The second derivative of equation (7.4.9) represents the acceleration:

$$\frac{d^2x}{dt^2} = -(s_0\omega^2 - q\rho)\cos\omega t - (v_0\omega - \frac{q\mu}{\tau\omega})\sin\omega t \quad \textbf{(7.4.11)}$$

Substituting into equations (7.4.9), (7.4.10), and (7.4.11) the time τ, we determine respectively the values of the displacement s, the velocity v, and the acceleration a at the end of the interval of time τ.

$$s = \frac{q(\rho + \mu)}{\omega^2} + (s_0 - \frac{q\rho}{\omega^2})\cos\omega\tau + (\frac{v_0}{\omega} - \frac{q\mu}{\tau\omega^3})\sin\omega\tau \quad \textbf{(7.4.12)}$$

$$v = \frac{q\mu}{\tau\omega^2} - (s_0\omega - \frac{q\rho}{\omega})\sin\omega\tau + (v_0 - \frac{q\mu}{\tau\omega^2})\cos\omega\tau \quad \textbf{(7.4.13)}$$

$$a = -(s_0\omega^2 - q\rho)\cos\omega\tau - (v_0\omega - \frac{q\mu}{\tau\omega})\sin\omega\tau \quad \textbf{(7.4.14)}$$

Multiplying equation (7.4.12) by the stiffness coefficient K and using notation (7.4.4), we calculate the maximum force R_{max} applied to the elastic link:

$$R_{max} = m[q(\rho + \mu) + (s_0\omega^2 - q\rho)\cos\omega\tau + \left(v_0\omega - \frac{q\mu}{\tau\omega}\right)\sin\omega\tau]$$
$$\textbf{(7.4.15)}$$

The force according to equation (7.4.15) should be used for the stress calculations of the elastic link.

7.4.2 Initial Displacement Equals Zero
The initial conditions of motion are:

$$\text{for} \quad t = 0 \quad x = 0; \quad \frac{dx}{dt} = v_0 \qquad (7.4.16)$$

The solution of differential equation of motion (7.4.1) with the initial conditions of motion (7.4.16) reads:

$$x = \frac{q\rho}{\omega^2} + \frac{q\mu t}{\tau\omega^2} - \frac{q\rho}{\omega^2}\cos\omega t + (\frac{v_0}{\omega} - \frac{q\mu}{\tau\omega^3})\sin\omega t \quad (7.4.17)$$

The first derivative of equation (7.4.17) represents the velocity:

$$\frac{dx}{dt} = \frac{q\mu}{\tau\omega^2} + \frac{q\rho}{\omega}\sin\omega t + (v_0 - \frac{q\mu}{\tau\omega^2})\cos\omega t \qquad (7.4.18)$$

Taking the second derivative of equation (7.4.17), we determine the acceleration:

$$\frac{d^2x}{dt^2} = q\rho\cos\omega t - (v_0\omega - \frac{q\mu}{\tau\omega})\sin\omega t \qquad (7.4.19)$$

Substituting the interval of time τ into equations (4.4.17), (7.4.18), and (7.4.19), we determine respectively the values of the displacement, the velocity, and the acceleration at the end of the predetermined time:

$$s = \frac{q(\rho + \mu)}{\omega^2} - \frac{q\rho}{\omega^2}\cos\omega\tau + (\frac{v_0}{\omega} - \frac{q\mu}{\tau\omega^3})\sin\omega\tau \quad (7.4.20)$$

$$v = \frac{q\mu}{\tau\omega^2} + \frac{q\rho}{\omega}\sin\omega\tau + (v_0 - \frac{q\mu}{\tau\omega^2})\cos\omega\tau \qquad (7.4.21)$$

$$a = q\rho\cos\omega\tau - (v_0\omega - \frac{q\mu}{\tau\omega})\sin\omega\tau \qquad (7.4.22)$$

Multiplying equation (7.4.20) by K and using notation (7.4.4), we determine the maximum force R_{max} applied to the elastic link:

$$R_{max} = m[q(\rho + \mu)\cos\omega\tau - q\rho\cos\omega\tau + (v_0\omega - \frac{q\mu}{\tau\omega})\sin\omega\tau] \ (7.4.23)$$

The stress calculations of the elastic link should be based on the force according to equation (7.4.23).

7.4.3 Initial Velocity Equals Zero

The initial conditions of motion are:

$$\text{for} \quad t = 0 \quad x = s_0; \quad \frac{dx}{dt} = 0 \qquad (7.4.24)$$

Solving differential equation of motion (7.4.1) with the initial conditions of motion (7.4.24), we obtain:

$$x = \frac{q\rho}{\omega^2} + \frac{q\mu t}{\tau\omega^2} + (s_0 - \frac{q\rho}{\omega^2})\cos\omega t - \frac{q\mu}{\tau\omega^3}\sin\omega t \qquad (7.4.25)$$

The velocity and acceleration represent the first and second derivatives of equation (7.4.25) respectively:

$$\frac{dx}{dt} = \frac{q\mu}{\tau\omega^2} - (s_0\omega - \frac{q\rho}{\omega})\sin\omega t - \frac{q\mu}{\tau\omega^2}\cos\omega t \qquad (7.4.26)$$

$$\frac{d^2x}{dt^2} = -(s_0\omega^2 - q\rho)\cos\omega t + \frac{q\mu}{\tau\omega}\sin\omega t \qquad (7.4.27)$$

Substituting the time τ into equations (7.4.25), (7.4.26), and (7.4.27), we determine respectively the values of the displacement, the velocity v, and the acceleration a at the end of the predetermined interval of time:

$$s = \frac{q(\rho + \mu)}{\omega^2} + (s_0 - \frac{q\rho}{\omega^2})\cos\omega\tau - \frac{q\mu}{\tau\omega^3}\sin\omega\tau \qquad (7.4.28)$$

$$v = \frac{q\mu}{\omega^2} - (s_0\omega - \frac{q\rho}{\omega})\sin\omega\tau - \frac{q\mu}{\tau\omega^2}\cos\omega\tau \qquad (7.4.29)$$

$$a = -(s_0\omega^2 - q\rho)\cos\omega\tau + \frac{q\mu}{\tau\omega}\sin\omega\tau \qquad (7.4.30)$$

Multiplying equation (7.4.28) by K and substituting notation (7.4.4), we calculate the maximum force R_{max} applied to the elastic link:

$$R_{max} = m[q(\rho + \mu) + (s_0\omega^2 - q\rho)\cos\omega\tau - \frac{q\mu}{\tau\omega}\sin\omega\tau] \qquad (7.4.31)$$

The stress calculations of the elastic link should be based on the force according to equation (7.4.31).

7.4.4 Both the Initial Displacement and Velocity Equal Zero

The initial conditions of motion are:

$$\text{for} \quad t = 0 \quad x = 0; \quad \frac{dx}{dt} = 0 \qquad (7.4.32)$$

The solution of differential equation of motion (7.4.1) with the initial conditions of motion (7.4.32) reads:

$$x = \frac{q\rho}{\omega^2} + \frac{q\mu t}{\tau\omega^2} - \frac{q\rho}{\omega^2}\cos\omega t - \frac{q\mu}{\tau\omega^3}\sin\omega t \qquad (7.4.33)$$

The first and second derivatives of equation (7.4.33) represent the velocity and the acceleration respectively:

$$\frac{dx}{dt} = \frac{q\mu}{\tau\omega^2} + \frac{q\rho}{\omega}\sin\omega t - \frac{q\mu}{\tau\omega^2}\cos\omega t \qquad (7.4.34)$$

$$\frac{d^2x}{dt^2} = q\rho\cos\omega t + \frac{q\mu}{\tau\omega}\sin\omega t \qquad (7.4.35)$$

Substituting the time τ into equations (7.4.33), (7.4.34), and (7.4.35), we determine the values of the displacement s, the velocity v, and the acceleration a at the end of the predetermined time respectively:

$$s = \frac{q(\rho + \mu)}{\omega^2} - \frac{q\rho}{\omega^2}\cos\omega\tau - \frac{q\mu}{\tau\omega^3}\sin\omega\tau \qquad (7.4.36)$$

$$v = \frac{q\mu}{\omega^2} + \frac{q\rho}{\omega}\sin\omega\tau - \frac{q\mu}{\tau\omega^2}\cos\omega\tau \qquad (7.4.37)$$

$$a = q\rho\cos\omega\tau + \frac{q\mu}{\tau\omega}\sin\omega\tau \qquad (7.4.38)$$

Multiplying equation (7.4.36) by K and substituting notation (7.4.4), we determine the force R_{max} applied to the elastic link:

$$R_{max} = m[q(\rho + \mu) - q\rho\cos\omega\tau - \frac{q\mu}{\tau\omega}\sin\omega\tau] \qquad (7.4.39)$$

The stress calculations of the elastic link should be based on the force according to equation (7.4.39).

7.5 Constant Force R and Harmonic Force $A\sin(\omega_1 t + \lambda)$

This section describes engineering systems subjected to the action of the force of inertia and the stiffness force as the resisting forces (Row 7 of Guiding Table 2.1) and the constant active force and the harmonic force as the active forces (Column 5). The current problems could be related to vibratory systems interacting with an elastoplastic or viscoelastoplastic medium. These types of media exert a stiffness force during the phase of elastic deformation (more related considerations are discussed in section 1.2).

The system is moving in the horizontal direction. We want to determine the basic parameters of motion. The model of a system subjected to the action of a constant active force, a harmonic force, and a stiffness force is shown in Figure 7.5.

With the considerations mentioned above and the model shown in Figure 7.5, we can compose the left and right sides of the system's differential equation of motion. The left side consists of the force of inertia and the stiffness force, while the right side includes the constant active force and the harmonic force. Therefore, the differential equation of motion reads:

$$m\frac{d^2x}{dt^2} + Kx = R + A\sin(\omega_1 t + \lambda) \qquad (7.5.1)$$

Differential equation of motion (7.5.1) has different solutions for various initial conditions of motion. These solutions and their analyses are presented below.

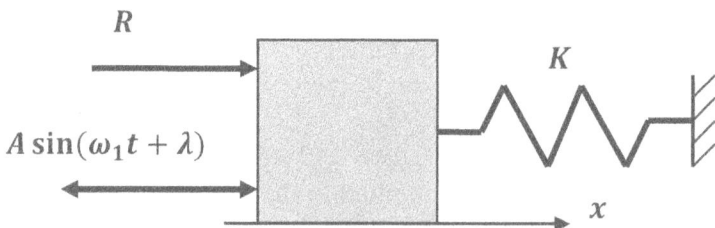

Figure 7.5 Model of a system subjected to a constant active force, a harmonic force, and a stiffness force

7.5.1 General Initial Conditions

The general initial conditions of motion are:

$$\text{for} \quad t = 0 \quad x = s_0; \quad \frac{dx}{dt} = v_0 \qquad (7.5.2)$$

where s_0 and v_0 are the initial displacement and initial velocity respectively.

Transforming the sinusoidal function in equation (7.5.1) and dividing the latter by m, we may write:

$$\frac{d^2x}{dt^2} + \omega^2 x = r + a \sin \omega_1 t \cos \lambda + a \cos \omega_1 t \sin \lambda \qquad (7.5.3)$$

where ω is the natural frequency of the system and:

$$\omega^2 = \frac{K}{m} \qquad (7.5.4)$$

$$r = \frac{R}{m} \qquad (7.5.5)$$

$$a = \frac{A}{m} \qquad (7.5.6)$$

Using Laplace Transform Pairs 3, 5, 1, 6, and 7 from Table 1.1, we convert differential equation of motion (7.5.3) with the initial conditions of motion (7.5.2) from the time domain into the Laplace domain. The resulting algebraic equation of motion in the Laplace domain reads:

$$l^2 x(l) - lv_0 - l^2 s_0 + \omega^2 x(l) = r + \frac{a\omega_1 l}{l^2 + \omega_1^2} \cos \lambda + \frac{al^2}{l^2 + \omega_1^2} \sin \lambda$$

$$(7.5.7)$$

Applying basic algebra to equation (7.5.7), we have:

$$x(l)(l^2 + \omega^2) = r + lv_0 + l^2 s_0 + \frac{a\omega_1 l}{l^2 + \omega_1^2} \cos \lambda + \frac{al^2}{l^2 + \omega_1^2} \sin \lambda$$

$$(7.5.8)$$

Solving equation (7.5.8) for the Laplace domain displacement $x(l)$, we may write:

$$x(l) = \frac{r}{l^2 + \omega^2} + \frac{lv_0}{l^2 + \omega^2} + \frac{l^2 s_0}{l^2 + \omega^2} + \frac{la\omega_1}{(l^2 + \omega^2)(l^2 + \omega_1^2)} \cos \lambda$$

$$+ \frac{l^2 a}{(l^2 + \omega^2)(l^2 + \omega_1^2)} \sin \lambda \qquad (7.5.9)$$

Inverting equation (7.5.9) from the Laplace domain into the time domain by applying pairs 1, 14, 5, 6, 7, 33, and 34 from Table 1.1, we obtain the solution of differential equation of motion (7.5.1) with the initial conditions of motion (7.5.2):

$$x = \frac{r}{\omega^2}(1 - \cos \omega t) + \frac{v_0}{\omega} \sin \omega t + s_0 \cos \omega t$$

$$+ \frac{a(\omega \sin \omega_1 t - \omega_1 \sin \omega t)\cos \lambda}{\omega(\omega^2 - \omega_1^2)} + \frac{a(\cos \omega_1 t - \cos \omega t)\sin \lambda}{\omega^2 - \omega_1^2}$$

$$(7.5.10)$$

Taking the first and second derivatives of equation (7.5.10), we determine respectively the velocity and the acceleration of the system:

$$\frac{dx}{dt} = \frac{r}{\omega} \sin \omega t + v_0 \cos \omega t - \omega s_0 \sin \omega t$$

$$+ \frac{a\omega_1(\cos \omega_1 t - \cos \omega t)\cos \lambda}{\omega^2 - \omega_1^2} - \frac{a(\omega_1 \sin \omega_1 t - \omega \sin \omega t)\sin \lambda}{\omega^2 - \omega_1^2}$$

$$(7.5.11)$$

$$\frac{d^2x}{dt^2} = r \cos \omega t - \omega v_0 \sin \omega - \omega^2 s_0 \cos \omega t$$

$$- \frac{a\omega_1(\omega_1 \sin \omega_1 t - \omega \sin \omega t)\cos \lambda}{\omega^2 - \omega_1^2}$$

$$- \frac{a(\omega_1^2 \cos \omega_1 t - \omega^2 \cos \omega t)\sin \lambda}{\omega^2 - \omega_1^2} \qquad (7.5.12)$$

7.5.2 Initial Displacement Equals Zero

The initial conditions of motion are:

$$\text{for} \quad t = 0 \quad x = 0; \quad \frac{dx}{dt} = v_0 \tag{7.5.13}$$

The solution of differential equation of motion (7.5.1) with the initial conditions of motion (7.5.13) reads:

$$x = \frac{r}{\omega^2}(1 - \cos \omega t) + \frac{v_0}{\omega} \sin \omega t + \frac{a(\omega \sin \omega_1 t - \omega_1 \sin \omega t) \cos \lambda}{\omega(\omega^2 - \omega_1^2)}$$

$$+ \frac{a(\cos \omega_1 t - \cos \omega t) \sin \lambda}{\omega^2 - \omega_1^2} \tag{7.5.14}$$

Taking the first derivative of equation (7.5.14), we determine the velocity:

$$\frac{dx}{dt} = \frac{r}{\omega} \sin \omega t + v_0 \cos \omega t + \frac{a\omega_1(\cos \omega_1 t - \cos \omega t) \cos \lambda}{\omega^2 - \omega_1^2}$$

$$- \frac{a(\omega_1 \sin \omega_1 t - \omega \sin \omega t) \sin \lambda}{\omega^2 - \omega_1^2} \tag{7.5.15}$$

The second derivative of equation (7.5.13) represents the acceleration:

$$\frac{d^2x}{dt^2} = r \cos \omega t - \omega v_0 \sin \omega t - \frac{a\omega_1(\omega_1 \sin \omega_1 t - \omega \sin \omega t) \cos \lambda}{\omega^2 - \omega_1^2}$$

$$- \frac{a(\omega_1^2 \cos \omega_1 t - \omega^2 \cos \omega t) \sin \lambda}{\omega^2 - \omega_1^2} \tag{7.5.16}$$

7.5.3 Initial Velocity Equals Zero

The initial conditions of motion are:

$$\text{for} \quad t = 0 \quad x = s_0; \quad \frac{dx}{dt} = 0 \tag{7.5.17}$$

Solving differential equation of motion (7.5.1) with the initial conditions of motion (7.3.17), we obtain:

$$x = \frac{r}{\omega^2}(1 - \cos \omega t) + s_0 \cos \omega t + \frac{a(\omega \sin \omega_1 t - \omega_1 \sin \omega t) \cos \lambda}{\omega(\omega^2 - \omega_1^2)}$$

$$+ \frac{a(\cos \omega_1 t - \cos \omega t) \sin \lambda}{\omega^2 - \omega_1^2} \tag{7.5.18}$$

The first derivative of equation (7.5.17) represents the velocity:

$$\frac{dx}{dt} = \frac{r}{\omega}\sin\omega t - \omega s_0 \sin\omega t + \frac{a\omega_1(\cos\omega_1 t - \cos\omega t)\cos\lambda}{\omega^2 - \omega_1^2}$$

$$-\frac{a(\omega_1 \sin\omega_1 t - \omega\sin\omega t)\sin\lambda}{\omega^2 - \omega_1^2}$$

$$(7.5.19)$$

Taking the second derivative of equation (7.5.17), we determine the acceleration:

$$\frac{d^2 x}{dt^2} = r\cos\omega t - \omega^2 s_0 \cos\omega t - \frac{a\omega_1(\omega_1 \sin\omega_1 t - \omega\sin\omega t)\cos\lambda}{\omega^2 - \omega_1^2}$$

$$-\frac{a(\omega_1^2 \cos\omega_1 t - \omega^2 \cos\omega t)\sin\lambda}{\omega^2 - \omega_1^2}$$

$$(7.5.20)$$

7.5.4 Both the Initial Displacement and Velocity Equal Zero

The initial conditions of motion are:

$$\text{for} \quad t = 0 \quad x = 0; \quad \frac{dx}{dt} = 0 \qquad (7.5.21)$$

The solution of differential equation of motion (7.5.1) with the initial conditions of motion (7.5.21) reads:

$$x = \frac{r}{\omega^2}(1 - \cos\omega t) + \frac{a(\omega\sin\omega_1 t - \omega_1 \sin\omega t)\cos\lambda}{\omega(\omega^2 - \omega_1^2)}$$

$$+\frac{a(\cos\omega_1 t - \cos\omega t)\sin\lambda}{\omega^2 - \omega_1^2}$$

$$(7.5.22)$$

Taking the first and second derivatives of equation (7.5.22), we determine the velocity and the acceleration respectively:

$$\frac{dx}{dt} = \frac{r}{\omega}\sin\omega t + \frac{a\omega_1(\cos\omega_1 t - \cos\omega t)\cos\lambda}{\omega^2 - \omega_1^2}$$

$$-\frac{a(\omega_1 \sin\omega_1 t - \omega\sin\omega t)\sin\lambda}{\omega^2 - \omega_1^2}$$

$$(7.5.23)$$

$$\frac{d^2x}{dt^2} = r\cos\omega t - \frac{a\omega_1(\omega_1 \sin\omega_1 t - \omega \sin\omega t)\cos\lambda}{\omega^2 - \omega_1^2}$$

$$- \frac{a(\omega_1^2 \cos\omega_1 t - \omega^2 \cos\omega t)\sin\lambda}{\omega^2 - \omega_1^2} \qquad (7.5.24)$$

7.6 Harmonic Force $A\sin(\omega_1 t + \lambda)$ and Time-Dependent Force $Q\left(\rho + \frac{\mu t}{\tau}\right)$

This section describes engineering systems subjected to the force of inertia and the stiffness force as the resisting forces (Row 7 in Guiding Table 2.1), and the harmonic force and the time-dependent force as the active forces (Column 6). The current problems could be related to vibratory systems intended for interaction with an elastoplastic or viscoelastoplastic medium. These media exert reactive stiffness forces on the phase of elastic deformation (more related information is discussed in section 1.2).

The system is moving in the horizontal direction. We want to determine the basic parameters of motion. The model of a system subjected to a harmonic force, a time-dependent force, and a stiffness force is shown in Figure 7.6.

Based on the considerations above and the model in Figure 7.6, we can assemble the left and right sides of the differential equation of motion. The left side consists of the force of inertia and the stiffness force, while the right side includes the harmonic force and the time-dependent force. Therefore, the differential equation of motion reads:

$$m\frac{d^2x}{dt^2} + Kx = A\sin(\omega_1 t + \lambda) + Q(\rho + \frac{\mu t}{\tau}) \qquad (7.6.1)$$

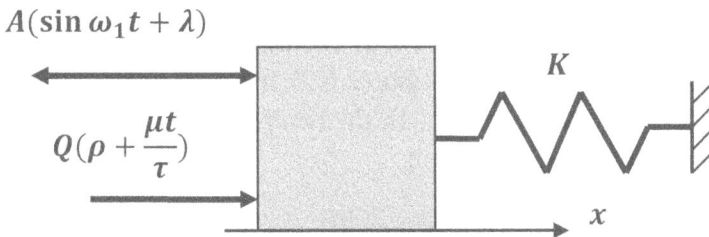

$A(\sin \omega_1 t + \lambda)$

$Q(\rho + \frac{\mu t}{\tau})$

K

x

Figure 7.6 Model of a system subjected to a harmonic force, a time-dependent force, and a stiffness force

Differential equation of motion (7.4.1) has different solutions for various initial conditions of motion. These solutions and their analyses are presented below.

7.6.1 General Initial Conditions

The general initial conditions of motion are:

$$\text{for} \quad t = 0 \quad x = s_0; \quad \frac{dx}{dt} = v_0 \tag{7.6.2}$$

where s_0 and v_0 are the initial displacement and initial velocity respectively.

Transforming equation (7.6.1), we have:

$$m\frac{d^2x}{dt^2} + Kx = Q\rho + Q\frac{\mu t}{\tau} + A\sin\omega_1 t\cos\lambda + A\cos\omega_1 t\sin\lambda \tag{7.6.3}$$

Dividing equation (7.6.3) by m, we may write:

$$\frac{d^2x}{dt^2} + \omega^2 x = q\rho + q\frac{\mu t}{\tau} + a\sin\omega_1 t\cos\lambda + a\cos\omega_1 t\sin\lambda \tag{7.6.4}$$

where ω is the natural frequency of the system and:

$$\omega^2 = \frac{K}{m} \tag{7.6.5}$$

$$q = \frac{Q}{m} \tag{7.6.6}$$

$$a = \frac{A}{a} \tag{7.6.7}$$

Using Laplace Transform Pairs 3, 1, 5, 2, 6, and 7 from Table 1.1, we convert differential equation of motion (7.6.4) with the initial conditions of motion (7.6.2) from the time domain into the Laplace domain, and obtain the resulting algebraic equation of motion in the Laplace domain:

$$l^2 x(l) - lv_0 - l^2 s_0 + \omega^2 x(l) = q\rho + \frac{q\mu}{\tau l} + \frac{a\omega_1 l}{l^2 + \omega_1^2}\cos\lambda + \frac{al^2}{l^2 + \omega_1^2}\sin\lambda \tag{7.6.8}$$

The solution of equation (7.6.8) for the Laplace domain displacement $x(l)$ reads:

$$x(l) = \frac{lv_0}{l^2 + \omega^2} + \frac{l^2 s_0}{l^2 + \omega^2} + \frac{q\rho}{l^2 + \omega^2} + \frac{q\mu}{\tau l(l^2 + \omega^2)}$$

$$+ \frac{la\omega_1}{(l^2 + \omega^2)(l^2 + \omega_1^2)} \cos \lambda + \frac{l^2 a}{(l^2 + \omega^2)(l^2 + \omega_1^2)} \sin \lambda \qquad \textbf{(7.6.9)}$$

Using pairs 1, 6, 7, 14, 16, 33, and 34, we invert equation (7.8.9) from the Laplace domain into the time domain. Upon inversion, we obtain the solution of differential equation of motion (7.6.1) with the initial conditions of motion (7.6.2):

$$x = \frac{v_0}{\omega} \sin \omega t + s_0 \cos \omega t + \frac{q\rho}{\omega^2}(1 - \cos \omega t) + \frac{q\mu}{\tau\omega^2}\left(t - \frac{1}{\omega}\sin \omega t\right)$$

$$+ \frac{a(\omega \sin \omega_1 t - \omega_1 \sin \omega t)\cos \lambda}{\omega(\omega^2 - \omega_1^2)} + \frac{a(\cos \omega_1 t - \cos \omega t)\sin \lambda}{\omega^2 - \omega_1^2}$$

$$\textbf{(7.6.10)}$$

Taking the first and second derivatives of equation (7.6.10), we determine respectively the velocity and the acceleration of the system:

$$\frac{dx}{dt} = v_0 \cos \omega t - s_0\omega \sin \omega t + \frac{q\rho}{\omega} \sin \omega t + \frac{q\mu}{\tau\omega^2} - \frac{q\mu}{\tau\omega^2}\cos \omega t$$

$$+ \frac{a\omega_1(\cos \omega_1 t - \cos \omega t)\cos \lambda}{\omega^2 - \omega_1^2} - \frac{a(\omega_1 \sin \omega_1 t - \omega \sin \omega t)\sin \lambda}{\omega^2 - \omega_1^2}$$

$$\textbf{(7.6.11)}$$

$$\frac{d^2 x}{dt^2} = -v_0\omega \sin \omega t - s_0\omega^2 \cos \omega t + q\rho \cos \omega t + \frac{q\mu}{\tau\omega} \sin \omega t$$

$$- \frac{a\omega_1(\omega_1 \sin \omega_1 t - \omega \sin \omega t)\cos \lambda}{\omega^2 - \omega_1^2}$$

$$- \frac{a(\omega_1^2 \cos \omega_1 t - \omega^2 \cos \omega t)\sin \lambda}{\omega^2 - \omega_1^2} \qquad \textbf{(7.6.12)}$$

7.6.2 Initial Displacement Equals Zero

The initial conditions of motion are:

$$\text{for} \quad t = 0 \quad x = 0; \quad \frac{dx}{dt} = v_0 \qquad (7.6.13)$$

Solving differential equation of motion (7.6.1) with the initial conditions of motion (7.6.13), we have:

$$x = \frac{v_0}{\omega} \sin \omega t + \frac{q\rho}{\omega^2}(1 - \cos \omega t) + \frac{q\mu}{\tau\omega^2}\left(t - \frac{1}{\omega}\sin \omega t\right)$$
$$+ \frac{a(\omega \sin \omega_1 t - \omega_1 \sin \omega t)\cos \lambda}{\omega(\omega^2 - \omega_1^2)}$$
$$+ \frac{a(\cos \omega_1 t - \cos \omega t)\sin \lambda}{\omega^2 - \omega_1^2} \qquad (7.6.14)$$

Taking the first and second derivatives of equation (7.6.10), we determine the velocity and the acceleration of the system respectively:

$$\frac{dx}{dt} = v_0 \cos \omega t + \frac{q\rho}{\omega}\sin \omega t + \frac{q\mu}{\tau\omega^2} - \frac{q\mu}{\tau\omega^2}\cos \omega t$$
$$+ \frac{a\omega_1(\cos \omega_1 t - \cos \omega t)\cos \lambda}{\omega^2 - \omega_1^2}$$
$$- \frac{a(\omega_1 \sin \omega_1 t - \omega \sin \omega t)\sin \lambda}{\omega^2 - \omega_1^2} \qquad (7.6.15)$$

$$\frac{d^2x}{dt^2} = -v_0\omega \sin \omega t + q\rho \cos \omega t + \frac{q\mu}{\tau\omega}\sin \omega t$$
$$- \frac{a\omega_1(\omega_1 \sin \omega_1 t - \omega \sin \omega t)\cos \lambda}{\omega^2 - \omega_1^2}$$
$$- \frac{a(\omega_1^2 \cos \omega_1 t - \omega^2 \cos \omega t)\sin \lambda}{\omega^2 - \omega_1^2} \qquad (7.6.16)$$

7.6.3 Initial Velocity Equals Zero

The initial conditions of motion are:

$$\text{for} \quad t = 0 \quad x = s_0; \quad \frac{dx}{dt} = 0 \qquad (7.6.16)$$

The solution of differential equation of motion (7.6.1) with the initial conditions of motion (7.6.16) reads:

$$x = s_0 \cos \omega t + \frac{q\rho}{\omega^2}(1 - \cos \omega t) + \frac{q\mu}{\tau\omega^2}(t - \frac{1}{\omega}\sin \omega t)$$
$$+ \frac{a(\omega \sin \omega_1 t - \omega_1 \sin \omega t)\cos \lambda}{\omega(\omega^2 - \omega_1^2)} + \frac{a(\cos \omega_1 t - \cos \omega t)\sin \lambda}{\omega^2 - \omega_1^2}$$

$$(7.6.17)$$

The velocity and the acceleration represent the first and second derivatives of equation (7.6.17) respectively:

$$\frac{dx}{dt} = -s_0\omega \sin \omega t + \frac{q\rho}{\omega}\sin \omega t + \frac{q\mu}{\tau\omega^2}\cos \omega t$$
$$+ \frac{a\omega_1(\cos \omega_1 t - \cos \omega t)\cos \lambda}{\omega^2 - \omega_1^2}$$
$$- \frac{a(\omega_1 \sin \omega_1 t - \omega \sin \omega t)\sin \lambda}{\omega^2 - \omega_1^2} \qquad (7.6.18)$$

$$\frac{d^2x}{dt^2} = -s_0\omega^2 \cos \omega t + q\rho \cos \omega t + \frac{q\mu}{\tau\omega}\sin \omega t$$
$$- \frac{a\omega_1(\omega_1 \sin \omega_1 t - \omega \sin \omega t)\cos \lambda}{\omega^2 - \omega_1^2}$$
$$- \frac{a(\omega_1^2 \cos \omega_1 t - \omega^2 \cos \omega t)\sin \lambda}{\omega^2 - \omega_1^2} \qquad (7.6.19)$$

7.6.4 Both the Initial Displacement and Velocity Equal Zero

The initial conditions of motion are:

$$\text{for} \quad t = 0 \quad x = 0; \quad \frac{dx}{dt} = 0 \qquad (7.6.20)$$

Solving differential equation of motion (7.6.1) with the initial conditions of motion (7.6.20), we obtain:

$$x = \frac{q\rho}{\omega^2}(1 - \cos \omega t) + \frac{q\mu}{\tau\omega^2}(t - \frac{1}{\omega}\sin \omega t)$$

$$+ \frac{a(\omega \sin \omega_1 t - \omega_1 \sin \omega t)\cos \lambda}{\omega(\omega^2 - \omega_1^2)}$$

$$+ \frac{a(\cos \omega_1 t - \cos \omega t)\sin \lambda}{\omega^2 - \omega_1^2} \tag{7.6.21}$$

Taking the first and second derivatives of equation (7.6.21), we determine the velocity and the acceleration of the system respectively:

$$\frac{dx}{dt} = \frac{q\rho}{\omega}\sin \omega t + \frac{q\mu}{\tau\omega^2} - \frac{q\mu}{\tau\omega^2}\cos \omega t$$

$$+ \frac{a\omega_1(\cos \omega_1 t - \cos \omega t)\cos \lambda}{\omega^2 - \omega_1^2}$$

$$- \frac{a(\omega_1 \sin \omega_1 t - \omega \sin \omega t)\sin \lambda}{\omega^2 - \omega_1^2} \tag{7.6.22}$$

$$\frac{d^2x}{dt^2} = q\rho \cos \omega t + \frac{q\mu}{\tau\omega}\sin \omega t$$

$$- \frac{a\omega_1(\omega_1 \sin \omega_1 t - \omega \sin \omega t)\cos \lambda}{\omega^2 - \omega_1^2}$$

$$- \frac{a(\omega_1^2 \cos \omega_1 t - \omega^2 \cos \omega t)\sin \lambda}{\omega^2 - \omega_1^2} \tag{7.6.23}$$

8

STIFFNESS AND FRICTION

This chapter addresses engineering systems subjected to the action of the force of inertia, the stiffness force, and the friction force as the resisting forces. These resisting forces are marked by the plus sign in Row 8 of Guiding Table 2.1. The intersections of this row with Columns 1 through 6 indicate the types of active forces applied to the systems and described in each section. Throughout this chapter, the left sides of the differential equations of motion are identical and represent the sum of the force inertia, the stiffness force, and the friction force. However, the right sides of these equations are different in each section and depend on the active forces applied to the systems.

The problems described in this chapter could be related to mechanical systems intended for interaction with an elastoplastic or viscoelastoplastic medium, or with specific elastic links. During the phase of elastic deformations, these media could simultaneously exert stiffness and friction forces. In cases of vibratory motion, the differential equations of motion are valid to the point when the system stops and the friction force instantaneously changes its direction. The presence of stiffness forces may cause the motion of the system in the opposite direction after the stop.

8.1 Active Force Equals Zero

According to Guiding Table 2.1, this section describes engineering system experiencing the action of the force of inertia, the stiffness force, and the friction force as the resisting forces (Row 8). No active forces are present in the problems related to this section (Column 1). The motion in this case is caused by the system's energy. (Considerations related to the motion of a system in the absence of active forces are discussed in section 1.3.)

The current problems could be associated with the working processes of systems intended for the interaction with an elasto-plastic or viscoelastoplastic medium during the phase of elastic deformations that exerts simultaneously stiffness and friction forces (more related information including the considerations regarding the behavior of a friction force applied to a vibratory system is discussed in section 1.2). These problems could be also related to the interaction of a system with a specific elastic link.

The system is moving in the horizontal direction. We want to determine the basic parameters of motion, their maximum values, and the characteristics of the forces applied to the elastic link. Figure 8.1 shows the model of a system subjected to the action of a stiffness force and a friction force.

Based on the considerations presented above and the model in Figure 8.1, we can compose the left and right sides of the differential equation of motion. The left side consists of the force of inertia, the stiffness force, and the friction force, while the right side equals zero. Thus, the differential equation of motion reads:

$$m\frac{d^2x}{dt^2} + Kx + F = 0 \qquad\qquad \textbf{(8.1.1)}$$

The differential equation of motion (8.1.1) has different solutions for various initial conditions of motion. These solutions and their analyses are presented below.

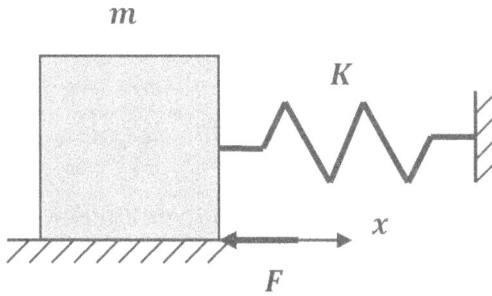

Figure 8.1 Model of a system subjected to a stiffness force and a friction force

8.1.1 General Initial Conditions

The general initial conditions of motion are:

$$\text{for} \quad t = 0 \quad x = s_0; \quad \frac{dx}{dt} = v_0 \tag{8.1.2}$$

where s_0 and v_0 are the initial displacement and initial velocity respectively.

Dividing equation (8.1.1) by m, we may write:

$$\frac{d^2x}{dt^2} + \omega^2 x + f = 0 \tag{8.1.3}$$

where ω is the natural frequency of the vibratory system and:

$$\omega^2 = \frac{K}{m} \tag{8.1.4}$$

$$f = \frac{F}{m} \tag{8.1.5}$$

Using Laplace Transform Pairs 3, 1, and 5 from Table 1.1, we convert differential equation of motion (8.1.3) with the initial conditions of motion (8.1.2) from the time domain into the Laplace domain, and obtain the resulting algebraic equation of motion in the Laplace domain:

$$l^2 x(l) - l v_0 - l^2 s_0 + \omega^2 x(l) + f = 0 \tag{8.1.6}$$

Solving equation (8.1.6) for the displacement $x(l)$ in the Laplace domain, we obtain:

$$x(l) = \frac{lv_0}{l^2 + \omega^2} + \frac{l^2 s_0}{l^2 + \omega^2} - \frac{f}{l^2 + \omega^2} \qquad (8.1.7)$$

Based on pairs 1, 5, 6, 7, and 14 from Table 1.1, we invert equation (8.1.7) from the Laplace domain into the time domain, and obtain the solution of differential equation of motion (8.1.1) with the initial conditions of motion (8.1.2):

$$x = \frac{v_0}{\omega} \sin \omega t + s_0 \cos \omega t - \frac{f}{\omega^2}(1 - \cos \omega t) \qquad (8.1.8)$$

In order to simplify the subsequent analysis, we apply some conventional algebraic procedures to the coefficients of the trigonometric functions in equation (8.1.8). Based on these procedures, we may write:

$$\frac{(s_0\omega^2 + f)^2}{(s_0\omega^2 + f)^2 + v_0^2\omega^2} + \frac{v_0^2\omega^2}{(s_0\omega^2 + f)^2 + v_0^2\omega^2}$$
$$= 1 = (\sin\alpha)^2 + (\cos\alpha)^2 \qquad (8.1.9)$$

where:

$$\sin\alpha = \frac{s_0\omega^2 + f}{\sqrt{(s_0\omega^2 + f)^2 + v_0^2\omega^2}} \qquad (8.1.10)$$

$$\cos\alpha = \frac{v_0\omega}{\sqrt{(s_0\omega^2 + f)^2 + v_0^2\omega^2}} \qquad (8.1.11)$$

Combining equations (8.1.10) and (8.1.11) with equation (8.1.8), we have:

$$x = \frac{1}{\omega^2}[\sqrt{(s_0\omega^2 + f)^2 + v_0^2\omega^2}$$
$$(\sin\omega t \cos\alpha + \cos\omega t \sin\alpha) - f] \qquad (8.1.12)$$

Therefore, we obtain:

$$x = \frac{1}{\omega^2}[\sqrt{(s_0\omega^2 + f)^2 + v_0^2\omega^2} \, \sin(\omega t + \alpha) - f] \qquad (8.1.13)$$

The first and second derivatives of equation (8.1.13) represent the velocity and the acceleration of the system respectively:

$$\frac{dx}{dt} = \frac{1}{\omega}\sqrt{(s_0\omega^2 + f)^2 + v_0^2\omega^2}\,\cos(\omega t + \alpha) \qquad \text{(8.1.14)}$$

$$\frac{d^2x}{dt^2} = -\sqrt{(s_0\omega^2 + f)^2 + v_0^2\omega^2}\,\sin(\omega t + \alpha) \qquad \text{(8.1.15)}$$

According to equation (8.1.14), the velocity becomes equal to zero at the moment of time when we have:

$$\cos(\omega t + \alpha) = 0 \qquad \text{(8.1.16)}$$

and, consequently, at the end of the first half of the vibratory cycle we have:

$$\sin(\omega t + \alpha) = 1 \qquad \text{(8.1.17)}$$

Combining equations (8.1.13) and (8.1.5) with equation (8.1.19), we determine the maximum values of the displacement s_{max} and the acceleration a_{max} respectively:

$$s_{max} = \frac{1}{\omega^2}[\sqrt{(s_0\omega^2 + f)^2 + v_0^2\omega^2} - f] \qquad \text{(8.1.18)}$$

$$a_{max} = -\sqrt{(s_0\omega^2 + f)^2 + v_0^2\omega^2} \qquad \text{(8.1.19)}$$

Multiplying equation (8.1.18) by K and substituting notations (8.1.4) and (8.1.5), we determine the maximum absolute value of the force R_{max} applied to the elastic link:

$$R_{max} = |\sqrt{(Ks_0 + F)^2 + mKv_0^2} - F \qquad \text{(8.1.20)}$$

It is justifiable to consider that the link is subjected to a completely reversed loading cycle having:

$$R_{min} = -R_{max} \qquad \text{(8.1.21)}$$

Therefore, the stress calculations should include fatigue considerations.

8.1.2 Initial Displacement Equals Zero

The initial conditions of motion are:

$$\text{for} \quad t = 0 \quad x = 0; \quad \frac{dx}{dt} = v_0 \qquad (8.1.22)$$

Solving differential equation of motion (8.1.1) with the initial conditions of motion (8.1.22), we obtain:

$$x = \frac{v_0}{\omega} \sin \omega t - \frac{f}{\omega^2} (1 - \cos \omega t) \qquad (8.1.23)$$

Applying the algebraic procedures similarly to the previous case, we may write:

$$x = \frac{\sqrt{v_0^2 \omega^2 + f^2}}{\omega^2} (\sin \omega t \cos \alpha_1 + \cos \omega t \sin \alpha_1) - \frac{f}{\omega^2} \qquad (8.1.24)$$

where:

$$\sin \alpha_1 = \frac{f}{\sqrt{v_0^2 \omega^2 + f^2}} \qquad (8.1.25)$$

$$\cos \alpha_1 = \frac{v_0 \omega}{\sqrt{v_0^2 \omega^2 + f^2}} \qquad (8.1.26)$$

Combining equations (8.1.25) and (8.1.26) with equation (8.1.24), we obtain:

$$x = \frac{1}{\omega^2} [\sqrt{v_0^2 \omega^2 + f^2} \sin(\omega t + \alpha_1) - f] \qquad (8.1.27)$$

Taking the first derivative of equation (8.1.27), we determine the velocity of the system:

$$\frac{dx}{dt} = \frac{\sqrt{v_0^2 \omega^2 + f^2}}{\omega} \cos(\omega t + \alpha_1) \qquad (8.1.28)$$

The second derivative of the equation (8.1.27) represents the acceleration:

$$\frac{d^2 x}{dt^2} = -\sqrt{v_0^2 \omega^2 + f^2} \sin(\omega t + \alpha_1) \qquad (8.1.29)$$

Equation (8.1.28) shows that the velocity becomes equal to zero at the moment when:

$$\cos(\omega t + \alpha_1) = 0 \tag{8.1.30}$$

and, consequently, at this time we have:

$$\sin(\omega t + \alpha_1) = 1 \tag{8.1.31}$$

Combining equations (8.1.27) and (8.1.29) with equation (8.1.31), we determine the maximum displacement and acceleration for this case respectively:

$$s_{max} = \frac{1}{\omega^2}[\sqrt{v_0^2\omega^2 + f^2} - f] \tag{8.1.32}$$

$$a_{max} = -\sqrt{v_0^2\omega^2 + f^2} \tag{8.1.33}$$

Multiplying equation (8.1.32) by K and substituting notations (8.1.4) and (8.1.5), we calculate the maximum absolute value of the force R_{max} applied to the elastic link:

$$R_{max} = |\sqrt{mKv_0^2 + F^2} - F| \tag{8.1.34}$$

As mentioned in the previous case, it may be accepted that the elastic link is subjected to a completely reversed loading cycle where the minimum force $R_{min} = -R_{max}$.

Therefore, the stress analysis should be based on fatigue calculations.

8.1.3 Initial Velocity Equals Zero

Figure 8.1 shows that in this case the system moves in a positive direction. This is possible if the initial displacement is negative. Thus, the initial conditions of motion are:

$$\text{for} \quad t = 0 \quad x = -s_0; \quad \frac{dx}{dt} = 0 \tag{8.1.35}$$

According to the initial conditions of motion, the system in this case possesses potential energy due to the deformation of the elastic link. The deformation is proportional to the initial negative displacement.

At the beginning of the process, the motion of the mass m is possible in those cases when the absolute value of the deformation force $|Ks_0|$ exceeds the friction force F. Assuming that the motion of the system in the positive direction is possible, we can solve differential equation of motion (8.1.1) with the initial conditions of motion (8.1.35). In this case, the term l^2s_0 in equation (8.1.6) should be taken with a positive sign. The solution of the differential equation (8.1.1) reads:

$$x = -s_0 \cos \omega t - \frac{f}{\omega^2}(1 - \cos \omega t) \tag{8.1.36}$$

Applying basic algebra to equation (8.1.36), we have:

$$x = \frac{1}{\omega^2}[(-s_0\omega^2 + f)\cos \omega t - f] \tag{8.1.37}$$

Taking the first and second derivatives of equation (8.1.37), we determine the velocity and the acceleration of the system respectively:

$$\frac{dx}{dt} = \frac{1}{\omega}(s_0\omega^2 - f)\sin \omega t \tag{8.1.38}$$

$$\frac{d^2x}{dt^2} = (s_0\omega^2 - f)\cos \omega t \tag{8.1.39}$$

According to equation (8.1.38), the velocity becomes equal to zero at the moment when:

$$\sin \omega t = 0$$

At this moment, we have:

$$\cos \omega t = -1 \tag{8.1.40}$$

Combining equation (8.1.40) with equations (8.1.37) and (8.1.39), we determine the maximum values of the displacement s_{max} and acceleration a_{max} respectively:

$$s_{max} = s_0 - \frac{2f}{\omega^2} \tag{8.1.41}$$

$$a_{max} = (s_0\omega^2 - f) \tag{8.1.42}$$

Multiplying equation (8.1.38) by the stiffness coefficient K and substituting notations (8.1.4) and (8.1.5), we calculate the maximum value of force R_{max} applied to the elastic link:

$$R_{max} = Ks_0 - 2F \tag{8.1.43}$$

For the stress analysis, we may accept that the elastic link is subjected to a completely reversed loading cycle having:

$$R_{min} = - Ks_0 + 2F$$

The stress analysis of the link should be based on fatigue considerations.

8.2 Constant Force R

This section describes engineering problems characterized by the force of inertia, the stiffness force, and the friction force as the resisting forces (Row 8 of Guiding Table 2.1). The constant active force, indicated in Column 2, is applied to the systems related to this section.

The current problems could be related to the working processes of engineering systems interacting with elastoplastic or viscoelastoplastic media or with a specific elastic link. These types of media exert stiffness and friction forces as a reaction to deformation (more related information including the considerations regarding the behavior of the friction force applied to a vibratory system is discussed in section 1.2).

The system is moving in the horizontal direction. The model of a system subjected to a constant active force, stiffness force, and friction force is shown in Figure 8.2.

Based on the considerations presented above and on the model in Figure 8.2 we can compose the left and right sides of the differential equation of motion of the system. The left side consists of the force of inertia, the stiffness force, and the friction force, while the right side includes a constant active force. Hence, the differential equation of motion reads:

$$m\frac{d^2x}{dt^2} + Kx + F = R \tag{8.2.1}$$

As explained in section 1.3, in some cases the force that depends on the displacement could play the role of an active force. For instance, in case of magnetic interaction, the force of attraction P_s may be characterized by the following equation:

$$P_s = R + kx \qquad (8.2.2)$$

where $R > 0$. In this case, the differential equation of motion reads:

$$m\frac{d^2x}{dt^2} - Kx + f = R \qquad (8.2.3)$$

We want to determine the basic parameters of motion; however, for the case according to equation (8.2.1), we also want to determine the extreme values of these parameters, and the characteristics of the forces applied to the elastic link.

Differential equations of motion (8.2.1) and (8.2.3) have different solutions for various initial conditions of motion. These solutions and their analyses are presented below.

8.2.1 General Initial Conditions

The general initial conditions of motion are:

$$\text{for} \quad t = 0 \quad x = s_0; \quad \frac{dx}{dt} = v_0 \qquad (8.2.4)$$

where s_0 and v_0 are the initial displacement and initial velocity respectively.

a) Equation (8.2.1)

Dividing equation (8.2.1) by m, we have:

$$\frac{d^2x}{dt^2} + \omega^2 x + f = r \qquad (8.2.5)$$

where ω is the natural frequency of the system and:

$$\omega^2 = \frac{K}{m} \qquad (8.2.6)$$

$$f = \frac{F}{m} \qquad (8.2.7)$$

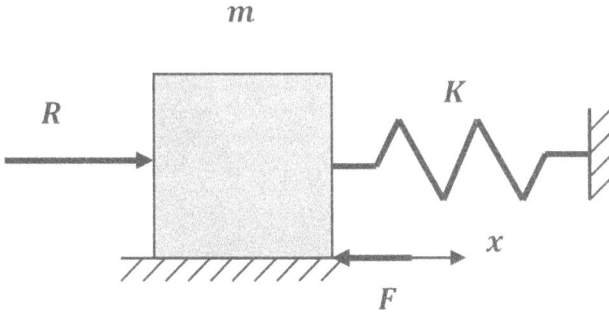

Figure 8.2 Model of a system subjected to a constant active force, a stiffness force, and a friction force

$$r = \frac{R}{m} \qquad (8.2.8)$$

Using Laplace Transform Pairs 3, 1, and 5 from Table 1.1, we convert differential equation of motion (8.2.5) with the initial conditions of motion (8.2.4) from the time domain into the Laplace domain, and obtain the resulting algebraic equation of motion in the Laplace domain:

$$l^2 x(l) - l v_0 - l^2 s_0 + \omega^2 x(l) + f = r \qquad (8.2.9)$$

Solving equation (8.2.8) for the displacement $x(l)$ in Laplace domain, we obtain:

$$x(l) = \frac{r - f}{l^2 + \omega^2} + \frac{l v_0}{l^2 + \omega^2} + \frac{l^2 s_0}{l^2 + \omega^2} \qquad (8.2.10)$$

By means of pairs 1, 5, 14, 6, and 7 from Table 1.1, we invert equation (8.2.10) from the Laplace domain into the time domain and obtain the solution of differential equation of motion (8.2.1) with the initial conditions of motion (8.2.4):

$$x = \frac{r - f}{\omega^2}(1 - \cos \omega t) + \frac{v_0}{\omega} \sin \omega t + s_0 \cos \omega t \qquad (8.2.11)$$

In order to simplify further analysis, we transform equation (8.2.11) to the following shape:

$$x = \frac{r - f}{\omega^2} + \frac{1}{\omega^2}[v_0 \omega \sin \omega t - (r - f - s_0 \omega^2) \cos \omega t] \qquad (8.2.12)$$

Using the coefficients of the trigonometric functions from equation (8.2.12), we compose the following expression:

$$\frac{(r-f-s_0\omega^2)^2}{(r-f-s_0\omega^2)^2+v_0^2\omega^2}+\frac{v_0^2\omega^2}{(r-f-s_0\omega^2)^2+v_0^2\omega^2}$$

$$=1=(\sin\beta)^2+(\cos\beta)^2 \qquad (8.2.13)$$

where

$$\sin\beta=-\frac{r-f-s_0\omega^2}{\sqrt{(r-f-s_0\omega^2)^2+v_0^2\omega^2}} \qquad (8.2.14)$$

$$\cos\beta=\frac{v_0\omega}{\sqrt{(r-f-s_0\omega^2)^2+v_0^2\omega^2}} \qquad (8.2.15)$$

Combining equations (8.2.14) and (8.2.15) with equation (8.2.12), we obtain:

$$x=\frac{r-f}{\omega^2}+\frac{1}{\omega^2}\sqrt{(r-f-s_0\omega^2)^2+v_0^2\omega^2}\,\sin(\omega t-\beta)] \qquad (8.2.16)$$

The first and second derivatives of equation (8.2.18) represent the velocity and the acceleration of the system respectively:

$$\frac{dx}{dt}=\frac{1}{\omega}\sqrt{(r-f-s_0\omega^2)^2+v_0^2\omega^2}\,\cos(\omega t-\beta) \qquad (8.2.17)$$

$$\frac{d^2x}{dt^2}=-\sqrt{(r-f-s_0\omega^2)^2+v_0^2\omega^2}\,\sin(\omega t-\beta) \qquad (8.2.18)$$

Equation (8.2.17) shows that the velocity equals zero at the moment when:

$$\cos(\omega t-\beta)=0 \qquad (8.2.19)$$

and, consequently, at this moment we have:

$$\sin(\omega t-\beta)=1 \qquad (8.2.20)$$

Combining equations (8.2.16) and (8.2.18) with equation (8.2.20), we determine the maximum values of the displacement s_{max} and acceleration a_{max} respectively:

$$S_{max} = \frac{r-f}{\omega^2} + \frac{1}{\omega^2}\sqrt{(r-f-s_0\omega^2)^2 + v_0^2\omega^2} \qquad (8.2.21)$$

$$a_{max} = -\sqrt{(r-f-s_0\omega^2)^2 + v_0^2\omega^2} \qquad (8.2.22)$$

Multiplying equation (8.2.21) by K and substituting notations (8.2.6), (8.2.7), and (8.2.8), we calculate the maximum force R_{max} applied to the elastic link:

$$R_{max} = R - F + \sqrt{(R-F-Ks_0)^2 + mKv_0^2} \qquad (8.2.23)$$

It is justifiable to accept that the link is subjected to a completely reversed loading cycle having the minimum force:

$$R_{min} = -R_{max} \qquad (8.2.24)$$

Therefore, the stress analysis of the link should be based on fatigue calculations.

b) **Equation (8.2.3)**

Dividing this equation by m, we obtain:

$$\frac{d^2x}{dt^2} - \omega^2 x + f = r \qquad (8.2.25)$$

Using pairs 3, 1 and 5 from Table 1.1, we convert differential equation of motion (8.2.25) with the initial conditions of motion (8.2.4) from the time domain into the Laplace domain, and obtain the corresponding algebraic equation of motion in the Laplace domain:

$$l^2x(l) - lv_0 - l^2s_0 - \omega^2x(l) + f = r \qquad (8.2.26)$$

Solving equation (8.2.26) for the displacement $x(l)$ in Laplace domain, we obtain:

$$x(l) = \frac{lv_0}{l^2 - \omega^2} + \frac{l^2s_0}{l^2 - \omega^2} + \frac{r-f}{l^2 - \omega^2} \qquad (8.2.27)$$

Based on pairs 1, 17, 18, and 21 from Table 1.1, we invert equation (8.2.27) from the Laplace domain into the time domain.

The inversion represents the solution of differential equation of motion (8.2.3) with the initial conditions of motion (8.2.4):

$$x = \frac{v_0}{\omega}\sinh\omega t + s_0\cosh\omega t + \frac{r-f}{\omega^2}(\cosh\omega t - 1) \quad \textbf{(8.2.28)}$$

Equation (8.2.28) describes the rectilinear translational motion of a system.

Taking the first and second derivatives of equation (8.2.28), we determine the velocity and the acceleration respectively:

$$\frac{dx}{dt} = v_0\cosh\omega t + s_0\omega\sinh\omega t + \frac{r-f}{\omega}\sinh\omega t \quad \textbf{(8.2.29)}$$

$$\frac{d^2x}{dt^2} = v_0\omega\sinh\omega t + s_0\cosh\omega t + (r-f)\cosh\omega t \quad \textbf{(8.2.30)}$$

Equations (8.2.28), (8.2.29), and (8.2.30) are increasing functions of time.

8.2.2 Initial Displacement Equals Zero

The initial conditions of motion are:

$$\text{for} \quad t = 0 \quad x = 0; \quad \frac{dx}{dt} = v_0 \quad \textbf{(8.2.31)}$$

a) Equation (8.2.1)

The solution of differential equation of motion (8.2.1) with the initial conditions of motion (8.2.31) reads:

$$x = \frac{r-f}{\omega^2}(1 - \cos\omega t) + \frac{v_0}{\omega}\sin\omega t \quad \textbf{(8.2.32)}$$

Transforming equation (8.2.31), we may write:

$$x = \frac{r-f}{\omega^2} + \frac{1}{\omega^2}[v_0\omega\sin\omega t - (r-f)\cos\omega t] \quad \textbf{(8.2.33)}$$

Combining the coefficients of the trigonometric functions in equation (8.2.33) similarly as in equation (8.2.12), we may write:

$$x = \frac{1}{\omega^2}[r + \sqrt{(r-f)^2 + v_0^2\omega^2}\,\sin(\omega t - \beta_1)] \quad \textbf{(8.2.34)}$$

where:

$$\sin \beta_1 = -\frac{r - f}{\sqrt{(r - f)^2 + v_0^2 \omega^2}} \qquad (8.2.35)$$

$$\cos \beta_1 = \frac{v_0 \omega}{\sqrt{(r - f)^2 + v_0^2 \omega^2}} \qquad (8.2.36)$$

Taking the first and second derivatives of equation (8.2.34), we determine the velocity and acceleration respectively:

$$\frac{dx}{dt} = \frac{1}{\omega} \sqrt{(r - f)^2 + v_0^2 \omega^2} \cos(\omega t - \beta_1) \qquad (8.2.37)$$

$$\frac{d^2 x}{dt^2} = -\sqrt{(r - f)^2 + v_0^2 \omega^2} \sin(\omega t - \beta_1) \qquad (8.2.38)$$

Equation (8.2.37) shows that at the moment when $\cos(\omega t - \beta_1) = 0$, the velocity equals zero. Consequently, at this moment we have:

$$\sin(\omega t - \beta_1) = 1 \qquad (8.2.39)$$

Substituting equation (8.2.39) into equations (8.2.34) and (8.2.38), we determine the maximum values of the displacement s_{max} and the acceleration a_{max} respectively:

$$s_{max} = \frac{1}{\omega^2} [r + \sqrt{(r - f)^2 + v_0^2 \omega^2}] \qquad (8.2.40)$$

$$a_{max} = -\sqrt{(r - f)^2 + v_0^2} \qquad (8.2.41)$$

Multiplying equation (8.2.40) by K and substituting notations (8.2.6), (8.2.7), and (8.2.8), we calculate the maximum force R_{max} applied to the elastic link:

$$R_{max} = R + \sqrt{(R - F)^2 + mKv_0^2} \qquad (8.2.42)$$

We may accept that the link is subjected to a completely reversed loading cycle having the minimum force $R_{min} = -R_{max}$. Therefore, the stress analysis of the link should be based on fatigue calculations.

b) Equation (8.2.3)

The solution of differential equation of motion (8.2.3) with the initial conditions of motion (8.2.31) reads:

$$x = \frac{v_0}{\omega}\sinh\omega t + \frac{r-f}{\omega^2}(\cosh\omega t - 1) \qquad (8.2.43)$$

Equation (8.2.43) shows that the system performs rectilinear translational motion.

Taking the first and the second derivatives of equation (8.2.43), we determine the velocity and the acceleration of the system respectively:

$$\frac{dx}{dt} = v_0\cosh\omega t + \frac{r-f}{\omega}\sinh\omega t$$

$$\frac{d^2x}{dt^2} = v_0\omega\sinh\omega t + (r-f)\cosh\omega t$$

8.2.3 Initial Velocity Equals Zero

The initial conditions of motion are:

$$\text{for}\quad t=0 \quad x=s_0;\quad \frac{dx}{dt}=0 \qquad (8.2.44)$$

a) Equation (8.2.1)

Solving differential equation of motion (8.2.1) with the initial conditions of motion (8.2.44), we obtain:

$$x = \frac{r-f}{\omega^2} - \frac{1}{\omega^2}(r-f-s_0\omega^2)\cos\omega t \qquad (8.2.45)$$

The first and second derivatives of equation (8.2.45) represent the velocity and acceleration respectively:

$$\frac{dx}{dt} = \frac{1}{\omega}(r-f-s_0\omega^2)\sin\omega t \qquad (8.2.46)$$

$$\frac{d^2x}{dt^2} = (r-f-s_0\omega^2)\cos\omega t \qquad (8.2.47)$$

According to equation (8.2.46), the velocity becomes equal to zero at the moment when

$$\sin \omega t = 0$$

This moment occurs after the system has completed one half of the vibratory cycle. Therefore, at this moment we have:

$$\cos \omega t = -1 \qquad \textbf{(8.2.48)}$$

Combining equation (8.2.48) with equations (8.2.45) and (8.2.47), we obtain the maximum values of the displacement s_{max} and acceleration a_{max} respectively:

$$s_{max} = \frac{2(r - f)}{\omega^2} - s_0 \qquad \textbf{(8.2.49)}$$

$$a_{max} = -(r - f - s_0\omega^2) \qquad \textbf{(8.2.50)}$$

Multiplying equation (8.2.49) by K and substituting notations (8.2.6), (8.2.7), and (8.2.8), we calculate the maximum value of the force R_{max} applied to the elastic link:

$$R_{max} = 2(R - F) - Ks_0 \qquad \textbf{(8.2.51)}$$

It is accepted that the link is subjected to a completely reversed loading cycle having the minimum force $R_{min} = - R_{max}$. Therefore, the stress analysis of the link should be based on fatigue calculations.

b) Equation (8.2.3)

Solving differential equation of motion (8.2.3) with the initial conditions of motion (8.2.44), we obtain:

$$x = s_0 \cosh \omega t + \frac{r - f}{\omega^2}(\cosh \omega t - 1) \qquad \textbf{(8.2.52)}$$

According to equation (8.2.52), the system is in rectilinear translational motion.

Taking the first and second derivatives of equation (8.2.52), we determine the velocity and the acceleration respectively:

$$\frac{dx}{dt} = s_0\omega \sinh \omega t + \frac{r-f}{\omega}\sinh \omega t$$

$$\frac{d^2x}{dt^2} = s_0 \cosh \omega t + (r-f)\cosh \omega t$$

8.2.4 Both the Displacement and Velocity Equal Zero

The initial conditions of motion are:

$$\text{for} \quad t = 0 \quad x = 0; \quad \frac{dx}{dt} = 0 \tag{8.2.53}$$

a) Equation (8.2.1)

The solution of differential equation of motion (8.2.1) with the initial conditions of motion (8.2.53) reads:

$$x = \frac{r-f}{\omega^2}(1 - \cos \omega t) \tag{8.2.54}$$

The first derivative of equation (8.2.54) represents the velocity:

$$\frac{dx}{dt} = \frac{r-f}{\omega}\sin \omega t \tag{8.2.55}$$

Taking the second derivative of equation (8.2. 54), we determine the acceleration:

$$\frac{d^2x}{dt^2} = (r-f)\cos \omega t \tag{8.2.56}$$

According to equation (8.2.55), the velocity equals zero when $\sin \omega t = 0$; therefore, we take for equations (8.2.54) and (8.2.56) that $\cos \omega t = -1$ and we calculate the maximum values of the displacement s_{max} and acceleration a_{max} respectively:

$$s_{max} = \frac{2(r-f)}{\omega^2} \tag{8.2.57}$$

$$a_{max} = -r + f \tag{8.2.58}$$

Multiplying equation (8.2.57) by K and substituting notations (8.2.6), (8.2.7), and (8.2.8), we determine the maximum force R_{max} applied to the link:

$$R_{max} = 2(R - F) \tag{8.2.59}$$

As in the previous cases, it is accepted that the link is subjected to a completely reversed loading cycle having the minimum force $R_{min} = -R_{max}$. Therefore, the stress analysis of the link should be based on fatigue calculations.

b) Equation (8.2.3)

Solving differential equation of motion (8.2.3) with the initial conditions of motion (8.2.53), we may write:

$$x = \frac{r - f}{\omega^2}(\cosh \omega t - 1) \tag{8.2.60}$$

Equation (8.2.60) shows that the system performs rectilinear translational motion.

Taking the first and second derivatives of equation (8.2.60), we determine the velocity and the acceleration respectively:

$$\frac{dx}{dt} = \frac{r - f}{\omega}\sinh \omega t$$

$$\frac{d^2x}{dt^2} = (r - f)\cosh \omega t$$

8.3 Harmonic Force $A \sin(\omega_1 t + \lambda)$

According to Guiding Table 2.1, this section describes engineering systems subjected to the action of the force of inertia, the stiffness force, and the friction force as the resisting forces (Row 8) and to the harmonic force (Column 3) as the active force. The problem described in this section could be associated with the working processes of vibratory systems intended for interaction with elastoplastic or viscoelastoplastic media that exert stiffness and friction forces during their deformation (more related information and considerations regarding the behavior of friction forces applied to vibratory systems are discussed in section 1.2).

The system is moving in the horizontal direction. We want to determine the basic parameters of motion. The model of a system subjected to the action of a harmonic force, a stiffness force, and a friction force is shown in Figure 8.3.

According to the considerations above and the model in Figure 8.3, we can compose the left and right sides of the differential equation of motion. Thus, the left side of this equation consists of the force of inertia, the stiffness force, and the friction force, while the right side includes the harmonic force. Therefore, the equation reads:

$$m\frac{d^2x}{dt^2} + Kx + F = A\sin(\omega_1 t + \lambda) \qquad (8.3.1)$$

Due to the presence of the friction force, equation (8.3.1) is valid to the moment when the system comes to a stop, causing the friction force to change its direction instantaneously. In this case, equation (8.3.1) is valid during the first half of the vibratory cycle.

Differential equation of motion (8.3.1) has different solutions for various initial conditions of motion. These solutions and their analyses are presented below.

8.3.1 General Initial Conditions
The general initial conditions of motion are:

$$\text{for} \quad t = 0 \quad x = s_0; \quad \frac{dx}{dt} = v_0 \qquad (8.3.2)$$

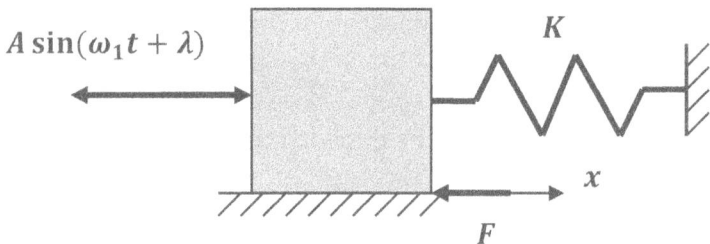

Figure 8.3 Model of a system subjected to a harmonic force, a stiffness force, and a friction force

where s_0 and v_0 are the initial displacement and initial velocity respectively.

Transforming the sinusoidal function in equation (8.3.1) and dividing the latter by m, we have:

$$\frac{d^2x}{dt^2} + \omega^2 x + f = a\sin\omega_1 t\cos\lambda + a\cos\omega_1 t\sin\lambda \qquad (8.3.3)$$

where ω is the natural frequency of the system and:

$$\omega^2 = \frac{K}{m} \qquad (8.3.4)$$

$$f = \frac{F}{m} \qquad (8.3.5)$$

$$a = \frac{A}{m} \qquad (8.3.6)$$

Using Laplace Transform Pairs 3, 1, 5, 6, and 7 from Table 1.1, we convert differential equation of motion (8.3.3) with the initial conditions of motion (8.3.2) from the time domain into the Laplace domain. The resulting algebraic equation of motion in Laplace domain reads:

$$l^2 x(l) - lv_0 - l^2 s_0 + \omega^2 x(l) + f = \frac{a\omega_1 l}{l^2 + \omega_1^2}$$

$$\cos\lambda + \frac{al^2}{l^2 + \omega_1^2}\sin\lambda \qquad (8.3.7)$$

Applying some transformations to equation (8.3.7), we have:

$$x(l)(l^2 + \omega^2) = lv_0 + l^2 s_0 - f + \frac{a\omega_1 l}{l^2 + \omega_1^2}$$

$$\cos\lambda + \frac{al^2}{l^2 + \omega_1^2}\sin\lambda \qquad (8.3.8)$$

Solving equation (8.3.8) for the Laplace domain displacement $x(l)$, we may write:

$$x(l) = \frac{lv_0}{l^2 + \omega^2} + \frac{l^2 s_0}{l^2 + \omega^2} - \frac{f}{l^2 + \omega^2} + \frac{la\omega_1}{(l^2 + \omega^2)(l^2 + \omega_1^2)}\cos\lambda$$

$$+ \frac{l^2 a}{(l^2 + \omega^2)(l^2 + \omega_1^2)}\sin\lambda \qquad (8.3.9)$$

By means of pairs 1, 6, 7, 14, 33, and 34 from Table 1.1, we invert equation (8.3.9) from the Laplace domain into the time domain, and obtain the solution of differential equation of motion (8.3.1) with the initial conditions of motion (8.3.2):

$$x = \frac{v_0}{\omega}\sin\omega t + s_0 \cos\omega t - \frac{f}{\omega^2}(1-\cos\omega t)$$

$$+ \frac{a(\omega\sin\omega_1 t - \omega_1\sin\omega t)\cos\lambda}{\omega(\omega^2-\omega_1^2)} + \frac{a(\cos\omega_1 t - \cos\omega t)\sin\lambda}{\omega^2-\omega_1^2}$$

$$(8.3.10)$$

The first derivative of equation (8.3.10) represents the velocity of the system:

$$\frac{dx}{dt} = v_0\cos\omega t - \omega s_0\sin\omega t - \frac{f}{\omega}\sin\omega t + \frac{a\omega_1(\cos\omega_1 t - \cos\omega t)\cos\lambda}{\omega^2-\omega_1^2}$$

$$- \frac{a(\omega_1\sin\omega_1 t - \omega\sin\omega t)\sin\lambda}{\omega^2-\omega_1^2}$$

$$(8.3.11)$$

Taking the second derivative of equation (8.3.10), we determine the acceleration of the system:

$$\frac{d^2x}{dt^2} = -\omega v_0\sin\omega t - f\cos\omega t - \omega^2 s_0\cos\omega t$$

$$- \frac{a\omega_1(\omega_1\sin\omega_1 t - \omega\sin\omega t)\cos\lambda}{\omega^2-\omega_1^2} - \frac{a(\omega_1^2\cos\omega_1 t - \omega^2\cos\omega t)\sin\lambda}{\omega^2-\omega_1^2}$$

$$(8.3.12)$$

8.3.2 Initial Displacement Equals Zero

The initial conditions of motion are:

$$\text{for} \quad t = 0 \quad x = 0; \quad \frac{dx}{dt} = v_0 \qquad (8.3.13)$$

Solving differential equation of motion (8.3.1) with the initial conditions of motion (8.3.13) we write:

$$x = \frac{v_0}{\omega}\sin\omega t - \frac{f}{\omega^2}(1-\cos\omega t) + \frac{a(\omega\sin\omega_1 t - \omega_1\sin\omega t)\cos\lambda}{\omega(\omega^2-\omega_1^2)}$$

$$+ \frac{a(\cos\omega_1 t - \cos\omega t)\sin\lambda}{\omega^2-\omega_1^2}$$

$$(8.3.14)$$

Taking the first and second derivatives of equation (8.3.14), we determine the velocity and acceleration of the system respectively:

$$\frac{dx}{dt} = v_0 \cos \omega t - \frac{f}{\omega} \sin \omega t + \frac{a\omega_1(\cos \omega_1 t - \cos \omega t)\cos \lambda}{\omega^2 - \omega_1^2}$$

$$- \frac{a(\omega_1 \sin \omega_1 t - \omega \sin \omega t)\sin \lambda}{\omega^2 - \omega_1^2} \tag{8.3.15}$$

$$\frac{d^2 x}{dt^2} = -\omega v_0 \sin \omega t - f \cos \omega t$$

$$- \frac{a\omega_1(\omega_1 \sin \omega_1 t - \omega \sin \omega t)\cos \lambda}{\omega^2 - \omega_1^2}$$

$$- \frac{a(\omega_1^2 \cos \omega_1 t - \omega^2 \cos \omega t)\sin \lambda}{\omega^2 - \omega_1^2} \tag{8.3.16}$$

8.3.3 Initial Velocity Equals Zero
The initial conditions of motion are:

$$\text{for} \quad t = 0 \quad x = s_0; \quad \frac{dx}{dt} = 0 \tag{8.3.17}$$

The solution of differential equation of motion (8.3.1) with the initial conditions of motion (8.3.17) reads:

$$x = s_0 \cos \omega t - \frac{f}{\omega^2}(1 - \cos \omega t) + \frac{a(\omega \sin \omega_1 t - \omega_1 \sin \omega t)\cos \lambda}{\omega(\omega^2 - \omega_1^2)}$$

$$+ \frac{a(\cos \omega_1 t - \cos \omega t)\sin \lambda}{\omega^2 - \omega_1^2} \tag{8.3.18}$$

The first derivative of equation (7.3.17) represents the velocity:

$$\frac{dx}{dt} = -\omega s_0 \sin \omega t - \frac{f}{\omega} \sin \omega t + \frac{a\omega_1(\cos \omega_1 t - \cos \omega t)\cos \lambda}{\omega^2 - \omega_1^2}$$

$$- \frac{a(\omega_1 \sin \omega_1 t - \omega \sin \omega t)\sin \lambda}{\omega^2 - \omega_1^2} \tag{8.3.19}$$

Taking the second derivative of equation (8.3.18), we obtain the acceleration:

$$\frac{d^2x}{dt^2} = -\omega^2 s_0 \cos\omega t - f\sin\omega t - \frac{a\omega_1(\omega_1\sin\omega_1 t - \omega\sin\omega t)\cos\lambda}{\omega^2 - \omega_1^2}$$

$$- \frac{a(\omega_1^2\cos\omega_1 t - \omega^2\cos\omega t)\sin\lambda}{\omega^2 - \omega_1^2} \tag{8.3.20}$$

8.3.4 Both the Initial Displacement and Velocity Equal Zero

The initial conditions of motion are:

$$\text{for} \quad t = 0 \quad x = 0; \quad \frac{dx}{dt} = 0 \tag{8.3.21}$$

Solving differential equation of motion (8.3.1) with the initial conditions of motion (8.3.21), we obtain:

$$x = -\frac{f}{\omega^2}(1 - \cos\omega t) + \frac{a(\omega\sin\omega_1 t - \omega_1\sin\omega t)\cos\lambda}{\omega(\omega^2 - \omega_1^2)}$$

$$+ \frac{a(\cos\omega_1 t - \cos\omega t)\sin\lambda}{\omega^2 - \omega_1^2} \tag{8.3.22}$$

The first derivative of equation (8.3.22) represents the velocity:

$$\frac{dx}{dt} = -\frac{f}{\omega}\sin\omega t + \frac{a\omega_1(\cos\omega_1 t - \cos\omega t)\cos\lambda}{\omega^2 - \omega_1^2}$$

$$- \frac{a(\omega_1\sin\omega_1 t - \omega\sin\omega t)\sin\lambda}{\omega^2 - \omega_1^2} \tag{8.3.23}$$

Taking the second derivative of equation (8.3.22), we determine the acceleration:

$$\frac{d^2x}{dt^2} = -f\cos\omega t - \frac{a\omega_1(\omega_1\sin\omega_1 t - \omega\sin\omega t)\cos\lambda}{\omega^2 - \omega_1^2}$$

$$- \frac{a(\omega_1^2\cos\omega_1 t - \omega^2\cos\omega t)\sin\lambda}{\omega^2 - \omega_1^2} \tag{8.3.24}$$

8.4 Time-Dependent Force $Q\left(\rho+\frac{\mu t}{\tau}\right)$

The current section describes engineering problems charac-terized by the action of the force of inertia, the stiffness force, and the friction force as the resisting forces (Row 8 in Guiding Table 2.1). The active force applied to the systems described in this sec-tion represents the time-dependent force (Column 4). The current problems could be related to engineering systems that interact with elastoplastic or viscoelastoplastic media, or with specific elastic links. These media exert stiffness and friction forces as a reaction to their deformation (see section 1.2 for more information regarding the deformation of these media and the behavior of a friction force applied to a vibratory system). In certain situations, the initial phase of the working process could be characterized by the action of a time-dependent force during a predetermined interval of time.

The system is moving in the horizontal direction. We want to determine the basic parameters of motion, their values at the end of the predetermined interval of time, and the characteristics of the forces applied to the elastic link at this time. Figure 8.4 shows the model of a system subjected to the action of a time-dependent force, a stiffness force, and a friction force.

Based on the considerations mentioned above and on the mod-el in Figure 8.4, we can assemble the left and right sides of the dif-ferential equation of motion. Thus, the left side consists of the force of inertia, the stiffness force, and the friction force. The right side

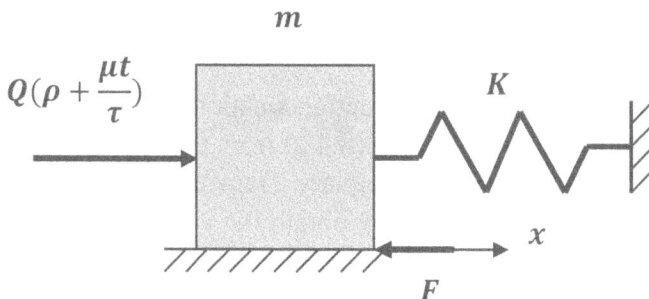

Figure 8.4 Model of a system subjected to a time-dependent force, a stiffness force, and a friction force

includes the time-dependent force. Therefore, the differential equation of motion reads:

$$m\frac{d^2x}{dt^2} + Kx + F = Q(\rho + \frac{\mu t}{\tau}) \qquad (8.4.1)$$

The differential equation of motion (8.4.1) has different solutions for various initial conditions of motion. These solutions and
their analyses are presented below.

8.4.1 General Initial Conditions

The general initial conditions of motion are:

$$\text{for} \quad t = 0 \quad x = s_0; \quad \frac{dx}{dt} = v_0 \qquad (8.4.2)$$

where s_0 and v_0 are the initial displacement and initial velocity
respectively.

Dividing equation (8.4.1) by m, we have:

$$\frac{d^2x}{dt^2} + \omega^2 x + f = q(\rho + \frac{\mu t}{\tau}) \qquad (8.4.3)$$

where ω is the natural frequency of the system and:

$$\omega^2 = \frac{K}{m} \qquad (8.4.4)$$

$$f = \frac{F}{m} \qquad (8.4.5)$$

$$q = \frac{Q}{m} \qquad (8.4.6)$$

In order to convert differential equation of motion (8.4.3) with
the initial conditions of motion (8.4.2) from the time domain into the
Laplace domain, we use the Laplace Transform Pairs 3, 1, 5, and 2
from Table 1.1. As a result, we obtain the corresponding algebraic
equation of motion in the Laplace domain:

$$l^2 x(l) - lv_0 - l^2 s_0 + \omega^2 x(l) + f = q\rho + \frac{q\mu}{\tau l} \qquad (8.4.7)$$

The solution of equation (8.4.7) for the Laplace domain displacement $x(l)$ reads:

$$x(l) = \frac{lv_0}{l^2 + \omega^2} + \frac{l^2 s_0}{l^2 + \omega^2} + \frac{q\rho - f}{l^2 + \omega^2} + \frac{q\mu}{\tau l(l^2 + \omega^2)} \tag{8.4.8}$$

Based on pairs 1, 6, 7, 14, and 16 from the Table 1.1, we invert equation (8.4.8) from the Laplace domain into the time domain and obtain the solution of differential equation of motion (8.4.1) with the initial conditions of motion (8.4.2):

$$x = \frac{v_0}{\omega}\sin\omega t + s_0\cos\omega t + \frac{q\rho - f}{\omega^2}(1 - \cos\omega t) + \frac{q\mu}{\tau\omega^2}(t - \frac{1}{\omega}\sin\omega t) \tag{8.4.9}$$

Transforming equation (8.4.9), we may write:

$$x = \frac{q\rho - f}{\omega^2} + \frac{q\mu t}{\tau\omega^2} + (s_0 - \frac{q\rho - f}{\omega^2})\cos\omega t + (\frac{v_0}{\omega} - \frac{q\mu}{\tau\omega^3})\sin\omega t \tag{8.4.10}$$

Taking the first derivative of equation (8.4.10), we determine the velocity:

$$\frac{dx}{dt} = \frac{q\mu}{\tau\omega^2} - (s_0\omega - \frac{q\rho - f}{\omega})\sin\omega t + (v_0 - \frac{q\mu}{\tau\omega^2})\cos\omega t \tag{8.4.11}$$

The second derivative of equation (8.4.10) represents the acceleration:

$$\frac{d^2x}{dt^2} = -(s_0\omega^2 - q\rho + f)\cos\omega t - (v_0\omega - \frac{q\mu}{\tau\omega})\sin\omega t \tag{8.4.12}$$

Substituting the time τ into equations (8.4.10), (8.4.11), and (8.4.12), we determine the values of the displacement s, the velocity v, and the acceleration a at the end of the predetermined interval of time respectively:

$$s = \frac{q(\rho + \mu) - f}{\omega^2} + (s_0\omega - \frac{q\rho - f}{\omega^2})\cos\omega\tau + (\frac{v_0}{\omega} - \frac{q\mu}{\tau\omega^2})\sin\omega\tau \tag{8.4.13}$$

$$v = \frac{q\mu}{\tau\omega^2} - (s_0\omega - \frac{q\rho - f}{\omega})\sin\omega\tau + (v_0 - \frac{q\mu}{\tau\omega^2})\cos\omega\tau \quad \textbf{(8.4.14)}$$

$$a = -(s_0\omega^2 - q\rho - f)\cos\omega\tau - (v_0\omega - \frac{q\mu}{\tau\omega})\sin\omega\tau \quad \textbf{(8.4.15)}$$

Multiplying equation (8.4.13) by the stiffness coefficient K and substituting notation (8.4.4), we calculate the maximum force R_{max} applied to the elastic link:

$$R_{max} = m[q(\rho + \mu) - f + (s_0\omega^2 - q\rho + f)\cos\omega\tau$$

$$+ (v_0\omega - \frac{q\mu}{\tau\omega})\sin\omega\tau] \quad \textbf{(8.4.16)}$$

The force according to equation (8.4.16) could be used for the stress calculations of the link.

8.4.2 Initial Displacement Equals Zero

The initial conditions of motion are:

$$\text{for} \quad t = 0 \quad x = 0; \quad \frac{dx}{dt} = v_0 \quad \textbf{(8.4.17)}$$

The solution of differential equation of motion (8.4.1) with the initial conditions of motion (8.4.17) reads:

$$x = \frac{v_0}{\omega}\sin\omega t + \frac{q\rho - f}{\omega^2}(1 - \cos\omega t) + \frac{q\mu}{\tau\omega^2}(t - \frac{1}{\omega}\sin\omega t) \quad \textbf{(8.4.18)}$$

Taking the first derivative of equation (8.4.18), we determine the velocity:

$$\frac{dx}{dt} = \frac{q\mu}{\tau\omega^2} + \frac{q\rho - f}{\omega}\sin\omega t + (v_0 - \frac{q\mu}{\tau\omega^2})\cos\omega t \quad \textbf{(8.4.19)}$$

The second derivative of equation (8.4.18) represents the acceleration:

$$\frac{d^2x}{dt^2} = (q\rho - f)\cos\omega t - (v_0\omega - \frac{q\mu}{\tau\omega})\sin\omega t \quad \textbf{(8.4.20)}$$

Substituting the interval of time τ into equations (8.4.18), (8.4.19), and (8.4.20), we determine the values of the displacement s, the velocity v, and the acceleration a at the end of the predetermined time respectively:

$$s = \frac{q(\rho + \mu) - f}{\omega^2} - \frac{q\rho - f}{\omega^2}\cos\omega\tau + (\frac{v_0}{\omega} - \frac{q\mu}{\tau\omega^3})\sin\omega\tau \quad (8.4.21)$$

$$v = \frac{q\mu}{\tau\omega^2} + \frac{q\rho - f}{\omega}\sin\omega\tau + (v_0 - \frac{q\mu}{\tau\omega^2})\cos\omega\tau \quad (8.4.22)$$

$$a = (q\rho - f)\cos\omega\tau - (v_0\omega - \frac{q\mu}{\tau\omega})\sin\omega\tau \quad (8.4.23)$$

Multiplying equation (8.4.21) by K and substituting notation (8.4.4), we calculate the maximum force applied to the elastic link:

$$R_{max} = m[q(\rho + \mu) - f - (q\rho - f)\cos\omega\tau + (v_0\omega - \frac{q\mu}{\tau\omega})\sin\omega\tau]$$

$$(8.4.24)$$

The stress calculations of the link could be based on the force according to equation (8.4.24).

8.4.3 Initial Velocity Equals Zero

The initial conditions of motion are:

$$\text{for} \quad t = 0 \quad x = s_0; \quad \frac{dx}{dt} = 0 \quad (8.4.25)$$

The solution of differential equation of motion (8.4.1) with the initial conditions of motion (8.4.25) reads:

$$x = \frac{q\rho - f}{\omega^2} + \frac{q\mu t}{\tau\omega^2} + (s_0 - \frac{q\rho - f}{\omega^2})\cos\omega t - \frac{q\mu}{\tau\omega^3}\sin\omega t \quad (8.4.26)$$

Taking the first and second derivatives of equation (8.4.26), we determine the velocity and the acceleration respectively:

$$\frac{dx}{dt} = \frac{q\mu}{\tau\omega^2} - (s_0\omega - \frac{q\rho - f}{\omega})\sin\omega t - \frac{q\mu}{\tau\omega^2}\cos\omega t \quad (8.4.27)$$

$$\frac{d^2x}{dt^2} = -(s_0\omega^2 - q\rho + f)\cos\omega t + \frac{q\mu}{\tau\omega}\sin\omega t \quad (8.4.28)$$

Substituting the time τ into equations (8.4.26), (8.4.27), and (8.4.28), we determine the values of the displacement s, the velocity v, and the acceleration a at the end of the interval of time τ respectively:

$$s = \frac{q(\rho + \mu) - f}{\omega^2} + (s_0 - \frac{q\rho - f}{\omega^2})\cos\omega\tau - \frac{q\mu}{\tau\omega^3}\sin\omega\tau \quad \textbf{(8.4.29)}$$

$$v = \frac{q\mu}{\omega^2} - (s_0\omega - \frac{q\rho - f}{\omega})\sin\omega\tau - \frac{q\mu}{\tau\omega^2}\cos\omega\tau \quad \textbf{(8.4.30)}$$

$$a = -(s_0\omega^2 - q\rho + f)\cos\omega\tau + \frac{q\mu}{\tau\omega}\sin\omega\tau \quad \textbf{(8.4.31)}$$

Multiplying equation (8.4.29) by K and substituting notation (8.4.4), we determine the maximum force R_{max} that is applied to the elastic link:

$$R_{max} = m[q(\rho + \mu) - f + (s_0\omega^2 - q\rho + f)\cos\omega\tau - \frac{q\mu}{\tau\omega}\sin\omega\tau]$$

$$\textbf{(8.4.32)}$$

The stress analysis of the link could be based on the force according to equation (8.4.32).

8.4.4 Both the Initial Displacement and Velocity Equal Zero

The initial conditions of motion are:

$$\text{for} \quad t = 0 \quad x = 0; \quad \frac{dx}{dt} = 0 \quad \textbf{(8.4.33)}$$

Solving differential equation of motion (8.4.1) with the initial conditions of motion (8.4.33), we obtain:

$$x = \frac{q\rho - f}{\omega^2} + \frac{q\mu t}{\tau\omega^2} - \frac{q\rho - f}{\omega^2}\cos\omega t - \frac{q\mu}{\tau\omega^3}\sin\omega t \quad \textbf{(8.4.34)}$$

The first and second derivatives of equation (8.4.34) represent the velocity and the acceleration respectively:

$$\frac{dx}{dt} = \frac{q\mu}{\tau\omega^2} + \frac{q\rho - f}{\omega}\sin\omega t - \frac{q\mu}{\tau\omega^2}\cos\omega t \quad \textbf{(8.4.35)}$$

$$\frac{d^2x}{dt^2} = (q\rho - f)\cos\omega t + \frac{q\mu}{\tau\omega}\sin\omega t \quad \textbf{(8.4.36)}$$

Substituting the time τ into equations (8.4.34), (8.4.35), and (8.4.36), we determine the values of the displacement s, the velocity v, and the acceleration a at the end of the predetermined time respectively:

$$s = \frac{q(\rho + \mu) - f}{\omega^2} - \frac{q\rho - f}{\omega^2} \cos \omega\tau - \frac{q\mu}{\tau\omega^3} \sin \omega\tau \qquad \textbf{(8.4.37)}$$

$$v = \frac{q\mu}{\omega^2} + \frac{q\rho - f}{\omega} \sin \omega\tau - \frac{q\mu}{\tau\omega^2} \cos \omega\tau \qquad \textbf{(8.4.38)}$$

$$a = (q\rho - f) \cos \omega\tau + \frac{q\mu}{\tau\omega} \sin \omega\tau \qquad \textbf{(8.4.39)}$$

Multiplying equation (8.4.37) by K and substituting notation (8.4.4), we calculate the maximum force R_{max} applied to the elastic link:

$$R_{max} = m[q(\rho + \mu) - f - (q\rho - f) \cos \omega\tau - \frac{q\mu}{\tau\omega} \sin \omega\tau]$$

$$\textbf{(8.4.40)}$$

The stress calculations of the link could be based on the force according to equation (8.4.40).

8.5 Constant Force R and Harmonic force $A \sin(\omega_1 t + \lambda)$

Guiding Table 2.1 indicates that this section focuses on engineering systems experiencing the action of the force of inertia, the stiffness force, and the friction force as the resisting forces (Row 8) and the sum of the constant active force and the harmonic force as the active forces (Column 5). The problem described in this section could be associated with a vibratory system intended for interaction with an elastoplastic or a viscoelastoplastic medium that exerts stiffness and friction forces as the reaction to deformation (see section 1.2 for more information related to the deformation of these media and to the behavior of a friction force applied to a vibratory system). This problem could be associated with an engineering system interacting with a specific elastic link.

The system is moving in the horizontal direction. We want to determine the basic parameters of motion. Figure 8.5 shows the model of a system subjected to the action of a constant active force, a harmonic force, a stiffness force, and a friction force.

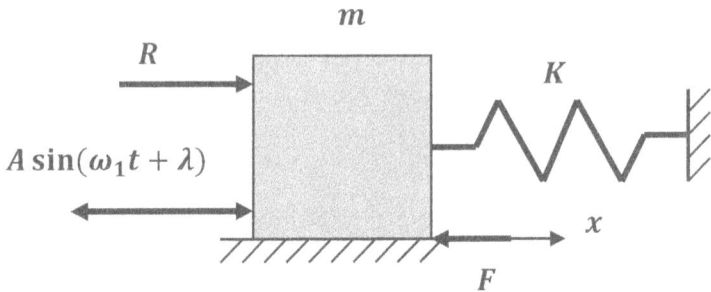

Figure 8.5 Model of a system subjected to a constant active force, a harmonic force, a stiffness force, and a friction force

With the considerations above and the model shown in Figure 8.5, we can compose the left and right sides of the differential equation of motion of the system. The left side consists of the force of inertia, the stiffness force, and the friction force, while the right side includes the constant active force and the harmonic force. Thus, the differential equation of motion reads:

$$m\frac{d^2x}{dt^2} + Kx + F = R + A\sin(\omega_1 t + \lambda) \qquad (8.5.1)$$

The differential equation of motion (8.5.1) has different solutions for various initial conditions of motion. These solutions and their analyses are presented below.

8.5.1 General Initial Conditions

The general initial conditions of motion are:

$$\text{for} \quad t = 0 \quad x = s_0; \quad \frac{dx}{dt} = v_0 \qquad (8.5.2)$$

where s_0 and v_0 are the initial displacement and initial velocity respectively.

Transforming the sinusoidal function in equation (8.5.1) and dividing the latter by m, we may write:

$$\frac{d^2x}{dt^2} + \omega^2 x + f = r + a\sin\omega_1 t \cos\lambda + a\cos\omega_1 t \sin\lambda \quad (8.5.3)$$

where ω is the natural frequency of the system and:

$$\omega^2 = \frac{K}{m} \tag{8.5.4}$$

$$f = \frac{F}{m} \tag{8.5.5}$$

$$r = \frac{R}{m} \tag{8.5.6}$$

$$a = \frac{A}{m} \tag{8.5.7}$$

In order to convert differential equation of motion (8.5.3) with the initial conditions of motion (8.5.2) from the time domain into the Laplace domain, we use Laplace Transform Pairs 3, 5, 1, 6, and 7 from Table 1.1. The resulting algebraic equation of motion in the Laplace domain reads:

$$l^2 x(l) - lv_0 - l^2 s_0 + \omega^2 x(l) + f = r + \frac{a\omega_1 l}{l^2 + \omega_1^2} \cos \lambda + \frac{al^2}{l^2 + \omega_1^2} \sin \lambda \tag{8.5.8}$$

Transforming equation (8.5.8), we may write:

$$x(l)(l^2 + \omega^2) = r - f + lv_0 + l^2 s_0 + \frac{a\omega_1 l}{l^2 + \omega_1^2} \cos \lambda + \frac{al^2}{l^2 + \omega_1^2} \sin \lambda \tag{8.5.9}$$

Solving equation (8.5.9) for the Laplace domain displacement $x(l)$, we have:

$$x(l) = \frac{r - f}{l^2 + \omega^2} + \frac{lv_0}{l^2 + \omega^2} + \frac{l^2 s_0}{l^2 + \omega^2}$$

$$+ \frac{la\omega_1}{(l^2 + \omega^2)(l^2 + \omega_1^2)} \cos \lambda + \frac{l^2 a}{(l^2 + \omega^2)(l^2 + \omega_1^2)} \sin \lambda \tag{8.5.10}$$

Using pairs 1, 14, 5, 6, 7, 33, and 34 from Table 1.1, we invert equation (8.5.10) and obtain the solution of differential equation of motion (8.5.1) with the initial conditions of motion (8.5.2):

$$x = \frac{r-f}{\omega^2}(1-\cos\omega t) + \frac{v_0}{\omega}\sin\omega t + s_0 \cos\omega t$$

$$+ \frac{a(\omega \sin\omega_1 t - \omega_1 \sin\omega t)\cos\lambda}{\omega(\omega^2 - \omega_1^2)} + \frac{a(\cos\omega_1 t - \cos\omega t)\sin\lambda}{\omega^2 - \omega_1^2} \quad \textbf{(8.5.11)}$$

Taking the first and second derivatives of equation (8.5.11), we determine the velocity and the acceleration of the system respectively:

$$\frac{dx}{dt} = \frac{r-f}{\omega}\sin\omega t + v_0 \cos\omega t - \omega s_0 \sin\omega t$$

$$+ \frac{a\omega_1(\cos\omega_1 t - \cos\omega t)\cos\lambda}{\omega^2 - \omega_1^2} - \frac{a(\omega_1 \sin\omega_1 t - \omega \sin\omega t)\sin\lambda}{\omega^2 - \omega_1^2}$$

$$\textbf{(8.5.12)}$$

$$\frac{d^2x}{dt^2} = (r-f)\cos\omega t - \omega v_0 \sin\omega t - \omega^2 s_0 \cos\omega t$$

$$- \frac{a\omega_1(\omega_1 \sin\omega_1 t - \omega \sin\omega t)\cos\lambda}{\omega^2 - \omega_1^2}$$

$$- \frac{a(\omega_1^2 \cos\omega_1 t - \omega^2 \cos\omega t)\sin\lambda}{\omega^2 - \omega_1^2} \quad \textbf{(8.5.13)}$$

8.5.2 Initial Displacement Equals Zero

The initial conditions of motion are:

$$\text{for} \quad t = 0 \quad x = 0; \quad \frac{dx}{dt} = v_0 \quad \textbf{(8.5.14)}$$

Solving differential equation of motion (8.5.1) with the initial conditions of motion (8.5.14), we have:

$$x = \frac{r-f}{\omega^2}(1-\cos\omega t) + \frac{v_0}{\omega}\sin\omega t + \frac{a(\omega \sin\omega_1 t - \omega_1 \sin\omega t)\cos\lambda}{\omega(\omega^2 - \omega_1^2)}$$

$$+ \frac{a(\cos\omega_1 t - \cos\omega t)\sin\lambda}{\omega^2 - \omega_1^2} \quad \textbf{(8.5.15)}$$

Taking the first derivative of the equation (8.5.15), we obtain the velocity:

$$\frac{dx}{dt} = \frac{r-f}{\omega}\sin\omega t + v_0 \cos\omega t + \frac{a\omega_1(\cos\omega_1 t - \cos\omega t)\cos\lambda}{\omega^2 - \omega_1^2}$$

$$-\frac{a(\omega_1 \sin\omega_1 t - \omega\sin\omega t)\sin\lambda}{\omega^2 - \omega_1^2} \qquad (8.5.16)$$

The second derivative of the equation (8.5.15) represents the acceleration:

$$\frac{d^2x}{dt^2} = (r-f)\cos\omega t - \omega v_0 \sin\omega t$$

$$-\frac{a\omega_1(\omega_1 \sin\omega_1 t - \omega\sin\omega t)\cos\lambda}{\omega^2 - \omega_1^2}$$

$$-\frac{a(\omega_1^2 \cos\omega_1 t - \omega^2 \cos\omega t)\sin\lambda}{\omega^2 - \omega_1^2} \qquad (8.5.17)$$

8.5.3 Initial Velocity Equals Zero

The initial conditions of motion are:

$$\text{for} \quad t = 0 \quad x = s_0; \quad \frac{dx}{dt} = 0 \qquad (8.5.18)$$

The solution of differential equation of motion (8.5.1) with the initial conditions of motion (8.5.18) reads:

$$x = \frac{r-f}{\omega^2}(1 - \cos\omega t) + s_0 \cos\omega t + \frac{a(\omega\sin\omega_1 t - \omega_1 \sin\omega t)\cos\lambda}{\omega(\omega^2 - \omega_1^2)}$$

$$+\frac{a(\cos\omega_1 t - \cos\omega t)\sin\lambda}{\omega^2 - \omega_1^2} \qquad (8.5.19)$$

The first derivative of equation (8.5.19) represents the velocity:

$$\frac{dx}{dt} = \frac{r-f}{\omega}\sin\omega t - \omega s_0 \sin\omega t + \frac{a\omega_1(\cos\omega_1 t - \cos\omega t)\cos\lambda}{\omega^2 - \omega_1^2}$$

$$-\frac{a(\omega_1 \sin\omega_1 t - \omega\sin\omega t)\sin\lambda}{\omega^2 - \omega_1^2} \qquad (8.5.20)$$

Taking the second derivative of equation (8.5.19), we determine the acceleration:

$$\frac{d^2x}{dt^2} = (r-f)\cos\omega t - \omega^2 s_0 \cos\omega t$$

$$-\frac{a\omega_1(\omega_1\sin\omega_1 t - \omega\sin\omega t)\cos\lambda}{\omega^2 - \omega_1^2}$$

$$-\frac{a(\omega_1^2\cos\omega_1 t - \omega^2\cos\omega t)\sin\lambda}{\omega^2 - \omega_1^2} \qquad (8.5.21)$$

8.5.4 Both the Displacement and Velocity Equal Zero

The initial conditions of motion are:

$$\text{for} \quad t = 0 \quad x = 0; \quad \frac{dx}{dt} = 0 \qquad (8.5.22)$$

The solution of differential equation motion of (8.5.1) with the initial conditions of motion (8.5.22) reads:

$$x = \frac{r-f}{\omega^2}(1-\cos\omega t) + \frac{a(\omega\sin\omega_1 t - \omega_1\sin\omega t)\cos\lambda}{\omega(\omega^2 - \omega_1^2)}$$

$$+\frac{a(\cos\omega_1 t - \cos\omega t)\sin\lambda}{\omega^2 - \omega_1^2} \qquad (8.5.23)$$

Taking the first and second derivatives of equation (8.5.23), we determine the velocity and the acceleration respectively:

$$\frac{dx}{dt} = \frac{r-f}{\omega}\sin\omega t + \frac{a\omega_1(\cos\omega_1 t - \cos\omega t)\cos\lambda}{\omega^2 - \omega_1^2}$$

$$-\frac{a(\omega_1\sin\omega_1 t - \omega\sin\omega t)\sin\lambda}{\omega^2 - \omega_1^2} \qquad (8.5.24)$$

$$\frac{d^2x}{dt^2} = (r-f)\cos\omega t - \frac{a\omega_1(\omega_1\sin\omega_1 t - \omega\sin\omega t)\cos\lambda}{\omega^2 - \omega_1^2}$$

$$-\frac{a(\omega_1^2\cos\omega_1 t - \omega^2\cos\omega t)\sin\lambda}{\omega^2 - \omega_1^2} \qquad (8.5.25)$$

8.6 Harmonic Force $A \sin(\omega_1 t + \lambda)$ and Time-Dependent Force $Q\left(\rho + \frac{\mu t}{\tau}\right)$

This section describes engineering problems characterized by the force of inertia, the stiffness force, and the friction force as the resisting forces (Row 8 in Guiding Table 2.1) and the sum of the harmonic force and the time-dependent force as the active forces (Column 6). The current problems could be associated with the working processes of vibratory systems that interact with elastoplastic or viscoelastoplastic media. Sometimes during the initial phase of the working process, it is reasonable to apply a time-dependent force that acts for a limited interval of time. The media exert stiffness and friction forces as a reaction to their deformation (see section 1.2 for more information regarding the deformation of the media and the behavior of the friction force applied to a vibratory system). These problems could be related to the interaction of an engineering system with a specific elastic link.

The system is moving in the horizontal direction. We want to determine the basic parameters of motion. Figure 8.6 shows the model of a system that is subjected to the action of the harmonic force, the time-dependent force, the stiffness force, and the friction force.

With the considerations above and the model in Figure 8.6, we can compose the left and right sides of the differential equation

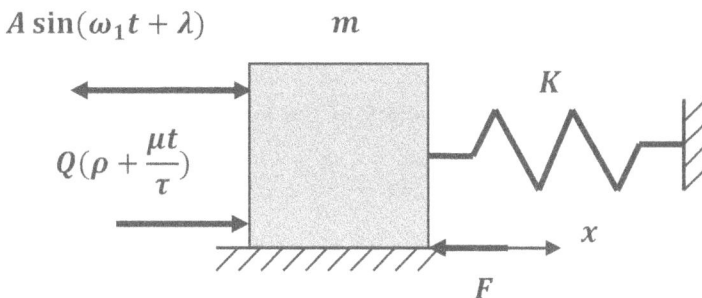

Figure 8.6 Model of a system subjected to a harmonic force, a time-dependent force, a stiffness force, and a friction force

of motion. The left side consists of the force of inertia, the stiffness force, and the friction force. The right side includes the harmonic force and the time-dependent force. Hence, the differential equation of motion reads:

$$m\frac{d^2x}{dt^2} + Kx + F = A\sin(\omega_1 t + \lambda) + Q(\rho + \frac{\mu t}{\tau}) \qquad (8.6.1)$$

Differential equation of motion (8.6.1) has different solutions for various initial conditions of motion. These solutions and their analyses are presented below.

8.6.1 General Initial Conditions

The general initial conditions of motion are:

$$\text{for} \quad t = 0 \quad x = s_0; \quad \frac{dx}{dt} = v_0 \qquad (8.6.2)$$

where s_0 and v_0 are the initial displacement and initial velocity respectively.

Transforming equation (8.6.1), we write:

$$m\frac{d^2x}{dt^2} + Kx + F = Q\rho + Q\frac{\mu t}{\tau} + A\sin\omega_1 t\cos\lambda + A\cos\omega_1 t\sin\lambda$$

$$(8.6.3)$$

Dividing equation (8.6.3) by m, we have:

$$\frac{d^2x}{dt^2} + \omega^2 x + f = q\rho + q\frac{\mu t}{\tau} + a\sin\omega_1 t\cos\lambda + a\cos\omega_1 t\sin\lambda$$

$$(8.6.4)$$

where ω is the natural frequency of the system and:

$$\omega^2 = \frac{K}{m} \qquad (8.6.5)$$

$$f = \frac{F}{m} \qquad (8.6.6)$$

$$q = \frac{Q}{m} \qquad (8.6.7)$$

$$a = \frac{A}{m} \qquad (8.6.8)$$

Using Laplace Transform Pairs 3, 5, 1, 2, 6, and 7 from Table 1.1, we convert the differential equation of motion (8.6.4) with the initial conditions of motion (8.6.2) from the time domain into the Laplace domain. The resulting algebraic equation of motion in Laplace domain reads:

$$l^2 x(l) - lv_0 - l^2 s_0 + \omega^2 x(l) + f = q\rho + \frac{q\mu}{\tau l}$$

$$+ \frac{a\omega_1 l}{l^2 + \omega_1^2} \cos\lambda + \frac{al^2}{l^2 + \omega_1^2} \sin\lambda \qquad (8.6.9)$$

Solving equation (8.6.9) for the Laplace domain displacement $x(l)$, we may write:

$$x(l) = \frac{lv_0}{l^2 + \omega^2} + \frac{l^2 s_0}{l^2 + \omega^2} + \frac{q\rho - f}{l^2 + \omega^2} + \frac{q\mu}{\tau l(l^2 + \omega^2)}$$

$$+ \frac{la\omega_1}{(l^2 + \omega^2)(l^2 + \omega_1^2)} \cos\lambda + \frac{l^2 a}{(l^2 + \omega^2)(l^2 + \omega_1^2)} \sin\lambda \qquad (8.6.10)$$

Based on pairs 1, 6, 7, 14, 16, 33, and 34 from Table 1.1, we invert equation (8.6.10) from the Laplace domain into the time domain. The inversion represents the solution of differential equation of motion (8.6.1) with the initial conditions of motion (8.6.2). Thus, we may write:

$$x = \frac{v_0}{\omega} \sin\omega t + s_0 \cos\omega t + \frac{q\rho - f}{\omega^2}(1 - \cos\omega t) + \frac{q\mu}{\tau\omega^2}(t - \frac{1}{\omega}\sin\omega t)$$

$$+ \frac{a(\omega \sin\omega_1 t - \omega_1 \sin\omega t)\cos\lambda}{\omega(\omega^2 - \omega_1^2)} + \frac{a(\cos\omega_1 t - \cos\omega t)\sin\lambda}{\omega^2 - \omega_1^2} \qquad (8.6.11)$$

Taking the first and second derivatives of equation (8.6.11), we determine the velocity and the acceleration of the system respectively:

$$\frac{dx}{dt} = -s_0 \omega \sin \omega t + v_0 \cos \omega t + \frac{q\rho - f}{\omega} \sin \omega t + \frac{q\mu}{\tau \omega^2} - \frac{q\mu}{\tau \omega^2} \cos \omega t$$

$$+ \frac{a\omega_1(\cos \omega_1 t - \cos \omega t)\cos \lambda}{\omega^2 - \omega_1^2} - \frac{a(\omega_1 \sin \omega_1 t - \omega \sin \omega t)\sin \lambda}{\omega^2 - \omega_1^2}$$

$$\textbf{(8.6.12)}$$

$$\frac{d^2 x}{dt^2} = -s_0 \omega^2 \cos \omega t - v_0 \omega \sin \omega t + (q\rho - f)\cos \omega t + \frac{q\mu}{\tau \omega} \sin \omega t$$

$$- \frac{a\omega_1(\omega_1 \sin \omega_1 t - \omega \sin \omega t)\cos \lambda}{\omega^2 - \omega_1^2}$$

$$- \frac{a(\omega_1^2 \cos \omega_1 t - \omega^2 \cos \omega t)\sin \lambda}{\omega^2 - \omega_1^2}$$

$$\text{(8.6.13)}$$

8.6.2 Initial Displacement Equals Zero

The initial conditions of motion are:

$$\text{for } t = 0 \quad x = 0; \quad \frac{dx}{dt} = v_0 \qquad \text{(8.6.14)}$$

Solving differential equation of motion (8.6.1) with the initial conditions of motion (8.6.14), we have:

$$x = \frac{v_0}{\omega} \sin \omega t + \frac{q\rho - f}{\omega^2}(1 - \cos \omega t) + \frac{q\mu}{\tau \omega^2}(t - \frac{1}{\omega} \sin \omega t)$$

$$+ \frac{a(\omega \sin \omega_1 t - \omega_1 \sin \omega t)\cos \lambda}{\omega(\omega^2 - \omega_1^2)} + \frac{a(\cos \omega_1 t - \cos \omega t)\sin \lambda}{\omega^2 - \omega_1^2}$$

$$\textbf{(8.6.15)}$$

The first and second derivatives of equation (8.6.15) represent the velocity and the acceleration of the system respectively:

$$\frac{dx}{dt} = v_0 \cos \omega t + \frac{q\rho - f}{\omega} \sin \omega t + \frac{q\mu}{\tau \omega^2} - \frac{q\mu}{\tau \omega^2} \cos \omega t$$

$$+ \frac{a\omega_1(\cos \omega_1 t - \cos \omega t)\cos \lambda}{\omega^2 - \omega_1^2} - \frac{a(\omega_1 \sin \omega_1 t - \omega \sin \omega t)\sin \lambda}{\omega^2 - \omega_1^2}$$

$$\textbf{(8.6.16)}$$

$$\frac{d^2x}{dt^2} = -v_0\omega \sin \omega t + (q\rho - f)\cos \omega t + \frac{q\mu}{\tau\omega}\sin \omega t$$

$$-\frac{a\omega_1(\omega_1 \sin \omega_1 t - \omega \sin \omega t)\cos \lambda}{\omega^2 - \omega_1^2}$$

$$-\frac{a(\omega_1^2 \cos \omega_1 t - \omega^2 \cos \omega t)\sin \lambda}{\omega^2 - \omega_1^2} \qquad \textbf{(8.6.17)}$$

8.6.3 Initial Velocity Equals Zero

The initial conditions of motion are:

$$\text{for} \quad t = 0 \quad x = s_0; \quad \frac{dx}{dt} = 0 \qquad \textbf{(8.6.18)}$$

The solution of the differential equation of motion (8.6.1) with the initial conditions of motion (8.6.18) reads:

$$x = s_0 \cos \omega t + \frac{q\rho - f}{\omega^2}(1 - \cos \omega t) + \frac{q\mu}{\tau\omega^2}(t - \frac{1}{\omega}\sin \omega t)$$

$$+ \frac{a(\omega \sin \omega_1 t - \omega_1 \sin \omega t)\cos \lambda}{\omega(\omega^2 - \omega_1^2)} + \frac{a(\cos \omega_1 t - \cos \omega t)\sin \lambda}{\omega^2 - \omega_1^2}$$

$$\textbf{(8.6.19)}$$

Taking the first and second derivatives of equation (8.6.19), we determine the velocity and acceleration respectively:

$$\frac{dx}{dt} = -s_0\omega \sin \omega t + \frac{q\rho - f}{\omega}\sin \omega t + \frac{q\mu}{\tau\omega^2}\cos \omega t$$

$$+ \frac{a\omega_1(\cos \omega_1 t - \cos \omega t)\cos \lambda}{\omega^2 - \omega_1^2} - \frac{a(\omega_1 \sin \omega_1 t - \omega \sin \omega t)\sin \lambda}{\omega^2 - \omega_1^2}$$

$$\textbf{(8.6.20)}$$

$$\frac{d^2x}{dt^2} = -s_0\omega^2 \cos \omega t + (q\rho - f)\cos \omega t + \frac{q\mu}{\tau\omega}\sin \omega t$$

$$-\frac{a\omega_1(\omega_1 \sin \omega_1 t - \omega \sin \omega t)\cos \lambda}{\omega^2 - \omega_1^2}$$

$$-\frac{a(\omega_1^2 \cos \omega_1 t - \omega^2 \cos \omega t)\sin \lambda}{\omega^2 - \omega_1^2} \qquad \textbf{(8.6.21)}$$

8.6.4 Both the Displacement and Velocity Equal Zero

The initial conditions of motion are:

$$\text{for} \quad t = 0 \quad x = 0; \quad \frac{dx}{dt} = 0 \tag{8.6.22}$$

Solving differential equation of motion equation (8.6.1) with the initial conditions of motion (8.6.22), we obtain:

$$x = \frac{q\rho - f}{\omega^2}(1 - \cos\omega t) + \frac{q\mu}{\tau\omega^2}\left(t - \frac{1}{\omega}\sin\omega t\right)$$

$$+ \frac{a(\omega\sin\omega_1 t - \omega_1\sin\omega t)\cos\lambda}{\omega(\omega^2 - \omega_1^2)} + \frac{a(\cos\omega_1 t - \cos\omega t)\sin\lambda}{\omega^2 - \omega_1^2}$$

$$\tag{8.6.23}$$

Taking the first and second derivatives of equation (8.6.23), we determine the velocity and the acceleration of the system respectively:

$$\frac{dx}{dt} = \frac{q\rho - f}{\omega}\sin\omega t + \frac{q\mu}{\tau\omega^2} - \frac{q\mu}{\tau\omega^2}\cos\omega t$$

$$+ \frac{a\omega_1(\cos\omega_1 t - \cos\omega t)\cos\lambda}{\omega^2 - \omega_1^2} - \frac{a(\omega_1\sin\omega_1 t - \omega\sin\omega t)\sin\lambda}{\omega^2 - \omega_1^2}$$

$$\tag{8.6.24}$$

$$\frac{d^2x}{dt^2} = (q\rho - f)\cos\omega t + \frac{q\mu}{\tau\omega}\sin\omega t$$

$$- \frac{a\omega_1(\omega_1\sin\omega_1 t - \omega\sin\omega t)\cos\lambda}{\omega^2 - \omega_1^2}$$

$$- \frac{a(\omega_1^2\cos\omega_1 t - \omega^2\cos\omega t)\sin\lambda}{\omega^2 - \omega_1^2} \tag{8.6.25}$$

9

STIFFNESS AND CONSTANT
RESISTANCE

This chapter contains descriptions of engineering systems subjected to the action of the force of inertia, the stiffness force, and the constant resisting force as the resisting forces. These forces are marked with a plus sign (+) in Row 9 of Guiding Table 2.1. The left sides of the differential equations of motion of all the systems described in this chapter are identical, consisting of these three resisting forces. However, the right sides of the differential equations of motion have different active forces, as indicated in Columns 1 through 6. The individual sections in this chapter reflect the characteristics of the active forces involved in the particular problem.

The problems described in this chapter could be related to the working processes of engineering systems intended to interact with some elastoplastic or viscoelastoplastic media that simultaneously exert stiffness and constant resisting forces as their reaction to residual deformations with strain-strengthening. Sometimes the interaction process of the system with the media consists of loading and unloading stages that occur during the vibratory motion of the system. The first half of the vibratory cycle represents the loading stage characterized by the stiffness and constant resisting forces. During the unloading stage, the constant resisting force disappears and the differential equation of motion becomes invalid for this stage. More information regarding the deformation of these media and the behavior of a constant resisting force applied to a vibratory system are described in section 1.2.

9.1 Active Force Equals Zero

As indicated by Guiding Table 2.1, this section describes the problems associated with the force of inertia, the stiffness force, and the constant resisting force as the resisting forces (Row 9), but also when the active force equals zero (Column 1). (The principles associated with the motion of a system in the absence of active forces are discussed in section 1.3.)

The current problem could be associated with the interaction of a system with elastoplastic or viscoelastoplastic media featuring plastic deformation with strain-strengthening and exerting simultaneously a stiffness force and a constant resisting force. This problem could also be related to the interaction of a system with a specific elastic link. At the end of the deformation process, the system may begin to move backward due to the elasticity of the media or the elastic link performing a vibratory cycle. Additional information related to the deformation of the media mentioned above and to the behavior of a constant resisting force applied to a vibratory system is described in section 1.2.

The system is moving in the horizontal direction. We want to determine the basic parameters of motion, their extreme values, and the characteristics of the forces applied to the elastic link. Figure 9.1 shows the model of a system subjected to the action of a stiffness force and a constant resisting force.

The considerations above and the model in Figure 9.1 let us compose the left and right sides of the differential equation of motion. The left side consists of the force of inertia, the stiffness force,

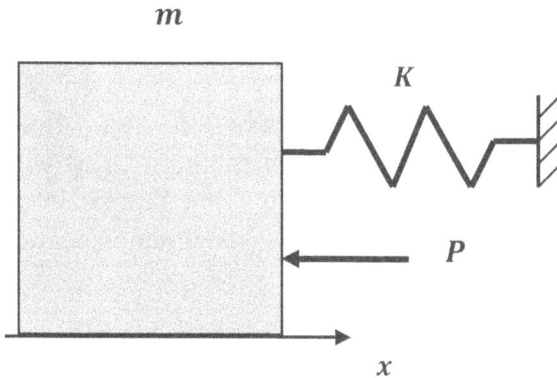

Figure 9.1 Model of a system subjected to a stiffness force and a constant resisting force

and the constant resisting force, while the right side equals zero. Hence, the differential equation of motion reads:

$$m\frac{d^2x}{dt^2} + Kx + P = 0 \qquad \text{(9.1.1)}$$

Differential equation of motion (9.1.1) has different solutions for various initial conditions of motion. These solutions and their analyses are presented below.

9.1.1 General Initial Conditions

The general initial conditions of motion are:

$$\text{for} \quad t = 0 \quad x = s_0; \quad \frac{dx}{dt} = v_0 \qquad \text{(9.1.2)}$$

where s_0 and v_0 are the initial displacement and initial velocity respectively.

Dividing equation (9.1.1) by m, we may write:

$$\frac{d^2x}{dt^2} + \omega^2 x + p = 0 \qquad \text{(9.1.3)}$$

where ω is the natural frequency of the vibratory system and:

$$\omega^2 = \frac{K}{m} \qquad \text{(9.1.4)}$$

$$p = \frac{P}{m} \qquad (9.1.5)$$

Using Laplace Transform Pairs 3, 1, and 5 from Table 1.1, we convert differential equation of motion (9.1.3) with the initial conditions of motion (9.1.2) from the time domain into the Laplace domain and obtain the corresponding algebraic equation of motion in the Laplace domain:

$$l^2 x(l) - lv_0 - l^2 s_0 + \omega^2 x(l) + p = 0 \qquad (9.1.6)$$

Solving equation (9.1.6) for the displacement $x(l)$ in the Laplace domain, we have:

$$x(l) = \frac{lv_0}{l^2 + \omega^2} + \frac{l^2 s_0}{l^2 + \omega^2} - \frac{p}{l^2 + \omega^2} \qquad (9.1.7)$$

In order to invert equation (9.1.7) from the Laplace domain into the time domain, we use pairs 1, 5, 6, 7, and 14 from Table 1.1. The inversion represents the solution of differential equation of motion (9.1.1) with the initial conditions of motion (9.1.2):

$$x = \frac{v_0}{\omega} \sin \omega t + s_0 \cos \omega t - \frac{p}{\omega^2}(1 - \cos \omega t) \qquad (9.1.8)$$

To simplify the analysis of equation (9.1.8), we apply some algebraic procedures to the coefficients at the trigonometric functions. Hence, we may write:

$$\frac{(s_0 \omega^2 + p)^2}{(s_0 \omega^2 + p)^2 + v_0^2 \omega^2} + \frac{v_0^2 \omega^2}{(s_0 \omega^2 + p)^2 + v_0^2 \omega^2} = 1$$

$$= (\sin \alpha)^2 + (\cos \alpha)^2 \qquad (9.1.9)$$

where:

$$\sin \alpha = \frac{s_0 \omega^2 + p}{\sqrt{(s_0 \omega^2 + p)^2 + v_0^2 \omega^2}} \qquad (9.1.10)$$

$$\cos \alpha = \frac{v_0 \omega}{\sqrt{(s_0 \omega^2 + p)^2 + v_0^2 \omega^2}} \qquad (9.1.11)$$

Combining equations (9.1.10) and (9.1.11) with equation (9.1.8), we have:

$$x = \frac{1}{\omega^2}[\sqrt{(s_0\omega^2 + p)^2 + v_0^2\omega^2}\,(\sin\omega t\cos\alpha + \cos\omega t\sin\alpha) - p]$$

(9.1.12)

Therefore, we may rewrite equation (9.1.12) in the following shape:

$$x = \frac{1}{\omega^2}[\sqrt{(s_0\omega^2 + p)^2 + v_0^2\omega^2}\,\sin(\omega t + \alpha) - p] \quad (9.1.13)$$

The first and second derivatives of equation (9.1.13) represent the velocity and acceleration of the system respectively:

$$\frac{dx}{dt} = \frac{1}{\omega}\sqrt{(s_0\omega^2 + p)^2 + v_0^2\omega^2}\,\cos(\omega t + \alpha) \quad (9.1.14)$$

$$\frac{d^2x}{dt^2} = -\sqrt{(s_0\omega^2 + p)^2 + v_0^2\omega^2}\,\sin(\omega t + \alpha) \quad (9.1.15)$$

According to equation (9.1.14), the velocity becomes equal zero at the moment of time when:

$$\cos\omega t = 0 \quad (9.1.16)$$

and, consequently:

$$\sin(\omega t + \alpha) = \pm 1 \quad (9.1.17)$$

Combining equations (9.1.13) and (9.1.15) with equation (9.1.17), we determine the extreme values of the displacement s_{ext} and the acceleration a_{ext} respectively:

$$s_{ext} = \pm\frac{1}{\omega^2}[\sqrt{(s_0\omega^2 + p)^2 + v_0^2\omega^2} - p] \quad (9.1.18)$$

$$a_{ext} = \pm\sqrt{(s_0\omega^2 + p)^2 + v_0^2\omega^2} \quad (9.1.19)$$

Multiplying equation (9.1.18) by K and substituting notations (9.1.4) and (9.1.5), we calculate the extreme values of the forces applied to the elastic link R_{ext}:

$$R_{ext} = \pm[\sqrt{(Ks_0 + P)^2 + mKv_0^2} - p] \quad (9.1.20)$$

According to equation (9.1.20), the link is subjected to a completely reversed loading cycle where the maximum and minimum forces are:

$$R_{max} = \sqrt{(Ks_0 + P)^2 + mKv_0^2} - P \qquad (9.1.21)$$

$$R_{min} = -R_{max}$$

The stress calculations of the link should include fatigue considerations.

9.1.2 Initial Displacement Equals Zero

The initial conditions of motion are:

$$\text{for} \quad t = 0 \quad x = 0; \quad \frac{dx}{dt} = v_0 \qquad (9.1.22)$$

The solution of differential equation of motion (9.1.1) with the initial conditions of motion (9.1.22) reads:

$$x = \frac{v_0}{\omega} \sin \omega t - \frac{p}{\omega^2}(1 - \cos \omega t) \qquad (9.1.23)$$

Applying to equation (9.1.23) the same procedures as in the previous case, we write:

$$x = \frac{\sqrt{v_0^2 \omega^2 + p^2}}{\omega^2}(\sin \omega t \cos \alpha_1 + \cos \omega t \sin \alpha_1) - \frac{p}{\omega^2} \qquad (9.1.24)$$

where

$$\sin \alpha_1 = \frac{p}{\sqrt{v_0^2 \omega^2 + p^2}} \qquad (9.1.25)$$

$$\cos \alpha_1 = \frac{v_0 \omega}{\sqrt{v_0^2 \omega^2 + p^2}} \qquad (9.1.26)$$

Equation (9.1.24) may be transformed into the following shape:

$$x = \frac{1}{\omega^2}[\sqrt{v_0^2 \omega^2 + p^2} \sin(\omega t + \alpha_1) - p] \qquad (9.1.27)$$

The first derivative of equation (9.1.27) represents the velocity of the system:

$$\frac{dx}{dt} = \frac{\sqrt{v_0^2\omega^2 + p^2}}{\omega}\cos(\omega t + \alpha_1) \qquad (9.1.28)$$

Taking the second derivative of equation (9.1.27), we determine the acceleration:

$$\frac{d^2x}{dt^2} = -\sqrt{v_0^2\omega^2 + p^2}\,\sin(\omega t + \alpha_1) \qquad (9.1.29)$$

Equation (9.1.28) shows that the velocity becomes equal to zero at the moment when:

$$\cos(\omega t + \alpha_1) = 0 \qquad (9.1.30)$$

and, consequently, at this moment of time we have:

$$\sin(\omega t + \alpha_1) = \pm 1 \qquad (9.1.31)$$

Combining equations (9.1.27) and (9.1.29) with equation (9.1.31), we calculate for this case the extreme values of the displacement and acceleration respectively:

$$S_{ext} = \frac{1}{\omega^2}[\sqrt{v_0^2\omega^2 + p^2} - p] \qquad (9.1.32)$$

$$a_{ext} = \pm[-\sqrt{v_0^2\omega^2 + p^2}] \qquad (9.1.33)$$

Multiplying equation (9.1.32) by K and substituting notations (9.1.4) and (9.1.5), we determine the values of the forces R_{ext} applied to the elastic link:

$$R_{ext} = \pm[\sqrt{mKv_0^2 + P^2} - P] \qquad (9.1.34)$$

The link is subjected to a completely reversed loading cycle having:

$$R_{max} = \sqrt{mKv_0^2 + P^2} - P \qquad (9.1.35)$$

$$R_{min} = -R_{max}$$

The stress analysis of the link should be based on fatigue calculations.

9.1.3 Initial Velocity Equals Zero

The model in Figure 9.1 shows that the system moves in the positive direction. This implies that in this case the initial displacement is negative. Therefore, the initial conditions of motion are:

$$\text{for} \quad t = 0 \quad x = -s_0; \quad \frac{dx}{dt} = 0 \qquad (9.1.36)$$

Consequently, in equation (9.1.6) the term $l^2 s_0$ should be taken with the positive sign. The solution of differential equation of motion (9.1.1) with the initial conditions of motion (9.1.36) reads:

$$x = -s_0 \cos \omega t - \frac{p}{\omega^2}(1 - \cos \omega t) \qquad (9.1.37)$$

Applying basic algebra to equation (9.1.37), we may write:

$$x = \frac{1}{\omega^2}[(-s_0\omega^2 + p)\cos \omega t - p] \qquad (9.1.38)$$

The first and second derivatives of the equation (9.1.38) represent the velocity and the acceleration of the system respectively:

$$\frac{dx}{dt} = \frac{1}{\omega}(s_0\omega^2 - p)\sin \omega t \qquad (9.1.39)$$

$$\frac{d^2x}{dt^2} = (s_0\omega^2 - p)\cos \omega t \qquad (9.1.40)$$

According to equation (9.1.39), the velocity becomes equal to zero at the moment when $\sin \omega t = 0$. For this moment of time, we take:

$$\cos \omega t = \pm 1 \qquad (9.1.41)$$

Combining equation (9.1.41) with equations (9.1.38) and (9.1.40), we determine the extreme values of the displacement S_{ext} and acceleration a_{ext} respectively:

$$S_{ext} = \frac{1}{\omega^2}[(-s_0\omega^2 + p)(\pm 1) - p] \qquad (9.1.42)$$

$$a_{ext} = \pm(s_0\omega^2 - p) \qquad (9.1.43)$$

Multiplying equation (9.1.42) by K and substituting notations (9.1.4) and (9.1.5), we calculate the maximum and minimum values of the force applied to the elastic link:

$$R_{max} = Ks_0 - 2P \qquad (9.1.44)$$

$$R_{min} = -Ks_0 \qquad (9.1.45)$$

The link is subjected to a random loading cycle, and the stress analysis should include fatigue considerations.

9.2 Constant Force R

This section represents the intersection of Row 9 and Column 2 in Guiding Table 2.1. As such, it describes engineering systems subjected to the action of the force of inertia, stiffness force, and a constant resisting force as the resisting forces, and to a constant active force.

The current problem could be related to the working process of systems intended for the interaction with elastoplastic or viscoelastoplastic media characterized by plastic deformations with strain-strengthening (see section 1.2 for more information regarding the deformation of the media and the behavior of a constant resisting force applied to a vibratory system). This problem could be also associated with a system that interacts with a certain specific elastic link. Due to the elasticity of the media or the elastic link, the system may perform a cycle of vibratory motion.

The system is moving in the horizontal direction. We want to determine the basic parameters of motion, their extreme values, and the characteristics of the forces applied to the elastic link. Figure 9.2 shows the model of a system subjected to the action of a constant active force, a stiffness force, and a constant resisting force.

The considerations above and the model in Figure 9.2 let us compose the left and the right sides of the differential equation of motion of this system. The left side consists of the force of inertia, the stiffness force, and the constant resisting force. The right side consists of the constant active force. Therefore, the differential equation of motion reads:

$$m\frac{d^2x}{dt^2} + Kx + P = R \qquad (9.2.1)$$

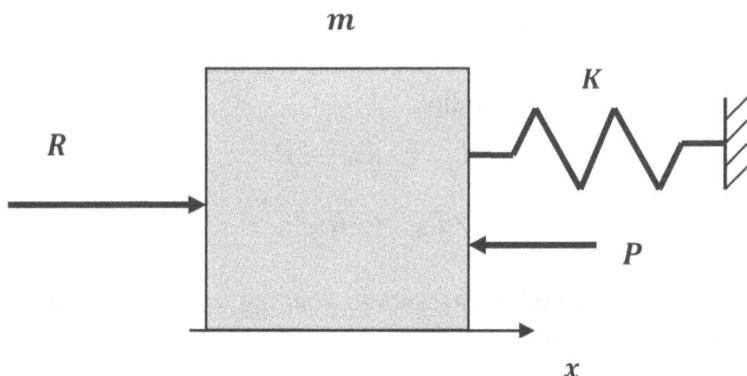

Figure 9.2 Model of a system subjected to a constant active force, a stiffness force, and a constant resisting force

Differential equation of motion (9.2.1) has different solutions for various initial conditions of motion. These solutions and their analyses are presented below.

9.2.1 General Initial Conditions

The general initial conditions of motion are:

$$\text{for} \quad t = 0 \quad x = s_0; \quad \frac{dx}{dt} = v_0 \qquad \textbf{(9.2.2)}$$

where s_0 and v_0 are the initial displacement and initial velocity respectively.

Dividing equation (9.2.1) by m, we have:

$$\frac{d^2x}{dt^2} + \omega^2 x + p = r \qquad \textbf{(9.2.3)}$$

where ω is the natural frequency of the system and:

$$\omega^2 = \frac{K}{m} \qquad \textbf{(9.2.4)}$$

$$p = \frac{P}{m} \qquad \textbf{(9.2.5)}$$

$$r = \frac{R}{m} \qquad \textbf{(9.2.6)}$$

Using Laplace Transform Pairs 3, 1, and 5 from Table 1.1, we convert equation (9.2.3) with the initial conditions of motion (9.2.2) from the time domain into the Laplace domain. The resulting algebraic equation of motion in the Laplace domain reads:

$$l^2 x(l) - lv_0 - l^2 s_0 + \omega^2 x(l) + p = r \qquad (9.2.7)$$

The solution of equation (9.2.7) for the displacement $x(l)$ in Laplace domain reads:

$$x(l) = \frac{r - p}{l^2 + \omega^2} + \frac{lv_0}{l^2 + \omega^2} + \frac{l^2 s_0}{l^2 + \omega^2} \qquad (9.2.8)$$

Using pairs 1, 5, 14, 6, and 7 from Table 1.1, we invert equation (9.2.8) from the Laplace domain into the time domain and obtain the solution of differential equation of motion (9.2.1) with the initial conditions of motion (9.2.2):

$$x = \frac{r - p}{\omega^2}(1 - \cos \omega t) + \frac{v_0}{\omega}\sin \omega t + s_0 \cos \omega t \qquad (9.2.9)$$

Applying conventional algebraic procedures to equation (9.2.9), we have:

$$x = \frac{r - p}{\omega^2} + \frac{1}{\omega^2}[v_0 \omega \sin \omega t - (r - p - s_0 \omega^2)\cos \omega t] \qquad (9.2.10)$$

Performing some appropriate algebraic actions with the coefficients of the trigonometric functions, we compose the following expression:

$$\frac{(r - p - s_0 \omega^2)^2}{(r - p - s_0 \omega^2)^2 + v_0^2 \omega^2} + \frac{v_0^2 \omega^2}{(r - p - s_0 \omega^2)^2 + v_0^2 \omega^2}$$
$$= 1 = (\sin \beta)^2 + (\cos \beta)^2 \qquad (9.2.11)$$

where:

$$\sin \beta = -\frac{r - p - s_0 \omega^2}{\sqrt{(r - p - s_0 \omega^2)^2 + v_0^2 \omega^2}} \qquad (9.2.12)$$

$$\cos \beta = \frac{v_0 \omega}{\sqrt{(r - p - s_0 \omega^2)^2 + v_0^2 \omega^2}} \qquad (9.2.13)$$

Combining equations (9.2.12) and (9.2.13) with equation (9.2.10), we write:

$$x = \frac{r-p}{\omega^2} + \frac{1}{\omega^2}\sqrt{(r-p-s_0\omega^2)^2 + v_0^2\omega^2}\,\sin(\omega t - \beta)] \qquad \textbf{(9.2.14)}$$

The first and second derivatives of equation (9.2.14) represent the velocity and the acceleration of the system respectively:

$$\frac{dx}{dt} = \frac{1}{\omega}\sqrt{(r-p-s_0\omega^2)^2 + v_0^2\omega^2}\,\cos(\omega t - \beta) \qquad \textbf{(9.2.15)}$$

$$\frac{d^2x}{dt^2} = -\sqrt{(r-p-s_0\omega^2)^2 + v_0^2\omega^2}\,\sin(\omega t - \beta) \qquad \textbf{(9.2.16)}$$

Equation (9.2.15) shows that the velocity becomes equal to zero at the moment when:

$$\cos(\omega t - \beta) = 0 \qquad \textbf{(9.2.17)}$$

and, consequently, at this moment we have:

$$\sin(\omega t - \beta) = 1 \qquad \textbf{(9.2.18)}$$

Combining equations (9.2.4 and (9.2.16) with equation (9.2.18), we determine the extreme values of the displacement s_{max} and acceleration a_{max} respectively:

$$s_{max} = \frac{r-p}{\omega^2} + \frac{1}{\omega^2}\sqrt{(r-p-s_0\omega^2)^2 + v_0^2\omega^2} \qquad \textbf{(9.2.19)}$$

$$a_{max} = -\sqrt{(r-p-s_0\omega^2)^2 + v_0^2\omega^2} \qquad \textbf{(9.2.20)}$$

Multiplying equation (9.2.29) by K and substituting notations (9.2.4), (9.2.5), and (9.2.6), we calculate the maximum value of the force R_{max} applied to the elastic link:

$$R_{max} = R - P + \sqrt{(R - P - Ks_0)^2 + mKv_0^2} \qquad \textbf{(9.2.21)}$$

It is reasonable to consider that the system is subjected to a completely reversed loading cycle having the minimum force R_{min}:

$$R_{min} = -R_{max} \qquad (9.2.22)$$

The stress analysis of the link should include fatigue considerations.

9.2.2 Initial Displacement Equals Zero

$$\text{for} \quad t = 0 \quad x = 0; \quad \frac{dx}{dt} = v_0 \qquad (9.2.23)$$

Solving differential equation of motion (9.2.1) with the initial conditions of motion (9.2.23), we may write:

$$x = \frac{r - p}{\omega^2}(1 - \cos \omega t) + \frac{v_0}{\omega} \sin \omega t \qquad (9.2.24)$$

Applying the appropriate transformations to equation (9.2.24), we have:

$$x = \frac{r - p}{\omega^2} + \frac{1}{\omega^2}[v_0 \omega \sin \omega t - (r - p)\cos \omega t] \qquad (9.2.25)$$

Based on similar procedures as in the previous case, we present equation (9.2.25) in the following shape:

$$x = \frac{1}{\omega^2}[r + \sqrt{(r - p)^2 + v_0^2 \omega^2} \, \sin(\omega t - \beta_1)] \qquad (9.2.26)$$

where:

$$\sin \beta_1 = -\frac{r - p}{\sqrt{(r - p)^2 + v_0^2 \omega^2}} \qquad (9.2.27)$$

$$\cos \beta_1 = \frac{v_0 \omega}{\sqrt{(r - p)^2 + v_0^2 \omega^2}} \qquad (9.2.28)$$

Taking the first and second derivatives of equation (9.2.26), we determine the velocity and acceleration respectively:

$$\frac{dx}{dt} = \frac{1}{\omega}\sqrt{(r - p)^2 + v_0^2 \omega^2} \, \cos(\omega t - \beta_1) \qquad (9.2.29)$$

$$\frac{d^2x}{dt^2} = -\sqrt{(r - p)^2 + v_0^2 \omega^2} \, \sin(\omega t - \beta_1) \qquad (9.2.30)$$

Equation (9.2.29) shows that at the moment when $\cos(\omega t - \beta_1) = 0$, the velocity equals zero. Consequently, at this moment, we have:

$$\sin(\omega t - \beta_1) = 1 \qquad (9.2.31)$$

Combining equation (9.2.31) with the equations (9.2.26) and (9.2.30), we determine the maximum values of the displacement s_{max} and the acceleration a_{max} respectively:

$$s_{max} = \frac{1}{\omega^2}[r + \sqrt{(r-p)^2 + v_0^2 \omega^2}] \qquad (9.2.32)$$

$$a_{max} = -\sqrt{(r-p)^2 + v_0^2} \qquad (9.2.33)$$

Multiplying equation (9.2.32) by K and substituting notations (9.2.4), (9.2.5) and (9.2.6), we calculate the maximum value of the force R_{max} applied to the elastic link:

$$R_{max} = R + \sqrt{(R-P)^2 + mKv_0^2} \qquad (9.2.34)$$

As in the previous case, it is accepted that the link is subjected to a completely reversed loading cycle having:

$$R_{min} = -R_{max}$$

The stress analysis of the link should include fatigue calculations.

9.2.3 Initial Velocity Equals Zero

The initial conditions of motion are:

$$\text{for} \quad t = 0 \quad x = s_0; \quad \frac{dx}{dt} = 0 \qquad (9.2.35)$$

The solution of differential equation of motion (9.2.1) with the initial conditions of motion (9.2.35) reads:

$$x = \frac{r-p}{\omega^2} - \frac{1}{\omega^2}(r - p - s_0\omega^2)\cos\omega t \qquad (9.2.36)$$

The first and second derivatives of equation (9.2.36) represent the velocity and acceleration respectively:

$$\frac{dx}{dt} = \frac{1}{\omega}(r - p - s_0\omega^2)\sin\omega t \qquad (9.2.37)$$

$$\frac{d^2x}{dt^2} = (r - p - s_0\omega^2)\cos\omega t \qquad (9.2.38)$$

Equation (9.2.37) shows that the velocity becomes equal to zero at the moment when $\sin\omega t = 0$, and, consequently, at this moment of time we have:

$$\cos\omega t = -1 \qquad (9.2.39)$$

Combining equation (9.2.39) with equations (9.2.36) and (9.2.38), we obtain the maximum values of the displacement s_{max} and acceleration a_{max} respectively:

$$s_{max} = \frac{2(r - p)}{\omega^2} - s_0 \qquad (9.2.40)$$

$$a_{max} = -(r - p - s_0\omega^2) \qquad (9.2.41)$$

Multiplying equation (9.2.40) by K and substituting notations (9.2.4), (9.2.5) and (9.2.6), we calculate the maximum value of the force R_{max} applied to the elastic link:

$$R_{max} = 2(R - P) - Ks_0 \qquad (9.2.42)$$

It is accepted that the stress analysis of the link is based on a completely reversed loading cycle having:

$$R_{max} = -R_{min}$$

The stress calculations of the link should include fatigue considerations.

9.2.4 Both the Initial Displacement and Velocity Equal Zero

The initial conditions of motion are:

$$\text{for} \quad t = 0 \quad x = 0; \quad \frac{dx}{dt} = 0 \tag{9.2.43}$$

Solving differential equation of motion (9.2.1) with the initial conditions of motion (9.2.43), we may write:

$$x = \frac{r - p}{\omega^2} (1 - \cos \omega t) \tag{9.2.44}$$

The first derivative of equation (9.2.44) represents the velocity:

$$\frac{dx}{dt} = \frac{r - p}{\omega} \sin \omega t \tag{9.2.45}$$

Taking the second derivative of equation (9.2. 44), we determine the acceleration:

$$\frac{d^2 x}{dt^2} = (r - p) \cos \omega t \tag{9.2.46}$$

According to equation (9.2.45), the velocity becomes equal to zero when $\sin \omega t = 0$. In this case, we have:

$$\cos \omega t = -1 \tag{9.2.47}$$

Combining equation (9.2.47) with equations (9.2.44) and (9.2.46), we obtain the maximum values of the displacement s_{max} and acceleration a_{max} respectively:

$$s_{max} = \frac{2(r - p)}{\omega^2} \tag{9.2.48}$$

$$a_{max} = -(r - p) \tag{9.2.49}$$

Multiplying equation (9.2.48) by K and substituting notations (9.2.4), (9.2.5) and (9.2.6), we calculate the maximum value of the force R_{max} applied to the elastic link:

$$R_{max} = 2(R - P) \tag{9.2.50}$$

For the stress calculations of the link, it is justifiable to assume that the system is subjected to a completely reversed loading cycle for which the minimum force is:

$$R_{min} = -R_{max}$$

The stress analysis of the link should include fatigue considerations.

9.3   Harmonic Force $A \sin(\omega_1 t + \lambda)$

The intersection of Row 9 and Column 3 in Guiding Table 2.1 indicates that the engineering systems described in this section are subjected to the action of the force of inertia, the stiffness force, and the constant resisting force as the resisting forces, and to the harmonic force as the active force.

The current problem could be associated with vibratory systems intended for interaction with elastoplastic or viscoelastoplastic media that feature strain-strengthening during the phase of plastic deformation. These media simultaneously exert stiffness and constant resisting forces as a reaction to the deformation (see section 1.2 for more information regarding the deformation of the media and the behavior of a constant resisting force applied to a vibratory system).

The system is moving in the horizontal direction. We want to determine the basic parameters of motion. Figure 9.3 shows the model of a system subjected to the action of a harmonic force, stiffness force, and a constant resisting force.

Based on the considerations above and the model in Figure 9.3, we can assemble the left and right sides of the differential equation of motion. The left side consists of the force of inertia, the stiffness force, and the constant resisting force, while the right side consists of the harmonic force. Thus, the equation reads:

$$m \frac{d^2 x}{dt^2} + Kx + P = A \sin(\omega_1 t + \lambda) \tag{9.3.1}$$

Differential equation of motion (9.3.1) has different solutions for various initial conditions of motion. These solutions and their analyses are presented below.

9.3.1 General Initial Conditions

The general initial conditions of motion are:

$$\text{for} \quad t = 0 \quad x = s_0; \quad \frac{dx}{dt} = v_0 \qquad \textbf{(9.3.2)}$$

where s_0 and v_0 are the initial displacement and initial velocity respectively.

Transforming the sinusoidal function in equation (9.3.1) and dividing this equation by m, we have:

$$\frac{d^2x}{dt^2} + \omega^2 x + p = a \sin \omega_1 t \cos \lambda + a \cos \omega_1 t \sin \lambda \qquad \textbf{(9.3.3)}$$

where ω is the natural frequency of the system and:

$$\omega^2 = \frac{K}{m} \qquad \textbf{(9.3.4)}$$

$$p = \frac{P}{m} \qquad \textbf{(9.3.5)}$$

$$a = \frac{A}{m} \qquad \textbf{(9.3.6)}$$

Using Laplace Transform Pairs 3, 5, 1, 6, and 7 from Table 1.1, we convert differential equation of motion (9.3.3) with the initial

$$m$$

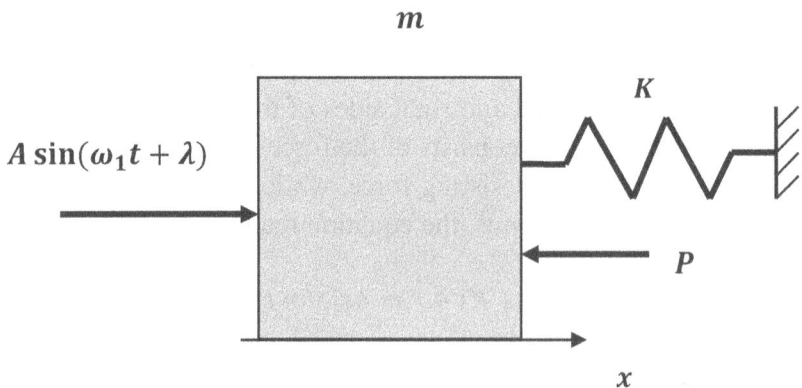

Figure 9.3 Model of a system subjected to a harmonic force, a stiffness force and a constant resisting force

conditions of motion (9.3.2) from the time domain into the Laplace domain. The resulting algebraic equation of motion in the Laplace domain reads:

$$l^2 x(l) - l v_0 - l^2 s_0 + \omega^2 x(l) + p = \frac{a \omega_1 l}{l^2 + \omega_1^2} \cos \lambda + \frac{a l^2}{l^2 + \omega_1^2} \sin \lambda$$

$$(9.3.7)$$

Applying appropriate algebraic actions to equation (9.3.7), we have:

$$x(l)(l^2 + \omega^2) = l v_0 + l^2 s_0 - p + \frac{a \omega_1 l}{l^2 + \omega_1^2} \cos \lambda + \frac{a l^2}{l^2 + \omega_1^2} \sin \lambda \quad (9.3.8)$$

Solving equation (9.3.8) for the Laplace domain displacement $x(l)$, we may write:

$$x(l) = \frac{l v_0}{l^2 + \omega^2} + \frac{l^2 s_0}{l^2 + \omega^2} - \frac{p}{l^2 + \omega^2} + \frac{l a \omega_1}{(l^2 + \omega^2)(l^2 + \omega_1^2)} \cos \lambda$$

$$+ \frac{l^2 a}{(l^2 + \omega^2)(l^2 + \omega_1^2)} \sin \lambda \quad (9.3.9)$$

Based on pairs 1, 5, 6, 7, 14, 33, and 34 from Table 1.1, we invert equation (9.3.9) from the Laplace domain into the time domain and obtain the solution of differential equation of motion (9.3.1) with the initial conditions of motion (9.3.2):

$$x = \frac{v_0}{\omega} \sin \omega t + s_0 \cos \omega t - \frac{p}{\omega^2}(1 - \cos \omega t)$$

$$+ \frac{a(\omega \sin \omega_1 t - \omega_1 \sin \omega t) \cos \lambda}{\omega(\omega^2 - \omega_1^2)} + \frac{a(\cos \omega_1 t - \cos \omega t) \sin \lambda}{\omega^2 - \omega_1^2} \quad (9.3.10)$$

The first derivative of equation (9.3.10) represents the velocity of the system:

$$\frac{dx}{dt} = v_0 \cos \omega t - \omega s_0 \sin \omega t - \frac{p}{\omega} \sin \omega t + \frac{a \omega_1 (\cos \omega_1 t - \cos \omega t) \cos \lambda}{\omega^2 - \omega_1^2}$$

$$- \frac{a(\omega_1 \sin \omega_1 t - \omega \sin \omega t) \sin \lambda}{\omega^2 - \omega_1^2} \quad (9.3.11)$$

Taking the second derivative of equation (9.3.10), we determine the acceleration of the system:

$$\frac{d^2x}{dt^2} = -\omega v_0 \sin\omega - \omega^2 s_0 \cos\omega t - p\cos\omega t$$

$$-\frac{a\omega_1(\omega_1 \sin\omega_1 t - \omega \sin\omega t)\cos\lambda}{\omega^2 - \omega_1^2}$$

$$-\frac{a(\omega_1^2 \cos\omega_1 t - \omega^2 \cos\omega t)\sin\lambda}{\omega^2 - \omega_1^2} \qquad (9.3.12)$$

9.3.2 Initial Displacement Equals Zero

The initial conditions of motion are:

$$\text{for} \quad t = 0 \quad x = 0; \quad \frac{dx}{dt} = v_0 \qquad (9.3.13)$$

The solution of differential equation of motion (9.3.1) with the initial conditions of motion (9.3.13) reads:

$$x = \frac{v_0}{\omega}\sin\omega t - \frac{p}{\omega^2}(1 - \cos\omega t) + \frac{a(\omega \sin\omega_1 t - \omega_1 \sin\omega t)\cos\lambda}{\omega(\omega^2 - \omega_1^2)}$$

$$+ \frac{a(\cos\omega_1 t - \cos\omega t)\sin\lambda}{\omega^2 - \omega_1^2} \qquad (9.3.14)$$

Taking the first and second derivatives of the equation (9.3.14), we determine the velocity and acceleration of the system respectively:

$$\frac{dx}{dt} = v_0 \cos\omega t - \frac{p}{\omega}\sin\omega t + \frac{a\omega_1(\cos\omega_1 t - \cos\omega t)\cos\lambda}{\omega^2 - \omega_1^2}$$

$$- \frac{a(\omega_1 \sin\omega_1 t - \omega \sin\omega t)\sin\lambda}{\omega^2 - \omega_1^2} \qquad (9.3.15)$$

$$\frac{d^2x}{dt^2} = -\omega v_0 \sin\omega t - p\cos\omega t - \frac{a\omega_1(\omega_1 \sin\omega_1 t - \omega \sin\omega t)\cos\lambda}{\omega^2 - \omega_1^2}$$

$$- \frac{a(\omega_1^2 \cos\omega_1 t - \omega^2 \cos\omega t)\sin\lambda}{\omega^2 - \omega_1^2} \qquad (9.1.16)$$

9.3.3 Initial Velocity Equals Zero

The initial conditions of motion are:

$$\text{for} \quad t = 0 \quad x = s_0; \quad \frac{dx}{dt} = 0 \tag{9.3.17}$$

Solving differential equation of motion (9.3.1) with the initial conditions of motion (9.3.17), we have:

$$x = s_0 \cos \omega t - \frac{p}{\omega^2}(1 - \cos \omega t) + \frac{a(\omega \sin \omega_1 t - \omega_1 \sin \omega t)\cos \lambda}{\omega(\omega^2 - \omega_1^2)}$$

$$+ \frac{a(\cos \omega_1 t - \cos \omega t)\sin \lambda}{\omega^2 - \omega_1^2} \tag{9.3.18}$$

The first derivative of the equation (9.3.18) represents the velocity:

$$\frac{dx}{dt} = -\omega s_0 \sin \omega t - \frac{p}{\omega}\sin \omega t + \frac{a\omega_1(\cos \omega_1 t - \cos \omega t)\cos \lambda}{\omega^2 - \omega_1^2}$$

$$- \frac{a(\omega_1 \sin \omega_1 t - \omega \sin \omega t)\sin \lambda}{\omega^2 - \omega_1^2} \tag{9.3.19}$$

Taking the second derivative of equation (9.3.18), we obtain the acceleration:

$$\frac{d^2x}{dt^2} = -\omega^2 s_0 \cos \omega t - p \sin \omega t - \frac{a\omega_1(\omega_1 \sin \omega_1 t - \omega \sin \omega t)\cos \lambda}{\omega^2 - \omega_1^2}$$

$$- \frac{a(\omega_1^2 \cos \omega_1 t - \omega^2 \cos \omega t)\sin \lambda}{\omega^2 - \omega_1^2} \tag{9.3.20}$$

9.3.4 Both the Initial Displacement and Velocity Equal Zero

The initial conditions of motion are:

$$\text{for} \quad t = 0 \quad x = 0; \quad \frac{dx}{dt} = 0 \tag{9.3.21}$$

The solution of differential equation of motion (9.3.1) with the initial conditions of motion (9.3.21) reads:

$$x = -\frac{p}{\omega^2}(1 - \cos \omega t) + \frac{a(\omega \sin \omega_1 t - \omega_1 \sin \omega t)\cos \lambda}{\omega(\omega^2 - \omega_1^2)}$$

$$+ \frac{a(\cos \omega_1 t - \cos \omega t)\sin \lambda}{\omega^2 - \omega_1^2} \tag{9.3.22}$$

The first derivative of equation (9.3.22) represents the velocity:

$$\frac{dx}{dt} = -\frac{p}{\omega}\sin\omega t + \frac{a\omega_1(\cos\omega_1 t - \cos\omega t)\cos\lambda}{\omega^2 - \omega_1^2}$$
$$- \frac{a(\omega_1\sin\omega_1 t - \omega\sin\omega t)\sin\lambda}{\omega^2 - \omega_1^2} \tag{9.3.23}$$

Taking the second derivative of equation (9.3.22), we determine the acceleration:

$$\frac{d^2x}{dt^2} = -p\cos\omega t - \frac{a\omega_1(\omega_1\sin\omega_1 t - \omega\sin\omega t)\cos\lambda}{\omega^2 - \omega_1^2}$$
$$- \frac{a(\omega_1^2\cos\omega_1 t - \omega^2\cos\omega t)\sin\lambda}{\omega^2 - \omega_1^2} \tag{9.3.24}$$

9.4 Time-Dependent Force $Q\left(\rho + \frac{\mu t}{\tau}\right)$

The intersection of Row 9 and Column 4 in Guiding Table 2.1 indicates that the engineering systems described in this section are subjected to the force of inertia, the stiffness force, and the constant resisting force as the resisting forces and the time-dependent force as the active force.

The current problem could be related to systems that interact with elastoplastic or viscoelastoplastic media characterized by plastic deformation with strain-strengthening. During this phase of deformation, the media simultaneously exerts stiffness forces and constant resisting forces as the reaction to their deformation (see section 1.2 for more information). This problem also could reflect interaction of a system with a specific elastic link. Sometimes during the initial stage of the working process, the system could be subjected to a time-dependent force that is acting for a predetermined interval of time.

The system is moving in the horizontal direction. We want to determine the basic parameters of motion, their values at the end of this interval of time, and the characteristic of forces applied to the elastic link. Figure 9.4 shows the model of a system subjected to the action of a time-dependent force, a stiffness force, and a constant resisting force.

Based on the considerations above and the model in Figure 9.4, we can compose the left and right sides of the differential equation of motion. Thus, the left side consists of the force of inertia, the stiffness force, and the constant resisting force. The right side consists of the time-dependent force. Therefore, the differential equation of motion reads:

$$m\frac{d^2x}{dt^2} + Kx + P = Q(\rho + \frac{\mu t}{\tau}) \qquad (9.4.1)$$

Differential equation of motion (9.4.1) has different solutions for various initial conditions of motion. These solutions and their analyses are presented below.

9.4.1 General Initial Conditions

The general initial conditions of motion are:

$$\text{for} \quad t = 0 \quad x = s_0; \quad \frac{dx}{dt} = v_0 \qquad (9.4.2)$$

where s_0 and v_0 are the initial displacement and initial velocity respectively.

Dividing equation (9.4.1) by m, we have:

$$\frac{d^2x}{dt^2} + \omega^2 x + p = q(\rho + \frac{\mu t}{\tau}) \qquad (9.4.3)$$

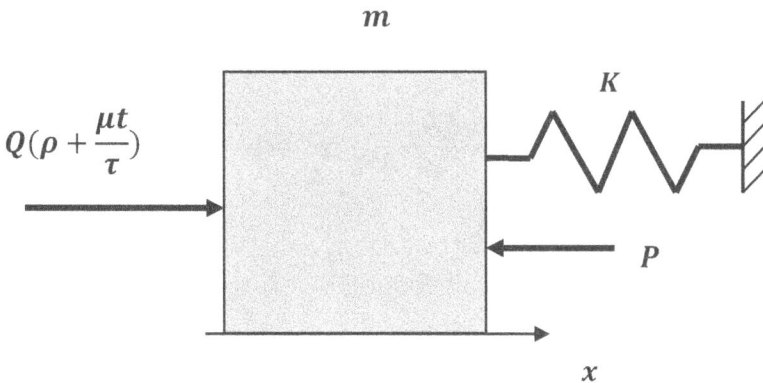

Figure 9.4 Model of a system subjected to a time-dependent force, a stiffness force, and a constant resisting force

where ω is the natural frequency of the system and:

$$\omega^2 = \frac{K}{m} \tag{9.4.4}$$

$$p = \frac{P}{m} \tag{9.4.5}$$

$$q = \frac{Q}{m} \tag{9.4.6}$$

Using Laplace Transform Pairs 3, 5, 1, and 2 from Table 1.1, we convert differential equation of motion (9.4.3) with the initial conditions of motion (9.4.2) from the time domain into the Laplace domain and obtain the corresponding algebraic equation of motion in the Laplace domain:

$$l^2 x(l) - lv_0 - l^2 s_0 + \omega^2 x(l) + p = qp + \frac{q\mu}{\tau l} \tag{9.4.7}$$

The solution of equation (9.4.7) for the Laplace domain displacement $x(l)$ reads:

$$x(l) = \frac{lv_0}{l^2 + \omega^2} + \frac{l^2 s_0}{l^2 + \omega^2} + \frac{qp - p}{l^2 + \omega^2} + \frac{q\mu}{\tau l(l^2 + \omega^2)} \tag{9.4.8}$$

Based on pairs 1, 6, 7, 14, and 16 from Table 1.1, we invert equation (9.4.8) from the Laplace domain into the time domain and obtain the solution of differential equation of motion (9.4.1) with the initial conditions of motion (9.4.2):

$$x = \frac{v_0}{\omega}\sin \omega t + s_0 \cos \omega t + \frac{qp - p}{\omega^2}(1 - \cos \omega t) + \frac{q\mu}{\tau \omega^2}(t - \frac{1}{\omega}\sin \omega t) \tag{9.4.9}$$

Applying the appropriate algebraic procedures to equation (9.4.9), we may write:

$$x = \frac{qp - p}{\omega^2} + \frac{q\mu t}{\tau \omega^2} + (s_0 - \frac{qp - p}{\omega^2})\cos \omega t + (\frac{v_0}{\omega} - \frac{q\mu}{\tau \omega^3})\sin \omega t \tag{9.4.10}$$

The first derivative of the equation (9.4.10) represents the velocity:

$$\frac{dx}{dt} = \frac{q\mu}{\tau\omega^2} - (s_0\omega - \frac{q\rho - p}{\omega})\sin\omega t + (v_0 - \frac{q\mu}{\tau\omega^2})\cos\omega t \quad \textbf{(9.4.11)}$$

Taking the second derivative of the equation (9.4.10), we determine the acceleration:

$$\frac{d^2x}{dt^2} = -(s_0\omega^2 - q\rho + p)\cos\omega t - (v_0\omega - \frac{q\mu}{\tau\omega})\sin\omega t \quad \textbf{(9.4.12)}$$

Substituting the time τ into equations (9.4.10), (9.4.11), and (9.4.12), we determine the values of the displacement s, the velocity v, and the acceleration a at the end of this interval of time respectively:

$$s = \frac{q(\rho + \mu) - p}{\omega^2} + (s_0 - \frac{q\rho - p}{\omega^2})\cos\omega\tau + (\frac{v_0}{\omega} - \frac{q\mu}{\tau\omega^3})\sin\omega\tau \quad \textbf{(9.4.13)}$$

$$v = \frac{q\mu}{\tau\omega^2} - (s_0\omega - \frac{q\rho - p}{\omega})\sin\omega\tau + (v_0 - \frac{q\mu}{\tau\omega^2})\cos\omega\tau \quad \textbf{(9.4.14)}$$

$$a = -(s_0\omega^2 - q\rho - p)\cos\omega\tau - (v_0\omega - \frac{q\mu}{\tau\omega})\sin\omega\tau \quad \textbf{(9.4.15)}$$

Multiplying equation (9.4.13) by the stiffness coefficient K and substituting notation (9.4.4), we calculate the maximum force R_{max} applied to the elastic link:

$$R_{max} = m[q(\rho + \mu) - p + (s_0\omega^2 - q\rho + p)\cos\omega\tau + (v_0\omega - \frac{q\mu}{\tau\omega})\sin\omega\tau] \quad \textbf{(9.4.16)}$$

The stress analysis of the link should be based on the force according to equation (9.4.16).

9.4.2 Initial Displacement Equals Zero
The initial conditions of motion are:

$$\text{for} \quad t = 0 \quad x = 0; \quad \frac{dx}{dt} = v_0 \quad \textbf{(9.4.17)}$$

The solution of differential equation of motion (9.4.1) with the initial conditions of motion (9.4.17) reads:

$$x = \frac{v_0}{\omega}\sin\omega t + \frac{q\rho - p}{\omega^2}(1 - \cos\omega t) + \frac{q\mu}{\tau\omega^2}(t - \frac{1}{\omega}\sin\omega t) \quad \textbf{(9.4.18)}$$

Taking the first derivative of the equation (9.4.18), we determine the velocity:

$$\frac{dx}{dt} = \frac{q\mu}{\tau\omega^2} + \frac{q\rho - p}{\omega}\sin\omega t + (v_0 - \frac{q\mu}{\tau\omega^2})\cos\omega t \quad \textbf{(9.4.19)}$$

The second derivative of the equation (9.4.18) represents the acceleration:

$$\frac{d^2 x}{dt^2} = (q\rho - p)\cos\omega t - (v_0\omega - \frac{q\mu}{\tau\omega})\sin\omega t \quad \textbf{(9.4.20)}$$

Substituting the interval of time τ into equations (9.4.18), (9.4.19), and (9.4.20), we determine the values of the displacement s, the velocity v, and the acceleration a at the end of the predetermined time τ respectively:

$$s = \frac{q(\rho + \mu) - p}{\omega^2} - \frac{q\rho - p}{\omega^2}\cos\omega\tau + (\frac{v_0}{\omega} - \frac{q\mu}{\tau\omega^3})\sin\omega\tau \quad \textbf{(9.4.21)}$$

$$v = \frac{q\mu}{\tau\omega^2} + \frac{q\rho - p}{\omega}\sin\omega\tau + (v_0 - \frac{q\mu}{\tau\omega^2})\cos\omega\tau \quad \textbf{(9.4.22)}$$

$$a = (q\rho - p)\cos\omega\tau - (v_0\omega - \frac{q\mu}{\tau\omega})\sin\omega\tau \quad \textbf{(9.4.23)}$$

Multiplying equation (9.4.21) by K and substituting notation (9.4.4), we calculate the maximum force applied to the elastic link:

$$R_{max} = m[q(\rho + \mu) - p - (q\rho - p)\cos\omega\tau + (v_0\omega - \frac{q\mu}{\tau\omega})\sin\omega\tau]$$

$$\textbf{(9.4.24)}$$

The stress analysis of the link should be based on the force according to equation (9.4.24).

9.4.3 Initial Velocity Equals Zero

The initial conditions of motion are:

$$\text{for} \quad t = 0 \quad x = s_0; \quad \frac{dx}{dt} = 0 \quad \text{(9.4.25)}$$

The solution of differential equation of motion (9.4.1) with the initial conditions of motion (9.4.25) reads:

$$x = \frac{q\rho - p}{\omega^2} + \frac{q\mu t}{\tau\omega^2} + (s_0 - \frac{q\rho - p}{\omega^2})\cos\omega t - \frac{q\mu}{\tau\omega^3}\sin\omega t \quad \text{(9.4.26)}$$

Taking the first and second derivatives of equation (9.4.26), we determine respectively the velocity and the acceleration:

$$\frac{dx}{dt} = \frac{q\mu}{\tau\omega^2} - (s_0\omega - \frac{q\rho - p}{\omega})\sin\omega t - \frac{q\mu}{\tau\omega^2}\cos\omega t \quad \text{(9.4.27)}$$

$$\frac{d^2x}{dt^2} = -(s_0\omega^2 - q\rho + p)\cos\omega t + \frac{q\mu}{\tau\omega}\sin\omega t \quad \text{(9.4.28)}$$

Substituting the time τ into equations (9.4.26), (9.4.27), and (9.4.28), we determine the values of the displacement s, the velocity v, and the acceleration a at the end of this interval of time respectively:

$$s = \frac{q(\rho + \mu) - p}{\omega^2} + (s_0 - \frac{q\rho - p}{\omega^2})\cos\omega\tau - \frac{q\mu}{\tau\omega^3}\sin\omega\tau \quad \text{(9.4.29)}$$

$$v = \frac{q\mu}{\omega^2} - (s_0\omega - q\rho + p)\sin\omega\tau - \frac{q\mu}{\tau\omega^2}\cos\omega\tau \quad \text{(9.4.30)}$$

$$a = -(s_0\omega^2 - q\rho + p)\cos\omega\tau + \frac{q\mu}{\tau\omega}\sin\omega\tau \quad \text{(9.4.31)}$$

Multiplying equation (9.4.29) by K and substituting notation (9.4.4), we determine the maximum force R_{max} applied to the elastic link:

$$R_{max} = m[(q(\rho + \mu) - p + (s_0\omega^2 - q\rho + p)\cos\omega\tau - \frac{q\mu}{\tau\omega}\sin\omega\tau]$$

$$\text{(9.4.32)}$$

The stress analysis of the link should be based on the force according to equation (9.4.32).

9.4.4 Both the Initial Displacement and Velocity Equal Zero

The initial conditions of motion are:

$$\text{for} \quad t = 0 \quad x = 0; \quad \frac{dx}{dt} = 0 \qquad (9.4.33)$$

Solving differential equation of motion (9.4.1) with the initial conditions of motion (9.4.33), we obtain:

$$x = \frac{q\rho - p}{\omega^2} + \frac{q\mu t}{\tau\omega^2} - \frac{q\rho - p}{\omega^2}\cos\omega t - \frac{q\mu}{\tau\omega^3}\sin\omega t \qquad (9.4.34)$$

The first and second derivatives of equation (9.4.34) represent the velocity and the acceleration respectively:

$$\frac{dx}{dt} = \frac{q\mu}{\tau\omega^2} + \frac{q\rho - p}{\omega}\sin\omega t - \frac{q\mu}{\tau\omega^2}\cos\omega t \qquad (9.4.35)$$

$$\frac{d^2x}{dt^2} = (q\rho - p)\cos\omega t + \frac{q\mu}{\tau\omega}\sin\omega t \qquad (9.4.36)$$

Substituting the time τ into equations (9.4.34), (9.4.35), and (9.4.36), we determine the values of the displacement s, the velocity v, and the acceleration a at the end of the predetermined time respectively:

$$s = \frac{q(\rho + \mu) - p}{\omega^2} - \frac{q\rho - p}{\omega^2}\cos\omega\tau - \frac{q\mu}{\tau\omega^3}\sin\omega\tau \qquad (9.4.37)$$

$$v = \frac{q\mu}{\omega^2} + \frac{q\rho - p}{\omega}\sin\omega\tau - \frac{q\mu}{\tau\omega^2}\cos\omega\tau \qquad (9.4.38)$$

$$a = (q\rho - p)\cos\omega\tau + \frac{q\mu}{\tau\omega}\sin\omega\tau \qquad (9.4.39)$$

By multiplying equation (9.4.37) by K and substituting notation (9.4.4), we calculate the maximum force R_{max} applied to the elastic link:

$$R_{max} = m[q(\rho + \mu) - p - (q\rho - p)\cos\omega\tau - \frac{q\mu}{\tau\omega}\sin\omega\tau]$$

The stress calculations should be performed based on the force according to equation (9.4.39).

9.5 Constant Force R and Harmonic Force $A \sin(\omega_1 t + \lambda)$

The intersection of Row 9 and Column 5 in Guiding Table 2.1 indicates that the engineering systems described in this section are subjected to the action of the force of inertia, the stiffness force, and the constant resisting force as the resisting forces, and to the constant active force and the harmonic force as the active forces.

The current problem could be related to the working process of a vibratory system that interacts with an elastoplastic or visco-elastoplastic medium exhibiting strain-strengthening on the phase of plastic deformation. During this phase of deformation, the medium simultaneously exerts a stiffness force and a constant resisting force (see section 1.2 for more information regarding the deformation of the media and the behavior of a constant resisting force applied to a vibratory system).

The system is moving in the horizontal direction. We want to determine the basic parameters of motion. Figure 9.5 shows the model of a system subjected to the action of a constant active force, a harmonic force, a stiffness force, and a constant resisting force.

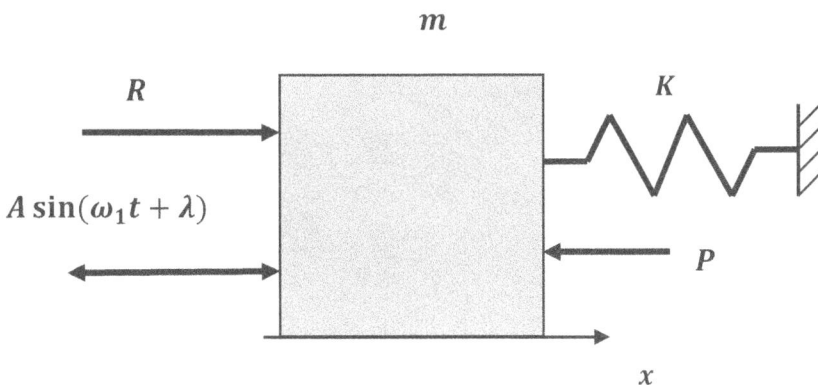

Figure 9.5 Model of a system subjected to a constant active force, a harmonic force, a stiffness force, and a constant resisting force

Based on the considerations above and the model shown in Figure 9.5, we can compose the left and right sides of the differential equation of motion of the system. The left side consists of the force of inertia, the stiffness force, and the constant resisting force, while the right side includes the constant active force and the harmonic force. Therefore, the differential equation of motion reads:

$$m\frac{d^2x}{dt^2} + Kx + P = R + A\sin(\omega_1 t + \lambda) \qquad (9.5.1)$$

Differential equation of motion (9.5.1) has different solutions for various initial conditions of motion. These solutions and their analyses are presented below.

9.5.1 General Initial Conditions

The general initial conditions of motion are:

$$\text{for} \quad t = 0 \quad x = s_0; \quad \frac{dx}{dt} = v_0 \qquad (9.5.2)$$

where s_0 and v_0 are the initial displacement and initial velocity respectively.

Transforming the sinusoidal function in the equation (9.5.1) and dividing this equation by m, we may write:

$$\frac{d^2x}{dt^2} + \omega^2 x + p = r + a\sin\omega_1 t\cos\lambda + a\cos\omega_1 t\sin\lambda \qquad (9.5.3)$$

where ω is the natural frequency of the system and:

$$\omega^2 = \frac{K}{m} \qquad (9.5.4)$$

$$p = \frac{P}{m} \qquad (9.5.5)$$

$$r = \frac{R}{m} \qquad (9.5.6)$$

$$a = \frac{A}{m} \qquad (9.5.7)$$

The conversion of differential equation of motion (9.5.3) with the initial conditions of motion (9.5.2) from the time domain into the Laplace domain is performed based on Laplace Transform Pairs 3, 5, 1, 6, and 7 from Table 1.1. The resulting algebraic equation of motion in the Laplace domain reads:

$$l^2 x(l) - l v_0 - l^2 s_0 + \omega^2 x(l) + p = r + \frac{a\omega_1 l}{l^2 + \omega_1^2} \cos \lambda + \frac{al^2}{l^2 + \omega_1^2} \sin \lambda$$
$$(9.5.8)$$

After transforming equation (9.5.8), we may write:

$$x(l)(l^2 + \omega^2) = r - p + l v_0 + l^2 s_0 + \frac{a\omega_1 l}{l^2 + \omega_1^2} \cos \lambda + \frac{al^2}{l^2 + \omega_1^2} \sin \lambda$$
$$(9.5.9)$$

Solving equation (9.5.9) for the Laplace domain displacement (l), we have:

$$x(l) = \frac{r - p}{l^2 + \omega^2} + \frac{l v_0}{l^2 + \omega^2} + \frac{l^2 s_0}{l^2 + \omega^2} + \frac{la\omega_1}{(l^2 + \omega^2)(l^2 + \omega_1^2)} \cos \lambda$$
$$+ \frac{l^2 a}{(l^2 + \omega^2)(l^2 + \omega_1^2)} \sin \lambda \qquad (9.5.10)$$

In order to invert equation (9.5.10) from the Laplace domain into the time domain we use pairs 1, 14, 5, 6, 7, 33, and 34 from Table 1.1. The inversion represents the solution of differential equation of motion (9.5.1) with the initial conditions of motion (9.5.2):

$$x = \frac{r - p}{\omega^2}(1 - \cos \omega t) + \frac{v_0}{\omega} \sin \omega t + s_0 \cos \omega t$$
$$+ \frac{a(\omega \sin \omega_1 t - \omega_1 \sin \omega t)\cos \lambda}{\omega(\omega^2 - \omega_1^2)} + \frac{a(\cos \omega_1 t - \cos \omega t)\sin \lambda}{\omega^2 - \omega_1^2}$$
$$(9.5.11)$$

Taking the first and second derivatives of equation (9.5.11), we determine the velocity and the acceleration of the system respectively:

$$\frac{dx}{dt} = \frac{r-p}{\omega}\sin\omega t + v_0\cos\omega t - \omega s_0\sin\omega t$$

$$+ \frac{a\omega_1(\cos\omega_1 t - \cos\omega t)\cos\lambda}{\omega^2 - \omega_1^2} - \frac{a(\omega_1\sin\omega_1 t - \omega\sin\omega t)\sin\lambda}{\omega^2 - \omega_1^2}$$

$$(9.5.12)$$

$$\frac{d^2x}{dt^2} = (r-p)\cos\omega t - \omega v_0\sin\omega t - \omega^2 s_0\cos\omega t$$

$$- \frac{a\omega_1(\omega_1\sin\omega_1 t - \omega\sin\omega t)\cos\lambda}{\omega^2 - \omega_1^2}$$

$$- \frac{a(\omega_1^2\cos\omega_1 t - \omega^2\cos\omega t)\sin\lambda}{\omega^2 - \omega_1^2}$$

$$(9.5.13)$$

9.5.2 Initial Displacement Equals Zero

The initial conditions of motion are:

$$\text{for} \quad t = 0 \quad x = 0; \quad \frac{dx}{dt} = v_0 \qquad (9.5.14)$$

Solving differential equation of motion (9.5.1) with the initial conditions of motion (9.5.14), we have:

$$x = \frac{r-p}{\omega^2}(1-\cos\omega t) + \frac{v_0}{\omega}\sin\omega t + \frac{a(\omega\sin\omega_1 t - \omega_1\sin\omega t)\cos\lambda}{\omega(\omega^2 - \omega_1^2)}$$

$$+ \frac{a(\cos\omega_1 t - \cos\omega t)\sin\lambda}{\omega^2 - \omega_1^2}$$

$$(9.5.15)$$

Taking the first derivative of equation (9.5.15), we determine the velocity:

$$\frac{dx}{dt} = \frac{r-p}{\omega}\sin\omega t + v_0\cos\omega t + \frac{a\omega_1(\cos\omega_1 t - \cos\omega t)\cos\lambda}{\omega^2 - \omega_1^2}$$

$$- \frac{a(\omega_1\sin\omega_1 t - \omega\sin\omega t)\sin\lambda}{\omega^2 - \omega_1^2}$$

$$(9.5.16)$$

The second derivative of equation (9.5.15) represents the acceleration:

$$\frac{d^2x}{dt^2} = (r-p)\cos\omega t - \omega v_0 \sin\omega t - \frac{a\omega_1(\omega_1 \sin\omega_1 t - \omega\sin\omega t)\cos\lambda}{\omega^2 - \omega_1^2}$$

$$-\frac{a(\omega_1^2 \cos\omega_1 t - \omega^2 \cos\omega t)\sin\lambda}{\omega^2 - \omega_1^2} \tag{9.5.17}$$

9.5.3 Initial Velocity Equals Zero

The initial conditions of motion are:

$$\text{for} \quad t = 0 \quad x = s_0; \quad \frac{dx}{dt} = 0 \tag{9.5.18}$$

The solution of differential equation of motion (9.5.1) with the initial conditions of motion (9.5.18) reads:

$$x = \frac{r-p}{\omega^2}(1-\cos\omega t) + s_0 \cos\omega t + \frac{a(\omega \sin\omega_1 t - \omega_1 \sin\omega t)\cos\lambda}{\omega(\omega^2 - \omega_1^2)}$$

$$+\frac{a(\cos\omega_1 t - \cos\omega t)\sin\lambda}{\omega^2 - \omega_1^2} \tag{9.5.19}$$

The first derivative of the equation (9.5.19) represents the velocity:

$$\frac{dx}{dt} = \frac{r-p}{\omega}\sin\omega t - \omega s_0 \sin\omega t + \frac{a\omega_1(\cos\omega_1 t - \cos\omega t)\cos\lambda}{\omega^2 - \omega_1^2}$$

$$-\frac{a(\omega_1 \sin\omega_1 t - \omega\sin\omega t)\sin\lambda}{\omega^2 - \omega_1^2} \tag{9.5.20}$$

Taking the second derivative of the equation (9.5.19), we determine the acceleration:

$$\frac{d^2x}{dt^2} = (r-p)\cos\omega t - \omega^2 s_0 \cos\omega t$$

$$-\frac{a\omega_1(\omega_1 \sin\omega_1 t - \omega\sin\omega t)\cos\lambda}{\omega^2 - \omega_1^2}$$

$$-\frac{a(\omega_1^2 \cos\omega_1 t - \omega^2 \cos\omega t)\sin\lambda}{\omega^2 - \omega_1^2} \tag{9.5.21}$$

9.5.4 Both the Initial Displacement and Velocity Equal Zero
The initial conditions of motion are:

$$\text{for} \quad t = 0 \quad x = 0; \quad \frac{dx}{dt} = 0 \qquad (9.5.22)$$

Solving differential equation of motion of (9.5.1) with the initial conditions of motion (9.5.22), we obtain:

$$x = \frac{r-p}{\omega^2}(1 - \cos\omega t) + \frac{a(\omega \sin\omega_1 t - \omega_1 \sin\omega t)\cos\lambda}{\omega(\omega^2 - \omega_1^2)}$$

$$+ \frac{a(\cos\omega_1 t - \cos\omega t)\sin\lambda}{\omega^2 - \omega_1^2} \qquad (9.5.23)$$

Taking the first and second derivatives of equation (9.5.23), we determine the velocity and the acceleration respectively:

$$\frac{dx}{dt} = \frac{r-p}{\omega}\sin\omega t + \frac{a\omega_1(\cos\omega_1 t - \cos\omega t)\cos\lambda}{\omega^2 - \omega_1^2}$$

$$- \frac{a(\omega_1 \sin\omega_1 t - \omega \sin\omega t)\sin\lambda}{\omega^2 - \omega_1^2} \qquad (9.5.24)$$

$$\frac{d^2 x}{dt^2} = (r - p)\cos\omega t - \frac{a\omega_1(\omega_1 \sin\omega_1 t - \omega \sin\omega t)\cos\lambda}{\omega^2 - \omega_1^2}$$

$$- \frac{a(\omega_1^2 \cos\omega_1 t - \omega^2 \cos\omega t)\sin\lambda}{\omega^2 - \omega_1^2} \qquad (9.5.25)$$

9.6 Harmonic Force $A \sin(\omega_1 t + \lambda)$ and Time-Dependent Force $Q\left(p + \frac{\mu t}{\tau}\right)$

The intersection of Row 9 and Column 11 of Guiding Table 2.1 indicates that this section describes engineering systems subjected to the action of the force of inertia, the stiffness force, and the constant resisting force as the resisting forces, and the harmonic force and the time-dependent force as the active forces..

The current problem could be associated with a vibratory system intended for interaction with elastoplastic or viscoelastoplastic media exhibiting strain-strengthening during the phase of plastic deformations (see section 1.2 for additional information regarding deformation of these media and for considerations related to the behavior of a constant resisting force applied to a vibratory system). This problem could also represent a system interacting with a specific elastic link.

In some situations, at the beginning of the vibratory working process, along with the harmonic force the system is also subjected to a time-dependent force that is acting a predetermined interval of time.

The system is moving in the horizontal direction. We want to determine the basic parameters of motion. Figure 9.6 shows the model of a system subjected to the action of a harmonic force, a time-dependent force, a stiffness force, and a constant resisting force.

Based on the considerations above and the model in Figure 9.6, we can compose the left and right sides of the differential equation of motion. The left side consists of the force of inertia, the stiffness force, and the constant resisting force. The right side includes the

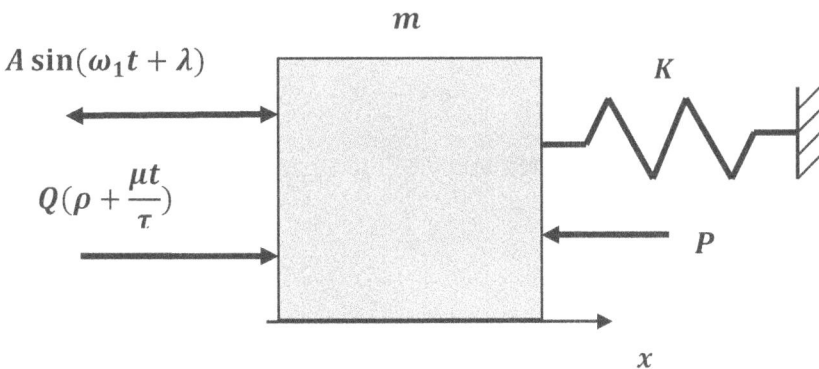

Figure 9.6 Model of a system subjected to a harmonic force, a time-dependent force, a stiffness force, and a constant resisting force

harmonic force and the time-dependent force. Hence, the differential equation of motion reads:

$$m\frac{d^2x}{dt^2} + Kx + P = A\sin(\omega_1 t + \lambda) + Q(\rho + \frac{\mu t}{\tau}) \qquad (9.6.1)$$

The differential equation of motion (9.6.1) has different solutions for various initial conditions of motion. These solutions and their analyses are presented below.

9.6.1 General Initial Conditions

The general initial conditions of motion are:

$$\text{for} \quad t = 0 \quad x = s_0; \quad \frac{dx}{dt} = v_0 \qquad (9.6.2)$$

where s_0 and v_0 are the initial displacement and initial velocity respectively.

Transforming the sinusoidal function in equation (9.6.1), we have:

$$m\frac{d^2x}{dt^2} + Kx + P = Q\rho + Q\frac{\mu t}{\tau} + A\sin\omega_1 t\cos\lambda + A\cos\omega_1 t\sin\lambda$$

$$(9.6.3)$$

Dividing equation (9.6.3) by m, we write:

$$\frac{d^2x}{dt^2} + \omega^2 x + p = q\rho + q\frac{\mu t}{\tau} + a\sin\omega_1 t\cos\lambda + a\cos\omega_1 t\sin\lambda \qquad (9.6.4)$$

where ω is the natural frequency of the system and:

$$\omega^2 = \frac{K}{m} \qquad (9.6.5)$$

$$p = \frac{P}{m} \qquad (9.6.6)$$

$$q = \frac{Q}{m} \qquad (9.6.7)$$

$$a = \frac{A}{m} \tag{9.6.8}$$

In order to convert differential equation of motion (9.6.4), we use Laplace Transform Pairs 3, 5, 1, 2, 6, and 7 from Table 1.1 and obtain the resulting algebraic equation of motion in the Laplace domain:

$$l^2 x(l) - l v_0 - l^2 s_0 + \omega^2 x(l) + p = q\rho + \frac{q\mu}{\tau l} + \frac{a\omega_1 l}{l^2 + \omega_1^2}\cos\lambda$$

$$+ \frac{a l^2}{l^2 + \omega_1^2}\sin\lambda \tag{9.6.9}$$

Applying basic algebra to equation (9.6.9), we have:

$$x(l)(l^2 + \omega^2) = l v_0 + l^2 s_0 - p + q\rho + \frac{q\mu}{\tau l} + \frac{a\omega_1 l}{l^2 + \omega_1^2}\cos\lambda$$

$$+ \frac{a l^2}{l^2 + \omega_1^2}\sin\lambda \tag{9.6.10}$$

Solving equation (9.6.10) for the Laplace domain displacement $x(l)$, we may write:

$$x(l) = \frac{l v_0}{l^2 + \omega^2} + \frac{l^2 s_0}{l^2 + \omega^2} + \frac{q\rho - p}{l^2 + \omega^2} + \frac{q\mu}{\tau l (l^2 + \omega^2)}$$

$$+ \frac{l a\omega_1}{(l^2 + \omega^2)(l^2 + \omega_1^2)}\cos\lambda + \frac{l^2 a}{(l^2 + \omega^2)(l^2 + \omega_1^2)}\sin\lambda \tag{9.6.11}$$

Using pairs 1, 6, 7, 14, 16, 33, and 34 from Table 1.1, we invert equation (9.8.11) from the Laplace domain into the time domain and obtain the solution of differential equation of motion (9.6.1) with the initial conditions of motion (9.6.2):

$$x = \frac{v_0}{\omega}\sin\omega t + s_0 \cos\omega t + \frac{q\rho - p}{\omega^2}(1 - \cos\omega t) + \frac{q\mu}{\tau\omega^2}\left(t - \frac{1}{\omega}\sin\omega t\right)$$

$$+ \frac{a(\omega\sin\omega_1 t - \omega_1 \sin\omega t)\cos\lambda}{\omega(\omega^2 - \omega_1^2)} + \frac{a(\cos\omega_1 t - \cos\omega t)\sin\lambda}{\omega^2 - \omega_1^2}$$

$$\tag{9.6.12}$$

Taking the first and second derivatives of equation (9.6.12), we determine the velocity and the acceleration of the system respectively:

$$\frac{dx}{dt} = v_0 \cos \omega t - s_0 \sin \omega t + \frac{q\rho - p}{\omega} \sin \omega t + \frac{q\mu}{\tau \omega^2} - \frac{q\mu}{\tau \omega^2} \cos \omega t$$

$$+ \frac{a\omega_1 (\cos \omega_1 t - \cos \omega t) \cos \lambda}{\omega^2 - \omega_1^2} - \frac{a(\omega_1 \sin \omega_1 t - \omega \sin \omega t) \sin \lambda}{\omega^2 - \omega_1^2}$$

$$\tag{9.6.13}$$

$$\frac{d^2 x}{dt^2} = -v_0 \omega \sin \omega t - s_0 \omega^2 \cos \omega t + (q\rho - p) \cos \omega t + \frac{q\mu}{\tau \omega} \sin \omega t$$

$$- \frac{a\omega_1 (\omega_1 \sin \omega_1 t - \omega \sin \omega t) \cos \lambda}{\omega^2 - \omega_1^2}$$

$$- \frac{a(\omega_1^2 \cos \omega_1 t - \omega^2 \cos \omega t) \sin \lambda}{\omega^2 - \omega_1^2}$$

$$\tag{9.6.14}$$

9.6.2 Initial Displacement Equals Zero
The initial conditions of motion are:

$$\text{for} \quad t = 0 \quad x = 0; \quad \frac{dx}{dt} = v_0 \tag{9.6.15}$$

Solving differential equation of motion (9.6.1) with the initial conditions of motion (9.6.15), we have:

$$x = \frac{v_0}{\omega} \sin \omega t + \frac{q\rho - p}{\omega^2} (1 - \cos \omega t) + \frac{q\mu}{\tau \omega^2} (t - \frac{1}{\omega} \sin \omega t)$$

$$+ \frac{a(\omega \sin \omega_1 t - \omega_1 \sin \omega t) \cos \lambda}{\omega(\omega^2 - \omega_1^2)} + \frac{a(\cos \omega_1 t - \cos \omega t) \sin \lambda}{\omega^2 - \omega_1^2}$$

$$\tag{9.6.16}$$

The first and second derivatives of equation (9.6.16) represent the velocity and the acceleration of the system respectively:

$$\frac{dx}{dt} = v_0 \cos \omega t + \frac{q\rho - p}{\omega} \sin \omega t + \frac{q\mu}{\tau \omega^2} - \frac{q\mu}{\tau \omega^2} \cos \omega t$$

$$+ \frac{a\omega_1 (\cos \omega_1 t - \cos \omega t) \cos \lambda}{\omega^2 - \omega_1^2} - \frac{a(\omega_1 \sin \omega_1 t - \omega \sin \omega t) \sin \lambda}{\omega^2 - \omega_1^2}$$

$$\tag{9.6.17}$$

$$\frac{d^2x}{dt^2} = -v_0\omega \sin\omega t + (q\rho - p)\cos\omega t + \frac{q\mu}{\tau\omega}\sin\omega t$$

$$-\frac{a\omega_1(\omega_1 \sin\omega_1 t - \omega \sin\omega t)\cos\lambda}{\omega^2 - \omega_1^2}$$

$$-\frac{a(\omega_1^2 \cos\omega_1 t - \omega^2 \cos\omega t)\sin\lambda}{\omega^2 - \omega_1^2} \qquad \textbf{(9.6.18)}$$

9.6.3 Initial Velocity Equals Zero

The initial conditions of motion are:

$$\text{for} \quad t = 0 \quad x = s_0; \quad \frac{dx}{dt} = 0 \qquad \textbf{(9.6.19)}$$

The solution of differential equation of motion (9.6.1) with the initial conditions of motion (9.6.19) reads:

$$x = s_0 \cos\omega t + \frac{q\rho - p}{\omega^2}(1 - \cos\omega t) + \frac{q\mu}{\tau\omega^2}\left(t - \frac{1}{\omega}\sin\omega t\right)$$

$$+\frac{a(\omega \sin\omega_1 t - \omega_1 \sin\omega t)\cos\lambda}{\omega(\omega^2 - \omega_1^2)} + \frac{a(\cos\omega_1 t - \cos\omega t)\sin\lambda}{\omega^2 - \omega_1^2}$$

$$\textbf{(9.6.20)}$$

Taking the first and second derivatives of equation (9.6.19), we determine the velocity and acceleration respectively:

$$\frac{dx}{dt} = -s_0\omega \sin\omega t + \frac{q\rho - p}{\omega}\sin\omega t + \frac{q\mu}{\tau\omega^2}\cos\omega t$$

$$+\frac{a\omega_1(\cos\omega_1 t - \cos\omega t)\cos\lambda}{\omega^2 - \omega_1^2} - \frac{a(\omega_1 \sin\omega_1 t - \omega \sin\omega t)\sin\lambda}{\omega^2 - \omega_1^2}$$

$$\textbf{(9.6.21)}$$

$$\frac{d^2x}{dt^2} = -s_0\omega^2 \cos\omega t + (q\rho - p)\cos\omega t + \frac{q\mu}{\tau\omega}\sin\omega t$$

$$-\frac{a\omega_1(\omega_1 \sin\omega_1 t - \omega \sin\omega t)\cos\lambda}{\omega^2 - \omega_1^2}$$

$$-\frac{a(\omega_1^2 \cos\omega_1 t - \omega^2 \cos\omega t)\sin\lambda}{\omega^2 - \omega_1^2} \qquad \textbf{(9.6.22)}$$

9.6.4 Both the Initial Displacement and Velocity Equal Zero

The initial conditions of motion are:

$$\text{for} \quad t = 0 \quad x = 0; \quad \frac{dx}{dt} = 0 \qquad \text{(9.6.23)}$$

Solving differential equation of motion (9.6.1) with the initial conditions of motion (9.6.23), we have:

$$x = \frac{q\rho - p}{\omega^2}(1 - \cos \omega t) + \frac{q\mu}{\tau\omega^2}(t - \frac{1}{\omega}\sin \omega t)$$

$$+ \frac{a(\omega \sin \omega_1 t - \omega_1 \sin \omega t)\cos \lambda}{\omega(\omega^2 - \omega_1^2)} + \frac{a(\cos \omega_1 t - \cos \omega t)\sin \lambda}{\omega^2 - \omega_1^2}$$

$$\text{(9.6.24)}$$

Taking the first and second derivatives of equation (9.6.23), we determine the velocity and the acceleration of the system respectively:

$$\frac{dx}{dt} = \frac{q\rho - p}{\omega}\sin \omega t + \frac{q\mu}{\tau\omega^2} - \frac{q\mu}{\tau\omega^2}\cos \omega t$$

$$+ \frac{a\omega_1(\cos \omega_1 t - \cos \omega t)\cos \lambda}{\omega^2 - \omega_1^2} - \frac{a(\omega_1 \sin \omega_1 t - \omega \sin \omega t)\sin \lambda}{\omega^2 - \omega_1^2}$$

$$\text{(9.6.25)}$$

$$\frac{d^2x}{dt^2} = (q\rho - p)\cos \omega t + \frac{q\mu}{\tau\omega}\sin \omega t$$

$$- \frac{a\omega_1(\omega_1 \sin \omega_1 t - \omega \sin \omega t)\cos \lambda}{\omega^2 - \omega_1^2}$$

$$- \frac{a(\omega_1^2 \cos \omega_1 t - \omega^2 \cos \omega t)\sin \lambda}{\omega^2 - \omega_1^2} \qquad \text{(9.6.26)}$$

10

STIFFNESS, CONSTANT RESISTANCE, AND FRICTION

This chapter discusses engineering systems subjected to the action of the force of inertia, the stiffness force, the constant resisting force, and the friction force as the resisting forces. These forces are marked by the plus signs (+) in Row 10 of Guiding Table 2.1. The intersection of this row with Columns 1 through 6 indicate the active forces associated with each system and discussed respective in this chapter's six sections. The left sides of the differential equations of motion in these sections are identical; however, the right sides of these equations are different depending on the active forces applied to the systems.

The problems described in this chapter could be related to engineering systems interacting with elastoplastic or viscoelasto-plastic media, or with specific elastic links. Certain types of these media exhibit strain-strengthening during the phase of plastic deformation. As a result, they may exert simultaneously the stiffness force, the constant resisting force, and the friction force (more related information is presented in section 1.2).

10.1 Active Force Equals Zero

According to Guiding Table 2.1, the engineering systems described in this section are associated with the force of inertia, the stiffness force, the constant resisting force, and the friction force as the resisting forces (Row 10), while the active force equals zero (Column 1). Considerations related to the motion of a system in the absence of active forces are discussed in section 1.3.

The current problem could be related to systems intended for the interaction with an elastoplastic or viscoelastoplastic medium featuring strain-strengthening during its plastic deformation. The presence of a stiffness force in this problem may cause vibratory motion of the system. Additional information related to the deformation of these types of media and the considerations related to the behavior of a constant resisting force and of a friction force applied to the vibratory system are presented in section 1.2. The current problem could be also related to a system that interacts with a specific elastic link.

The system is moving into the horizontal direction. We want to determine the basic parameters of motion, their maximum values, and the characteristics of the forces applied to the elastic link. Figure 10.1 shows the model of a system subjected to the action of a stiffness force, a constant resisting force, and a friction force.

Based on the considerations above and the model in Figure 10.1, we assemble the left and right sides of the differential equation of motion. The left side consists of the force of inertia, the stiffness force, the constant resisting force, and the friction force, while the right side of the equation equals zero. Therefore, the differential equation of motion reads:

$$m\frac{d^2x}{dt^2} + Kx + P + F = 0 \qquad (10.1.1)$$

Differential equation of motion (10.1.1) has different solutions for various initial conditions of motion. These solutions and their analyses are presented below.

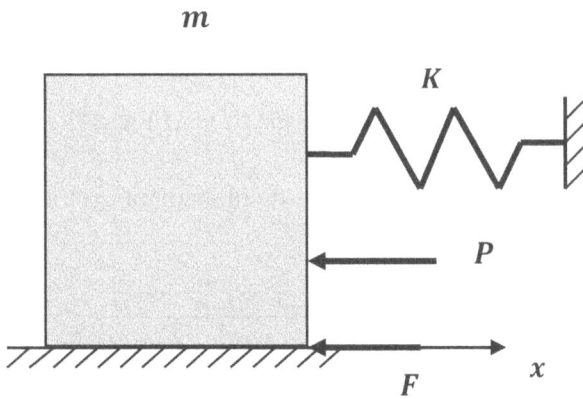

Figure 10.1 Model of a system subjected to a stiffness force, a constant resisting force, and a friction force

10.1.1 General Initial Conditions

The general initial conditions of motion are:

$$\text{for} \quad t = 0 \quad x = s_0; \quad \frac{dx}{dt} = v_0 \qquad \text{(10.1.2)}$$

where s_0 and v_0 are the initial displacement and initial velocity respectively.

Dividing equation (10.1.1) by m, we may write:

$$\frac{d^2x}{dt^2} + \omega^2 x + p + f = 0 \qquad \text{(10.1.3)}$$

where ω is the natural frequency of the vibratory system and:

$$\omega^2 = \frac{K}{m} \qquad \text{(10.1.4)}$$

$$p = \frac{P}{m} \qquad \text{(10.1.5)}$$

$$f = \frac{F}{m} \qquad \text{(10.1.6)}$$

Using Laplace Transform Pairs 3, 1, and 5 from Table 1.1, we convert differential equation of motion (10.1.3) with the initial conditions of motion (10.1.2) from the time domain into the Laplace domain and obtain the corresponding algebraic equation of motion in the Laplace domain:

$$l^2 x(l) - l v_0 - l^2 s_0 + \omega^2 x(l) + p + f = 0 \qquad (10.1.7)$$

Solving equation (10.1.7) for the displacement $x(l)$ in the Laplace domain, we have:

$$x(l) = \frac{l v_0}{l^2 + \omega^2} + \frac{l^2 s_0}{l^2 + \omega^2} - \frac{p + f}{l^2 + \omega^2} \qquad (10.1.8)$$

To invert equation (10.1.8) from the Laplace domain into the time domain, we apply pairs 1, 5, 6, 7, and 14 from Table 1.1. The inversion represents the solution of differential equation of motion (10.1.1) with the initial conditions of motion (10.1.2):

$$x = \frac{v_0}{\omega} \sin \omega t + s_0 \cos \omega t - \frac{p + f}{\omega^2} (1 - \cos \omega t) \qquad (10.1.9)$$

Applying algebraic procedures to the coefficients at the trigonometric functions of equation (10.1.9), we write:

$$\frac{(s_0 \omega^2 + p + f)^2}{(s_0 \omega^2 + p + f)^2 + v_0^2 \omega^2} + \frac{v_0^2 \omega^2}{(s_0 \omega^2 + p + f)^2 + v_0^2 \omega^2}$$

$$= 1 = (\sin \alpha)^2 + (\cos \alpha)^2 \qquad (10.1.10)$$

where:

$$\sin \alpha = \frac{s_0 \omega^2 + p + f}{\sqrt{(s_0 \omega^2 + p + f)^2 + v_0^2 \omega^2}} \qquad (10.1.11)$$

$$\cos \alpha = \frac{v_0 \omega}{\sqrt{(s_0 \omega^2 + p + f)^2 + v_0^2 \omega^2}} \qquad (10.1.12)$$

Combining equations (10.1.11) and (10.1.12) with equation (10.1.9), we obtain:

$$x = \frac{1}{\omega^2} [\sqrt{(s_0 \omega^2 + p + f)^2 + v_0^2 \omega^2} \ (\sin \omega t \cos \alpha + \cos \omega t \sin \alpha) - p - f]$$

$$(10.1.13)$$

Transforming equation (10.1.13), we have:

$$x = \frac{1}{\omega^2}[\sqrt{(s_0\omega^2 + p)^2 + v_0^2\omega^2} \sin(\omega t + \alpha) - p - f] \quad (10.1.14)$$

The first and second derivatives of the equation (10.1.14) represent the velocity and the acceleration of the system respectively:

$$\frac{dx}{dt} = \frac{1}{\omega}\sqrt{(s_0\omega^2 + p + f)^2 + v_0^2\omega^2} \cos(\omega t + \alpha) \quad (10.1.15)$$

$$\frac{d^2x}{dt^2} = -\sqrt{(s_0\omega^2 + p + f)^2 + v_0^2\omega^2} \sin(\omega t + \alpha) \quad (10.1.16)$$

Equation (10.1.15) shows that the velocity becomes equal to zero when:

$$\cos(\omega t + \alpha) = 0 \quad (10.1.17)$$

and, consequently, at this moment of time we have:

$$\sin(\omega t + \alpha) = 1 \quad (10.1.18)$$

Combining equations (10.1.14) and (10.1.16) with equation (10.1.18), we determine the maximum values of the displacement s_{max} and the acceleration a_{max} respectively:

$$s_{max} = \frac{1}{\omega^2}[\sqrt{(s_0\omega^2 + p + f)^2 + v_0^2\omega^2} - p - f] \quad (10.1.19)$$

$$a_{max} = \sqrt{(s_0\omega^2 + p + f)^2 + v_0^2\omega^2} \quad (10.1.20)$$

Multiplying equation (10.1.19) by K and substituting notations (10.1.4), (10.1.5), (10.1.6), we calculate the maximum value of the force R_{max} applied to the elastic link:

$$R_{max} = \sqrt{(Ks_0 + P + F)^2 + mKv_0^2} - P - F \quad (10.1.21)$$

We may accept that the link is subjected to a completely reversed loading cycle having $R_{min} = -R_{max}$. The stress calculations should include fatigue considerations.

10.1.2 Initial Displacement Equals Zero

The initial conditions of motion are:

$$\text{for}\quad t = 0\quad x = 0;\quad \frac{dx}{dt} = v_0 \qquad (10.1.22)$$

Solving differential equation of motion (10.1.1) with the initial conditions of motion (10.1.22), we have:

$$x = \frac{v_0}{\omega}\sin\omega t - \frac{p+f}{\omega^2}(1-\cos\omega t) \qquad (10.1.23)$$

Combining the coefficients at the trigonometric functions in equation (10.1.23) by using the same procedures like in the previous case, we write:

$$x = \frac{\sqrt{v_0^2\omega^2 + (p+f)^2}}{\omega^2}(\sin\omega t\cos\alpha_1 + \cos\omega t\sin\alpha_1) - \frac{p+f}{\omega^2}$$
$$(10.1.24)$$

where:

$$\sin\alpha_1 = \frac{p+f}{\sqrt{v_0^2\omega^2 + (p+f)^2}} \qquad (10.1.25)$$

$$\cos\alpha_1 = \frac{v_0\omega}{\sqrt{v_0^2\omega^2 + (p+f)^2}} \qquad (10.1.26)$$

Transforming equation (10.1.24), we may write:

$$x = \frac{1}{\omega^2}[\sqrt{v_0^2\omega^2 + (p+f)^2}\,\sin(\omega t + \alpha_1) - p - f] \qquad (10.1.27)$$

Taking the first derivative of equation (10.1.27), we determine the velocity of the system:

$$\frac{dx}{dt} = \frac{\sqrt{v_0^2\omega^2 + (p+f)^2}}{\omega}\cos(\omega t + \alpha_1) \qquad (10.1.28)$$

The second derivative of equation (10.1.27) represents the acceleration:

$$\frac{d^2x}{dt^2} = -\sqrt{v_0^2\omega^2 + (p+f)^2}\,\sin(\omega t + \alpha_1) \qquad (10.1.29)$$

Equation (10.1.28) indicates that the velocity becomes equal to zero at the moment when:

$$\cos(\omega t + \alpha_1) = 0 \qquad \qquad \textbf{(10.1.30)}$$

and, consequently, at this moment of time we have:

$$\sin(\omega t + \alpha_1) = 1 \qquad \qquad \textbf{(10.1.31)}$$

Combining equations (10.1.27) and (10.1.29) with the equation (10.1.31), we calculate the maximum values of the displacement and acceleration for this case respectively:

$$s_{max} = \frac{1}{\dot{u}^2}[\sqrt{v_0^2\omega^2 + (p+f)^2} - p - f] \qquad \qquad \textbf{(10.1.32)}$$

$$a_{max} = -\sqrt{v_0^2\omega^2 + (p+f)^2} \qquad \qquad \textbf{(10.1.33)}$$

Multiplying equation (10.1.32) by K and substituting notations (10.1.4), (10.1.5), and (10.1.6), we determine the value of the force R_{max} applied to the elastic link:

$$R_{max} = \sqrt{mKv_0^2 + (P+F)^2} - P - F \qquad \qquad \textbf{(10.1.34)}$$

The considerations related to the stress analysis are the same as for the previous case.

10.1.3 Initial Velocity Equals Zero

Figure 10.1 shows that on the beginning of the process, the system moves into the positive direction. This is possible if the initial displacement has a negative value. Therefore, the initial conditions of motion in this case are:

$$\text{for} \quad t = 0 \quad x = -s_0; \quad \frac{dx}{dt} = 0 \qquad \qquad \textbf{(10.1.35)}$$

Hence, for this case the term $l^2 s_0$ in equation (10.1.7) should be taken with the positive sign.

Solving differential equation of motion (10.1.1) with the initial conditions of motion (10.1.35), we obtain:

$$x = -s_0 \cos \omega t - \frac{p+f}{\omega^2}(1 - \cos \omega t) \qquad \textbf{(10.1.36)}$$

Transforming equation (10.1.36), we may write:

$$x = \frac{1}{\omega^2}[(-s_0\omega^2 + p + f)\cos \omega t - p - f] \qquad \textbf{(10.1.37)}$$

The first and second derivatives of the equation (10.1.37) represent the velocity and the acceleration of the system respectively:

$$\frac{dx}{dt} = \frac{1}{\omega}(s_0\omega^2 - p - f)\sin \omega t \qquad \textbf{(10.1.38)}$$

$$\frac{d^2x}{dt^2} = (s_0\omega^2 - p - f)\cos \omega t \qquad \textbf{(10.1.39)}$$

Equation (10.1.38) shows that the velocity becomes equal to zero at the moment when $\sin \omega t = 0$. Consequently, at this moment of time we have:

$$\cos \omega t = -1 \qquad \textbf{(10.1.40)}$$

Combining equation (10.1.40) with equations (10.1.37) and (10.1.39), we determine the maximum values of the displacement s_{max} and acceleration a_{max} respectively:

$$s_{max} = s_0 - \frac{2(p+f)}{\omega^2} \qquad \textbf{(10.1.41)}$$

$$a_{max} = (s_0\omega^2 - p - f) \qquad \textbf{(10.1.42)}$$

Multiplying equation (10.1.41) by K we calculate the maximum value of the force R_{max} applied to the elastic link:

$$R_{max} = Ks_0 - 2(P + F) \qquad \textbf{(10.1.43)}$$

However, at the beginning of the process when $t = 0$, the link is deformed by the force R_1 that is equal to the product of multiplying K by the initial displacement $-s_0$:

$$R_1 = -Ks_0 \qquad (10.1.44)$$

The absolute value of the force according to equation (10.1.44) exceeds the value of the absolute value of the force according to equation (10.1.43). Therefore, it is acceptable to perform the stress calculations of the link based on the force according to equation (10.1.44), assuming as in the previous case that the link is subjected to a completely reversed loading cycle having the following maximum and minimum forces:

$$R_{max} = |Ks_0| \text{ and } R_{min} = -R_{max}$$

The stress analysis should include fatigue considerations.

10.2 Constant Force R

The intersection of Row 10 and Column 2 in Guiding Table 2.1 indicates that this section focuses on engineering systems subjected to the action of the force of inertia, the stiffness force, the constant resisting force, and the friction force as the resisting forces, and to a constant active force.

This problem could be related to the working process of a system interacting with an elastoplastic or viscoelastic medium that is characterized by the strain-strengthening during the phase of plastic deformation. This type of a medium simultaneously exerts the stiffness force, the constant resisting force, and the friction force as a reaction to its deformation. The presence of a stiffness force may cause vibratory motion of the system. (Additional information related to the deformation of elastoplastic and viscoelastoplastic media as well as the considerations regarding the behavior of the constant resisting force and the friction force applied to a vibratory system can be found in section 1.2.) The current problem could be also related to the interaction of a system with a specific elastic link.

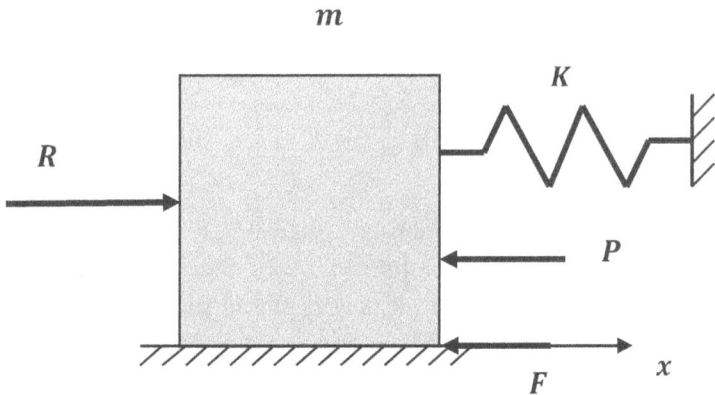

Figure 10.2 Model of a system subjected to a constant active force, a stiffness force, a constant resisting force, and a friction force

The system is moving in the horizontal direction. We want to determine the basic parameters of motion, their values, and the characteristics of forces applied to the elastic link. Figure 10.2 shows the model of a system subjected to the action of a constant active force, a stiffness force, a constant resisting force, and a friction force.

Based on the considerations above and the model in Figure 10.2, we compose the left and right sides of the differential equation of motion of this system. The left side consists of the force of inertia, the stiffness force, the constant resisting force, and the friction force, while the right side of this equation consists of a constant active force. Hence, the differential equation of motion reads:

$$m\frac{d^2x}{dt^2} + Kx + P + F = R \qquad (10.2.1)$$

Differential equation of motion (10.2.1) has different solutions for various initial conditions of motion. These solutions and their analyses are presented below.

10.2.1 General Initial Condition

The general initial conditions of motion are:

$$\text{for} \quad t = 0 \quad x = s_0; \quad \frac{dx}{dt} = v_0 \qquad \textbf{(10.2.2)}$$

where s_0 and v_0 are the initial displacement and initial velocity respectively.

Dividing the equation (10.2.1) by m, we have:

$$\frac{d^2x}{dt^2} + \omega^2 x + p + f = r \qquad \textbf{(10.2.3)}$$

where ω is the natural frequency of the system and:

$$\omega^2 = \frac{K}{m} \qquad \textbf{(10.2.4)}$$

$$p = \frac{P}{m} \qquad \textbf{(10.2.5)}$$

$$f = \frac{F}{m} \qquad \textbf{(10.2.6)}$$

$$r = \frac{R}{m} \qquad \textbf{(10.2.7)}$$

Using Laplace Transform Pairs 3, 1, and 5 from Table 1.1, we convert differential equation of motion (10.2.3) with the initial conditions of motion (10.2.2) from the time domain into the Laplace domain. The resulting algebraic equation of motion in the Laplace domain reads:

$$l^2 x(l) - lv_0 - l^2 s_0 + \omega^2 x(l) + p + f = r \qquad \textbf{(10.2.8)}$$

The solution of equation (10.2.7) for the displacement $x(l)$ in Laplace domain has the following shape:

$$x(l) = \frac{r - p - f}{l^2 + \omega^2} + \frac{lv_0}{l^2 + \omega^2} + \frac{l^2 s_0}{l^2 + \omega^2} \qquad \textbf{(10.2.9)}$$

Applying pairs 1, 5, 14, 6, and 7 from Table 1.1, we invert equation (10.2.9) from the Laplace domain into the time domain. The inversion represents the solution of differential equation of motion (10.2.1) with the initial conditions of motion (10.2.2):

$$x = \frac{r-p-f}{\omega^2}(1-\cos\omega t) + \frac{v_0}{\omega}\sin\omega t + s_0\cos\omega t \quad (10.2.10)$$

Transforming equation (10.2.10), we may write:

$$x = \frac{r-p-f}{\omega^2} + \frac{1}{\omega^2}[v_0\omega\sin\omega t$$

$$- (r-p-f-s_0\omega^2)\cos\omega t] \quad (10.2.11)$$

Performing algebraic actions with the trigonometric function's coefficients in equation (10.2.11), we have:

$$\frac{(r-p-f-s_0\omega^2)^2}{(r-p-f-s_0\omega^2)^2+v_0^2\omega^2} + \frac{v_0^2\omega^2}{(r-p-f-s_0\omega^2)^2+v_0^2\omega^2}$$

$$= 1 = (\sin\beta)^2 + (\cos\beta)^2 \quad (10.2.12)$$

where:

$$\sin\beta = -\frac{r-p-f-s_0\omega^2}{\sqrt{(r-p-f-s_0\omega^2)^2+v_0^2\omega^2}} \quad (10.2.13)$$

$$\cos\beta = \frac{v_0\omega}{\sqrt{(r-p-f-s_0\omega^2)^2+v_0^2\omega^2}} \quad (10.2.14)$$

Combining equations (10.2.15) and (10.2.16) with equation (10.2.11), we obtain:

$$x = \frac{r-p-f}{\omega^2}$$

$$+ \frac{1}{\omega^2}\sqrt{(r-p-f-s_0\omega^2)^2+v_0^2\omega^2}\,\sin(\omega t - \beta)] \quad (10.2.15)$$

The first and second derivatives of the equation (10.2.15) represent the velocity and the acceleration of the system respectively:

$$\frac{dx}{dt} = \frac{1}{\omega}\sqrt{(r-p-f-s_0\omega^2)^2 + v_0^2\omega^2}\cos(\omega t - \beta) \quad \textbf{(10.2.16)}$$

$$\frac{d^2x}{dt^2} = -\sqrt{(r-p-f-s_0\omega^2)^2 + v_0^2\omega^2}\sin(\omega t - \beta) \quad \textbf{(10.2.17)}$$

According to equation (10.2.16), the velocity becomes equal to zero at the moment when:

$$\cos(\omega t - \beta) = 0 \quad \textbf{(10.2.18)}$$

and, consequently, at this moment we have:

$$\sin(\omega t - \beta) = 1 \quad \textbf{(10.2.19)}$$

Combining equations (10.2.15) and (10.2.17) with equation (10.2.19), we determine the maximum values of the displacement s_{max} and acceleration a_{max} respectively:

$$s_{max} = \frac{r-p-f}{\omega^2} + \frac{1}{\omega^2}\sqrt{(r-p-f-s_0\omega^2)^2 + v_0^2\omega^2} \quad \textbf{(10.2.20)}$$

$$a_{max} = \sqrt{(r-p-f-s_0\omega^2)^2 + v_0^2\omega^2} \quad \textbf{(10.2.21)}$$

Multiplying equation (10.2.20) by K and substituting notations (10.2.4), (10.2.5), (10.2.6), and (10.2.7), we calculate the maximum value of the force R_{max} applied to the elastic link:

$$R_{max} = R - P - F + \sqrt{(R-P-F-Ks_0)^2 + mKv_0^2} \quad \textbf{(10.2.22)}$$

It is reasonable to accept that the link is subjected to a completely reversed loading cycle having the following minimum force:

$$R_{min} = -R_{max} \quad \textbf{(10.2.23)}$$

The stress analysis of the link should include fatigue considerations.

10.2.2 Initial Displacement Equals Zero

The initial conditions of motion are:

$$\text{for} \quad t = 0 \quad x = 0; \quad \frac{dx}{dt} = v_0 \quad\quad\quad \textbf{(10.2.24)}$$

The solution of differential equation of motion (10.2.1) with the initial conditions of motion (10.2.24) reads:

$$x = \frac{r - p - f}{\omega^2}(1 - \cos \omega t) + \frac{v_0}{\omega}\sin \omega t \quad\quad \textbf{(10.2.25)}$$

Transforming equation (10.2.25), we have:

$$x = \frac{r - p - f}{\omega^2} + \frac{1}{\omega^2}[v_0 \omega \sin \omega t - (r - p - f)\cos \omega t] \quad \textbf{(10.2.26)}$$

Applying algebraic procedures to equation (10.2.26) as in the previous case, we may write:

$$x = \frac{1}{\omega^2}[r + \sqrt{(r - p - f)^2 + v_0^2 \omega^2}\ \sin(\omega t - \beta_1)] \quad \textbf{(10.2.27)}$$

where:

$$\sin \beta_1 = -\frac{r - p - f}{\sqrt{(r - p - f)^2 + v_0^2 \omega^2}} \quad\quad\quad \textbf{(10.2.28)}$$

$$\cos \beta_1 = \frac{v_0 \omega}{\sqrt{(r - p - f)^2 + v_0^2 \omega^2}} \quad\quad\quad \textbf{(10.2.29)}$$

Taking the first and second derivatives of equation (9.2.27), we determine the velocity and acceleration respectively:

$$\frac{dx}{dt} = \frac{1}{\omega}\sqrt{(r - p - f)^2 + v_0^2 \omega^2}\ \cos(\omega t - \beta_1) \quad \textbf{(10.2.30)}$$

$$\frac{d^2 x}{dt^2} = -\sqrt{(r - p - f)^2 + v_0^2 \omega^2}\ \sin(\omega t - \beta_1) \quad \textbf{(10.2.31)}$$

Equation (10.2.30) shows that at the moment when $\cos(\omega t - \beta_1) = 0$, the velocity is equal to zero. Consequently, at this moment we have:

$$\sin(\omega t - \beta_1) = 1 \quad\quad\quad\quad \textbf{(10.2.32)}$$

Combining equation (10.2.32) with equations (10.2.27) and (10.2.31), we determine the maximum values of the displacement s_{max} and the acceleration a_{max} respectively:

$$s_{max} = \frac{1}{\omega^2}[r + \sqrt{(r-p-f)^2 + v_0^2\omega^2}\,] \qquad \textbf{(10.2.33)}$$

$$a_{max} = -\sqrt{(r-p-f)^2 + v_0^2\,\omega^2} \qquad \textbf{(10.2.34)}$$

Multiplying equation (10.2.33) by K and substituting notations (10.2.4), (10.2.5), (10.2.6), and (10.2.7), we calculate the maximum value of the force R_{max} applied to the elastic link:

$$R_{max} = R + \sqrt{(R-P-F)^2 + mKv_0^2} \qquad \textbf{(10.2.35)}$$

As in the previous case, we accept that the link is subjected to a completely reversed loading cycle having the following minimum force:

$$R_{min} = -R_{max}$$

The stress analysis of the link should include fatigue calculations.

10.2.3 Initial Velocity Equals Zero

The initial conditions of motion are:

$$\text{for} \quad t = 0 \quad x = s_0; \quad \frac{dx}{dt} = 0 \qquad \textbf{(10.2.36)}$$

The solution of differential equation of motion (10.2.1) with the initial conditions of motion (10.2.36) reads:

$$x = \frac{r-p-f}{\omega^2} - \frac{1}{\omega^2}(r-p-f-s_0\omega^2)\cos\omega t \qquad \textbf{(10.2.37)}$$

The first and second derivatives of equation (10.2.37) represent the velocity and acceleration respectively:

$$\frac{dx}{dt} = \frac{1}{\omega}(r-p-f-s_0\omega^2)\sin\omega t \qquad \textbf{(10.2.38)}$$

$$\frac{d^2x}{dt^2} = (r-p-f-s_0\omega^2)\cos\omega t \qquad \textbf{(10.2.39)}$$

Based on equation (10.2.38), it can be seen that the velocity becomes equal to zero at the moment when $\sin \omega t = 0$ and, consequently, at this moment of time we have:

$$\cos \omega t = 1 \qquad (10.2.40)$$

Combining equation (10.2.40) with equations (10.2.37) and (10.2.39), we obtain the maximum values of the displacement s_{max} and acceleration a_{max} respectively:

$$s_{max} = \frac{2(r - p - f)}{\omega^2} - s_0 \qquad (10.2.41)$$

$$a_{max} = r - p - f - s_0\omega^2 \qquad (10.2.42)$$

Multiplying equation (9.2.40) by K and substituting notations (10.2.4), (10.2.5), (10.2.6), and (10.2.7), we determine the maximum value of the force R_{max} applied to the elastic link:

$$R_{max} = 2(R - P - F) - Ks_0 \qquad (10.2.43)$$

It is accepted, as in the previous cases, that the link is subjected to a completely reversed loading cycle having the following minimum force:

$$R_{min} = - R_{max}$$

The stress analysis of the link should be based on fatigue considerations.

10.2.4 Both the Initial Displacement and Velocity Equal Zero

The initial conditions of motion are:

$$\text{for} \quad t = 0 \quad x = 0; \quad \frac{dx}{dt} = 0 \qquad (10.2.44)$$

The solution of differential equation of motion (10.2.1) with the initial conditions of motion (10.2.44) reads:

$$x = \frac{r - p - f}{\omega^2}(1 - \cos \omega t) \qquad (10.2.45)$$

Taking the first derivative of the equation (10.2.45), we determine the velocity of the system:

$$\frac{dx}{dt} = \frac{r - p - f}{\omega} \sin \omega t \qquad (10.2.46)$$

The second derivative of the equation (10.2.45) represents the acceleration:

$$\frac{d^2 x}{dt^2} = (r - p - f) \cos \omega t \qquad (10.2.47)$$

According to equation (10.2.46), the velocity becomes equal to zero when $\sin \omega t = 0$. In this case, we have:

$$\cos \omega t = -1 \qquad (10.2.48)$$

Combining equation (10.2.48) with equations (10.2.45) and (10.2.47), we obtain the maximum values of the displacement s_{max} and acceleration a_{max} respectively:

$$s_{max} = \frac{2(r - p - f)}{\omega^2} \qquad (10.2.49)$$

$$a_{max} = -(r - p - f) \qquad (10.2.50)$$

Multiplying equation (10.2.49) by K and substituting notations (10.2.4), (10.2.5), (10.2.6), and (10.2.7), we determine the maximum value of the force R_{max} applied to the elastic link:

$$R_{max} = 2(R - P - F) \qquad (10.2.51)$$

It is accepted, as in the previous cases, that the link is subjected to a completely reversed loading cycle having the following minimum force:

$$R_{min} = - R_{max}$$

The stress analysis of the link should include fatigue calculations.

10.3 Harmonic Force $A \sin(\omega_1 t + \lambda)$

The intersection of Row 10 and Column 3 of Guiding Table 2.1 indicates that the engineering systems discussed in this section are subjected to the action of the force of inertia, the stiffness force, the constant resisting force, and the friction force as the resisting forces, and to the harmonic force as the active force.

The current problem could be associated with the working processes of vibratory systems subjected to the action of a harmonic force and intended for deformation of an elastoplastic or viscoelastic medium featuring strain-strengthening during its plastic deformation. In certain conditions, these media simultaneously exert stiffness, constant resisting, and friction forces as a result of their deformation (more information regarding the deformation of these media and the behavior of the constant resisting force and friction force applied to a vibratory system is provided in section 1.2).

The system is moving in the horizontal direction. We want to determine the basic parameters of motion. Figure 10.3 shows the model of a system subjected to the action of a harmonic force, a stiffness force, a constant resisting force, and a friction force.

The considerations above and the model in Figure 10.3 allow to describe the left and right sides of the differential equation of motion. The left side consists of the force of inertia, the stiffness force,

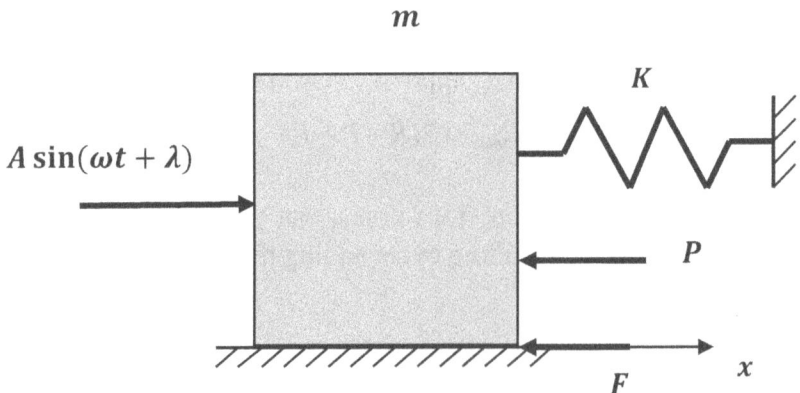

Figure 10.3 Model of a system subjected to a harmonic force, a stiffness force, a constant resisting force, and a friction force

the constant resisting force, and the friction force, while the right side includes the harmonic force. Thus, the differential equation of motion reads:

$$m\frac{d^2x}{dt^2} + Kx + P + F = A\sin(\omega_1 t + \lambda) \qquad (10.3.1)$$

Differential equation of motion (10.3.1) has different solutions for various initial conditions of motion. These solutions and their analyses are presented below.

10.3.1 General Initial Conditions

The general initial conditions of motion are:

$$\text{for} \quad t = 0 \quad x = s_0; \quad \frac{dx}{dt} = v_0 \qquad (10.3.2)$$

where s_0 and v_0 are the initial displacement and initial velocity respectively.

Transforming the sinusoidal function in equation (10.3.1) and dividing this equation by m, we have:

$$\frac{d^2x}{dt^2} + \omega^2 x + p + f = a\sin\omega_1 t \cos\lambda + a\cos\omega_1 t \sin\lambda \qquad (10.3.3)$$

where ω is the natural frequency of the system and:

$$\omega^2 = \frac{K}{m} \qquad (10.3.4)$$

$$p = \frac{P}{m} \qquad (10.3.5)$$

$$f = \frac{F}{m} \qquad (10.3.6)$$

$$a = \frac{A}{m} \qquad (10.3.7)$$

Based on Laplace Transform Pairs 3, 5, 1, 6, and 7 from Table 1.1, we convert differential equation of motion (10.3.3) with the initial conditions of motion (10.3.2) from the time domain into the Laplace domain. The resulting algebraic equation of motion in the Laplace domain reads:

$$l^2 x(l) - lv_0 - l^2 s_0 + \omega^2 x(l) + p + f$$

$$= \frac{a\omega_1 l}{l^2 + \omega_1^2} \cos \lambda + \frac{al^2}{l^2 + \omega_1^2} \sin \lambda \qquad (10.3.8)$$

Applying basic algebra to equation (10.3.8), we have:

$$x(l)(l^2 + \omega^2)$$

$$= lv_0 + l^2 s_0 - p - f + \frac{a\omega_1 l}{l^2 + \omega_1^2} \cos \lambda + \frac{al^2}{l^2 + \omega_1^2} \sin \lambda \quad (10.3.9)$$

Solving equation (10.3.8) for the Laplace domain displacement (l), we may write:

$$x(l) = \frac{lv_0}{l^2 + \omega^2} + \frac{l^2 s_0}{l^2 + \omega^2} - \frac{p + f}{l^2 + \omega^2} + \frac{la\omega_1}{(l^2 + \omega^2)(l^2 + \omega_1^2)} \cos \lambda$$

$$+ \frac{l^2 a}{(l^2 + \omega^2)(l^2 + \omega_1^2)} \sin \lambda \qquad (10.3.10)$$

Using pairs $1, 5, 6, 7, 14, 33$, and 34 from Table 1.1, we invert equation (10.3.10) from the Laplace domain into the time domain and obtain the solution of differential equation of motion (10.3.1) with the initial conditions of motion (10.3.2):

$$x = \frac{v_0}{\omega} \sin \omega t + s_0 \cos \omega t - \frac{p + f}{\omega^2} (1 - \cos \omega t)$$

$$+ \frac{a(\omega \sin \omega_1 t - \omega_1 \sin \omega t) \cos \lambda}{\omega(\omega^2 - \omega_1^2)}$$

$$+ \frac{a(\cos \omega_1 t - \cos \omega t) \sin \lambda}{\omega^2 - \omega_1^2} \qquad (10.3.11)$$

Taking the first derivative of equation (10.3.11), we determine the velocity of the system:

$$\frac{dx}{dt} = v_0 \cos \omega t - \omega s_0 \sin \omega t - \frac{p+f}{\omega} \sin \omega t$$

$$+ \frac{a\omega_1(\cos \omega_1 t - \cos \omega t)\cos \lambda}{\omega^2 - \omega_1^2}$$

$$- \frac{a(\omega_1 \sin \omega_1 t - \omega \sin \omega t)\sin \lambda}{\omega^2 - \omega_1^2} \qquad (10.3.12)$$

The second derivative of equation (10.3.11) represents the acceleration of the system:

$$\frac{d^2x}{dt^2} = -\omega v_0 \sin \omega t - -\omega^2 s_0 \cos \omega t - (p+f)\cos \omega t$$

$$- \frac{a\omega_1(\omega_1 \sin \omega_1 t - \omega \sin \omega t)\cos \lambda}{\omega^2 - \omega_1^2}$$

$$- \frac{a(\omega_1^2 \cos \omega_1 t - \omega^2 \cos \omega t)\sin \lambda}{\omega^2 - \omega_1^2} \qquad (10.3.13)$$

10.3.2 Initial Displacement Equals Zero
The initial conditions of motion are:

$$\text{for} \quad t = 0 \quad x = 0; \quad \frac{dx}{dt} = v_0 \qquad (10.3.14)$$

Solving differential equation of motion (10.3.1) with the initial conditions of motion (10.3.14), we may write:

$$x = \frac{v_0}{\omega} \sin \omega t - \frac{p+f}{\omega^2}(1 - \cos \omega t) + \frac{a(\omega \sin \omega_1 t - \omega_1 \sin \omega t)\cos \lambda}{\omega(\omega^2 - \omega_1^2)}$$

$$+ \frac{a(\cos \omega_1 t - \cos \omega t)\sin \lambda}{\omega^2 - \omega_1^2} \qquad (10.3.15)$$

The first and second derivatives of equation (10.3.14) represent respectively the velocity and acceleration of the system:

$$\frac{dx}{dt} = v_0 \cos \omega t - \frac{p+f}{\omega} \sin \omega t + \frac{a\omega_1(\cos \omega_1 t - \cos \omega t)\cos \lambda}{\omega^2 - \omega_1^2}$$

$$- \frac{a(\omega_1 \sin \omega_1 t - \omega \sin \omega t)\sin \lambda}{\omega^2 - \omega_1^2} \qquad (10.3.16)$$

$$\frac{d^2x}{dt^2} = -\omega v_0 \sin \omega t - (p+f)\cos \omega t$$

$$- \frac{a\omega_1(\omega_1 \sin \omega_1 t - \omega \sin \omega t)\cos \lambda}{\omega^2 - \omega_1^2}$$

$$- \frac{a(\omega_1^2 \cos \omega_1 t - \omega^2 \cos \omega t)\sin \lambda}{\omega^2 - \omega_1^2} \qquad (10.3.17)$$

10.3.3 Initial Velocity Equals Zero

The initial conditions of motion are:

$$\text{for} \quad t = 0 \quad x = s_0; \quad \frac{dx}{dt} = 0 \qquad (10.3.18)$$

The solution of differential equation of motion (10.3.1) with the initial conditions of motion (10.3.18) reads:

$$x = s_0 \cos \omega t - \frac{p+f}{\omega^2}(1 - \cos \omega t) + \frac{a(\omega \sin \omega_1 t - \omega_1 \sin \omega t)\cos \lambda}{\omega(\omega^2 - \omega_1^2)}$$

$$+ \frac{a(\cos \omega_1 t - \cos \omega t)\sin \lambda}{\omega^2 - \omega_1^2} \qquad (10.3.19)$$

Taking the first derivative of equation (10.3.19), we determine the velocity of the system:

$$\frac{dx}{dt} = -\omega s_0 \sin \omega t - \frac{p+f}{\omega} \sin \omega t + \frac{a\omega_1(\cos \omega_1 t - \cos \omega t)\cos \lambda}{\omega^2 - \omega_1^2}$$

$$- \frac{a(\omega_1 \sin \omega_1 t - \omega \sin \omega t)\sin \lambda}{\omega^2 - \omega_1^2} \qquad (10.3.20)$$

The second derivative of the equation (10.3.19) represents the acceleration:

$$\frac{d^2x}{dt^2} = -\omega^2 s_0 \cos\omega t - (p+f)\sin\omega t$$

$$-\frac{a\omega_1(\omega_1 \sin\omega_1 t - \omega \sin\omega t)\cos\lambda}{\omega^2 - \omega_1^2}$$

$$-\frac{a(\omega_1^2 \cos\omega_1 t - \omega^2 \cos\omega t)\sin\lambda}{\omega^2 - \omega_1^2} \qquad \textbf{(10.3.21)}$$

10.3.4 Both the Initial Displacement and Velocity Equals Zero

The initial conditions of motion are:

$$\text{for} \quad t = 0 \quad x = 0; \quad \frac{dx}{dt} = 0 \qquad \textbf{(10.3.22)}$$

The solution of differential equation of motion (10.3.1) with the initial conditions of motion (10.3.22) reads:

$$x = -\frac{p+f}{\omega^2}(1-\cos\omega t) + \frac{a(\omega \sin\omega_1 t - \omega_1 \sin\omega t)\cos\lambda}{\omega(\omega^2 - \omega_1^2)}$$

$$+\frac{a(\cos\omega_1 t - \cos\omega t)\sin\lambda}{\omega^2 - \omega_1^2} \qquad \textbf{(10.3.23)}$$

Taking the first derivative of equation (10.3.23), we determine the velocity of the system:

$$\frac{dx}{dt} = -\frac{p+f}{\omega}\sin\omega t + \frac{a\omega_1(\cos\omega_1 t - \cos\omega t)\cos\lambda}{\omega^2 - \omega_1^2}$$

$$-\frac{a(\omega_1 \sin\omega_1 t - \omega \sin\omega t)\sin\lambda}{\omega^2 - \omega_1^2} \qquad \textbf{(10.3.24)}$$

The second derivative of the equation (10.3.23) represents the acceleration:

$$\frac{d^2x}{dt^2} = -(p+f)\cos\omega t - \frac{a\omega_1(\omega_1 \sin\omega_1 t - \omega \sin\omega t)\cos\lambda}{\omega^2 - \omega_1^2}$$

$$-\frac{a(\omega_1^2 \cos\omega_1 t - \omega^2 \cos\omega t)\sin\lambda}{\omega^2 - \omega_1^2} \qquad \textbf{(10.3.25)}$$

10.4 Time-Dependent Force $Q\left(\rho + \frac{\mu t}{\tau}\right)$

This section describes engineering systems associated with the force of inertia, the stiffness force, the constant resisting force, and the friction force as the resisting forces (Row 10 of Guiding Table 2.1) and the time-dependent force as the active force (Column 4).

The current problem could be related to a system that interacts with an elastoplastic or viscoelastoplastic medium exhibiting strain-strengthening during the plastic deformations and exerting stiffness, constant resisting, and friction forces as the reaction to the deformation (more information is provided in section 1.2). The current problem could also represent a system interacting with a specific elastic link. Sometimes during the initial stage of the working process, the system is subjected to a time-dependent force that acts for a predetermined interval of time.

The system is moving in the horizontal direction. We want to determine the basic parameters of motion, their values at the end of this interval of time, and the characteristics of forces applied to the elastic link. The model of a system subjected to the action of a time-dependent force, a stiffness force, a constant resisting force, and a friction force is shown in Figure 10.4.

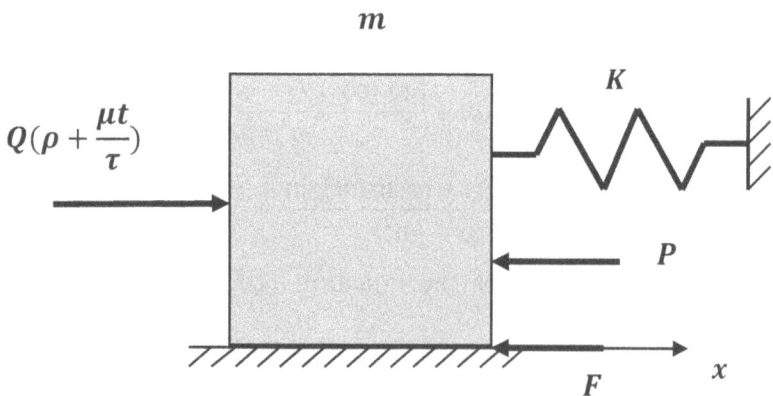

Figure 10.4 Model of a system subjected to a time-dependent force, a stiffness force, a constant resisting force, and a friction force

The considerations above and the model in Figure 10.4 let us compose the left and right sides of the differential equation of motion. The left side consists of the force of inertia, the stiffness force, the constant resisting force, and the friction force, while the right side includes the time-dependent force. Therefore, the differential equation of motion reads:

$$m\frac{d^2x}{dt^2} + Kx + P + F = Q(\rho + \frac{\mu t}{\tau}) \tag{10.4.1}$$

The differential equation of motion (10.4.1) has different solutions for various initial conditions of motion. These solutions and their analyses are presented below.

10.4.1 General Initial Conditions

The general initial conditions of motion are:

$$\text{for} \quad t = 0 \quad x = s_0; \quad \frac{dx}{dt} = v_0 \tag{10.4.2}$$

where s_0 and v_0 are the initial displacement and initial velocity respectively.

Dividing the equation (10.4.1) by m, we have:

$$\frac{d^2x}{dt^2} + \omega^2 x + p + f = q(\rho + \frac{\mu t}{\tau}) \tag{10.4.3}$$

where ω is the natural frequency of the system and:

$$\omega^2 = \frac{K}{m} \tag{10.4.4}$$

$$p = \frac{P}{m} \tag{10.4.5}$$

$$f = \frac{F}{m} \tag{10.4.6}$$

$$q = \frac{Q}{m} \tag{10.4.7}$$

Applying Laplace Transform Pairs 3, 5, 1, and 2 from Table 1.1, we convert the differential equation of motion (10.4.3) with the initial conditions of motion (10.4.2) from the time domain into the Laplace domain, and obtain the corresponding algebraic equation of motion in the Laplace domain:

$$l^2 x(l) - lv_0 - l^2 s_0 + \omega^2 x(l) + p + f = q\rho + \frac{q\mu}{\tau l} \quad \textbf{(10.4.8)}$$

The solution of equation (10.4.8) for the Laplace domain displacement $x(l)$ reads:

$$x(l) = \frac{lv_0}{l^2 + \omega^2} + \frac{l^2 s_0}{l^2 + \omega^2} + \frac{q\rho - p - f}{l^2 + \omega^2} + \frac{q\mu}{\tau l(l^2 + \omega^2)} \quad \textbf{(10.4.9)}$$

Using pairs 1, 6, 7, 14, and 16 from Table 1.1, we invert equation (10.4.9) from the Laplace domain into the time domain and obtain the solution of differential equation of motion (10.4.1) with the initial conditions of motion (10.4.2):

$$x = \frac{v_0}{\omega} \sin \omega t + s_0 \cos \omega t + \frac{q\rho - p - f}{\omega^2}(1 - \cos \omega t)$$

$$+ \frac{q\mu}{\tau \omega^2}(t - \frac{1}{\omega} \sin \omega t) \quad \textbf{(10.4.10)}$$

Applying basic algebra to equation (10.4.10), we have:

$$x = \frac{q\rho - p - f}{\omega^2} + \frac{q\mu t}{\tau \omega^2} + (s_0 - \frac{q\rho - p - f}{\omega^2}) \cos \omega t$$

$$+ (\frac{v_0}{\omega} - \frac{q\mu}{\tau \omega^3}) \sin \omega t \quad \textbf{(10.4.11)}$$

Taking the first derivative of the equation (10.4.11), we determine the velocity of the system:

$$\frac{dx}{dt} = \frac{q\mu}{\tau \omega^2} - (s_0 \omega - \frac{q\rho - p - f}{\omega}) \sin \omega t + (v_0 - \frac{q\mu}{\tau \omega^2}) \cos \omega t \quad \textbf{(10.4.12)}$$

The second derivative of equation (10.4.11) represents the acceleration:

$$\frac{d^2x}{dt^2} = -(s_0\omega^2 - q\rho + p + f)\cos\omega t - (v_0\omega - \frac{q\mu}{\tau\omega})\sin\omega t \qquad \textbf{(10.4.13)}$$

Substituting time interval τ into equations (10.4.11), (10.4.12), and (10.4.13), we determine the values of the displacement s, the velocity v, and the acceleration a at the end of this interval of time respectively:

$$s = \frac{q(\rho + \mu) - p - f}{\omega^2} + (s_0 - \frac{q\rho - p - f}{\omega^2})\cos\omega\tau$$

$$+(\frac{v_0}{\omega} - \frac{q\mu}{\tau\omega^3})\sin\omega\tau \qquad \textbf{(10.4.14)}$$

$$v = \frac{q\mu}{\tau\omega^2} - (s_0\omega - \frac{q\rho - p - f}{\omega})\sin\omega\tau + (v_0 - \frac{q\mu}{\tau\omega^2})\cos\omega\tau \qquad \textbf{(10.4.15)}$$

$$a = -(s_0\omega^2 - q\rho - p - f)\cos\omega\tau - (v_0\omega - \frac{q\mu}{\tau\omega})\sin\omega\tau \qquad \textbf{(10.4.16)}$$

Multiplying equation (10.4.14) by the stiffness coefficient K and substituting notation (10.4.4), we calculate the force R_1 applied to the elastic link at the end of this time interval:

$$R_1 = m[q(\rho + \mu) - p - f + (s_0\omega^2 - q\rho + p + f)\cos\omega\tau$$

$$+\left(v_0\omega - \frac{q\mu}{\tau\omega}\right)\sin\omega\tau] \qquad \textbf{(10.4.17)}$$

The stress calculations of the link could be based on the force according to equation (10.4.17).

10.4.2 Initial Displacement Equals Zero

The initial conditions of motion are:

$$\text{for} \quad t = 0 \quad x = 0; \quad \frac{dx}{dt} = v_0 \qquad \textbf{(10.4.18)}$$

Solving differential equation of motion (10.4.1) with the initial conditions of motion (10.4.18), we have:

$$x = \frac{v_0}{\omega}\sin\omega t + \frac{q\rho-p-f}{\omega^2}(1-\cos\omega t) + \frac{q\mu}{\tau\omega^2}(t-\frac{1}{\omega}\sin\omega t) \quad \textbf{(10.4.19)}$$

The first derivative of equation (10.4.19) represents the velocity of the system:

$$\frac{dx}{dt} = \frac{q\mu}{\tau\omega^2} + \frac{q\rho-p=f}{\omega}\sin\omega t + (v_0 - \frac{q\mu}{\tau\omega^2})\cos\omega t \qquad \textbf{(10.4.20)}$$

Taking the second derivative of equation (10.4.19), we determine the acceleration:

$$\frac{d^2x}{dt^2} = (q\rho-p-f)\cos\omega t - (v_0\omega - \frac{q\mu}{\tau\omega})\sin\omega t \quad \textbf{(10.4.21)}$$

Substituting time interval τ into equations (10.4.19), (10.4.20), and (10.4.21), we determine the values of the displacement s, the velocity v, and the acceleration a at the end of this interval of time respectively:

$$s = \frac{q(\rho+\mu)-p-f}{\omega^2} - \frac{q\rho-p-f}{\omega^2}\cos\omega\tau$$

$$+(\frac{v_0}{\omega} - \frac{q\mu}{\tau\omega^3})\sin\omega\tau \qquad \textbf{(10.4.22)}$$

$$v = \frac{q\mu}{\tau\omega^2} + \frac{q\rho-p-f}{\omega}\sin\omega\tau + (v_0 - \frac{q\mu}{\tau\omega^2})\cos\omega\tau \quad \textbf{(10.4.23)}$$

$$a = (q\rho-p-f)\cos\omega\tau - (v_0\omega - \frac{q\mu}{\tau\omega})\sin\omega\tau \quad \textbf{(10.4.24)}$$

Multiplying equation (10.4.22) by K and substituting notation (10.4.4), we calculate the force R_1 applied to the elastic link at the end of this interval of time:

$$R_1 = m[q(\rho+\mu)-p-f-(q\rho-p-f)\cos\omega\tau$$

$$+(v_0\omega - \frac{q\mu}{\tau\omega})\sin\omega\tau] \qquad \textbf{(10.4.25)}$$

The stress analysis of the link could be performed using the force according to equation (10.4.25).

10.4.3 Initial Velocity Equals Zero

The initial conditions of motion are:

$$\text{for} \quad t = 0 \quad x = s_0; \quad \frac{dx}{dt} = 0 \qquad \textbf{(10.4.26)}$$

Solving differential equation of motion (10.4.1) with the initial conditions of motion (10.4.26), we have:

$$x = \frac{q\rho - p - f}{\omega^2} + \frac{q\mu t}{\tau\omega^2} + \left(s_0 - \frac{q\rho - p - f}{\omega^2}\right)\cos\omega t - \frac{q\mu}{\tau\omega^3}\sin\omega t$$

$$\textbf{(10.4.27)}$$

Taking the first and second derivatives of equation (10.4.27), we determine the velocity and the acceleration respectively:

$$\frac{dx}{dt} = \frac{q\mu}{\tau\omega^2} - \left(s_0\omega - \frac{q\rho - p - f}{\omega}\right)\sin\omega t - \frac{q\mu}{\tau\omega^2}\cos\omega t \quad \textbf{(10.4.28)}$$

$$\frac{d^2x}{dt^2} = -(s_0\omega^2 - q\rho + p - f)\cos\omega t + \frac{q\mu}{\tau\omega}\sin\omega t$$

Substituting the time τ into equations (10.4.27), (10.4.28), and (10.4.29), we determine the values of the displacement s, the velocity v, and the acceleration a at the end of the of the predetermined interval of time respectively:

$$s = \frac{q(\rho + \mu) - p - f}{\omega^2} + \left(s_0 - \frac{q\rho - p - f}{\omega^2}\right)\cos\omega\tau$$

$$- \frac{q\mu}{\tau\omega^3}\sin\omega\tau \qquad \textbf{(10.4.30)}$$

$$v = \frac{q\mu}{\omega^2} - (s_0\omega - q\rho + p + f)\sin\omega\tau - \frac{q\mu}{\tau\omega^2}\cos\omega\tau \quad \textbf{(10.4.31)}$$

$$a = -(s_0\omega^2 - q\rho + p + f)\cos\omega\tau + \frac{q\mu}{\tau\omega}\sin\omega\tau \quad \textbf{(10.4.32)}$$

Multiplying equation (10.4.30) by K and substituting notation (10.4.4), we determine the force R_1 applied to the elastic link:

$$R_1 = m[(q(\rho + \mu) - p - f$$

$$+ (s_0\omega^2 - q\rho + p + f)\cos\omega\tau - \frac{q\mu}{\tau\omega}\sin\omega\tau] \qquad \textbf{(10.4.33)}$$

The stress calculations of the link could be based on the force according to equation (10.4.33).

10.4.4 Both the Initial Displacement and Velocity Equal Zero
The initial conditions of motion are:

$$\text{for} \quad t = 0 \quad x = 0; \quad \frac{dx}{dt} = 0 \qquad \textbf{(10.4.34)}$$

The solution of differential equation of motion (10.4.1) with the initial conditions of motion (10.4.34) reads:

$$x = \frac{q\rho - p - f}{\omega^2} + \frac{q\mu t}{\tau\omega^2} - \frac{q\rho - p - f}{\omega^2}\cos\omega t - \frac{q\mu}{\tau\omega^3}\sin\omega t \qquad \textbf{(10.4.35)}$$

The first and second derivatives of equation (10.4.35) represent the velocity and the acceleration of the system respectively:

$$\frac{dx}{dt} = \frac{q\mu}{\tau\omega^2} + \frac{q\rho - p = f}{\omega}\sin\omega t - \frac{q\mu}{\tau\omega^2}\cos\omega t \qquad \textbf{(10.4.36)}$$

$$\frac{d^2x}{dt^2} = (q\rho - p - f)\cos\omega t + \frac{q\mu}{\tau\omega}\sin\omega t \qquad \textbf{(10.4.37)}$$

Substituting time interval τ into equations (10.4.35), (10.4.36), and (10.4.37), we determine the values of the displacement s, the velocity v, and the acceleration a at the end of the of the predetermined time interval respectively:

$$s = \frac{q(\rho + \mu) - p - f}{\omega^2} - \frac{q\rho - p - f}{\omega^2}\cos\omega\tau - \frac{q\mu}{\tau\omega^3}\sin\omega\tau \qquad \textbf{(10.4.38)}$$

$$v = \frac{q\mu}{\omega^2} + \frac{q\rho - p - f}{\omega}\sin\omega\tau - \frac{q\mu}{\tau\omega^2}\cos\omega\tau \qquad \textbf{(10.4.39)}$$

$$a = (q\rho - p - f)\cos\omega\tau + \frac{q\mu}{\tau\omega}\sin\omega\tau \qquad \textbf{(10.4.40)}$$

Multiplying equation (10.4.38) by K and substituting notation (10.4.4), we calculate the force R_1 applied to the elastic link:

$$R_1 = m[q(\rho + \mu) - p - f - (q\rho - p - f)$$

$$\cos\omega\tau - \frac{q\mu}{\tau\omega}\sin\omega\tau] \qquad \textbf{(10.4.41)}$$

The stress analysis of the link could be performed using the force according to equation (10.4.41).

10.5   Constant Force R and Harmonic Force $A \sin(\omega_1 t + \lambda)$

The intersection of Row 10 and Column 5 in Guiding Table 2.1 indicates that this section describes engineering systems subjected to the action of the force of inertia, the stiffness force, the constant resisting force, and the friction force as the resisting forces, and to the constant active force and the harmonic force as the active forces.

The problem could be related to the working process of a vibratory system intended for the interaction with an elastoplastic or viscoelastoplastic medium that exhibits strain-strengthening during its plastic deformation. During this phase of deformation, the medium simultaneously exerts a stiffness force, a constant resisting force, and a friction force (more information regarding the deformation of these media and the behavior of the constant resisting force and friction force applied to a vibratory system is provided in section 1.2).

The system is moving in the horizontal direction. We want to determine the basic parameters of motion. Figure 10.5 shows the model of a system subjected to the action of a constant active force, a harmonic force, a stiffness force, a constant resisting force, and a friction force.

Based on the considerations above and the model in Figure 10.5, we can assemble the left and right sides of the differential equation of motion of the system. The left side consists of the force of inertia,

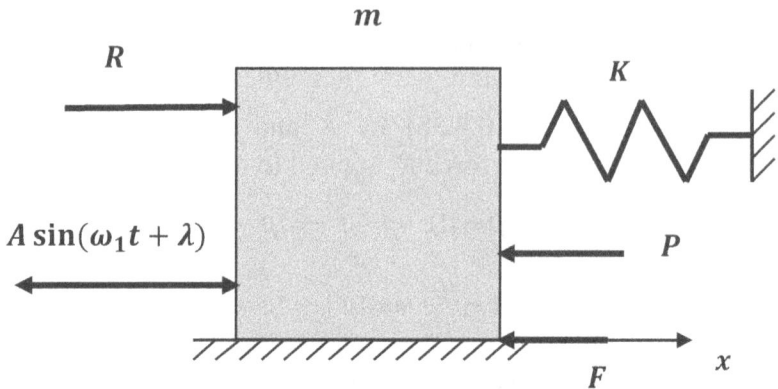

Figure 10.5 Model of a system subjected to a constant active force, a harmonic force, a stiffness force, a constant resisting force, and a friction force

the stiffness force, the constant resisting force, and the friction force, while the right side includes the constant active force and the harmonic force. Thus, the differential equation of motion reads:

$$m\frac{d^2x}{dt^2} + Kx + P + F = R + A\sin(\omega_1 t + \lambda) \qquad (10.5.1)$$

Differential equation of motion (10.5.1) has different solutions for various initial conditions of motion. These solutions and their analyses are presented below.

10.5.1 General Initial Conditions

The general initial conditions of motion are:

$$\text{for} \quad t = 0 \quad x = s_0; \quad \frac{dx}{dt} = v_0 \qquad (10.5.2)$$

where s_0 and v_0 are the initial displacement and initial velocity respectively.

Transforming the sinusoidal function in equation (10.5.1) and dividing this equation by m, we may write:

$$\frac{d^2x}{dt^2} + \omega^2 x + pf = r + a\sin\omega_1 t \cos\lambda + a\cos\omega_1 t \sin\lambda \qquad (10.5.3)$$

where ω is the natural frequency of the system and:

$$\omega^2 = \frac{K}{m} \qquad (10.5.4)$$

$$p = \frac{P}{m} \qquad (10.5.5)$$

$$f = \frac{F}{m} \qquad (10.5.6)$$

$$r = \frac{R}{m} \qquad (10.5.7)$$

$$a = \frac{A}{m} \qquad (10.5.8)$$

To convert differential equation of motion (10.5.3) with the initial conditions of motion (10.5.2) from the time domain into the Laplace domain, we use Laplace Transform Pairs 3, 5, 1, 6, and 7 from Table 1.1. The resulting algebraic equation of motion in the Laplace domain reads:

$$l^2 x(l) - l v_0 - l^2 s_0 + \omega^2 x(l) + p + f$$

$$= r + \frac{a \omega_1 l}{l^2 + \omega_1^2} \cos \lambda + \frac{a l^2}{l^2 + \omega_1^2} \sin \lambda \qquad (10.5.9)$$

Applying basic algebra to equation (10.5.9), we have:

$$x(l)(l^2 + \omega^2) = r - p - f + l v_0 + l^2 s_0$$

$$+ \frac{a \omega_1 l}{l^2 + \omega_1^2} \cos \lambda + \frac{a l^2}{l^2 + \omega_1^2} \sin \lambda \qquad (10.5.10)$$

Solving equation (10.5.10) for the Laplace domain displacement, $x(l)$ we write:

$$x(l) = \frac{r - p - f}{l^2 + \omega^2} + \frac{l v_0}{l^2 + \omega^2} + \frac{l^2 s_0}{l^2 + \omega^2} + \frac{l a \omega_1}{(l^2 + \omega^2)(l^2 + \omega_1^2)} \cos \lambda$$

$$+ \frac{l^2 a}{(l^2 + \omega^2)(l^2 + \omega_1^2)} \sin \lambda \qquad (10.5.11)$$

Based on pairs 1, 14, 5, 6, 7, 33, and 34 from Table 1.1, we invert equation (10.5.11) from the Laplace domain into the time domain. The inversion represents the solution of differential equation of motion (10.5.1) with the initial conditions of motion (10.5.2):

$$x = \frac{r-p-f}{\omega^2}(1-\cos\omega t)+\frac{v_0}{\omega}\sin\omega t + s_0\cos\omega t$$

$$+\frac{a(\omega\sin\omega_1 t - \omega_1\sin\omega t)\cos\lambda}{\omega(\omega^2-\omega_1^2)}+\frac{a(\cos\omega_1 t - \cos\omega t)\sin\lambda}{\omega^2-\omega_1^2}$$

$$(10.5.12)$$

Taking the first and second derivatives of equation (10.5.12), we determine the velocity and the acceleration of the system respectively:

$$\frac{dx}{dt}=\frac{r-p-f}{\omega}\sin\omega t + v_0\cos\omega t - \omega s_0\sin\omega t$$

$$+\frac{a\omega_1(\cos\omega_1 t - \cos\omega t)\cos\lambda}{\omega^2-\omega_1^2}-\frac{a(\omega_1\sin\omega_1 t - \omega\sin\omega t)\sin\lambda}{\omega^2-\omega_1^2}$$

$$(10.5.13)$$

$$\frac{d^2x}{dt^2}=(r-p-f)\cos\omega t - \omega v_0\sin\omega t - \omega^2 s_0\cos\omega t$$

$$-\frac{a\omega_1(\omega_1\sin\omega_1 t - \omega\sin\omega t)\cos\lambda}{\omega^2-\omega_1^2}$$

$$-\frac{a(\omega_1^2\cos\omega_1 t - \omega^2\cos\omega t)\sin\lambda}{\omega^2-\omega_1^2}$$

$$(10.5.14)$$

10.5.2 Initial Displacement Equals Zero

The initial conditions of motion are:

$$\text{for}\quad t=0\quad x=0;\quad \frac{dx}{dt}=v_0 \qquad (10.5.15)$$

The solution of differential equation of motion (10.5.1) with the initial conditions of motion (10.5.15) reads:

$$x=\frac{r-p-f}{\omega^2}(1-\cos\omega t)+\frac{v_0}{\omega}\sin\omega t + \frac{a(\omega\sin\omega_1 t - \omega_1\sin\omega t)\cos\lambda}{\omega(\omega^2-\omega_1^2)}$$

$$+\frac{a(\cos\omega_1 t - \cos\omega t)\sin\lambda}{\omega^2-\omega_1^2} \qquad (10.5.16)$$

The first derivative of equation (10.5.16) represents the velocity of the system:

$$\frac{dx}{dt} = \frac{r-p-f}{\omega}\sin\omega t + v_0\cos\omega t + \frac{a\omega_1(\cos\omega_1 t - \cos\omega t)\cos\lambda}{\omega^2 - \omega_1^2}$$

$$- \frac{a(\omega_1\sin\omega_1 t - \omega\sin\omega t)\sin\lambda}{\omega^2 - \omega_1^2} \qquad \textbf{(10.5.17)}$$

Taking the second derivative of equation (10.5.16), we determine the acceleration:

$$\frac{d^2x}{dt^2} = (r-p-f)\cos\omega t - \omega v_0\sin\omega t$$

$$- \frac{a\omega_1(\omega_1\sin\omega_1 t - \omega\sin\omega t)\cos\lambda}{\omega^2 - \omega_1^2}$$

$$- \frac{a(\omega_1^2\cos\omega_1 t - \omega^2\cos\omega t)\sin\lambda}{\omega^2 - \omega_1^2} \qquad \textbf{(10.5.18)}$$

10.5.3 Initial Velocity Equals Zero

The initial conditions of motion are:

$$\text{for} \quad t=0 \quad x = s_0; \quad \frac{dx}{dt} = 0 \qquad \textbf{(10.5.19)}$$

Solving differential equation of motion (10.5.1) with the initial conditions of motion (10.5.19), we may write:

$$x = \frac{r-p-f}{\omega^2}(1-\cos\omega t) + s_0\cos\omega t + \frac{a(\omega\sin\omega_1 t - \omega_1\sin\omega t)\cos\lambda}{\omega(\omega^2 - \omega_1^2)}$$

$$+ \frac{a(\cos\omega_1 t - \cos\omega t)\sin\lambda}{\omega^2 - \omega_1^2} \qquad \textbf{(10.5.20)}$$

Taking the first derivative of equation (10.5.20), we determine the velocity of the system:

$$\frac{dx}{dt} = \frac{r-p-f}{\omega}\sin\omega t - \omega s_0\sin\omega t + \frac{a\omega_1(\cos\omega_1 t - \cos\omega t)\cos\lambda}{\omega^2 - \omega_1^2}$$

$$- \frac{a(\omega_1\sin\omega_1 t - \omega\sin\omega t)\sin\lambda}{\omega^2 - \omega_1^2}$$

$$\textbf{(10.5.21)}$$

The second derivative of equation (10.5.20) represents the acceleration:

$$\frac{d^2x}{dt^2} = (r - p - f)\cos\omega t - \omega^2 s_0 \cos\omega t$$

$$- \frac{a\omega_1(\omega_1 \sin\omega_1 t - \omega\sin\omega t)\cos\lambda}{\omega^2 - \omega_1^2}$$

$$- \frac{a(\omega_1^2 \cos\omega_1 t - \omega^2 \cos\omega t)\sin\lambda}{\omega^2 - \omega_1^2} \qquad (10.5.22)$$

10.5.4 Both the Initial Displacement and Velocity Equal Zero

The initial conditions of motion are:

$$\text{for} \quad t = 0 \quad x = 0; \quad \frac{dx}{dt} = 0 \qquad (10.5.23)$$

The solution of differential equation motion (10.5.1) with the initial conditions of motion (10.5.23) reads:

$$x = \frac{r - p - f}{\omega^2}(1 - \cos\omega t) + \frac{a(\omega \sin\omega_1 t - \omega_1 \sin\omega t)\cos\lambda}{\omega(\omega^2 - \omega_1^2)}$$

$$+ \frac{a(\cos\omega_1 t - \cos\omega t)\sin\lambda}{\omega^2 - \omega_1^2} \qquad (10.5.24)$$

Taking the first and second derivatives of equation (10.5.24), we determine the velocity and the acceleration of the system respectively:

$$\frac{dx}{dt} = \frac{r - p - f}{\omega}\sin\omega t + \frac{a\omega_1(\cos\omega_1 t - \cos\omega t)\cos\lambda}{\omega^2 - \omega_1^2}$$

$$- \frac{a(\omega_1 \sin\omega_1 t - \omega\sin\omega t)\sin\lambda}{\omega^2 - \omega_1^2} \qquad (10.5.25)$$

$$\frac{d^2x}{dt^2} = (r - p - f)\cos\omega t - \frac{a\omega_1(\omega_1 \sin\omega_1 t - \omega\sin\omega t)\cos\lambda}{\omega^2 - \omega_1^2}$$

$$- \frac{a(\omega_1^2 \cos\omega_1 t - \omega^2 \cos\omega t)\sin\lambda}{\omega^2 - \omega_1^2} \qquad (10.5.26)$$

10.6 Harmonic Force $A \sin(\omega_1 t + \lambda)$ and Time-Dependent Force $Q\left(\rho + \frac{\mu t}{\tau}\right)$

The intersection of Row 10 and Column 6 in Guiding Table 2.1 indicates that this section describes problems associated with the action of the force of inertia, the stiffness force, the constant resistance force, and the friction force as the resisting forces, and the harmonic force and a time-dependent force as the active forces.

The current problem could be related to the working process of a vibratory system that interacts with an elastoplastic or visco-elastoplastic medium featuring strain-strengthening during its plastic deformation. As a reaction to the deformation, these types of media exert a stiffness force, a constant resisting force, and a friction force that are acting simultaneously. (Additional information related to the deformation of these media and to the behavior of a constant resisting force and a friction force applied to a vibratory system is provided in section 1.2.) Sometimes, at the beginning of the working process, the system is subjected to a harmonic force along with a time-dependent force that acts for a predetermined interval of time.

The system is moving in the horizontal direction. We want to determine the basic parameters of motion. The model of a system subjected to the action of a harmonic force, a time-dependent force, a stiffness force, a constant resisting force, and a friction force is shown in Figure 10.6.

According to the considerations above and the model in Figure 10.6, we can assemble the left and right sides of the differential equation of motion. The left side consists of the force of inertia, the stiffness force, the constant resisting force, and the friction force. The right side includes the harmonic force and the time-dependent force. Therefore, the differential equation of motion reads:

$$m\frac{d^2x}{dt^2} + Kx + P + F = A\sin(\omega_1 t + \lambda) + Q(\rho + \frac{\mu t}{\tau}) \quad \textbf{(10.6.1)}$$

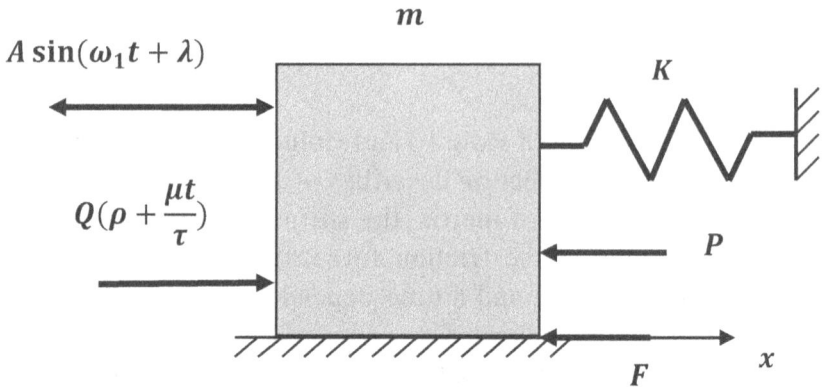

Figure 10.6 Model of a system subjected to a harmonic force, a time-dependent force, a stiffness force, a constant resisting force, and a friction force

Differential equation of motion (10.6.1) has different solutions for various initial conditions of motion. These solutions and their analyses are presented below.

10.6.1 General Initial Conditions

The general initial conditions of motion are:

$$\text{for} \quad t = 0 \quad x = s_0; \quad \frac{dx}{dt} = v_0 \qquad (10.6.2)$$

where s_0 and v_0 are the initial displacement and initial velocity respectively.

Transforming equation (10.6.1), we may write:

$$m\frac{d^2x}{dt^2} + Kx + P + F = Q\rho + Q\frac{\mu t}{\tau}$$
$$+ A\sin\omega_1 t \cos\lambda + A\cos\omega_1 t \sin\lambda \qquad (10.6.3)$$

Dividing equation (10.6.3) by m, we have:

$$\frac{d^2x}{dt^2} + \omega^2 x + p + f = q\rho + q\frac{\mu t}{\tau}$$
$$+ a\sin\omega_1 t \cos\lambda + a\cos\omega_1 t \sin\lambda \qquad (10.6.4)$$

where ω is the natural frequency and:

$$\omega^2 = \frac{K}{m} \tag{10.6.5}$$

$$p = \frac{P}{m} \tag{10.6.6}$$

$$f = \frac{F}{m} \tag{10.6.7}$$

$$q = \frac{Q}{m} \tag{10.6.8}$$

$$a = \frac{A}{m} \tag{10.6.9}$$

Using Laplace Transform Pairs 3, 5, 1, 2, 6, and 7 from Table 1.1, we convert differential equation of motion (10.6.4) with the initial conditions of motion (10.6.2) from the time domain into the Laplace domain. Upon conversion, we obtain the resulting algebraic equation of motion in the Laplace domain:

$$l^2 x(l) - l v_0 - l^2 s_0 + \omega^2 x(l) + p + f = q p + \frac{q\mu}{\tau l}$$

$$+ \frac{a\omega_1 l}{l^2 + \omega_1^2} \cos \lambda + \frac{a l^2}{l^2 + \omega_1^2} \sin \lambda \tag{10.6.10}$$

Solving equation (10.6.10) for the Laplace domain displacement $x(l)$, we may write:

$$x(l) = \frac{l v_0}{l^2 + \omega^2} + \frac{l^2 s_0}{l^2 + \omega^2} + \frac{q p - p - f}{l^2 + \omega^2}$$

$$+ \frac{q\mu}{\tau l(l^2 + \omega^2)} + \frac{l a \omega_1}{(l^2 + \omega^2)(l^2 + \omega_1^2)} \cos \lambda$$

$$+ \frac{l^2 a}{(l^2 + \omega^2)(l^2 + \omega_1^2)} \sin \lambda \tag{10.6.11}$$

Based on pairs $1, 6, 7, 14, 16, 33$, and 34 from Table 1.1, we invert equation (10.6.11) from the Laplace domain into the time domain and obtain the solution of differential equation of motion (10.6.1) with the initial conditions of motion (10.6.2):

$$x = \frac{v_0}{\omega}\sin\omega t + s_0\cos\omega t + \frac{q\rho - p - f}{\omega^2}(1-\cos\omega t)$$

$$+\frac{q\mu}{\tau\omega^2}(t-\frac{1}{\omega}\sin\omega t) + \frac{a(\omega\sin\omega_1 t - \omega_1\sin\omega t)\cos\lambda}{\omega(\omega^2 - \omega_1^2)}$$

$$+\frac{a(\cos\omega_1 t - \cos\omega t)\sin\lambda}{\omega^2 - \omega_1^2} \tag{10.6.12}$$

Taking the first and second derivatives of equation (10.6.12), we determine the velocity and the acceleration of the system respectively:

$$\frac{dx}{dt} = v_0\cos\omega t - s_0\omega\cos\omega t + \frac{q\rho - p - f}{\omega}\sin\omega t + \frac{q\mu}{\tau\omega^2}$$

$$-\frac{q\mu}{\tau\omega^2}\cos\omega t + \frac{a\omega_1(\cos\omega_1 t - \cos\omega t)\cos\lambda}{\omega^2 - \omega_1^2}$$

$$-\frac{a(\omega_1\sin\omega_1 t - \omega\sin\omega t)\sin\lambda}{\omega^2 - \omega_1^2} \tag{10.6.13}$$

$$\frac{d^2x}{dt^2} = -v_0\omega\sin\omega t - s_0\omega^2\cos\omega t + (q\rho - p - f)\cos\omega t$$

$$+\frac{q\mu}{\tau\omega}\sin\omega t - \frac{a\omega_1(\omega_1\sin\omega_1 t - \omega\sin\omega t)\cos\lambda}{\omega^2 - \omega_1^2}$$

$$-\frac{a(\omega_1^2\cos\omega_1 t - \omega^2\cos\omega t)\sin\lambda}{\omega^2 - \omega_1^2} \tag{10.6.14}$$

10.6.2 Initial Displacement Equals Zero

The initial conditions of motion are:

$$\text{for} \quad t = 0 \quad x = 0; \quad \frac{dx}{dt} = v_0 \tag{10.6.15}$$

The solution of differential equation of motion (10.6.1) with the initial conditions of motion (10.6.15) reads:

$$x = \frac{v_0}{\omega}\sin\omega t + \frac{q\rho - p - f}{\omega^2}(1 - \cos\omega t) + \frac{q\mu}{\tau\omega^2}\left(t - \frac{1}{\omega}\sin\omega t\right)$$

$$+ \frac{a(\omega\sin\omega_1 t - \omega_1\sin\omega t)\cos\lambda}{\omega(\omega^2 - \omega_1^2)}$$

$$+ \frac{a(\cos\omega_1 t - \cos\omega t)\sin\lambda}{\omega^2 - \omega_1^2} \tag{10.6.16}$$

The first and second derivatives of equation (10.6.16) represent the velocity and the acceleration of the system respectively:

$$\frac{dx}{dt} = v_0\cos\omega t + \frac{q\rho - p - f}{\omega}\sin\omega t + \frac{q\mu}{\tau\omega^2} - \frac{q\mu}{\tau\omega^2}\cos\omega t$$

$$+ \frac{a\omega_1(\cos\omega_1 t - \cos\omega t)\cos\lambda}{\omega^2 - \omega_1^2}$$

$$- \frac{a(\omega_1\sin\omega_1 t - \omega\sin\omega t)\sin\lambda}{\omega^2 - \omega_1^2} \tag{10.6.17}$$

$$\frac{d^2x}{dt^2} = -v_0\omega\sin\omega t + (q\rho - p - f)\cos\omega t + \frac{q\mu}{\tau\omega}\sin\omega t$$

$$- \frac{a\omega_1(\omega_1\sin\omega_1 t - \omega\sin\omega t)\cos\lambda}{\omega^2 - \omega_1^2}$$

$$- \frac{a(\omega_1^2\cos\omega_1 t - \omega^2\cos\omega t)\sin\lambda}{\omega^2 - \omega_1^2} \tag{10.6.18}$$

10.6.3 Initial Velocity Equals Zero

The initial conditions of motion are:

$$\text{for} \quad t = 0 \quad x = s_0; \quad \frac{dx}{dt} = 0 \tag{10.6.19}$$

Solving differential equation of motion (10.6.1) with the initial conditions of motion (10.6.19), we may write:

$$x = s_0 \cos \omega t + \frac{q\rho - p - f}{\omega^2}(1 - \cos \omega t) + \frac{q\mu}{\tau\omega^2}\left(t - \frac{1}{\omega}\sin \omega t\right)$$

$$+ \frac{a(\omega \sin \omega_1 t - \omega_1 \sin \omega t)\cos \lambda}{\omega(\omega^2 - \omega_1^2)}$$

$$+ \frac{a(\cos \omega_1 t - \cos \omega t)\sin \lambda}{\omega^2 - \omega_1^2} \qquad (10.6.20)$$

Taking the first and second derivatives of equation (10.6.20), we determine the velocity and acceleration of the system respectively:

$$\frac{dx}{dt} = -s_0\omega \sin \omega t + \frac{q\rho - p - f}{\omega}\sin \omega t + \frac{q\mu}{\tau\omega^2}\cos \omega t$$

$$+ \frac{a\omega_1(\cos \omega_1 t - \cos \omega t)\cos \lambda}{\omega^2 - \omega_1^2}$$

$$- \frac{a(\omega_1 \sin \omega_1 t - \omega \sin \omega t)\sin \lambda}{\omega^2 - \omega_1^2} \qquad (10.6.21)$$

$$\frac{d^2x}{dt^2} = -s_0\omega^2 \cos \omega t + (q\rho - p - f)\cos \omega t + \frac{q\mu}{\tau\omega}\sin \omega t$$

$$- \frac{a\omega_1(\omega_1 \sin \omega_1 t - \omega \sin \omega t)\cos \lambda}{\omega^2 - \omega_1^2}$$

$$- \frac{a(\omega_1^2 \cos \omega_1 t - \omega^2 \cos \omega t)\sin \lambda}{\omega^2 - \omega_1^2} \qquad (10.6.22)$$

10.6.4 Both the Initial Displacement and Velocity Equal Zero

The initial conditions of motion are:

$$\text{for} \quad t = 0 \quad x = 0; \quad \frac{dx}{dt} = 0 \qquad (10.6.23)$$

The solution of differential equation of motion (10.6.1) with the initial conditions of motion (10.6.23) reads:

$$x = \frac{q\rho - p - f}{\omega^2}(1 - \cos\omega t) + \frac{q\mu}{\tau\omega^2}(t - \frac{1}{\omega}\sin\omega t)$$

$$+ \frac{a(\omega\sin\omega_1 t - \omega_1\sin\omega t)\cos\lambda}{\omega(\omega^2 - \omega_1^2)}$$

$$+ \frac{a(\cos\omega_1 t - \cos\omega t)\sin\lambda}{\omega^2 - \omega_1^2} \qquad (10.6.24)$$

Taking the first and second derivatives of equation (10.6.24), we determine the velocity and the acceleration of the system respectively:

$$\frac{dx}{dt} = \frac{q\rho - p - f}{\omega}\sin\omega t + \frac{q\mu}{\tau\omega^2} - \frac{q\mu}{\tau\omega^2}\cos\omega t$$

$$+ \frac{a\omega_1(\cos\omega_1 t - \cos\omega t)\cos\lambda}{\omega^2 - \omega_1^2}$$

$$- \frac{a(\omega_1\sin\omega_1 t - \omega\sin\omega t)\sin\lambda}{\omega^2 - \omega_1^2} \qquad (10.6.25)$$

$$\frac{d^2 x}{dt^2} = (q\rho - p - f)\cos\omega t + \frac{q\mu}{\tau\omega}\sin\omega t$$

$$- \frac{a\omega_1(\omega_1\sin\omega_1 t - \omega\sin\omega t)\cos\lambda}{\omega^2 - \omega_1^2}$$

$$- \frac{a(\omega_1^2\cos\omega_1 t - \omega^2\cos\omega t)\sin\lambda}{\omega^2 - \omega_1^2} \qquad (10.6.26)$$

11

DAMPING

This chapter covers engineering systems subjected to the force of inertia and the damping force as the resisting forces. These forces are marked by the plus signs (+) in Row 11 of Guiding Table 2.1. The intersections of Columns 1 through 6 with this row indicate the active forces applied to the engineering systems described in the chapter's six sections. The titles of these sections reflect the active forces that characterize the problems.

The left sides of the differential equations of motion for the systems presented in this chapter are identical and consist of the force of inertia and the damping force. The right sides of these equations consist of the active forces applied to the systems.

In real conditions of motion, the vast majority of movable systems experience damping forces that represent the reaction of fluid media to their deformation.

11.1 Active Force Equals Zero

According to Guiding Table 2.1, this section describes the problems associated with the force of inertia and the damping force as the resisting forces (Row 11) at the absence of active forces (Column 1). The considerations related to the motion of a system in the absence of active forces are described in section 1.3. The current problem could be associated with the motion of a system subjected to the resistance of a fluid medium or a specific hydraulic link.

The system moves in the horizontal direction. We want to determine the basic parameters of motion and the maximum values of the displacement, the acceleration, and the force applied to the system. Figure 11.1 shows the model of a system subjected to the action of a damping force.

The considerations above and the model in Figure 11.1 let us compose the left and the right sides of the differential equation of motion. The left side consists of the force of inertia and the damping force, while the right side of the equation equals zero. Therefore, the differential equation of motion reads:

$$m\frac{d^2x}{dt^2} + C\frac{dx}{dt} = 0 \qquad (11.1.1)$$

Differential equation of motion (11.1.1) describes the deceleration process and has different solutions for various initial conditions of motion. These solutions and their analyses are presented below.

m

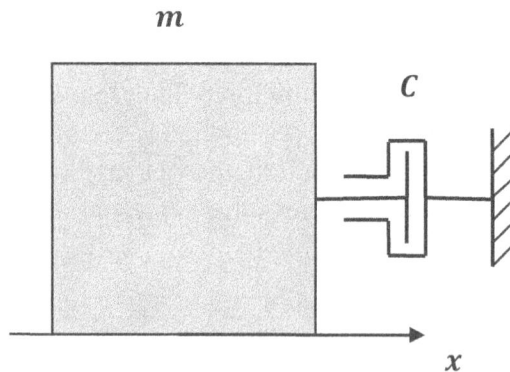

C

x

Figure 11.1 Model of a system subjected to a damping force

11.1.1 General Initial Conditions

The general initial conditions of motion are:

$$\text{for} \quad t = 0 \quad x = s_0; \quad \frac{dx}{dt} = v_0 \qquad \textbf{(11.1.2)}$$

where s_0 and v_0 are the initial displacement and initial velocity respectively.

Dividing equation (11.1.1) by m, we may write:

$$\frac{d^2x}{dt^2} + 2n\frac{dx}{dt} = 0 \qquad \textbf{(11.1.3)}$$

where n is the damping factor and:

$$2n = \frac{C}{m} \qquad \textbf{(11.1.4)}$$

Using Laplace Transform Pairs 3 and 4 from Table 1.1, we convert differential equation of motion (11.1.3) with the initial conditions of motion (11.1.2) from the time domain into the Laplace domain and obtain the corresponding algebraic equation of motion in the Laplace domain:

$$l^2 x(l) - l v_0 - l^2 s_0 + 2nlx(l) - 2nls_0 = 0 \qquad \textbf{(11.1.5)}$$

Solving equation (11.1.5) for the displacement $x(l)$ in the Laplace domain, we have:

$$x(l) = \frac{v_0 + 2ns_0}{l + 2n} + \frac{ls_0}{l + 2n} \qquad \textbf{(11.1.6)}$$

In order to invert equation (11.1.6) from the Laplace domain into the time domain, we use pairs 1, 10, and 12 from Table 1.1. The inversion represents the solution of differential equation of motion (11.1.1) with the initial conditions of motion (11.1.2):

$$x = \frac{v_0 + 2ns_0}{2n}(1 - e^{-2nt}) + s_0 e^{-2nt} \qquad \textbf{(11.1.7)}$$

Applying basic algebra to equation (11.1.7), we obtain:

$$x = \frac{1}{2n}(v_0 + 2ns_0 - v_0 e^{-2nt}) \qquad \textbf{(11.1.8)}$$

Taking the first and second derivatives of equation (11.1.8), we determine the velocity and the deceleration of the system respectively:

$$\frac{dx}{dt} = v_0 e^{-2nt} \tag{11.1.9}$$

$$\frac{d^2x}{dt^2} = -2nv_0 e^{-2nt} \tag{11.1.10}$$

Equating the velocity according to equation (11.1.9) to zero, we determine the time T that the process may last. Equation (11.1.9) shows that the term e^{-2nT} tends to zero when the time T tends to infinity. Therefore, the velocity tends to zero when:

$$e^{-2nT} \to 0 \tag{11.1.11}$$

Combining expression (11.1.11) with equations (11.1.8) and (11.1.10) and substituting notation (11.1.4), we calculate the maximum displacement s_{max} and acceleration (in this case deceleration) a_{max} of the system respectively:

$$s_{max} = \frac{C}{m} v_0 + s_0 \tag{11.1.12}$$

$$a_{max} = -\frac{C}{m} v_0 \tag{11.1.13}$$

Multiplying equation (11.1.13) by m, we determine the absolute value of the maximum force R_{max} applied to the system:

$$R_{max} = |Cv_0| \tag{11.1.14}$$

For a transportation system, the maximum value of the acceleration should be verified with the norms of public health and safety. The maximum force applied to the system is used for the corresponding stress calculations.

11.1.2 Initial Displacement Equals Zero
The initial conditions of motion are:

$$\text{for} \quad t = 0 \quad x = 0; \quad \frac{dx}{dt} = v_0 \tag{11.1.15}$$

The solution of differential equation of motion (11.1.1) with the initial conditions of motion (11.1.15) reads:

$$x = \frac{v_0}{2n}(1 - e^{-2nt}) \tag{11.1.16}$$

The rest of the parameters are the same as in the previous case.

11.2 Constant Force R

The intersection of Row 11 and Column 2 in Guiding Table 2.1 indicates that this section describes a system that is subjected to the force of inertia and the damping force as the resisting forces, and the constant active force.

In general, the current problem is related to the interaction of mechanical systems with a fluid medium or a specific hydraulic link. These systems also could represent means of ground, water, or air transportation. The working processes of a variety of hydraulic and pneumatic machines are characterized by the interaction of fluid media with movable elements subjected to constant active forces (more related information is presented in section 1.2).

The system moves in the horizontal direction. We want to determine the basic parameters of motion, the maximum values of the velocity and the acceleration, the maximum force applied to the system, and the power of the energy source. Figure 11.2 shows the

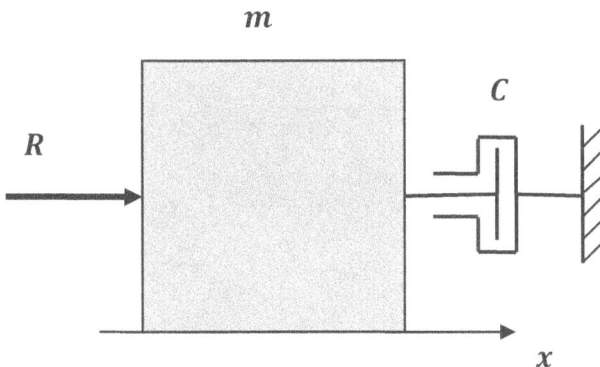

Figure 11.2 Model of a system subjected to a constant active force and a damping force

model of a system subjected to the action of a constant active force and a damping force.

The considerations above and the model in Figure 11.2 let us compose the left and the right sides of the differential equation of motion of the system. The left side includes the force of inertia and the damping force, while the right side consists of a constant active force. Therefore, the differential equation of motion reads:

$$m\frac{d^2x}{dt^2} + C\frac{dx}{dt} = R \qquad (11.2.1)$$

Differential equation of motion (11.2.1) has different solutions for various initial conditions of motion. These solutions and their analyses are presented below.

11.2.1 General Initial Conditions

The general initial conditions of motion are:

$$\text{for} \quad t = 0 \quad x = s_0; \quad \frac{dx}{dt} = v_0 \qquad (11.2.2)$$

where s_0 and v_0 are the initial displacement and initial velocity respectively.

Dividing equation (11.2.1) by m, we have:

$$\frac{d^2x}{dt^2} + 2n\frac{dx}{dt} = r \qquad (11.2.3)$$

where n is the damping factor and:

$$2n = \frac{C}{m} \qquad (11.2.4)$$

$$r = \frac{R}{m} \qquad (11.2.5)$$

Applying Laplace Transform Pairs 3, 4 and 5 from Table 1.1, we convert differential equation of motion (11.2.3) with the initial conditions of motion (11.2.2) from the time domain into the Laplace domain. The resulting algebraic equation of motion in the Laplace domain has the following shape:

$$l^2 x(l) - l v_0 - l^2 s_0 + 2nl x(l) - 2nl s_0 = r \qquad (11.2.6)$$

The solution of equation (11.2.6) for the displacement $x(l)$ in Laplace domain reads:

$$x(l) = \frac{r}{l(l+2n)} + \frac{v_0 + 2n s_0}{l+2n} + \frac{l s_0}{l+2n} \qquad (11.2.7)$$

Using pairs 1, 13, 10, and 12 from Table 1.1, we invert equation (11.2.7) from the Laplace domain into the time domain and obtain the solution of differential equation of motion (11.2.1) with the initial conditions of motion (11.2.2):

$$x = \frac{r}{2n}[t + \frac{1}{2n}(e^{-2nt} - 1)] + \frac{v_0 + 2n s_0}{2n}(1 - e^{-2nt}) + s_0 e^{-2nt} \qquad (11.2.8)$$

Transforming equation (11.2.8), we may write:

$$x = \frac{1}{2n}[rt + (\frac{r}{2n} - v_0)e^{-2nt} - \frac{r}{2n} + v_0 + 2n s_0] \qquad (11.2.9)$$

The first derivative of equation (11.2.9) represents the velocity of the system:

$$\frac{dx}{dt} = \frac{r}{2n}(1 - e^{-2nt}) + v_0 e^{-2nt} \qquad (11.2.10)$$

Taking the second derivative of equation (11.2.9), we determine the acceleration:

$$\frac{d^2 x}{dt^2} = (r - 2n v_0)e^{-2nt} \qquad (11.2.11)$$

In this case, the velocity approaches its maximum value when the acceleration tends to zero. Because $r - 2n v_0 > 0$, consequently, according to equation (11.2.11), the acceleration tends to zero when:

$$e^{-2nt} \to 0 \qquad (11.2.12)$$

Therefore, combining expression (11.2.12) with equation (11.2.10) and using notations (11.2.4) and (11.2.5), we determine the maximum value v_{max} that the velocity is tending to:

$$v_{max} \to \frac{R}{C} \qquad (11.2.13)$$

The acceleration has the maximum value at the beginning of the process of motion, when $t = 0$. Hence, at this moment, according to equation (11.2.11), the maximum value of the acceleration a_{max} reads:

$$a_{max} = r - 2nv_0 \qquad (11.2.14)$$

For public transportation systems, the maximum value of the acceleration according to equation (11.2.14) should be in compliance with the norms of public health and safety.

In general, the maximum force applied to the system equals the sum of the force of inertia and the resisting forces. However, at the time when the velocity is approaching its maximum value, the acceleration approaches zero. Therefore, in this case, the maximum force R_{max} applied to the system equals the maximum value of the resisting force, which in its turn equals the product of multiplying the damping coefficient C by the velocity according to expression (11.2.13):

$$R_{max} = R \qquad (11.2.15)$$

Multiplying the maximum force according to equation (11.2.15) by the maximum velocity according to expression (11.2.13), we determine the power N of the energy source:

$$N = \frac{R^2}{C} \qquad (11.2.16)$$

The stress calculations of the system should be based on the maximum force applied to the system according to equation (11.2.15).

11.2.2 Initial Displacement Equals Zero

The initial conditions of motion are:

$$\text{for} \quad t = 0 \quad x = 0; \quad \frac{dx}{dt} = v_0 \qquad (11.2.17)$$

Solving differential equation of motion (11.2.1) with the initial conditions of motion (11.2.17), we obtain:

$$x = \frac{1}{2n}[rt + (\frac{r}{2n} - v_0)e^{-2nt} - \frac{r}{2n} + v_0] \qquad (11.2.18)$$

The rest of the parameters and their analysis are the same as in the previous case.

11.2.3 Initial Velocity Equals Zero

The initial conditions of motion are:

$$\text{for} \quad t = 0 \quad x = s_0; \quad \frac{dx}{dt} = 0 \qquad (11.2.19)$$

The solution of differential equation of motion (11.2.1) with the initial conditions of motion (11.2.19) reads:

$$x = \frac{1}{2n}(rt + \frac{r}{2n}e^{-2nt} - \frac{r}{2n} + 2ns_0) \qquad (11.2.20)$$

The first and the second derivatives of the equation (11.2.20) represent the velocity and the acceleration respectively:

$$\frac{dx}{dt} = \frac{r}{2n}(1 - e^{-2nt}) \qquad (11.2.21)$$

$$\frac{d^2x}{dt^2} = re^{-2nt} \qquad (11.2.22)$$

Equation (11.2.22) shows that when $e^{-2nt} \to 0$, the acceleration is approaching zero and, consequently, the velocity approaches its maximum value that can be calculated according to expression (11.2.13). The value of the maximum force and the value of the power of the energy source are the same as in the previous cases and do not depend on the initial velocity.

11.2.4 Both the Initial Displacement and Velocity Equal Zero

The initial conditions of motion are:

$$\text{for} \quad t = 0 \quad x = 0; \quad \frac{dx}{dt} = 0 \qquad (11.2.23)$$

Solving differential equation of motion (11.2.1) with the initial conditions of motion (11.2.23), we obtain:

$$x = \frac{1}{2n}(rt + \frac{r}{2n}e^{-2nt} - \frac{r}{2n}) \qquad (11.2.24)$$

The rest of the parameters and their analysis are the same as in the previous case.

11.3 Harmonic Force $A \sin(\omega_1 t + \lambda)$

The intersection of Row 11 and Column 3 in Guiding Table 2.1 indicates that this section describes a system subjected to the action of the force of inertia and the damping force as the resisting forces, and the harmonic force as the active force.

The current problem could be related to the interaction of a vibratory system with fluid media or a hydraulic link. The system is moving in the horizontal direction. We want to determine the basic parameters of motion. Figure 11.3 shows the model of a system subjected to the action of a harmonic force and a damping force.

Based on considerations above and on the model in Figure 11.3, we can write the left and the right sides of the differential equation of motion of the system. The left side consists of the force of inertia and the damping force, while the right side includes the harmonic force. Therefore, the differential equation of motion reads:

$$m\frac{d^2x}{dt^2} + C\frac{dx}{dt} = A\sin(\omega_1 t + \lambda) \qquad (11.3.1)$$

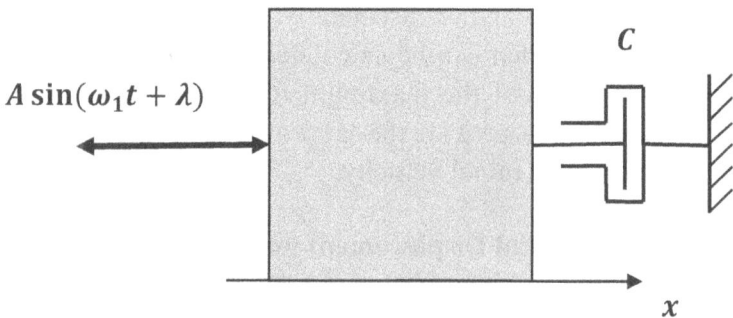

Figure 11.3 Model of a system subjected to a harmonic force and a damping force

Differential equation of motion (11.3.1) has different solutions for various initial conditions of motion. These solutions and their analyses are presented below.

11.3.1 General Initial Conditions

The general initial conditions of motion are:

$$\text{for} \quad t = 0 \quad x = s_0; \quad \frac{dx}{dt} = v_0 \qquad (11.3.2)$$

where s_0 and v_0 are the initial displacement and initial velocity respectively.

Transforming the sinusoidal function in the equation (11.3.1) and dividing the latter by m, we have:

$$\frac{d^2x}{dt^2} + 2n\frac{dx}{dt} = a\sin\omega_1 t\cos\lambda + a\cos\omega_1 t\sin\lambda \qquad (11.3.3)$$

where n is the damping factor and:

$$2n = \frac{C}{m} \qquad (11.3.4)$$

$$a = \frac{A}{m} \qquad (11.3.5)$$

Using Laplace Transform Pairs 3, 4, 6, and 7 from Table 1.1, we convert differential equation of motion (11.3.3) with the initial conditions of motion (11.3.2) from the time domain into the Laplace domain. The resulting algebraic equation of motion in the Laplace domain reads:

$$l^2 x(l) - lv_0 - l^2 s_0 + 2nlx(l) - 2nls_0 = \frac{a\omega_1 l}{l^2 + \omega_1^2}\cos\lambda + \frac{al^2}{l^2 + \omega_1^2}\sin\lambda$$

$$(11.3.6)$$

Applying some basic algebra to equation (11.3.6), we may write:

$$x(l)l(l + 2n) = l(2ns_0 + v_0) + s_0 l^2 + \frac{a\omega_1 l}{l^2 + \omega_1^2}\cos\lambda + \frac{al^2}{l^2 + \omega_1^2}\sin\lambda$$

$$(11.3.7)$$

Solving equation (11.3.7) for the Laplace domain displacement $x(l)$, we obtain:

$$x(l) = \frac{2ns_0 + v_0}{l + 2n} + \frac{ls_0}{l + 2n} + \frac{a\omega_1}{(l + 2n)(l^2 + \omega_1^2)} \cos \lambda$$

$$+ \frac{la}{(l + 2n)(l^2 + \omega_1^2)} \sin \lambda \qquad (11.3.8)$$

In order to invert equation (11.3.8) from the Laplace domain into the time domain we use pairs 1, 10, 12, 19, and 20 from Table 1.1 and obtain the solution of differential equation of motion (11.3.1) with the initial conditions of motion (11.3.2):

$$x = \frac{2ns_0 + v_0}{2n}(1 - e^{-2nt}) + s_0 e^{-2nt} + \frac{2na(1 - \cos\omega_1 t)\cos\lambda}{\omega_1(\omega_1^2 + 4n^2)}$$

$$- \frac{a\sin\omega_1 t \cos\lambda}{\omega_1^2 + 4n^2} + \frac{a\omega_1(1 - e^{-2nt})\cos\lambda}{2n(\omega_1^2 + 4n^2)} + \frac{a(1 - \cos\omega_1 t)\sin\lambda}{\omega_1^2 + 4n^2}$$

$$+ \frac{2na\sin\omega_1 t \sin\lambda}{\omega_1(\omega_1^2 + 4n^2)} - \frac{a(1 - e^{-2nt})\sin\lambda}{\omega_1^2 + 4n^2} \qquad (11.3.9)$$

Applying the appropriate algebraic procedures to equation (11.3.9), we have:

$$x = s_0 + \frac{v_0}{2n}(1 - e^{-2nt}) + \frac{2na(1 - \cos\omega_1 t)\cos\lambda}{\omega_1(\omega_1^2 + 4n^2)} - \frac{a\sin\omega_1 t \cos\lambda}{\omega_1^2 + 4n^2}$$

$$+ \frac{a\omega_1(1 - e^{-2nt})\cos\lambda}{2n(\omega_1^2 + 4n^2)} + \frac{a(1 - \cos\omega_1 t)\sin\lambda}{\omega_1^2 + 4n^2}$$

$$+ \frac{2na\sin\omega_1 t \sin\lambda}{\omega_1(\omega_1^2 + 4n^2)} - \frac{a(1 - e^{-2nt})\sin\lambda}{\omega_1^2 + 4n^2} \qquad (11.3.10)$$

Taking the first derivative of equation (11.3.10), we determine the velocity of the system:

$$\frac{dx}{dt} = v_0 e^{-2nt} + \frac{2na\sin\omega_1 t \cos\lambda}{\omega_1^2 + 4n^2}$$

$$- \frac{a\omega_1 \cos\omega_1 t \cos\lambda}{\omega_1^2 + 4n^2} + \frac{a\omega_1 e^{-2nt} \cos\lambda}{\omega_1^2 + 4n^2} + \frac{a\omega_1 \sin\omega_1 t \sin\lambda}{\omega_1^2 + 4n^2}$$

$$+\frac{2na\cos\omega_1 t\sin\lambda}{\omega_1^2+4n^2}-\frac{2nae^{-2nt}\sin\lambda}{\omega_1^2+4n^2} \qquad (11.3.11)$$

The second derivative of equation (11.3.10) represents the acceleration of the system:

$$\frac{d^2x}{dt^2}=-2nv_0e^{-2nt}+\frac{2na\omega_1\cos\omega_1 t\cos\lambda}{\omega_1^2+4n^2}+\frac{a\omega_1^2\sin\omega_1 t\cos\lambda}{\omega_1^2+4n^2}$$

$$-\frac{2na\omega_1e^{-2nt}\cos\lambda}{\omega_1^2+4n^2}+\frac{a\omega_1^2\cos\omega_1 t\sin\lambda}{\omega_1^2+4n^2}$$

$$-\frac{2na\omega_1\sin\omega_1 t\sin\lambda}{\omega_1^2+4n^2}+\frac{4n^2ae^{-2nt}\sin\lambda}{\omega_1^2+4n^2} \qquad (11.3.12)$$

11.3.2 Initial Displacement Equals Zero

The initial condition of motion are:

$$\text{for}\quad t=0\quad x=0;\quad \frac{dx}{dt}=v_0 \qquad (11.3.13)$$

The solution of differential equation of motion (11.3.1) with the initial conditions of motion (11.3.13) reads:

$$x=\frac{v_0}{2n}(1-e^{-2nt})+\frac{2na(1-\cos\omega_1 t)\cos\lambda}{\omega_1(\omega_1^2+4n^2)}-\frac{a\sin\omega_1 t\cos\lambda}{\omega_1^2+4n^2}$$

$$+\frac{a\omega_1(1-e^{-2nt})\cos\lambda}{2n(\omega_1^2+4n^2)}+\frac{a(1-\cos\omega_1 t)\sin\lambda}{\omega_1^2+4n^2}$$

$$+\frac{2na\sin\omega_1 t\sin\lambda}{\omega_1(\omega_1^2+4n^2)}-\frac{a(1-e^{-2nt})\sin\lambda}{\omega_1^2+4n^2} \qquad (11.3.14)$$

The equations for the velocity and the acceleration are the same as in the previous case.

11.3.3 Initial Velocity Equals Zero

The initial conditions of motion are:

$$\text{for}\quad t=0\quad x=s_0;\quad \frac{dx}{dt}=0 \qquad (11.3.15)$$

Solving differential equation of motion (11.3.1) with the initial conditions of motion (11.3.15), we obtain:

$$x = s_0 + \frac{2na(1 - \cos\omega_1 t)\cos\lambda}{\omega_1(\omega_1^2 + 4n^2)} - \frac{a\sin\omega_1 t\cos\lambda}{\omega_1^2 + 4n^2} + \frac{a\omega_1(1 - e^{-2nt})\cos\lambda}{2n(\omega_1^2 + 4n^2)}$$

$$+ \frac{a(1 - \cos\omega_1 t)\sin\lambda}{\omega_1^2 + 4n^2} + \frac{2na\sin\omega_1 t\sin\lambda}{\omega_1(\omega_1^2 + 4n^2)} - \frac{a(1 - e^{-2nt})\sin\lambda}{\omega_1^2 + 4n^2}$$

$$(11.3.16)$$

Taking the first and the second derivatives of equation (11.3.16), we determine the velocity and the acceleration of the system respectively:

$$\frac{dx}{dt} = \frac{2na\sin\omega_1 t\cos\lambda}{\omega_1^2 + 4n^2} - \frac{a\omega_1\cos\omega_1 t\cos\lambda}{\omega_1^2 + 4n^2} + \frac{a\omega_1 e^{-2nt}\cos\lambda}{\omega_1^2 + 4n^2}$$

$$+ \frac{a\omega_1\sin\omega_1 t\sin\lambda}{\omega_1^2 + 4n^2} + \frac{2na\cos\omega_1 t\sin\lambda}{\omega_1^2 + 4n^2} - \frac{2nae^{-2nt}\sin\lambda}{\omega_1^2 + 4n^2} \quad (11.3.17)$$

$$\frac{d^2x}{dt^2} = \frac{2na\omega_1\cos\omega_1 t\cos\lambda}{\omega_1^2 + 4n^2} + \frac{a\omega_1^2\sin\omega_1 t\cos\lambda}{\omega_1^2 + 4n^2} - \frac{2na\omega_1 e^{-2nt}\cos\lambda}{\omega_1^2 + 4n^2}$$

$$+ \frac{a\omega_1^2\cos\omega_1 t\sin\lambda}{\omega_1^2 + 4n^2} - \frac{2na\omega_1\sin\omega_1 t\sin\lambda}{\omega_1^2 + 4n^2} + \frac{4n^2 ae^{-2nt}\sin\lambda}{\omega_1^2 + 4n^2}$$

$$(11.3.18)$$

11.3.4 Both the Initial Displacement and Velocity Equal Zero

The initial conditions of motion are:

$$\text{for} \quad t = 0 \quad x = 0; \quad \frac{dx}{dt} = 0 \quad (11.3.19)$$

The solution of differential equation of motion (11.3.1) with the initial conditions of motion (11.3.19) reads:

$$x = \frac{2na(1 - \cos\omega_1 t)\cos\lambda}{\omega_1(\omega_1^2 + 4n^2)} - \frac{a\sin\omega_1 t\cos\lambda}{\omega_1^2 + 4n^2} + \frac{a\omega_1(1 - e^{-2nt})\cos\lambda}{2n(\omega_1^2 + 4n^2)}$$

$$+ \frac{a(1 - \cos\omega_1 t)\sin\lambda}{\omega_1^2 + 4n^2} + \frac{2na\sin\omega_1 t\sin\lambda}{\omega_1(\omega_1^2 + 4n^2)} - \frac{a(1 - e^{-2nt})\sin\lambda}{\omega_1^2 + 4n^2}$$

$$(11.3.20)$$

The equations for the velocity and acceleration are the same as for the previous case.

11.4 Time-Dependent Force $Q\left(\rho + \frac{\mu t}{\tau}\right)$

The intersection of Row 11 and Column 4 in Guiding Table 2.1 indicates that this section describes problems associated with the force of inertia and damping force as the resisting forces, and the time-dependent force as the active force.

The current problem could represent the initial stage of the acceleration process of a transportation system interacting with fluid media. This stage lasts for a predetermined interval of time during which the active force represents an increasing function of time.

The system is moving in the horizontal direction. We want to determine the basic parameters of motion, the values of the velocity, the acceleration, the force applied to the system, and the power of the energy source at the end of the predetermined interval of time. The model of a system subjected to the action of a time-dependent force and a damping force is shown in Figure 11.4.

Based on the considerations above and on the model in Figure 11.4, we can compose the left and the right sides of the differential equation of motion. The left side includes the force of inertia and the

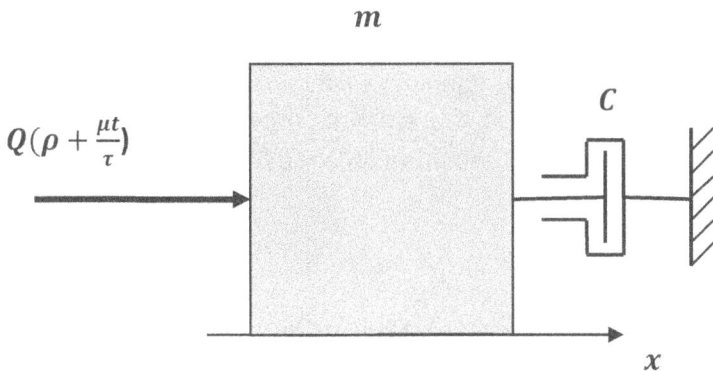

Figure 11.4 Model of a system subjected to a time-dependent force and a damping force

damping force, while the right side consists of the time-dependent force. Hence, the differential equation of motion reads:

$$m\frac{d^2x}{dt^2} + C\frac{dx}{dt} = Q(\rho + \frac{\mu t}{\tau}) \tag{11.4.1}$$

Differential equation of motion (11.4.1) has different solutions for various initial conditions of motion. These solutions and their analyses are presented below.

11.4.1 General Initial Conditions

The general initial conditions of motion are:

$$\text{for} \quad t = 0 \quad x = s_0; \quad \frac{dx}{dt} = v_0 \tag{11.4.2}$$

where s_0 and v_0 are the initial displacement and initial velocity respectively.

Dividing equation (11.4.1) by m, we obtain:

$$\frac{d^2x}{dt^2} + 2n\frac{dx}{dt} = q(\rho + \frac{\mu t}{\tau}) \tag{11.4.3}$$

where n is the damping factor and:

$$2n = \frac{C}{m} \tag{11.4.4}$$

$$q = \frac{Q}{m} \tag{11.4.5}$$

Using Laplace Transform Pairs 3, 4, 5, and 2 from Table 1.1, we convert differential equation of motion (11.4.3) with the initial conditions of motion (11.4.2) from the time domain into the Laplace domain, and obtain the resulting algebraic equation of motion in the Laplace domain:

$$l^2x(l) - lv_0 - l^2s_0 + 2nlx(l) - 2nls_0 = q\rho + \frac{q\mu}{\tau l} \tag{11.4.6}$$

The solution of equation (11.4.6) for the Laplace domain displacement $x(l)$ reads:

$$x(l) = \frac{v_0 + 2ns_0}{l + 2n} + \frac{ls_0}{l + 2n} + \frac{q\rho}{l(l + 2n)} + \frac{q\mu}{\tau l^2(l + 2n)} \tag{11.4.7}$$

Based on pairs 1, 10, 12, 13, and 31 from Table 1.1, we invert equation (11.4.7) from the Laplace domain into the time domain and obtain the solution of differential equation of motion (11.4.1) with the initial conditions of motion (11.4.2):

$$x = \frac{v_0 + 2ns_0}{2n}(1 - e^{-2nt}) + s_0 e^{-2nt} + \frac{q\rho}{2n}\left[t + \frac{1}{2n}(e^{-2nt} - 1)\right]$$

$$+ \frac{q\mu}{4n\tau}[t^2 - \frac{1}{n}t - \frac{1}{2n^2}(e^{-2nt} - 1)] \tag{11.4.8}$$

Performing some conventional algebraic procedures with equation (11.4.8), we obtain:

$$x = \frac{1}{2n}(v_0 + 2ns_0 - v_0 e^{-2nt})$$

$$+ \frac{q}{2n}[(\rho - \frac{\mu}{2n\tau})t + \frac{\mu}{2\tau}t^2 + (\frac{\rho}{2n} - \frac{\mu}{4n^2\tau})(e^{-2nt} - 1)] \tag{11.4.9}$$

The first derivative of equation (11.4.9) represents the velocity of the system:

$$\frac{dx}{dt} = v_0 e^{-2nt} + \frac{q}{2n}[\rho - \frac{\mu}{2n\tau} + \frac{\mu t}{\tau} - (\rho - \frac{\mu}{2n\tau})e^{-2nt}] \tag{11.4.10}$$

Taking the second derivative of equation (11.4.9), we determine the acceleration:

$$\frac{d^2x}{dt^2} = -2nv_0 e^{-2nt} + \frac{q\mu}{2n\tau} + q(\rho - \frac{\mu}{2n\tau})e^{-2nt} \tag{11.4.11}$$

Substituting the time τ into equations (11.4.10), (11.4.11), we determine the velocity v and acceleration a at end of the predetermined interval of time respectively:

$$v = v_0 e^{-2n\tau} + \frac{q}{2n}[\rho - \frac{\mu}{2n\tau} + \mu - (\rho - \frac{\mu}{2n\tau})e^{-2n\tau}] \tag{11.4.12}$$

$$a = -2nv_0 e^{-2n\tau} + \frac{q\mu}{2n\tau} + q(\rho - \frac{\mu}{2n\tau})e^{-2n\tau} \tag{11.4.13}$$

For a public transportation system, the acceleration according to equation (11.4.13) should be in compliance with the norms of public health and safety.

The force applied to the system equals the sum of the force of inertia and the resisting forces. The force of inertia equals the product of multiplying the acceleration according to equation (11.4.13) by the mass m. The resisting force, represented by the damping force, equals the product of multiplying C by the velocity according to equation (11.4.12). Adding these two forces together and recalling notations (11.4.4) and (11.4.5), we determine the force R_1 applied to the system at the end of the predetermined interval of time:

$$R_1 = Q(\rho + \mu) \tag{11.4.14}$$

The stress calculations of the system should be based on the force according to equation (11.4.14).

The power of the energy source N equals the product of multiplying the force according to equation (11.4.14) by the velocity according to equation (11.4.12):

$$N = R_1 v \tag{11.4.15}$$

11.4.2 Initial Displacement Equals Zero

The initial conditions of motion are:

$$\text{for} \quad t = 0 \quad x = 0; \quad \frac{dx}{dt} = v_0 \tag{11.4.16}$$

The solution of differential equation of motion (11.4.1) with the initial conditions of motion (11.4.16) reads:

$$x = \frac{v_0}{2n}(1 - e^{-2nt}) + \frac{q}{2n}[(\rho - \frac{\mu}{2n\tau})t + \frac{\mu}{2\tau}t^2 + (\frac{\rho}{2n} - \frac{\mu}{4n^2\tau})(e^{-2nt} - 1)] \tag{11.4.17}$$

The rest of the parameters and their analysis are the same as for the previous case.

11.4.3 Initial Velocity Equals Zero

The initial conditions of motion are:

$$\text{for} \quad t = 0 \quad x = s_0; \quad \frac{dx}{dt} = 0 \tag{11.4.18}$$

Solving differential equation of motion (11.4.1) with the initial conditions of motion (11.4.18), we have:

$$x = s_0 + \frac{q}{2n}[(\rho - \frac{\mu}{2n\tau})t + \frac{\mu}{2\tau}t^2 + (\frac{\rho}{2n} - \frac{\mu}{4n^2\tau})(e^{-2nt} - 1)] \tag{11.4.19}$$

Taking the first and second derivatives of equation (11.4.19), we determine the velocity and acceleration of the system respectively:

$$\frac{dx}{dt} = \frac{q}{2n}[\rho - \frac{\mu}{2n\tau} + \frac{\mu t}{\tau} - (\rho - \frac{\mu}{2n\tau})e^{-2nt}] \tag{11.4.20}$$

$$\frac{d^2x}{dt^2} = \frac{q\mu}{2n\tau} + q(\rho - \frac{\mu}{2n\tau})e^{-2nt} \tag{11.4.21}$$

Substituting the time τ into equations (11.4.20) and (11.4.21), we determine the values of the velocity v and the acceleration a at the end of the predetermined interval of time respectively:

$$v = \frac{q}{2n}[\rho - \frac{\mu}{2n\tau} + \mu - (\rho - \frac{\mu}{2n\tau})e^{-2n\tau}] \tag{11.4.22}$$

$$a = \frac{q\mu}{2n\tau} + q(\rho - \frac{\mu}{2n\tau})e^{-2n\tau} \tag{11.4.23}$$

The force applied to the system can be calculated according to equation (11.4.14). Multiplying the force according to equation (11.4.14) by the velocity according to equation (11.4.22), and recalling notations (11.4.4) and (11.4.5), we calculate the power of the energy system:

$$N = \frac{Q^2}{C}(\rho + \mu)\frac{q}{2n}[\rho - \frac{\mu}{2n\tau} + \mu - (\rho - \frac{\mu}{2n\tau})e^{-2n\tau}]$$

11.4.4 Both the Initial Displacement and Velocity Equal Zero
The initial conditions are:

$$\text{for} \quad t = 0 \quad x = 0; \quad \frac{dx}{dt} = 0 \qquad (11.4.24)$$

Solving differential equation of motion (11.4.1) with the initial conditions of motion (11.4.24), we have:

$$x = \frac{q}{2n}[(\rho - \frac{\mu}{2n\tau})t + \frac{\mu}{2\tau}t^2 + (\frac{\rho}{2n} - \frac{\mu}{4n^2\tau})(e^{-2nt} - 1)] \quad (11.4.25)$$

The rest of the parameters and their analysis are the same as for the previous case.

11.5 Constant Force R and Harmonic Force $A\sin(\omega_1 t + \lambda)$
The intersection of Row 11 and Column 5 in Guiding Table 2.1 indicates that this section describes engineering systems subjected to the action of the force of inertia and the damping force as the resisting forces, and to the action of a constant active force and a harmonic force as the active forces. The current problems could be related to some vibratory systems interacting with a fluid medium or a hydraulic link.

The system is moving in the horizontal direction. We want to determine the basic parameters of motion. Figure 11.5 shows the

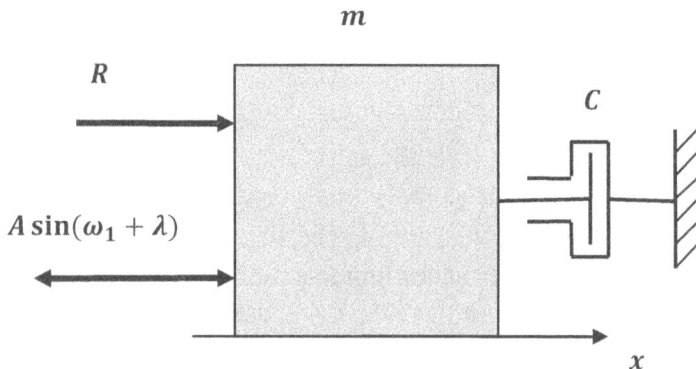

Figure 11.5 Model of a system subjected to a constant active force, a harmonic force, and a damping force

model of a system subjected to the action of a constant active force, a harmonic force, and a damping force.

The considerations above and the model shown in Figure 11.5 let us compose the left and right sides of the differential equation of motion of the system. The left side consists of the force of inertia and the damping force, while the right side includes the constant active force and the harmonic force. Therefore, the differential equation of motion reads:

$$m\frac{d^2x}{dt^2} + C\frac{dx}{dt} = R + A\sin(\omega_1 t + \lambda) \qquad (11.5.1)$$

Differential equation of motion (11.5.1) has different solutions for various initial conditions of motion. These solutions and their analyses are presented below.

11.5.1 General Initial Conditions

The general initial conditions of motion are:

$$\text{for} \quad t = 0 \quad x = s_0; \quad \frac{dx}{dt} = v_0 \qquad (11.5.2)$$

Transforming the sinusoidal function in equation (11.5.1) and dividing the latter by m, we obtain:

$$\frac{d^2x}{dt^2} + 2n\frac{dx}{dt} = r + a\sin\omega_1 t\cos\lambda + a\cos\omega_1 t\sin\lambda \quad (11.5.3)$$

where n is the damping factor and:

$$2n = \frac{C}{m} \qquad (11.5.4)$$

$$r = \frac{R}{m} \qquad (11.5.5)$$

$$a = \frac{A}{m} \qquad (11.5.6)$$

Using Laplace Transform Pairs 3, 4, 5, 6, and 7 from Table 1.1, we convert differential equation of motion (11.5.3) with the initial conditions of motion (11.5.2) from the time domain into the Laplace domain, and obtain the corresponding algebraic equation of motion in the Laplace domain:

$$l^2 x(l) - l v_0 - l^2 s_0 + 2nlx(l) - 2nls_0 = r + \frac{a\omega_1 l \cos\lambda}{l^2 + \omega_1^2} + \frac{al^2 \sin\lambda}{l^2 + \omega_1^2}$$

$$(11.5.7)$$

Applying basic algebra to equation (11.5.7), we may write:

$$x(l)l(l+2n) = r + l(v_0 + 2ns_0) + l^2 s_0 + \frac{a\omega_1 l \cos\lambda}{l^2 + \omega_1^2} + \frac{al^2 \sin\lambda}{l^2 + \omega_1^2}$$

$$(11.5.8)$$

The solution of equation (11.5.8) for the displacement $x(l)$ in Laplace domain reads:

$$x(l) = \frac{r}{l(l+2n)} + \frac{v_0 + 2ns_0}{l+2n} + \frac{ls_0}{l+2n}$$
$$+ \frac{a\omega_1 \cos\lambda}{(l+2n)(l^2+\omega_1^2)} + \frac{al \sin\lambda}{(l+2n)(l^2+\omega_1^2)} \quad (11.5.9)$$

Based on pairs 1, 13, 10, 12, 19, and 20 from Table 1.1, we invert equation (11.5.9) from the Laplace domain into the time domain and obtain the solution of differential equation of motion (11.5.1) with the initial conditions of motion (11.5.2):

$$x = \frac{r}{2n}[t + \frac{1}{2n}(e^{-2nt}-1)] + \frac{v_0+2ns_0}{2n}(1-e^{-2nt}) + s_0 e^{-2nt}$$
$$+ \frac{2na(1-\cos\omega_1 t)\cos\lambda}{\omega_1(\omega_1^2+4n^2)} - \frac{a\sin\omega_1 t\cos\lambda}{\omega_1^2+4n^2} + \frac{a\omega_1(1-e^{-2nt})\cos\lambda}{2n(\omega_1^2+4n^2)}$$
$$+ \frac{a(1-\cos\omega_1 t)\sin\lambda}{\omega_1^2+4n^2} + \frac{2na\sin\omega_1 t\sin\lambda}{\omega_1(\omega_1^2+4n^2)} - \frac{a(1-e^{-2nt})\sin\lambda}{\omega_1^2+4n^2}$$

$$(11.5.10)$$

Transforming equation (11.5.10), we may write:

$$x = s_0 + \frac{v_0}{2n}(1-e^{-2nt}) + \frac{r}{2n}[t + \frac{1}{2n}(e^{-2nt}-1)] + \frac{2na(1-\cos\omega_1 t)\cos\lambda}{\omega_1(\omega_1^2+4n^2)}$$

$$-\frac{a\sin\omega_1 t\cos\lambda}{\omega_1^2+4n^2}+\frac{a\omega_1(1-e^{-2nt})\cos\lambda}{2n(\omega_1^2+4n^2)}+\frac{a(1-\cos\omega_1 t)\sin\lambda}{\omega_1^2+4n^2}$$

$$+\frac{2na\sin\omega_1 t\sin\lambda}{\omega_1(\omega_1^2+4n^2)}-\frac{a(1-e^{-2nt})\sin\lambda}{\omega_1^2+4n^2}\qquad (11.5.11)$$

Taking the first derivative of equation (11.5.11), we determine the velocity of the system:

$$\frac{dx}{dt}=v_0 e^{-2nt}+\frac{r}{2n}(1-e^{-2nt})+\frac{2na\sin\omega_1 t\cos\lambda}{\omega_1^2+4n^2}-\frac{a\omega_1\cos\omega_1 t\cos\lambda}{\omega_1^2+4n^2}$$

$$+\frac{a\omega_1 e^{-2nt}\cos\lambda}{\omega_1^2+4n^2}+\frac{a\omega_1\sin\omega_1 t\sin\lambda}{\omega_1^2+4n^2}$$

$$+\frac{2na\cos\omega_1 t\sin\lambda}{\omega_1^2+4n^2}-\frac{2nae^{-2nt}\sin\lambda}{\omega_1^2+4n^2}\qquad (11.5.12)$$

The second derivative of equation (11.5.11) represents the acceleration:

$$\frac{d^2x}{dt^2}=(r-2nv_0)e^{-2nt}+\frac{2na\omega_1\cos\omega_1 t\cos\lambda}{\omega_1^2+4n^2}+\frac{a\omega_1^2\sin\omega_1 t\cos\lambda}{\omega_1^2+4n^2}$$

$$-\frac{2na\omega_1 e^{-2nt}\cos\lambda}{\omega_1^2+4n^2}+\frac{a\omega_1^2\cos\omega_1 t\sin\lambda}{\omega_1^2+4n^2}$$

$$-\frac{2na\omega_1\sin\omega_1 t\sin\lambda}{\omega_1^2+4n^2}+\frac{4n^2ae^{-2nt}\sin\lambda}{\omega_1^2+4n^2}\qquad (11.5.13)$$

11.5.2 Initial Displacement Equals Zero

The initial conditions of motion are:

$$\text{for}\quad t=0\quad x=0;\quad \frac{dx}{dt}=v_0\qquad (11.5.14)$$

Solving differential equation of motion (11.5.1) with the initial conditions of motion (11.5.14), we may write:

$$x=\frac{v_0}{2n}(1-e^{-2nt})+\frac{r}{2n}[t+\frac{1}{2n}(e^{-2nt}-1)]+\frac{2na(1-\cos\omega_1 t)\cos\lambda}{\omega_1(\omega_1^2+4n^2)}$$

$$-\frac{a\sin\omega_1 t\cos\lambda}{\omega_1^2+4n^2}+\frac{a\omega_1(1-e^{-2nt})\cos\lambda}{2n(\omega_1^2+4n^2)}+\frac{a(1-\cos\omega_1 t)\sin\lambda}{\omega_1^2+4n^2}$$

$$+\frac{2na\sin\omega_1 t\sin\lambda}{\omega_1(\omega_1^2+4n^2)}-\frac{a(1-e^{-2nt})\sin\lambda}{\omega_1^2+4n^2} \qquad \textbf{(11.5.15)}$$

The equations for the velocity and acceleration are the same as in the previous case.

11.5.3 Initial Velocity Equals Zero

The initial conditions of motion are:

$$\text{for}\quad t=0\quad x=s_0;\quad \frac{dx}{dt}=0 \qquad \textbf{(11.5.16)}$$

The solution of differential equation of motion (11.5.1) with the initial conditions of motion (11.5.16) reads:

$$x=s_0+\frac{r}{2n}[t+\frac{1}{2n}(e^{-2nt}-1)]+\frac{2na(1-\cos\omega_1 t)\cos\lambda}{\omega_1(\omega_1^2+4n^2)}$$

$$-\frac{a\sin\omega_1 t\cos\lambda}{\omega_1^2+4n^2}+\frac{a\omega_1(1-e^{-2nt})\cos\lambda}{2n(\omega_1^2+4n^2)}+\frac{a(1-\cos\omega_1 t)\sin\lambda}{\omega_1^2+4n^2}$$

$$+\frac{2na\sin\omega_1 t\sin\lambda}{\omega_1(\omega_1^2+4n^2)}-\frac{a(1-e^{-2nt})\sin\lambda}{\omega_1^2+4n^2} \qquad \textbf{(11.5.17)}$$

Taking the first derivative of equation (11.5.17), we determine the velocity of the system:

$$\frac{dx}{dt}=\frac{r}{2n}(1-e^{-2nt})+\frac{2na\sin\omega_1 t\cos\lambda}{\omega_1^2+4n^2}-\frac{a\omega_1\cos\omega_1 t\cos\lambda}{\omega_1^2+4n^2}$$

$$+\frac{a\omega_1 e^{-2nt}\cos\lambda}{\omega_1^2+4n^2}+\frac{a\omega_1\sin\omega_1 t\sin\lambda}{\omega_1^2+4n^2}$$

$$+\frac{2na\cos\omega_1 t\sin\lambda}{\omega_1^2+4n^2}-\frac{2nae^{-2nt}\sin\lambda}{\omega_1^2+4n^2} \qquad \textbf{(11.5.18)}$$

The second derivative of equation (11.5.17) represents the acceleration:

$$\frac{d^2x}{dt^2} = re^{-2nt} + \frac{2na\omega_1 \cos\omega_1 t \cos\lambda}{\omega_1^2 + 4n^2} + \frac{a\omega_1^2 \sin\omega_1 t \cos\lambda}{\omega_1^2 + 4n^2}$$

$$-\frac{2na\omega_1 e^{-2nt} \cos\lambda}{\omega_1^2 + 4n^2} + \frac{a\omega_1^2 \cos\omega_1 t \sin\lambda}{\omega_1^2 + 4n^2}$$

$$-\frac{2na\omega_1 \sin\omega_1 t \sin\lambda}{\omega_1^2 + 4n^2} + \frac{4n^2 ae^{-2nt} \sin\lambda}{\omega_1^2 + 4n^2} \qquad \textbf{(11.5.19)}$$

11.5.4 Both the Initial Displacement and Velocity Equal Zero
The initial conditions of motion are:

$$\text{for} \quad t = 0 \quad x = 0; \quad \frac{dx}{dt} = 0 \qquad \textbf{(11.5.20)}$$

We obtain:

$$x = \frac{r}{2n}[t + \frac{1}{2n}(e^{-2nt} - 1)] + \frac{2na(1 - \cos\omega_1 t)\cos\lambda}{\omega_1(\omega_1^2 + 4n^2)} - \frac{a\sin\omega_1 t \cos\lambda}{\omega_1^2 + 4n^2}$$

$$+\frac{a\omega_1(1 - e^{-2nt})\cos\lambda}{2n(\omega_1^2 + 4n^2)} + \frac{a(1 - \cos\omega_1 t)\sin\lambda}{\omega_1^2 + 4n^2}$$

$$+\frac{2na\sin\omega_1 t \sin\lambda}{\omega_1(\omega_1^2 + 4n^2)} - \frac{a(1 - e^{-2nt})\sin\lambda}{\omega_1^2 + 4n^2} \qquad \textbf{(11.5.21)}$$

The equations for the velocity and acceleration are the same as in the previous case.

11.6 Harmonic Force $A \sin(\omega_1 t + \lambda)$
and Time-Dependent Force $Q\left(\rho + \frac{\mu t}{t}\right)$
Guiding Table 2.1 indicates that this section describes engineering systems subjected to the action of the force of inertia and damping force as the resisting forces (Row 11), and the harmonic force and the time-dependent force as the active forces (Column 6).

The current problem could be related to a vibratory system interacting with fluid media or with a specific hydraulic link, while on the initial stage of the working process this system utilizes, in addition to the harmonic force, the time-dependent force that is acting for a predetermined interval of time.

The system is moving in the horizontal direction. We want to determine the basic parameters of motion. The model of a system subjected to the action of a harmonic force, a time-dependent force, and a damping force is shown in Figure 11.6.

Based on the considerations above and on the model in Figure 11.6, we can compose the left and the right sides of the differential equation of motion. The left side includes the force of inertia and the damping force. The right side consists of the harmonic force and the time-dependent force. Hence, the differential equation of motion reads:

$$m\frac{d^2x}{dt^2} + C\frac{dx}{dt} = A\sin(\omega_1 t + \lambda) + Q(\rho + \frac{\mu t}{\tau}) \qquad (11.6.1)$$

Differential equation of motion (11.6.1) has different solutions for various initial conditions of motion. These solutions and their analyses are presented below.

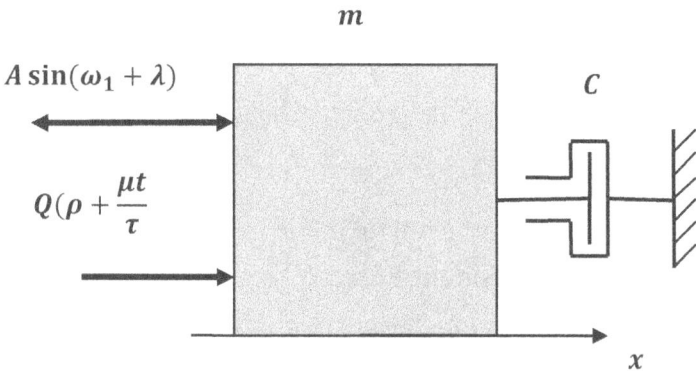

Figure 11.6 Model of a system subjected to a harmonic force, a time-dependent force, and a damping force

11.6.1 General Initial Conditions

The general initial conditions of motion are:

$$\text{for} \quad t = 0 \quad x = s_0; \quad \frac{dx}{dt} = v_0 \tag{11.6.2}$$

where s_0 and v_0 are the initial displacement and initial velocity respectively.

Transforming equation (11.6.1), we have:

$$m\frac{d^2x}{dt^2} + C\frac{dx}{dt} = Q\rho + Q\frac{\mu t}{\tau} + A\sin\omega_1 t\cos\lambda + A\cos\omega_1 t\sin\lambda \tag{11.6.3}$$

Dividing equation (11.6.3) by m, we may write:

$$\frac{d^2x}{dt^2} + 2n\frac{dx}{dt} = q\rho + q\frac{\mu t}{\tau} + a\sin\omega_1 t\cos\lambda + a\cos\omega_1 t\sin\lambda \tag{11.6.4}$$

where n is the damping factor and:

$$2n = \frac{C}{m} \tag{11.6.5}$$

$$q = \frac{Q}{m} \tag{11.6.6}$$

$$a = \frac{A}{m} \tag{11.6.7}$$

Based on Laplace Transform Pairs 3, 4, 5, 2, 6, and 7 from Table 1.1, we convert differential equation of motion (11.6.4) with the initial conditions of motion (11.6.2) from the time domain into the Laplace domain. The resulting algebraic equation of motion in the Laplace domain reads:

$$l^2x(l) - lv_0 - l^2s_0 + 2nlx(l) - 2nls_0$$

$$= q\rho + \frac{q\mu}{\tau l} + \frac{a\omega_1 l}{l^2 + \omega_1^2}\cos\lambda + \frac{al^2}{l^2 + \omega_1^2}\sin\lambda \tag{11.6.8}$$

Applying to equation (11.6.8) some basic algebra, we have:

$$x(l)l(l+2n)=l(v_0+2ns_0)+l^2 s_0+q\rho+\frac{q\mu}{\tau l}$$

$$+\frac{a\omega_1 l}{l^2+\omega_1^2}\cos\lambda+\frac{al^2}{l^2+\omega_1^2}\sin\lambda \qquad (11.6.9)$$

Solving equation (11.6.9) for the Laplace domain displacement $x(l)$ we may write:

$$x(l)=\frac{v_0+2ns_0}{l+2n}+\frac{ls_0}{l+2n}+\frac{q\rho}{l(l+2n)}+\frac{q\mu}{\tau l^2(l+2n)}$$

$$+\frac{a\omega_1\cos\lambda}{(l+2n)(l^2+\omega_1^2)}+\frac{al\sin\lambda}{(l+2n)(l^2+\omega_1^2)} \qquad (11.6.10)$$

Using pairs 1, 10, 12, 13, 31, 19, and 20 from Table 1.1, we invert equation (11.6.10) from the Laplace domain into the time domain and obtain the solution of differential equation of motion (11.6.1) with the initial conditions of motion (11.6.2):

$$x=\frac{v_0+2ns_0}{2n}(1-e^{-2nt})+s_0 e^{-2nt}+\frac{q\rho}{2n}\left[t+\frac{1}{2n}\left(e^{-2nt}-1\right)\right]$$

$$+\frac{q\mu}{4n\tau}\left[t^2-\frac{1}{n}t-\frac{1}{2n^2}(e^{-2nt}-1)\right]+\frac{2na(1-\cos\omega_1 t)\cos\lambda}{\omega_1(\omega_1^2+4n^2)}$$

$$-\frac{a\sin\omega_1 t\cos\lambda}{\omega_1^2+4n^2}+\frac{a\omega_1(1-e^{-2nt})\cos\lambda}{2n(\omega_1^2+4n^2)}+\frac{a(1-\cos\omega_1 t)\sin\lambda}{\omega_1^2+4n^2}$$

$$+\frac{2na\sin\omega_1 t\sin\lambda}{\omega_1(\omega_1^2+4n^2)}-\frac{a(1-e^{-2nt})\sin\lambda}{\omega_1^2+4n^2} \qquad (11.6.11)$$

Transforming equation (11.6.11) to the shape more suitable for the analysis, we write:

$$x=s_0+\frac{v_0}{2n}(1-e^{-2nt})+\frac{q}{2n}\left(\rho+\frac{\mu}{2n\tau}\right)\left(t+\frac{1}{2n}e^{-2nt}-\frac{1}{2n}\right)$$

$$+\frac{q\mu}{2n\tau}t^2+\frac{2na(1-\cos\omega_1 t)\cos\lambda}{\omega_1(\omega_1^2+4n^2)}-\frac{a\sin\omega_1 t\cos\lambda}{\omega_1^2+4n^2}$$

$$+\frac{a\omega_1(1-e^{-2nt})\cos\lambda}{2n(\omega_1^2+4n^2)}+\frac{a(1-\cos\omega_1 t)\sin\lambda}{\omega_1^2+4n^2}$$

$$+\frac{2na\sin\omega_1 t\sin\lambda}{\omega_1(\omega_1^2+4n^2)}-\frac{a(1-e^{-2nt})\sin\lambda}{\omega_1^2+4n^2} \qquad \textbf{(11.6.12)}$$

Taking the first derivative of equation (11.6.12), we determine the velocity of the system:

$$\frac{dx}{dt}=v_0 e^{-2nt}+\frac{q}{2n}\left[\left(\rho+\frac{\mu}{2n\tau}\right)(1-e^{-2nt})+\frac{\mu}{\tau}t\right]$$

$$+\frac{2na\sin\omega_1 t\cos\lambda}{\omega_1^2+4n^2}-\frac{a\omega_1\cos\omega_1 t\cos\lambda}{\omega_1^2+4n^2}+\frac{a\omega_1 e^{-2nt}\cos\lambda}{\omega_1^2+4n^2}$$

$$+\frac{a\omega_1\sin\omega_1 t\sin\lambda}{\omega_1^2+4n^2}+\frac{2na\cos\omega_1 t\sin\lambda}{\omega_1^2+4n^2}-\frac{2nae^{-2nt}\sin\lambda}{\omega_1^2+4n^2}$$

$$\textbf{(11.6.13)}$$

The second derivative represents the acceleration:

$$\frac{d^2x}{dt^2}=-2nv_0 e^{-2nt}+\frac{q}{2n}[(2n\rho+\frac{\mu}{\tau})e^{-2nt}+\frac{\mu}{\tau}]+\frac{2na\omega_1\cos\omega_1 t\cos\lambda}{\omega_1^2+4n^2}$$

$$+\frac{a\omega_1^2\sin\omega_1 t\cos\lambda}{\omega_1^2+4n^2}-\frac{2na\omega_1 e^{-2nt}\cos\lambda}{\omega_1^2+4n^2}+\frac{a\omega_1^2\cos\omega_1 t\sin\lambda}{\omega_1^2+4n^2}$$

$$-\frac{2na\omega_1\sin\omega_1 t\sin\lambda}{\omega_1^2+4n^2}+\frac{4n^2ae^{-2nt}\sin\lambda}{\omega_1^2+4n^2} \qquad \textbf{(11.6.14)}$$

11.6.2 Initial Displacement Equals Zero

The initial conditions of motion are:

$$\text{for}\quad t=0\quad x=0;\quad \frac{dx}{dt}=v_0 \qquad \textbf{(11.6.15)}$$

Solving differential equation of motion (11.6.1) with the initial conditions of motion (11.6.15), we obtain:

$$x=\frac{v_0}{2n}(1-e^{-2nt})+\frac{q}{2n}(\rho+\frac{\mu}{2n\tau})(t+\frac{1}{2n}e^{-2nt}-\frac{1}{2n})+\frac{q\mu}{2n\tau}t^2$$

$$+\frac{2na(1-\cos\omega_1 t)\cos\lambda}{\omega_1(\omega_1^2+4n^2)}-\frac{a\sin\omega_1 t\cos\lambda}{\omega_1^2+4n^2}+\frac{a\omega_1(1-e^{-2nt})\cos\lambda}{2n(\omega_1^2+4n^2)}$$

$$+\frac{a(1-\cos\omega_1 t)\sin\lambda}{\omega_1^2+4n^2}+\frac{2na\sin\omega_1 t\sin\lambda}{\omega_1(\omega_1^2+4n^2)}-\frac{a(1-e^{-2nt})\sin\lambda}{\omega_1^2+4n^2}$$

$$\textbf{(11.6.16)}$$

The rest of the parameters are the same as in the previous case.

11.6.3 Initial Velocity Equals Zero

The initial conditions of motion are:

$$\text{for}\quad t=0\quad x=s_0;\quad \frac{dx}{dt}=0 \qquad \textbf{(11.6.17)}$$

The solution of differential equation of motion (11.6.1) with the initial conditions of motion (11.6.17) reads:

$$x=s_0+\frac{q}{2n}(\rho+\frac{\mu}{2n\tau})(t+\frac{1}{2n}e^{-2nt}-\frac{1}{2n})+\frac{q\mu}{2n\tau}t^2$$

$$+\frac{2na(1-\cos\omega_1 t)\cos\lambda}{\omega_1(\omega_1^2+4n^2)}-\frac{a\sin\omega_1 t\cos\lambda}{\omega_1^2+4n^2}$$

$$+\frac{a\omega_1(1-e^{-2nt})\cos\lambda}{2n(\omega_1^2+4n^2)}+\frac{a(1-\cos\omega_1 t)\sin\lambda}{\omega_1^2+4n^2}$$

$$+\frac{2na\sin\omega_1 t\sin\lambda}{\omega_1(\omega_1^2+4n^2)}-\frac{a(1-e^{-2nt})\sin\lambda}{\omega_1^2+4n^2} \qquad \textbf{(11.6.18)}$$

The first derivative of equation (11.6.18) represents the velocity of the system:

$$\frac{dx}{dt}=\frac{q}{2n}\left[(\rho+\frac{\mu}{2n\tau})(1-e^{-2nt})+\frac{\mu}{\tau}t\right]+\frac{2na\sin\omega_1 t\cos\lambda}{\omega_1^2+4n^2}$$

$$-\frac{a\omega_1\cos\omega_1 t\cos\lambda}{\omega_1^2+4n^2}+\frac{a\omega_1 e^{-2nt}\cos\lambda}{\omega_1^2+4n^2}+\frac{a\omega_1\sin\omega_1 t\sin\lambda}{\omega_1^2+4n^2}$$

$$+\frac{2na\cos\omega_1 t\sin\lambda}{\omega_1^2+4n^2}-\frac{2nae^{-2nt}\sin\lambda}{\omega_1^2+4n^2} \qquad \textbf{(11.6.19)}$$

Taking the second derivative of equation (11.6.18), we determine the acceleration:

$$\frac{d^2x}{dt^2} = \frac{q}{2n}\left[(2n\rho + \frac{\mu}{\tau})e^{-2nt} + \frac{\mu}{\tau}\right] + \frac{2na\omega_1\cos\omega_1 t\cos\lambda}{\omega_1^2 + 4n^2}$$

$$+ \frac{a\omega_1^2\sin\omega_1 t\cos\lambda}{\omega_1^2 + 4n^2} - \frac{2na\omega_1 e^{-2nt}\cos\lambda}{\omega_1^2 + 4n^2} + \frac{a\omega_1^2\cos\omega_1 t\sin\lambda}{\omega_1^2 + 4n^2}$$

$$- \frac{2na\omega_1\sin\omega_1 t\sin\lambda}{\omega_1^2 + 4n^2} + \frac{4n^2 a e^{-2nt}\sin\lambda}{\omega_1^2 + 4n^2} \qquad \textbf{(11.6.20)}$$

11.6.4 Both the Initial Displacement and Velocity Equal Zero

The initial conditions of motion are:

$$\text{for} \quad t = 0 \quad x = 0; \quad \frac{dx}{dt} = 0 \qquad \textbf{(11.6.21)}$$

Solving differential equation of motion (11.6.1) with the initial conditions of motion (11.6.21), we obtain:

$$x = \frac{q}{2n}(\rho + \frac{\mu}{2n\tau})(t + \frac{1}{2n}e^{-2nt} - \frac{1}{2n}) + \frac{q\mu}{2n\tau}t^2$$

$$+ \frac{2na(1 - \cos\omega_1 t)\cos\lambda}{\omega_1(\omega_1^2 + 4n^2)} - \frac{a\sin\omega_1 t\cos\lambda}{\omega_1^2 + 4n^2} + \frac{a\omega_1(1 - e^{-2nt})\cos\lambda}{2n(\omega_1^2 + 4n^2)}$$

$$+ \frac{a(1 - \cos\omega_1 t)\sin\lambda}{\omega_1^2 + 4n^2} + \frac{2na\sin\omega_1 t\sin\lambda}{\omega_1(\omega_1^2 + 4n^2)} - \frac{a(1 - e^{-2nt})\sin\lambda}{\omega_1^2 + 4n^2}$$

$$\textbf{(11.6.22)}$$

The equations for the velocity and the acceleration are the same as in the previous case.

12

DAMPING AND FRICTION

This chapter discusses engineering systems subjected to the force of inertia, the damping force, and the friction force as the resisting forces. These forces are marked by the plus signs (+) in Row 12 of Guiding Table 2.1. The active forces applied to the systems are displayed in Columns 1 through to 6. Throughout this chapter, the left sides of the differential equations of motion are identical and represent the resisting forces mentioned above. The right sides of these equations consist of the active forces applied to the particular systems, and vary from section to section. The titles of the sections reflect the active forces applied to the systems,

The problems described in this chapter could be associated with systems moving on a horizontal frictional surface and interacting with fluid media or specific hydraulic links.

12.1 Active Force Equals Zero

This section describes problems associated with the force of inertia, the damping force, and the friction force as the resisting

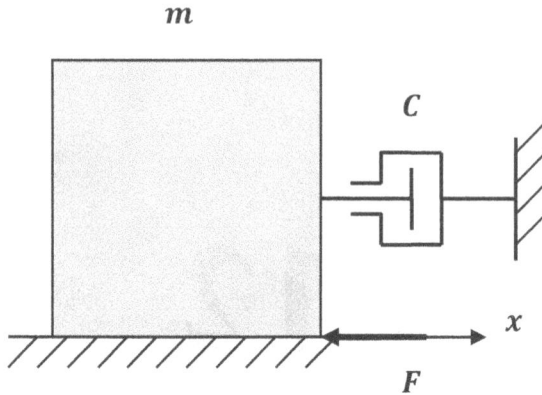

**Figure 12.1 Model of a system subjected to a damping force
and a friction force**

forces (Row 12 in Guiding Table 2.1), when the active force equals
zero (Column 1). The considerations related to the motion of a system in the absence of active forces are described in section 1.3.

 The deceleration process of a ground transportation system
moving in the horizontal direction is an example of the current problem. We want to determine the basic parameters of motion and the
maximum values of the displacement, the acceleration, and the force
applied to the system. Figure 12.1 shows the model of a system subjected to the action of a damping force and friction force.

 Based on the considerations above and accounting the model
in Figure 12.1, we compose the left and the right sides of the differential equation of motion. The left side consists of the force of
inertia, the damping force, and the friction force, while the right
side of the equation equals zero. Hence, the differential equation of
motion reads:

$$m\frac{d^2x}{dt^2} + C\frac{dx}{dt} + F = 0 \qquad (12.1.1)$$

 Differential equation of motion (12.1.1) has different solutions for various initial conditions of motion. These solutions and
their analyses are presented below.

12.1.1 General Initial Conditions

The general initial conditions of motion are:

$$\text{for} \quad t = 0 \quad x = s_0; \quad \frac{dx}{dt} = v_0 \tag{12.1.2}$$

where s_0 and v_0 are the initial displacement and initial velocity respectively.

Dividing equation (12.1.1) by m, we may write:

$$\frac{d^2x}{dt^2} + 2n\frac{dx}{dt} + f = 0 \tag{12.1.3}$$

where n is the damping factor and:

$$2n = \frac{C}{m} \tag{12.1.4}$$

$$f = \frac{F}{m} \tag{12.1.5}$$

Based on Laplace Transform Pairs 3, 4, and 5 from Table 1.1, we convert differential equation of motion (12.1.3) with the initial conditions of motion (12.1.2) from the time domain into the Laplace domain and obtain the corresponding algebraic equation of motion in the Laplace domain:

$$l^2x(l) - lv_0 - l^2s_0 + 2nlx(l) - 2nls_0 + f = 0 \tag{12.1.6}$$

Solving equation (12.1.6) for the displacement $x(l)$ in the Laplace domain, we have:

$$x(l) = \frac{v_0 + 2ns_0}{l + 2n} + \frac{ls_0}{l + 2n} - \frac{f}{l(l + 2n)} \tag{12.1.7}$$

Using pairs 1, 10, 12, and 13 from Table 1.1, we invert equation (12.1.7) from the Laplace domain into the time domain and obtain

the solution of differential equation of motion (12.1.1) with the initial conditions of motion (12.1.2):

$$x = \frac{v_0 + 2ns_0}{2n}(1 - e^{-2nt}) + s_0 e^{-2nt} - \frac{f}{2n}[t + \frac{1}{2n}(e^{-2nt} - 1)] \qquad \textbf{(12.1.8)}$$

Applying conventional algebraic procedures to equation (12.1.8), we may write:

$$x = \frac{1}{2n}[2ns_0 + v_0 + \frac{f}{2n} - ft - \left(v_0 + \frac{f}{2n}\right)e^{-2nt}] \qquad \textbf{(12.1.9)}$$

Taking the first and second derivatives of equation (12.1.9), we determine the velocity and the acceleration of the system respectively:

$$\frac{dx}{dt} = (v_0 + \frac{f}{2n})e^{-2nt} - \frac{f}{2n} \qquad \textbf{(12.1.10)}$$

$$\frac{d^2x}{dt^2} = -2n(v_0 + \frac{f}{2n})e^{-2nt} \qquad \textbf{(12.1.11)}$$

Because the system is in a deceleration process, we equate the velocity according to the equation (12.1.10) to zero and determine the time T that the process lasts. Actually, we may write:

$$(v_0 + \frac{f}{2n})e^{-2nT} - \frac{f}{2n} = 0 \qquad \textbf{(12.1.12)}$$

According to equation (12.1.12), we have:

$$e^{-2nT} = \frac{f}{2nv_0 + f} \qquad \textbf{(12.1.13)}$$

Solving equation (12.1.13) for the time T, we obtain:

$$T = \frac{1}{2n}\ln\frac{2nv_0 + f}{f} \qquad \textbf{(12.1.14)}$$

Combining equations (12.1.13) and (12.1.14) with equation (12.1.9), we determine the maximum displacement s_{max} of the system:

$$s_{max} = \frac{1}{2n}(2ns_0 + v_0 - \frac{f}{2n} \ln \frac{2nv_0 + f}{f})$$ (12.1.15)

Eliminating from equation (12.1.15) the initial displacement s_0, we calculate the braking distance s_{br}:

$$s_{br} = \frac{1}{2n}(v_0 - \frac{f}{2n} \ln \frac{2nv_0 + f}{f})$$ (12.1.16)

Combining equations (12.1.13) and (12.1.11), we determine the maximum acceleration (deceleration) a_{max}:

$$a_{max} = -f$$ (12.1.17)

However equation (12.1.11) shows that at the beginning of the process when $t = 0$, the deceleration a_0 equals:

$$a_0 = -2nv_0 - f$$ (12.1.18)

The absolute value of a_0 according to equation (12.1.18) exceeds the absolute value of a_{max} according to equation (12.1.17); therefore, for public transportation systems, the verification of the acceleration with the norms of public health and safety should be based on the absolute value a_0 according to equation (12.1.18):

$$a_0 = |2nv_0 + f|$$ (12.1.19)

Multiplying equation (12.1.19) by m and substituting notations (12.1.4) and (12.1.5), we determine the maximum force R_{max} applied to the system:

$$R_{max} = Cv_0 + F$$ (12.1.20)

The stress calculations of the system should be based on the force according to equation (12.1.20).

12.1.2 Initial Displacement Equals Zero

The initial conditions of motion are:

$$\text{for} \quad t = 0 \quad x = 0; \quad \frac{dx}{dt} = v_0 \qquad (12.1.21)$$

The solution of differential equation of motion (12.1.1) with the initial conditions of motion (12.1.21) reads:

$$x = \frac{1}{2n}[v_0 + \frac{f}{2n} - ft - \left(v_0 + \frac{f}{2n}\right)e^{-2nt}] \qquad (12.1.22)$$

The rest of the parameters and their analysis are the same as for the previous case, keeping in mind that in this case $s_{max} = s_{br}$.

12.2 Constant Force R

As indicated by the intersection of Row 12 with Column 2 in Guiding Table 2.1, this section describes engineering systems subjected to the action of the force of inertia, the damping force, and the friction force as resisting forces, and the constant active force.

The current problem could be related to the acceleration of ground transportation systems. This problem could be also associated with some pneumatic machines characterized by the decreasing of the active force proportionally to the increasing of the velocity of a corresponding component (piston) of the machine. Section 1.3 contains a detailed description of this situation. Equation (1.3.1) reflects the behavior of this active force and shows that the problems of these pneumatic machines are actually identical to the problems associated with the transportations systems.

Considering that the system moves in the horizontal direction, we want to determine the basic parameters of motion, the maximum values of the velocity and the acceleration of the system, the maximum force applied to the system, and the power of the energy source. Figure 12.2 shows the model of a system subjected to the action of a constant active force, a damping force, and a friction force.

Based on the considerations above and the model in Figure 12.2, we can assemble the left and the right sides of the

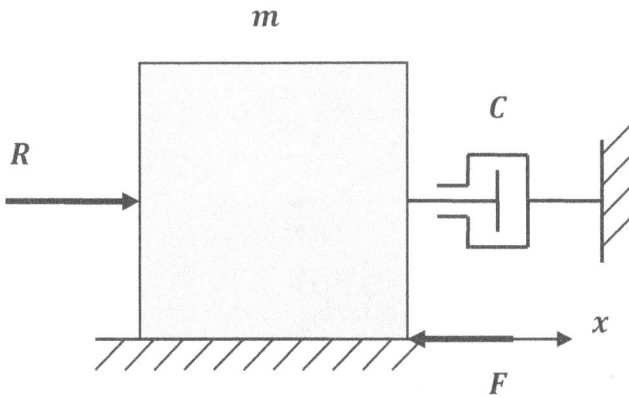

Figure 12.2 Model of a system subjected to a constant active force, a damping force, and a friction force

differential equation of motion of this system. The left side consists of the force of inertia, the damping force, and the friction force, while the right side consists of the constant active force. Hence, the differential equation of motion reads:

$$m\frac{d^2x}{dt^2} + C\frac{dx}{dt} + F = R \qquad (12.2.1)$$

Differential equation of motion (12.2.1) has different solutions for various initial conditions of motion. These solutions and their analyses are presented below.

12.2.1 General Initial Conditions

The general initial conditions of motion are:

$$\text{for} \quad t = 0 \quad x = s_0; \quad \frac{dx}{dt} = v_0 \qquad (12.2.2)$$

where s_0 and v_0 are the initial displacement and initial velocity respectively.

Dividing equation (12.2.1) by m, we have:

$$\frac{d^2x}{dt^2} + 2n\frac{dx}{dt} + f = r \qquad (12.2.3)$$

where n is the damping factor and:

$$2n = \frac{C}{m} \qquad (12.2.4)$$

$$f = \frac{F}{m} \qquad (12.2.5)$$

$$r = \frac{R}{m} \qquad (12.2.6)$$

Using Laplace Transform Pairs 3, 4, and 5 from Table 1.1, we convert differential equation of motion (12.2.3) with the initial conditions of motion (12.2.2) from the time domain into the Laplace domain. The resulting algebraic equation of motion in the Laplace domain reads:

$$l^2 x(l) - l v_0 - l^2 s_0 + 2nlx(l) - 2nls_0 + f = r \qquad (12.2.7)$$

Solving equation (12.2.7) for the displacement $x(l)$ in Laplace domain, we may write:

$$x(l) = \frac{r - f}{l(l + 2n)} + \frac{v_0 + 2ns_0}{l + 2n} + \frac{ls_0}{l + 2n} \qquad (12.2.8)$$

Based on pairs 1, 13, 10, and 12 from Table 1.1, we invert equation (12.2.8) from the Laplace domain into the time domain and obtain the solution of differential equation of motion (12.2.1) with the initial conditions of motion (12.2.2):

$$x = \frac{r - f}{2n}[t + \frac{1}{2n}(e^{-2nt} - 1)] + \frac{v_0 + 2ns_0}{2n}(1 - e^{-2nt}) + s_0 e^{-2nt} \qquad (12.2.9)$$

Applying basic algebra to equation (12.2.9), we have:

$$x = \frac{1}{2n}[(r - f)t + \left(\frac{r - f}{2n} - v_0\right)e^{-2nt} - \frac{r - f}{2n} + v_0 + 2ns_0] \qquad (12.2.10)$$

Taking the first derivative of equation (12.2.10), we determine the velocity of the system:

$$\frac{dx}{dt} = \frac{r-f}{2n}(1 - e^{-2nt}) + v_0 e^{-2nt} \qquad (12.2.11)$$

The second derivative of equation (12.2.10) represents the acceleration:

$$\frac{d^2 x}{dt^2} = (r - f - 2nv_0)e^{-2nt} \qquad (12.2.12)$$

In this case, the velocity approaches its maximum value when the acceleration approaches zero. Because $r - f - 2nv_0 > 0$, then according to equation (12.2.12) the acceleration approaches zero when:

$$e^{-2nT} \to 0 \qquad (12.2.13)$$

where T is the time that the process lasts (actually T approaches infinity).

Combining expression (12.2.13) with equation (12.2.11) and substituting notations (12.2.4), (12.2.5), and (12.2.6), we determine the maximum value of the velocity v_{max}:

$$v_{max} \to \frac{R - F}{C} \qquad (12.2.14)$$

The acceleration has its maximum value at the beginning of the process when $t = 0$. Therefore, at this moment, the maximum value of the acceleration a_{max} according to equation (12.2.12) equals:

$$a_{max} = r - f - 2nv_0 \qquad (12.2.15)$$

For public transportation systems, the maximum value of the acceleration should comply with the norms of public health and safety.

The maximum force applied to the system represents the sum of the values of the force of inertia and the resisting forces at any given time. At the time when the velocity approaches its maximum value, the acceleration approaches zero. Consequently, the force of inertia at this time also approaches zero. Therefore, in this case, the maximum force applied to the system equals the sum of the resisting forces. Thus, multiplying expression (12.2.14) by the C, we determine the value of the damping force R_d:

$$R_d = R - F \qquad (12.2.16)$$

Adding the friction force F to the damping force according to equation (12.2.16), we determine the maximum value of the force R_{max} applied to the system:

$$R_{max} = R \qquad (12.2.17)$$

The value of the maximum force should be used for the stress analysis of the system.

The power N of the energy source equals the product of multiplying the maximum force according to equation (12.2.17) by the maximum velocity according to expression (12.2.14):

$$N = \frac{R(R - F)}{C} \qquad (12.2.18)$$

12.2.2 Initial Displacement Equals Zero

The initial conditions of motion are:

$$\text{for} \quad t = 0 \quad x = 0; \quad \frac{dx}{dt} = v_0 \qquad (12.2.19)$$

The solution of differential equation of motion (12.2.1) with the initial conditions of motion (12.2.19) reads:

$$x = \frac{1}{2n}[(r - f)t + \left(\frac{r - f}{2n} - v_0 \right)e^{-2nt} - \frac{r - f}{2n} + v_0] \,(12.2.20)$$

The rest of the parameters and their analysis are the same as in the previous case.

12.2.3 Initial Velocity Equals Zero

The initial conditions of motion are:

$$\text{for} \quad t = 0 \quad x = s_0; \quad \frac{dx}{dt} = 0 \qquad (12.2.21)$$

Solving differential equation of motion (12.2.1) with the initial conditions of motion (12.2.21), we obtain:

$$x = \frac{1}{2n}[(r-f)t + \frac{r-f}{2n}e^{-2nt} - \frac{r-f}{2n} + 2ns_0] \qquad (12.2.22)$$

The first and the second derivatives of equation (12.2.22) represent the velocity and acceleration respectively:

$$\frac{dx}{dt} = \frac{r-f}{2n}(1-e^{-2nt}) \qquad (12.2.23)$$

$$\frac{d^2x}{dt^2} = (r-f)e^{-2nt} \qquad (12.2.24)$$

The analysis of equation (12.2.23) shows that the maximum velocity in this case is the same as in the previous cases; it does not depend on the initial velocity and can be calculated according to expression (12.2.14). The maximum acceleration a_{max} in this case occurs at the beginning of the process and equals:

$$a_{max} = \frac{F}{m} \qquad (12.2.25)$$

The rest of the parameters and their analysis are also the same as in the previous cases.

12.2.4 Both the Initial Displacement and Velocity Equal Zero

The initial conditions of motion are:

$$\text{for} \quad t = 0 \quad x = 0; \quad \frac{dx}{dt} = 0 \qquad (12.2.26)$$

The solution of differential equation of motion (12.2.1) with the initial conditions of motion (12.2.26) reads:

$$x = \frac{1}{2n}[(r-f)t + \frac{r-f}{2n}e^{-2nt} - \frac{r-f}{2n}] \qquad (12.2.26)$$

The rest of the parameters and their analysis are the same as in the previous case.

12.3 Harmonic Force $A\sin(\omega_1 t + \lambda)$

This section discusses engineering systems subjected to the force of inertia, the damping force, and the friction force as the resisting forces (Row 12 in Guiding Table 2.1), and to the harmonic force as the active force (Column 3). This problem could be related to the working processes of vibratory systems interacting with fluid media or with a specific hydraulic link.

The system is moving on a frictional horizontal surface. We want to determine the basic parameters of motion. Considerations related to the behavior of a friction force subjected to a vibratory system are described in section 1.2. Figure 12.3 shows the model of a system subjected to the action of a harmonic force, a damping force, and a friction force.

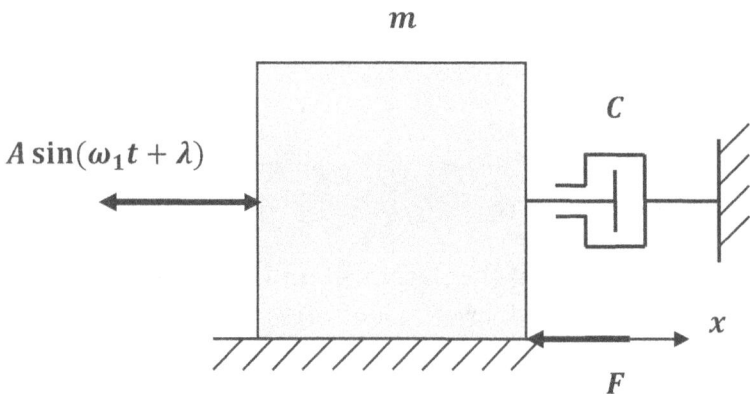

Figure 12.3 Model of a system subjected to a harmonic force, a damping force, and a friction force

The considerations above and the model in the Figure 12.3 let us compose the left and right sides of the differential equation of motion. The left side consists of the force of inertia, the damping force, and the friction force, while the right side consists of the harmonic force. Hence, the differential equation of motion reads:

$$m\frac{d^2x}{dt^2} + C\frac{dx}{dt} + F = A\sin(\omega_1 t + \lambda) \qquad (12.3.1)$$

Differential equation of motion (12.3.1) has different solutions for various initial conditions of motion. These solutions and their analyses are presented below.

12.3.1 General Initial Conditions

The general initial conditions of motion are:

$$\text{for} \quad t = 0 \quad x = s_0; \quad \frac{dx}{dt} = v_0 \qquad (12.3.2)$$

where s_0 and v_0 are the initial displacement and initial velocity respectively.

Transforming the sinusoidal function in equation (12.3.1) and dividing the latter by m, we have:

$$\frac{d^2x}{dt^2} + 2n\frac{dx}{dt} + f = a\sin\omega_1 t \cos\lambda + a\cos\omega_1 t \sin\lambda$$

where n is the damping factor and:

$$2n = \frac{C}{m} \qquad (12.3.4)$$

$$f = \frac{F}{m} \qquad (12.3.5)$$

$$a = \frac{A}{m} \qquad (12.3.6)$$

Applying Laplace Transform Pairs 3, 4, 5, 6, and 7 from Table 1.1, we convert differential equation of motion (12.3.3) with the initial conditions of motion (12.3.2) from the time domain into the Laplace domain. The resulting algebraic equation of motion in the Laplace domain reads:

$$l^2 x(l) - l v_0 - l^2 s_0 + 2nlx(l) - 2nls_0 + f$$

$$= \frac{a\omega_1 l}{l^2 + \omega_1^2} \cos \lambda + \frac{al^2}{l^2 + \omega_1^2} \sin \lambda \qquad (12.3.7)$$

Applying some basic algebra to equation (12.3.7), we may write:

$$x(l)l(l + 2n) = l(2ns_0 + v_0) + s_0 l^2 - f$$

$$+ \frac{a\omega_1 l}{l^2 + \omega_1^2} \cos \lambda + \frac{al^2}{l^2 + \omega_1^2} \sin \lambda \qquad (12.3.8)$$

Solving equation (12.3.8) for the Laplace domain displacement $x(l)$, we obtain:

$$x(l) = \frac{2ns_0 + v_0}{l + 2n} + \frac{ls_0}{l + 2n} - \frac{f}{l(l + 2n)} + \frac{a\omega_1}{(l + 2n)(l^2 + \omega_1^2)} \cos \lambda$$

$$+ \frac{la}{(l + 2n)(l^2 + \omega_1^2)} \sin \lambda \qquad (12.3.9)$$

Using pairs 1, 10, 12, 13, 19, and 20 from Table 1.1, we invert equation (12.3.9) from the Laplace domain into the time domain and obtain the solution of differential equation of motion (12.3.1) with the initial conditions of motion (12.3.2):

$$x = \frac{2ns_0 + v_0}{2n}(1 - e^{-2nt}) + s_0 e^{-2nt} - \frac{f}{2n}\left[t + \frac{1}{2n}(e^{-2nt} - 1) \right]$$

$$+ \frac{2na(1 - \cos \omega_1 t)\cos \lambda}{\omega_1(\omega_1^2 + 4n^2)} - \frac{a \sin \omega_1 t \cos \lambda}{\omega_1^2 + 4n^2} + \frac{a\omega_1(1 - e^{-2nt})\cos \lambda}{2n(\omega_1^2 + 4n^2)}$$

$$+ \frac{a(1 - \cos \omega_1 t)\sin \lambda}{\omega_1^2 + 4n^2} + \frac{2na \sin \omega_1 t \sin \lambda}{\omega_1(\omega_1^2 + 4n^2)} - \frac{a(1 - e^{-2nt})\sin \lambda}{\omega_1^2 + 4n^2}$$

$$(12.3.10)$$

Transforming equation (12.3.10), we may write:

$$x = s_0 + \frac{v_0}{2n}(1 - e^{-2nt}) - \frac{f}{2n}\left[t + \frac{1}{2n}(e^{-2nt} - 1)\right] + \frac{2na(1 - \cos\omega_1 t)\cos\lambda}{\omega_1(\omega_1^2 + 4n^2)}$$

$$- \frac{a\sin\omega_1 t\cos\lambda}{\omega_1^2 + 4n^2} + \frac{a\omega_1(1 - e^{-2nt})\cos\lambda}{2n(\omega_1^2 + 4n^2)} + \frac{a(1 - \cos\omega_1 t)\sin\lambda}{\omega_1^2 + 4n^2}$$

$$+ \frac{2na\sin\omega_1 t\sin\lambda}{\omega_1(\omega_1^2 + 4n^2)} - \frac{a(1 - e^{-2nt})\sin\lambda}{\omega_1^2 + 4n^2} \qquad (12.3.11)$$

The first derivative of equation (12.3.11) represents the velocity of the system:

$$\frac{dx}{dt} = v_0 e^{-2nt} - \frac{f}{2n}(1 - e^{-2nt}) + \frac{2na\sin\omega_1 t\cos\lambda}{\omega_1^2 + 4n^2}$$

$$- \frac{a\omega_1\cos\omega_1 t\cos\lambda}{\omega_1^2 + 4n^2} + \frac{a\omega_1 e^{-2nt}\cos\lambda}{\omega_1^2 + 4n^2} + \frac{a\omega_1\sin\omega_1 t\sin\lambda}{\omega_1^2 + 4n^2}$$

$$+ \frac{2na\cos\omega_1 t\sin\lambda}{\omega_1^2 + 4n^2} - \frac{2nae^{-2nt}\sin\lambda}{\omega_1^2 + 4n^2} \qquad (12.3.12)$$

Taking the second derivative of equation (12.3.11), we determine the acceleration of the system:

$$\frac{d^2x}{dt^2} = -2nv_0 e^{-2nt} - fe^{-2nt} + \frac{2na\omega_1\cos\omega_1 t\cos\lambda}{\omega_1^2 + 4n^2}$$

$$+ \frac{a\omega_1^2\sin\omega_1 t\cos\lambda}{\omega_1^2 + 4n^2} - \frac{2na\omega_1\cos\omega_1 t\cos\lambda}{\omega_1^2 + 4n^2} + \frac{a\omega_1^2\cos\omega_1 t\sin\lambda}{\omega_1^2 + 4n^2}$$

$$- \frac{2na\omega_1\sin\omega_1 t\sin\lambda}{\omega_1^2 + 4n^2} + \frac{4n^2ae^{-2nt}\sin\lambda}{\omega_1^2 + 4n^2} \qquad (12.3.13)$$

12.3.2 Initial Displacement Equals Zero

The initial condition of motion are:

$$\text{for} \quad t = 0 \quad x = 0; \quad \frac{dx}{dt} = v_0 \qquad (12.3.14)$$

Solving differential equation of motion (12.3.1) with the initial conditions of motion (12.3.14), we obtain:

$$x = \frac{v_0}{2n}(1 - e^{-2nt}) - \frac{f}{2n}\left[t + \frac{1}{2n}(e^{-2nt} - 1)\right]$$
$$+ \frac{2na(1 - \cos\omega_1 t)\cos\lambda}{\omega_1(\omega_1^2 + 4n^2)} - \frac{a\sin\omega_1 t\cos\lambda}{\omega_1^2 + 4n^2}$$
$$+ \frac{a\omega_1(1 - e^{-2nt})\cos\lambda}{2n(\omega_1^2 + 4n^2)} + \frac{a(1 - \cos\omega_1 t)\sin\lambda}{\omega_1^2 + 4n^2}$$
$$+ \frac{2na\sin\omega_1 t\sin\lambda}{\omega_1(\omega_1^2 + 4n^2)} - \frac{a(1 - e^{-2nt})\sin\lambda}{\omega_1^2 + 4n^2} \qquad \textbf{(12.3.15)}$$

The equations for the velocity and the acceleration are the same as in the previous case.

12.3.3 Initial Velocity Equals Zero

The initial conditions of motion are:

$$\text{for} \quad t = 0 \quad x = s_0; \quad \frac{dx}{dt} = 0 \qquad \textbf{(12.3.16)}$$

The solution of differential equation of motion (12.3.1) with the initial conditions of motion (12.3.16) reads:

$$x = s_0 - \frac{f}{2n}\left[t + \frac{1}{2n}(e^{-2nt} - 1)\right] + \frac{2na(1 - \cos\omega_1 t)\cos\lambda}{\omega_1(\omega_1^2 + 4n^2)}$$
$$- \frac{a\sin\omega_1 t\cos\lambda}{\omega_1^2 + 4n^2} + \frac{a\omega_1(1 - e^{-2nt})\cos\lambda}{2n(\omega_1^2 + 4n^2)} + \frac{a(1 - \cos\omega_1 t)\sin\lambda}{\omega_1^2 + 4n^2}$$
$$+ \frac{2na\sin\omega_1 t\sin\lambda}{\omega_1(\omega_1^2 + 4n^2)} - \frac{a(1 - e^{-2nt})\sin\lambda}{\omega_1^2 + 4n^2} \qquad \textbf{(12.3.17)}$$

Taking the first and second derivatives of equation (12.3.17), we determine the velocity and the acceleration of the system respectively:

$$\frac{dx}{dt} = \frac{f}{2n}(e^{-2nt} - 1) + \frac{2na \sin \omega_1 t \cos \lambda}{\omega_1^2 + 4n^2}$$

$$- \frac{a\omega_1 \cos \omega_1 t \cos \lambda}{\omega_1^2 + 4n^2} + \frac{a\omega_1 e^{-2nt} \cos \lambda}{\omega_1^2 + 4n^2} + \frac{a\omega_1 \sin \omega_1 t \sin \lambda}{\omega_1^2 + 4n^2}$$

$$+ \frac{2na \cos \omega_1 t \sin \lambda}{\omega_1^2 + 4n^2} - \frac{2nae^{-2nt} \sin \lambda}{\omega_1^2 + 4n^2} \qquad \textbf{(12.3.18)}$$

$$\frac{d^2 x}{dt^2} = - fe^{-2nt} + \frac{2na\omega_1 \cos \omega_1 t \cos \lambda}{\omega_1^2 + 4n^2} + \frac{a\omega_1^2 \sin \omega_1 t \cos \lambda}{\omega_1^2 + 4n^2}$$

$$- \frac{2na\omega_1 \cos \omega_1 t \cos \lambda}{\omega_1^2 + 4n^2} + \frac{a\omega_1^2 \cos \omega_1 t \sin \lambda}{\omega_1^2 + 4n^2}$$

$$- \frac{2na\omega_1 \sin \omega_1 t \sin \lambda}{\omega_1^2 + 4n^2} + \frac{4n^2 ae^{-2nt} \sin \lambda}{\omega_1^2 + 4n^2} \qquad \textbf{(12.3.19)}$$

12.3.4 Both the Initial Displacement and Velocity Equal Zero

The initial conditions of motion are:

$$\text{for} \quad t = 0 \quad x = 0; \quad \frac{dx}{dt} = 0 \qquad \textbf{(12.3.20)}$$

Solving differential equation of motion (12.3.1) with the initial conditions of motion (12.3.20), we obtain:

$$x = \frac{f}{2n}\left[\frac{1}{2n}(1 - e^{-2nt}) - t\right] + \frac{2na(1 - \cos \omega_1 t)\cos \lambda}{\omega_1(\omega_1^2 + 4n^2)} - \frac{a \sin \omega_1 t \cos \lambda}{\omega_1^2 + 4n^2}$$

$$+ \frac{a\omega_1(1 - e^{-2nt})\cos \lambda}{2n(\omega_1^2 + 4n^2)} + \frac{a(1 - \cos \omega_1 t)\sin \lambda}{\omega_1^2 + 4n^2}$$

$$+ \frac{2na \sin \omega_1 t \sin \lambda}{\omega_1(\omega_1^2 + 4n^2)} - \frac{a(1 - e^{-2nt})\sin \lambda}{\omega_1^2 + 4n^2} \qquad \textbf{(12.3.21)}$$

The equations for the velocity and acceleration are the same as for the previous case.

12.4 Time-Dependent Force $Q\left(\rho + \dfrac{\mu t}{\tau}\right)$

The intersection of Row 12 with Column No 4 in Guiding Table 2.1 indicates that the engineering system described in this section is subjected to the action of the force of inertia, the damping force, and the friction force as the resisting forces, and to the time-dependent force as the active force.

The current problem could be related to the acceleration of a ground transportation system in its initial phase of motion. In this phase, the active force is increasing during a predetermined interval of time, while the system reaches a certain velocity.

The system is moving in the horizontal direction. We want to determine the basic parameters of motion and the values of the velocity and acceleration at the end of the predetermined interval of time. Figure 12.4 shows the model of a system subjected to the action of a time-dependent force, a damping force, and a friction force.

The considerations above and the model in Figure 12.4 let us compose the left and right sides of the differential equation of motion. The left side consists of the force of inertia, the damping force, and the friction force, while the right side includes the time-dependent force. Therefore, the differential equation of motion reads:

$$m\frac{d^2x}{dt^2} + C\frac{dx}{dt} + F = Q(\rho + \frac{\mu t}{\tau}) \qquad \textbf{(12.4.1)}$$

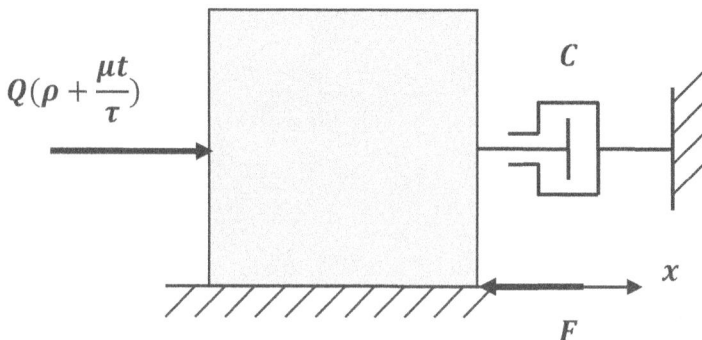

Figure 12.4 Model of a system subjected to a time-dependent force, a damping force, and a friction force

Differential equation of motion (12.4.1) has different solutions for various initial conditions of motion. These solutions and their analyses are presented below.

12.4.1 General Initial Conditions
The general initial conditions of motion are:

$$\text{for} \quad t = 0 \quad x = s_0; \quad \frac{dx}{dt} = v_0 \tag{12.4.2}$$

where s_0 and v_0 are the initial displacement and initial velocity respectively.

Dividing equation (12.4.1) by m, we obtain:

$$\frac{d^2 x}{dt^2} + 2n \frac{dx}{dt} + f = q(\rho + \frac{\mu t}{\tau}) \tag{12.4.3}$$

where n is the damping factor and:

$$2n = \frac{C}{m} \tag{12.4.4}$$

$$f = \frac{F}{m} \tag{12.4.5}$$

$$q = \frac{Q}{m} \tag{12.4.6}$$

Applying Laplace Transform Pairs 3, 4, 5, and 2 from Table 1.1, we convert differential equation of motion (12.4.3) with the initial conditions of motion (12.4.2) from the time domain into the Laplace domain, and obtain the resulting algebraic equation of motion in the Laplace domain:

$$l^2 x(l) - l v_0 - l^2 s_0 + 2nlx(l) - 2nls_0 + f = q\rho + \frac{q\mu}{\tau l} \tag{12.4.7}$$

The solution of equation (12.4.7) for the Laplace domain displacement $x(l)$ reads:

$$x(l) = \frac{v_0 + 2ns_0}{l + 2n} + \frac{ls_0}{l + 2n} + \frac{q\rho - f}{l(l + 2n)} + \frac{q\mu}{\tau l^2 (l + 2n)} \tag{12.4.8}$$

Using pairs 1, 10, 12, 13, and 31 from Table 1.1, we invert equation (12.4.8) from the Laplace domain into the time domain and obtain the solution of differential equation of motion (12.4.1) with the initial conditions of motion (12.4.2):

$$x = \frac{1}{2n}(v_0 + 2ns_0 - v_0 e^{-2nt}) + \frac{q\rho - f}{2n}\left[t + \frac{1}{2n}(e^{-2nt} - 1)\right]$$

$$+ \frac{q\mu}{4n\tau}[t^2 - \frac{1}{n}t - \frac{1}{2n^2}(e^{-2nt} - 1)] \qquad (12.4.9)$$

Applying algebraic procedures to equation (12.4.9), we have:

$$x = \frac{1}{2n}\{v_0 + 2ns_0 - v_0 e^{-2nt} + \frac{q\mu}{2\tau}t^2$$

$$+ \left(q\rho - f - \frac{q\mu}{2n\tau}\right)[t + \frac{1}{2n}(e^{-2nt} - 1)]\} \quad (12.4.10)$$

Taking the first derivative of equation (12.4.10), we determine the velocity of the system:

$$\frac{dx}{dt} = v_0 e^{-2nt} + \frac{q\mu}{2n\tau}t + \frac{1}{2n}\left(q\rho - f - \frac{q\mu}{2n\tau}\right)(1 - e^{-2nt}) \qquad (12.4.11)$$

The second derivative of equation (12.4.10) represents the acceleration:

$$\frac{d^2x}{dt^2} = -2nv_0 e^{-2nt} + \frac{q\mu}{2n\tau} + (q\rho - f - \frac{q\mu}{2n\tau})e^{-2nt} \qquad (12.4.12)$$

Substituting the time τ into equations (12.4.11) and (12.4.12), we determine the velocity v and acceleration a at end of the predetermined interval of time respectively:

$$v = v_0 e^{-2n\tau} + \frac{q\mu}{2n} + \frac{1}{2n}\left(q\rho - f - \frac{q\mu}{2n\tau}\right)(1 - e^{-2n\tau}) \quad (12.4.13)$$

$$a = -2nv_0 e^{-2n\tau} + \frac{q\mu}{2n\tau} + (q\rho - f - \frac{q\mu}{2n\tau})e^{-2n\tau} \qquad (12.4.14)$$

12.4.2 Initial Displacement Equals Zero

The initial conditions of motion are:

$$\text{for} \quad t = 0 \quad x = 0; \quad \frac{dx}{dt} = v_0 \qquad \textbf{(12.4.15)}$$

Solving differential equation of motion (12.4.1) with the initial conditions of motion (12.4.15), we obtain:

$$x = \frac{1}{2n}\{v_0 - v_0 e^{-2nt} + \frac{q\mu}{2\tau}t^2 + \left(q\rho - f - \frac{q\mu}{2n\tau}\right)[t + \frac{1}{2n}(e^{-2nt} - 1)]\}$$
$$\textbf{(12.4.16)}$$

The rest of the parameters are the same as in the previous case.

12.4.3 Initial Velocity Equals Zero

The initial conditions of motion are:

$$\text{for} \quad t = 0 \quad x = s_0; \quad \frac{dx}{dt} = 0 \qquad \textbf{(12.4.17)}$$

The solution of differential equation of motion (12.4.1) with the initial conditions of motion (12.4.17) reads:

$$x = \frac{1}{2n}\{2ns_0 - v_0 e^{-2nt} + \frac{q\mu}{2\tau}t^2 + \left(q\rho - f - \frac{q\mu}{2n\tau}\right)[t + \frac{1}{2n}(e^{-2nt} - 1)]\}$$
$$\textbf{(12.4.18)}$$

Taking the first and the second derivatives of equation (12.4.18), we determine the velocity and the acceleration of the system respectively:

$$\frac{dx}{dt} = \frac{q\mu}{2n\tau}t + \frac{1}{2n}\left(q\rho - f - \frac{q\mu}{2n\tau}\right)(1 - e^{-2nt}) \qquad \textbf{(12.4.19)}$$

$$\frac{d^2x}{dt^2} = \frac{q\mu}{2n\tau} + (q\rho - f - \frac{q\mu}{2n\tau})e^{-2nt} \qquad \textbf{(12.4.20)}$$

Substituting the time τ into equations (12.4.19) and (12.4.20), we calculate the values of the velocity v and the acceleration a at the end of the predetermined interval of time respectively:

$$\frac{dx}{dt} = \frac{q\mu}{2n\tau}\tau + \frac{1}{2n}\left(q\rho - f - \frac{q\mu}{2n\tau}\right)(1 - e^{-2n\tau}) \quad \textbf{(12.4.21)}$$

$$\frac{d^2x}{dt^2} = \frac{q\mu}{2n\tau} + (q\rho - f - \frac{q\mu}{2n\tau})e^{-2n\tau} \quad \textbf{(12.4.22)}$$

12.4.4 Both the Initial Displacement and Velocity Equal Zero
The initial conditions of motion are:

$$\text{for} \quad t = 0 \quad x = 0; \quad \frac{dx}{dt} = 0 \quad \textbf{(12.4.23)}$$

Solving differential equation of motion (12.4.1) with the initial conditions of motion (12.4.23), we have:

$$x = \frac{1}{2n}\{\frac{q\mu}{2\tau}t^2 + \left(q\rho - f - \frac{q\mu}{2n\tau}\right)[t + \frac{1}{2n}(e^{-2nt} - 1)]\} \quad \textbf{(12.4.24)}$$

The rest of the parameters are the same as for the previous case.

12.5 Constant Force R and Harmonic Force $A\sin(\omega_1 t + \lambda)$
The intersection of Row 12 and Column 5 in Guiding Table 2.1 indicates that this section describes engineering problems associated with the force of inertia, the damping force, and the friction force as the resisting forces, and the constant active force and the harmonic force as the active forces. The current problem could be related to the working process of vibratory systems interacting with a fluid medium or a specific hydraulic link.

The system is moving on a horizontal frictional surface. We want to determine the basic parameters of motion. The considerations related to the behavior of a friction force applied to a vibratory system are described in section 1.2. Figure 12.5 shows the model of a system subjected to the action of a constant active force, a harmonic force, a damping force, and a friction force.

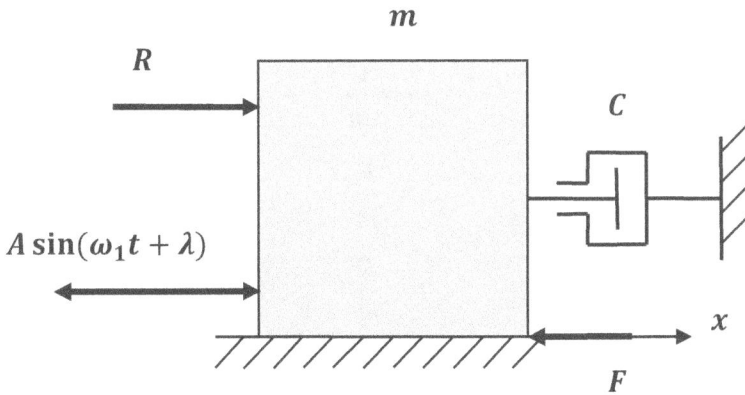

Figure 12.5 Model of a system subjected to a constant active force, a harmonic force, a damping force, and a friction force

Accounting for the considerations above and the model shown in Figure 12.5, we can compose the left and right sides of the differential equation of motion of the system. The left side consists of the force of inertia, the damping force, and the friction force, while the right side includes the constant active force and the harmonic force. Hence, the differential equation of motion reads:

$$m\frac{d^2x}{dt^2} + C\frac{dx}{dt} + F = R + A\sin(\omega_1 t + \lambda) \qquad (12.5.1)$$

Differential equation of motion (12.5.1) has different solutions for various initial conditions of motion. These solutions and their analyses are presented below.

12.5.1 General Initial Conditions

The general initial conditions of motion are:

$$\text{for} \quad t = 0 \quad x = s_0; \quad \frac{dx}{dt} = v_0 \qquad (12.5.2)$$

Transforming the sinusoidal function in equation (12.5.1) and dividing the latter by m, we obtain:

$$\frac{d^2x}{dt^2} + 2n\frac{dx}{dt} + f = r + a\sin\omega_1 t\cos\lambda + a\cos\omega_1 t\sin\lambda \qquad (12.5.3)$$

where n is the damping factor and:

$$2n = \frac{C}{m} \qquad (12.5.4)$$

$$f = \frac{F}{m} \qquad (12.5.5)$$

$$r = \frac{R}{m} \qquad (12.5.6)$$

$$a = \frac{A}{m} \qquad (12.5.7)$$

Using Laplace Transform Pairs 3, 4, 5, 6, and 7 from Table 1.1, we convert differential equation of motion (12.5.3) with the initial conditions of motion (12.5.2) from the time domain into the Laplace domain, and obtain the corresponding algebraic equation of motion in the Laplace domain:

$$l^2 x(l) - l v_0 - l^2 s_0 + 2nlx(l) - 2nls_0 + f = r + \frac{a\omega_1 l \cos\lambda}{l^2 + \omega_1^2} + \frac{al^2 \sin\lambda}{l^2 + \omega_1^2} \qquad (12.5.8)$$

Applying conventional algebraic procedures to equation (12.5.8), we may write:

$$x(l)l(l + 2n) = r - f + l(v_0 + 2ns_0) + l^2 s_0 + \frac{a\omega_1 l \cos\lambda}{l^2 + \omega_1^2} + \frac{al^2 \sin\lambda}{l^2 + \omega_1^2} \qquad (12.5.9)$$

The solution of equation (12.5.9) for the displacement $x(l)$ in Laplace domain reads:

$$x(l) = \frac{r - f}{l(l + 2n)} + \frac{v_0 + 2ns_0}{l + 2n} + \frac{ls_0}{l + 2n}$$

$$+ \frac{a\omega_1 \cos\lambda}{(l + 2n)(l^2 + \omega_1^2)} + \frac{al \sin\lambda}{(l + 2n)(l^2 + \omega_1^2)} \qquad (12.5.10)$$

Using pairs 1, 13, 10, 12, 19, and 20 from Table 1.1, we invert equation (12.5.10) from the Laplace domain into the time domain and obtain the solution of differential equation of motion (12.5.1) with the initial conditions of motion (12.5.2):

$$x = \frac{r-f}{2n}[t + \frac{1}{2n}(e^{-2nt} - 1)] + \frac{v_0 + 2ns_0}{2n}(1 - e^{-2nt}) + s_0 e^{-2nt}$$
$$+ \frac{2na(1 - \cos\omega_1 t)\cos\lambda}{\omega_1(\omega_1^2 + 4n^2)} - \frac{a\sin\omega_1 t\cos\lambda}{\omega_1^2 + 4n^2} + \frac{a\omega_1(1 - e^{-2nt})\cos\lambda}{2n(\omega_1^2 + 4n^2)}$$
$$+ \frac{a(1 - \cos\omega_1 t)\sin\lambda}{\omega_1^2 + 4n^2} + \frac{2na\sin\omega_1 t\sin\lambda}{\omega_1(\omega_1^2 + 4n^2)} - \frac{a(1 - e^{-2nt})\sin\lambda}{\omega_1^2 + 4n^2}$$
$$\tag{12.5.11}$$

Using basic algebra, we transform equation (12.5.11) into the following shape:

$$x = s_0 + \frac{v_0}{2n}(1 - e^{-2nt}) + \frac{r-f}{2n}[t + \frac{1}{2n}(e^{-2nt} - 1)]$$
$$+ \frac{2na(1 - \cos\omega_1 t)\cos\lambda}{\omega_1(\omega_1^2 + 4n^2)} - \frac{a\sin\omega_1 t\cos\lambda}{\omega_1^2 + 4n^2} + \frac{a\omega_1(1 - e^{-2nt})\cos\lambda}{2n(\omega_1^2 + 4n^2)}$$
$$+ \frac{a(1 - \cos\omega_1 t)\sin\lambda}{\omega_1^2 + 4n^2} + \frac{2na\sin\omega_1 t\sin\lambda}{\omega_1(\omega_1^2 + 4n^2)} - \frac{a(1 - e^{-2nt})\sin\lambda}{\omega_1^2 + 4n^2}$$
$$\tag{12.5.12}$$

The first derivative of equation (12.5.12) represents the velocity of the system:

$$\frac{dx}{dt} = v_0 e^{-2nt} + \frac{r-f}{2n}(1 - e^{-2nt}) + \frac{2na\sin\omega_1 t\cos\lambda}{\omega_1^2 + 4n^2}$$
$$- \frac{a\omega_1\cos\omega_1 t\cos\lambda}{\omega_1^2 + 4n^2} + \frac{a\omega_1 e^{-2nt}\cos\lambda}{\omega_1^2 + 4n^2}$$
$$+ \frac{a\omega_1\sin\omega_1 t\sin\lambda}{\omega_1^2 + 4n^2} + \frac{2na\cos\omega_1 t\sin\lambda}{\omega_1^2 + 4n^2} - \frac{2nae^{-2nt}\sin\lambda}{\omega_1^2 + 4n^2}$$
$$\tag{12.5.13}$$

Taking the second derivative of equation (12.5.12), we determine the acceleration:

$$\frac{d^2x}{dt^2} = (r - f - 2nv_0)e^{-2nt} + \frac{2na\omega_1 \cos\omega_1 t \cos\lambda}{\omega_1^2 + 4n^2} + \frac{a\omega_1^2 \sin\omega_1 t \cos\lambda}{\omega_1^2 + 4n^2}$$

$$- \frac{2na\omega_1 e^{-2nt} \cos\lambda}{\omega_1^2 + 4n^2} + \frac{a\omega_1^2 \cos\omega_1 t \sin\lambda}{\omega_1^2 + 4n^2}$$

$$- \frac{2na\omega_1 \sin\omega_1 t \sin\lambda}{\omega_1^2 + 4n^2} + \frac{4n^2 ae^{-2nt} \sin\lambda}{\omega_1^2 + 4n^2} \qquad (12.5.14)$$

12.5.2 Initial Displacement Equals Zero

The initial conditions of motion are:

$$\text{for} \quad t = 0 \quad x = 0; \quad \frac{dx}{dt} = v_0 \qquad (12.5.15)$$

The solution of differential equation of motion (12.5.1) with the initial conditions of motion (12.5.15) reads:

$$x = \frac{v_0}{2n}(1 - e^{-2nt}) + \frac{r - f}{2n}[t + \frac{1}{2n}(e^{-2nt} - 1)]$$

$$+ \frac{2na(1 - \cos\omega_1 t)\cos\lambda}{\omega_1(\omega_1^2 + 4n^2)} - \frac{a\sin\omega_1 t \cos\lambda}{\omega_1^2 + 4n^2}$$

$$+ \frac{a\omega_1(1 - e^{-2nt})\cos\lambda}{2n(\omega_1^2 + 4n^2)} + \frac{a(1 - \cos\omega_1 t)\sin\lambda}{\omega_1^2 + 4n^2}$$

$$+ \frac{2na\sin\omega_1 t \sin\lambda}{\omega_1(\omega_1^2 + 4n^2)} - \frac{a(1 - e^{-2nt})\sin\lambda}{\omega_1^2 + 4n^2} \qquad (12.5.16)$$

The equations for the velocity and acceleration are the same as in the previous case.

12.5.3 Initial Velocity Equals Zero

The initial conditions of motion are:

$$\text{for} \quad t = 0 \quad x = s_0; \quad \frac{dx}{dt} = 0 \qquad (12.5.17)$$

Solving differential equation of motion (12.5.1) with the initial conditions of motion (12.5.17), we obtain:

$$x = s_0 + \frac{r-f}{2n}\left[t + \frac{1}{2n}(e^{-2nt} - 1)\right] + \frac{2na(1 - \cos\omega_1 t)\cos\lambda}{\omega_1(\omega_1^2 + 4n^2)}$$

$$-\frac{a\sin\omega_1 t\cos\lambda}{\omega_1^2 + 4n^2} + \frac{a\omega_1(1 - e^{-2nt})\cos\lambda}{2n(\omega_1^2 + 4n^2)} + \frac{a(1 - \cos\omega_1 t)\sin\lambda}{\omega_1^2 + 4n^2}$$

$$+\frac{2na\sin\omega_1 t\sin\lambda}{\omega_1(\omega_1^2 + 4n^2)} - \frac{a(1 - e^{-2nt})\sin\lambda}{\omega_1^2 + 4n^2} \qquad \textbf{(12.5.18)}$$

The first derivative of equation (12.5.18) represents the velocity of the system:

$$\frac{dx}{dt} = \frac{r-f}{2n}(1 - e^{-2nt}) + \frac{2na\sin\omega_1 t\cos\lambda}{\omega_1^2 + 4n^2} - \frac{a\omega_1\cos\omega_1 t\cos\lambda}{\omega_1^2 + 4n^2}$$

$$+\frac{a\omega_1 e^{-2nt}\cos\lambda}{\omega_1^2 + 4n^2} + \frac{a\omega_1\sin\omega_1 t\sin\lambda}{\omega_1^2 + 4n^2}$$

$$+\frac{2na\cos\omega_1 t\sin\lambda}{\omega_1^2 + 4n^2} - \frac{2nae^{-2nt}\sin\lambda}{\omega_1^2 + 4n^2} \qquad \textbf{(12.5.19)}$$

Taking the second derivative of equation (12.5.18), we determine the acceleration:

$$\frac{d^2x}{dt^2} = (r-f)e^{-2nt} + \frac{2na\omega_1\cos\omega_1 t\cos\lambda}{\omega_1^2 + 4n^2} + \frac{a\omega_1^2\sin\omega_1 t\cos\lambda}{\omega_1^2 + 4n^2}$$

$$-\frac{2na\omega_1 e^{-2nt}\cos\lambda}{\omega_1^2 + 4n^2} + \frac{a\omega_1^2\cos\omega_1 t\sin\lambda}{\omega_1^2 + 4n^2}$$

$$-\frac{2na\omega_1\sin\omega_1 t\sin\lambda}{\omega_1^2 + 4n^2} + \frac{4n^2 ae^{-2nt}\sin\lambda}{\omega_1^2 + 4n^2} \qquad \textbf{(12.5.20)}$$

12.5.4 Both the Initial Displacement and Velocity Equals Zero

The initial conditions of motion are:

$$\text{for } t = 0 \quad x = 0; \quad \frac{dx}{dt} = 0 \qquad \textbf{(12.5.21)}$$

The solution of differential equation motion (12.5.1) with the initial conditions of motion (12.5.21) reads:

$$x = \frac{r-f}{2n}[t + \frac{1}{2n}(e^{-2nt} - 1)] + \frac{2na(1 - \cos\omega_1 t)\cos\lambda}{\omega_1(\omega_1^2 + 4n^2)}$$

$$-\frac{a\sin\omega_1 t\cos\lambda}{\omega_1^2 + 4n^2} + \frac{a\omega_1(1 - e^{-2nt})\cos\lambda}{2n(\omega_1^2 + 4n^2)} + \frac{a(1 - \cos\omega_1 t)\sin\lambda}{\omega_1^2 + 4n^2}$$

$$+\frac{2na\sin\omega_1 t\sin\lambda}{\omega_1(\omega_1^2 + 4n^2)} - \frac{a(1 - e^{-2nt})\sin\lambda}{\omega_1^2 + 4n^2} \qquad (12.5.22)$$

The equations for the velocity and acceleration are the same as in the previous case.

12.6 Harmonic Force $A \sin(\omega_1 t + \lambda)$, and Time-Dependent Force $Q\left(\rho + \frac{\mu t}{\tau}\right)$

According to Guiding Table 2.1, this section describes engi-neering systems subjected to the action of the force of inertia, the damping force, and the friction force as the resisting forces (Row 12), and to the action of the harmonic force and the time-dependent force as the active forces (Column 6).

The current problem could be associated with the initial phase of the working process for some vibratory systems interacting with fluid media or specific hydraulic links. During this phase, the system is utilizing the harmonic force along with the time-dependent force that is acting a predetermined interval of time.

The system is moving on a frictional horizontal surface. We want to determine the basic parameters of motion. Figure 12.6 shows the model of a system subjected to the action of a harmonic force, a time-dependent force, a damping force, and a friction force. The discussion related to the behavior of a friction force applied to a vibratory system is described in section 1.2.

The considerations above and the model in Figure 12.6 let us assemble the left and right sides of the differential equation of mo-tion. The left side consists of the force of inertia, the damping force,

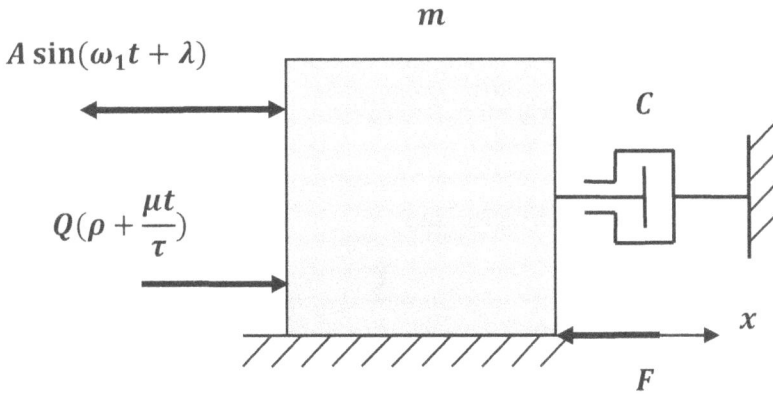

Figure 12.6 Model of a system subjected to a harmonic force, a time-dependent force, a damping force, and a friction force

and the friction force. The right side includes the harmonic force and the time-dependent force. Therefore, the differential equation of motion reads:

$$m\frac{d^2x}{dt^2}+C\frac{dx}{dt}+F = A\sin(\omega_1 t + \lambda)+Q(\rho+\frac{\mu t}{\tau}) \quad \textbf{(12.6.1)}$$

The differential equation of motion (12.6.1) has different solutions for various initial conditions of motion. These solutions and their analyses are presented below.

12.6.1 General Initial Conditions

The general initial conditions of motion are:

$$\text{for} \quad t=0 \quad x=s_0; \quad \frac{dx}{dt}=v_0 \quad \textbf{(12.6.2)}$$

where s_0 and v_0 are the initial displacement and initial velocity respectively.

Transforming equation (12.6.1), we may write:

$$m\frac{d^2x}{dt^2}+C\frac{dx}{dt}+F = Q\rho+Q\frac{\mu t}{\tau}+A\sin\omega_1 t\cos\lambda+A\cos\omega_1 t\sin\lambda$$
$$\textbf{(12.6.3)}$$

Dividing equation (12.6.3) by m, we have:

$$\frac{d^2x}{dt^2} + 2n\frac{dx}{dt} = q\rho + q\frac{\mu t}{\tau} - f + a\sin\omega_1 t\cos\lambda + a\cos\omega_1 t\sin\lambda$$
$$\text{(12.6.4)}$$

where n is the damping factor and:

$$2n = \frac{C}{m} \qquad\qquad \text{(12.6.5)}$$

$$f = \frac{F}{m} \qquad\qquad \text{(12.6.6)}$$

$$q = \frac{Q}{m} \qquad\qquad \text{(12.6.7)}$$

$$a = \frac{A}{m} \qquad\qquad \text{(12.6.8)}$$

Using Laplace Transform Pairs 3, 4, 5, 6, and 7 from Table 1.1, we convert differential equation of motion (12.6.4) with the initial conditions of motion (12.6.2) from the time domain into the Laplace domain. The resulting algebraic equation of motion in the Laplace domain reads:

$$l^2x(l) - lv_0 - l^2s_0 + 2nlx(l) - 2nls_0$$
$$= q\rho + \frac{q\mu}{\tau l} - f + \frac{a\omega_1 l}{l^2 + \omega_1^2}\cos\lambda + \frac{al^2}{l^2 + \omega_1^2}\sin\lambda$$
$$\text{(12.6.9)}$$

Applying algebraic procedures to equation (12.6.9), we obtain:

$$x(l)l(l + 2n) = l(v_0 + 2ns_0) + l^2s_0$$
$$+ q\rho + \frac{q\mu}{\tau l} + \frac{a\omega_1 l}{l^2 + \omega_1^2}\cos\lambda + \frac{al^2}{l^2 + \omega_1^2}\sin\lambda \quad \text{(12.6.10)}$$

Solving equation (12.6.8) for the Laplace domain displacement $x(l)$ we may write:

$$x(l) = \frac{v_0 + 2ns_0}{l + 2n} + \frac{ls_0}{l + 2n} + \frac{q\rho - f}{l(l + 2n)} + \frac{q\mu}{l^2\tau(l + 2n)}$$

$$+ \frac{a\omega_1\cos\lambda}{(l + 2n)(l^2 + \omega_1^2)} + \frac{al\sin\lambda}{(l + 2n)(l^2 + \omega_1^2)} \qquad (12.6.11)$$

Based on pairs 1, 10, 12, 13, 31, 19, and 20 from Table 1.1, we invert equation (12.6.11) from the Laplace domain into the time domain and obtain the solution of differential equation of motion (12.6.1) with the initial conditions of motion (12.6.2):

$$x = \frac{v_0 + 2ns_0}{2n}(1 - e^{-2nt}) + s_0 e^{-2nt} + \frac{q\rho - f}{2n}\left[t + \frac{1}{2n}(e^{-2nt} - 1)\right]$$

$$+ \frac{q\mu}{4n\tau}\left[t^2 - \frac{1}{n}t - \frac{1}{2n^2}(e^{-2nt} - 1)\right] + \frac{2na(1 - \cos\omega_1 t)\cos\lambda}{\omega_1(\omega_1^2 + 4n^2)}$$

$$- \frac{a\sin\omega_1 t\cos\lambda}{\omega_1^2 + 4n^2} + \frac{a\omega_1(1 - e^{-2nt})\cos\lambda}{2n(\omega_1^2 + 4n^2)} + \frac{a(1 - \cos\omega_1 t)\sin\lambda}{\omega_1^2 + 4n^2}$$

$$+ \frac{2na\sin\omega_1 t\sin\lambda}{\omega_1(\omega_1^2 + 4n^2)} - \frac{a(1 - e^{-2nt})\sin\lambda}{\omega_1^2 + 4n^2} \qquad (12.6.12)$$

The first derivative of equation (12.6.12) represents the velocity of the system:

$$\frac{dx}{dt} = (v_0 - \frac{q\rho - f}{2n} + \frac{q\mu}{4n^2\tau})e^{-2nt} + \frac{q\rho - f}{2n} + \frac{q\mu}{2n\tau}(t - \frac{1}{2n})$$

$$+ \frac{2na\sin\omega_1 t\cos\lambda}{\omega_1^2 + 4n^2} - \frac{a\omega_1\cos\omega_1 t\cos\lambda}{\omega_1^2 + 4n^2} + \frac{a\omega_1 e^{-2nt}\cos\lambda}{\omega_1^2 + 4n^2}$$

$$+ \frac{a\omega_1\sin\omega_1 t\sin\lambda}{\omega_1^2 + 4n^2} + \frac{2na\cos\omega_1 t\sin\lambda}{\omega_1^2 + 4n^2} - \frac{2nae^{-2nt}\sin\lambda}{\omega_1^2 + 4n^2}$$

$$(12.6.13)$$

Taking the second derivative of equation (12.6.12), we determine the acceleration:

$$\frac{d^2x}{dt^2} = \left(-2nv_0 + q\rho - f - \frac{q\mu}{2n\tau}\right)e^{-2nt} + \frac{q\mu}{2n\tau}$$

$$+ \frac{2na\omega_1 \cos\omega_1 t \cos\lambda}{\omega_1^2 + 4n^2} + \frac{a\omega_1^2 \sin\omega_1 t \cos\lambda}{\omega_1^2 + 4n^2}$$

$$- \frac{2na\omega_1 e^{-2nt} \cos\lambda}{\omega_1^2 + 4n^2} + \frac{a\omega_1^2 \cos\omega_1 t \sin\lambda}{\omega_1^2 + 4n^2}$$

$$- \frac{2na\omega_1 \sin\omega_1 t \sin\lambda}{\omega_1^2 + 4n^2} + \frac{4n^2 a e^{-2nt} \sin\lambda}{\omega_1^2 + 4n^2} \qquad \textbf{(12.6.14)}$$

12.6.2 Initial Displacement Equals Zero

The initial conditions of motion are:

$$\text{for} \quad t = 0 \quad x = 0; \quad \frac{dx}{dt} = v_0 \qquad \textbf{(12.6.15)}$$

The solution of differential equation of motion (12.6.1) with the initial conditions of motion (12.6.15) reads:

$$x = \frac{v_0 + 2ns_0}{2n}(1 - e^{-2nt}) + \frac{q\rho - f}{2n}\left[t + \frac{1}{2n}(e^{-2nt} - 1)\right]$$

$$+ \frac{q\mu}{4n\tau}\left[t^2 - \frac{1}{n}t - \frac{1}{2n^2}(e^{-2nt} - 1)\right] + \frac{2na(1 - \cos\omega_1 t)\cos\lambda}{\omega_1(\omega_1^2 + 4n^2)}$$

$$- \frac{a\sin\omega_1 t \cos\lambda}{\omega_1^2 + 4n^2} + \frac{a\omega_1(1 - e^{-2nt})\cos\lambda}{2n(\omega_1^2 + 4n^2)} + \frac{a(1 - \cos\omega_1 t)\sin\lambda}{\omega_1^2 + 4n^2}$$

$$+ \frac{2na\sin\omega_1 t \sin\lambda}{\omega_1(\omega_1^2 + 4n^2)} - \frac{a(1 - e^{-2nt})\sin\lambda}{\omega_1^2 + 4n^2} \qquad \textbf{(12.6.16)}$$

The rest of the parameters are the same as in the previous case.

12.6.3 Initial Velocity Equals Zero

The initial conditions of motion are:

$$\text{for} \quad t = 0 \quad x = s_0; \quad \frac{dx}{dt} = 0 \qquad \textbf{(12.6.17)}$$

Solving differential equation of motion (12.6.1) with the initial conditions of motion (12.6.17), we obtain:

$$x = s_0 + \frac{q\rho - f}{2n}\left[t + \frac{1}{2n}(e^{-2nt} - 1)\right] + \frac{q\mu}{4n\tau}\left[t^2 - \frac{1}{n}t - \frac{1}{2n^2}(e^{-2nt} - 1)\right]$$

$$+ \frac{2na(1 - \cos\omega_1 t)\cos\lambda}{\omega_1(\omega_1^2 + 4n^2)} - \frac{a\sin\omega_1 t \cos\lambda}{\omega_1^2 + 4n^2} + \frac{a\omega_1(1 - e^{-2nt})\cos\lambda}{2n(\omega_1^2 + 4n^2)}$$

$$+ \frac{a(1 - \cos\omega_1 t)\sin\lambda}{\omega_1^2 + 4n^2} + \frac{2na\sin\omega_1 t \sin\lambda}{\omega_1(\omega_1^2 + 4n^2)} - \frac{a(1 - e^{-2nt})\sin\lambda}{\omega_1^2 + 4n^2}$$

$$\text{(12.6.18)}$$

Taking the first derivative of equation (12.6.18), we determine the velocity of the system:

$$\frac{dx}{dt} = \frac{1}{2n}\left(q\rho - f - \frac{q\mu}{2n\tau}\right)(1 - e^{-2nt}) + \frac{q\mu}{2n\tau}t + \frac{2na\sin\omega_1 t \cos\lambda}{\omega_1^2 + 4n^2}$$

$$- \frac{a\omega_1 \cos\omega_1 t \cos\lambda}{\omega_1^2 + 4n^2} + \frac{a\omega_1 e^{-2nt}\cos\lambda}{\omega_1^2 + 4n^2} + \frac{a\omega_1 \sin\omega_1 t \sin\lambda}{\omega_1^2 + 4n^2}$$

$$+ \frac{2na\cos\omega_1 t \sin\lambda}{\omega_1^2 + 4n^2} - \frac{2nae^{-2nt}\sin\lambda}{\omega_1^2 + 4n^2} \qquad \text{(12.6.19)}$$

The second derivative of equation (12.6.18) represents the acceleration:

$$\frac{d^2x}{dt^2} = \left(q\rho - f - \frac{q\mu}{2n\tau}\right)e^{-2nt} + \frac{q\mu}{2n\tau} + \frac{2na\omega_1 \cos\omega_1 t \cos\lambda}{\omega_1^2 + 4n^2}$$

$$+ \frac{a\omega_1^2 \sin\omega_1 t \cos\lambda}{\omega_1^2 + 4n^2} - \frac{2na\omega_1 e^{-2nt}\cos\lambda}{\omega_1^2 + 4n^2} + \frac{a\omega_1^2 \cos\omega_1 t \sin\lambda}{\omega_1^2 + 4n^2}$$

$$- \frac{2na\omega_1 \sin\omega_1 t \sin\lambda}{\omega_1^2 + 4n^2} + \frac{4n^2 a e^{-2nt}\sin\lambda}{\omega_1^2 + 4n^2} \qquad \text{(12.6.20)}$$

12.6.4 Both the Initial Displacement and Velocity Equal Zero

The initial conditions of motion are:

$$\text{for} \quad t = 0 \quad x = 0; \quad \frac{dx}{dt} = 0 \qquad \text{(12.6.21)}$$

The solution of differential equation of motion (12.6.1) with the initial conditions of motion (12.6.21) reads:

$$
x = \frac{q\rho - f}{2n}\left[t + \frac{1}{2n}(e^{-2nt} - 1)\right] + \frac{q\mu}{4n\tau}\left[t^2 - \frac{1}{n}t - \frac{1}{2n^2}(e^{-2nt} - 1)\right]
$$

$$
+ \frac{2na(1 - \cos\omega_1 t)\cos\lambda}{\omega_1(\omega_1^2 + 4n^2)} - \frac{a\sin\omega_1 t\cos\lambda}{\omega_1^2 + 4n^2} + \frac{a\omega_1(1 - e^{-2nt})\cos\lambda}{2n(\omega_1^2 + 4n^2)}
$$

$$
+ \frac{a(1 - \cos\omega_1 t)\sin\lambda}{\omega_1^2 + 4n^2} + \frac{2na\sin\omega_1 t\sin\lambda}{\omega_1(\omega_1^2 + 4n^2)} - \frac{a(1 - e^{-2nt})\sin\lambda}{\omega_1^2 + 4n^2}
$$

$$
(12.6.22)
$$

The equations for the velocity and the acceleration are the same as in the previous case.

13

DAMPING AND CONSTANT RESISTANCE

This chapter discusses engineering systems subjected to the action of the force of inertia, the damping force, and the constant resisting force as the resisting forces, as indicated by the plus signs on Row 13 in Guiding Table 2.1. The intersection of this row with Columns 1 through 6 displays the numbers of this chapter's six sections and indicate the active forces applied to the systems. Throughout this chapter, the left sides of the differential equations of motion for the engineering systems are identical. The right sides of these equations differ, corresponding to the active forces applied to the systems in each section

The problems described in this chapter could be related to the working processes of engineering systems interacting with specific hydraulic links or viscoelastioplastic media (the considerations related to the deformation of these media are presented in section 1.2).

13.1 Active Force Equals Zero

Row 13 in Guiding Table 2.1 indicates that the engineering systems described in this section are associated with the force of inertia, the damping force, and the constant resisting force as the resisting forces. Because this section corresponds with Column 1, the systems experience these resisting forces in the absence of active forces. The considerations related to the motion of a system in the absence of active forces are described in section 1.3. The current problem could be associated with the upward motion of a system subjected to the damping resistance of the air and the constant resisting force related to the gravity. Alternatively, the problem could be related to the deceleration process of a system moving in the horizontal direction where the air resistance plays the role of the damping force, while the dry friction force represents a constant resisting force. The current problem could be also related to the interaction of a system with a viscoelastoplastic medium that exerts a damping and a constant resisting force during the stage of plastic deformation (more related information is presented in section 1.2).

The system is moving in the horizontal direction. We want to determine the basic parameters of motion and the maximum values of the displacement, the acceleration, and the force applied to the system. Figure 13.1 shows the model of a system subjected to the action of a damping force and a constant resisting force.

Based on the considerations above and the model in Figure 13.1, we can compose the left and right sides of the differential equation of motion. The left side consists of the force of inertia, the damping force, and the constant resisting force, while the right side equals zero. Therefore, the differential equation of motion reads:

$$m\frac{d^2x}{dt^2} + C\frac{dx}{dt} + P = 0 \qquad (13.1.1)$$

Differential equation of motion (13.1.1) has different solutions for various initial conditions of motion. These solutions and their analyses are presented below.

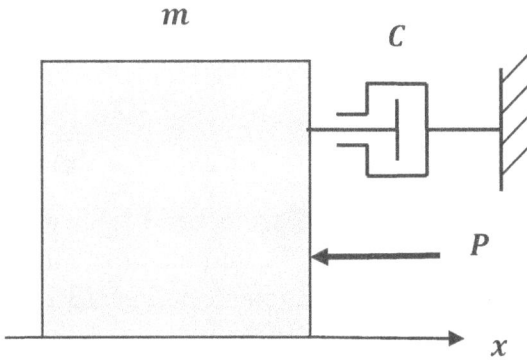

Figure 13.1 Model of a system subjected to a damping force and a constant resisting force

13.1.1 General Initial Conditions

The general initial conditions of motion are:

$$\text{for} \quad t = 0 \quad x = s_0; \quad \frac{dx}{dt} = v_0 \tag{13.1.2}$$

where s_0 and v_0 are the initial displacement and initial velocity respectively.

Dividing equation (13.1.1) by m, we may write:

$$\frac{d^2x}{dt^2} + 2n\frac{dx}{dt} + p = 0 \tag{13.1.3}$$

where n is the damping factor and:

$$2n = \frac{C}{m} \tag{13.1.4}$$

$$p = \frac{P}{m} \tag{13.1.5}$$

Using Laplace Transform Pairs 3, 4, and 5 from Table 1.1, we convert differential equation of motion (13.1.3) with the initial conditions of motion (13.1.2) from the time domain into the Laplace domain and obtain the corresponding algebraic equation of motion in the Laplace domain:

$$l^2x(l) - lv_0 - l^2s_0 + 2nlx(l) - 2nls_0 + p = 0 \tag{13.1.6}$$

Solving equation (13.1.6) for the displacement $x(l)$ in the Laplace domain, we have:

$$x(l) = \frac{v_0 + 2ns_0}{l + 2n} + \frac{ls_0}{l + 2n} - \frac{p}{l(l + 2n)} \qquad (13.1.7)$$

Based on pairs 1, 10, 12, and 13 from Table 1.1, we invert equation (13.1.7) from the Laplace domain into the time domain and obtain the solution of differential equation of motion (13.1.1) with the initial conditions of motion (13.1.2):

$$x = \frac{v_0 + 2ns_0}{2n}(1 - e^{-2nt}) + s_0 e^{-2nt} - \frac{p}{2n}[t + \frac{1}{2n}(e^{-2nt} - 1)] \qquad (13.1.8)$$

Using conventional algebraic procedures, we present equation (13.1.8) in the following shape:

$$x = \frac{1}{2n}[2ns_0 + v_0 + \frac{p}{2n} - pt - \left(v_0 + \frac{p}{2n}\right)e^{-2nt}] \qquad (13.1.9)$$

Taking the first and the second derivatives of equation (13.1.9), we determine the velocity and acceleration of the system respectively:

$$\frac{dx}{dt} = (v_0 + \frac{p}{2n})e^{-2nt} - \frac{p}{2n} \qquad (13.1.10)$$

$$\frac{d^2x}{dt^2} = -2n(v_0 + \frac{p}{2n})e^{-2nt} \qquad (13.1.11)$$

Equating the velocity according to equation (13.1.10) to zero, we may write:

$$(v_0 + \frac{p}{2n})e^{-2nT} - \frac{p}{2n} = 0 \qquad (13.1.12)$$

where T is the time that the process lasts.

From equation (13.1.12), we have:

$$e^{-2nT} = \frac{p}{2nv_0 + p} \qquad (13.1.13)$$

Equation (13.1.13) allows us to determine the time T:

$$T = \frac{1}{2n} ln \frac{2nv_0 + p}{p} \tag{13.1.14}$$

Combining equations (13.1.13) and (13.1.14) with equation (13.1.9), we calculate the maximum displacement s_{max} of the system:

$$s_{max} = \frac{1}{2n}(2ns_0 + v_0 - \frac{p}{2n} ln \frac{2nv_0 + p}{p}) \tag{13.1.15}$$

The braking distance does not include the initial displacement s_0; therefore, eliminating the initial displacement from equation (13.1.15), we obtain the equation for the braking distance s_{br} (in case if it is a braking process):

$$s_{br} = \frac{1}{2n}(v_0 - \frac{p}{2n} ln \frac{2nv_0 + p}{p}) \tag{13.1.16}$$

Combining equations (13.1.13) and (13.1.11), we determine the maximum acceleration (deceleration) a_{max}:

$$a_{max} = -p \tag{13.1.17}$$

However, according to equation (13.1.11), at the beginning of the process of motion when $t = 0$, the deceleration a_0 equals:

$$a_0 = -2nv_0 - p \tag{13.1.18}$$

The absolute value of a_0 according to equation (13.1.18) exceeds the absolute value of a_{max} according to equation (13.1.17). Hence, the maximum value of the acceleration in this case represents the absolute value a_1 according to equation (13.1.18):

$$a_1 = |2nv_0 + p| \tag{13.1.19}$$

For public transportation systems, the value of a_1 should comply with the norms of public health and safety.

By multiplying the equation (13.1.19) by m and substituting the notations (13.1.4) and (13.1.5), we determine the maximum force R_{max} applied to the system:

$$R_{max} = Cv_0 + P \tag{13.1.20}$$

The stress calculations of the system should be based on the force according to equation (13.1.20).

13.1.2 Initial Displacement Equals Zero

The initial conditions of motion are:

$$\text{for} \quad t = 0 \quad x = 0; \quad \frac{dx}{dt} = v_0 \qquad (13.1.21)$$

Solving differential equation of motion (13.1.1) with the initial conditions of motion (13.1.21), we obtain:

$$x = \frac{1}{2n}[v_0 + \frac{p}{2n} - pt - \left(v_0 + \frac{p}{2n} \right)e^{-2nt}] \qquad (13.1.22)$$

The rest of the parameters and their analysis are the same as for the previous case, keeping in mind that in this case $s_{max} = s_{br}$.

13.2 Constant Force R

According to Guiding Table 2.1, this section describes engineering systems associated with the force of inertia, the damping force, and constant resisting force as the resisting forces (Row 13) and the constant active force (Column 2). The current problem could be related to the acceleration of an upward moving system experiencing air resistance as a damping force and a constant resisting force as the weight of the system. Alternatively, this problem could be associated with the acceleration process of a horizontally moving system subjected to the air resistance and any constant resistance force. The current problem could also represent the working process of a system intended for the interaction with a viscoelastoplastic medium that exerts a damping and a constant resisting force during the stage of plastic deformation (additional information can be found in section 1.2). The current problem could be associated with certain pneumatically operated machines characterized by the decreasing of the air pressure force proportionally to the increasing of the velocity of a specific component (piston) of the machine. More related information is discussed in section 1.3, where equation (1.3.1) shows that the

decreasing of the air pressure force can be accounted by a corresponding damping force included in the left side of the differential
equation of motion.

The system moves in the horizontal direction. We want to
determine the basic parameters of motion, the maximum values
of the velocity and the acceleration of the system, the maximum
force applied to the system, and the power of the energy source.
Figure 13.2 shows the model of a system subjected to the action of
a constant active force, a damping force, and a constant resisting
force.

The considerations above and the model in Figure 13.2 let us
compose the left and right sides of the differential equation of motion of this system. The left side consists of the force of inertia, the
damping force, and the constant resisting force, while the right side
consists of a constant active force. Therefore, the differential equation of motion reads:

$$m\frac{d^2x}{dt^2} + C\frac{dx}{dt} + P = R \qquad (13.2.1)$$

Differential equation of motion (13.2.1) has different solutions for various initial conditions of motion. These solutions and
their analyses are presented below.

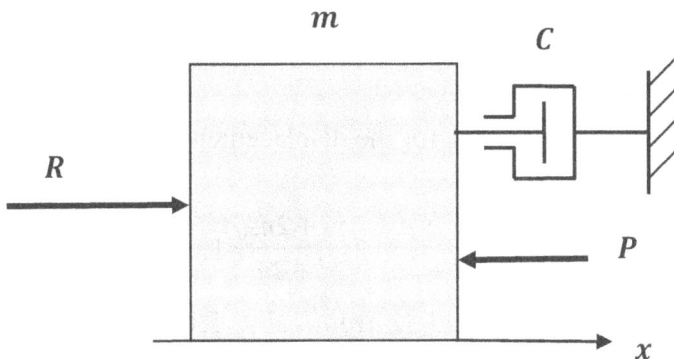

**Figure 13.2 Model of a system subjected to a constant active
force, a damping force, and a constant resisting force**

13.2.1 General Initial Conditions

The general initial conditions of motion are:

$$\text{for} \quad t = 0 \quad x = s_0; \quad \frac{dx}{dt} = v_0 \qquad \textbf{(13.2.2)}$$

where s_0 and v_0 are the initial displacement and initial velocity respectively.

Dividing equation (13.2.1) by m, we have:

$$\frac{d^2x}{dt^2} + 2n\frac{dx}{dt} + p = r \qquad \textbf{(13.2.3)}$$

where n is the damping factor and:

$$2n = \frac{C}{m} \qquad \textbf{(13.2.4)}$$

$$p = \frac{P}{m} \qquad \textbf{(13.2.5)}$$

$$r = \frac{R}{m} \qquad \textbf{(13.2.6)}$$

Based on Laplace Transform Pairs 3, 4, and 5 from Table 1.1, we convert differential equation of motion (13.2.3) with the initial conditions of motion (13.2.2) from the time domain into the Laplace domain. The resulting algebraic equation of motion in the Laplace domain reads:

$$l^2x(l) - lv_0 - l^2s_0 + 2nlx(l) - 2nls_0 + p = r \qquad \textbf{(13.2.7)}$$

Solving equation (13.2.7) for the displacement $x(l)$ in Laplace domain, we may write:

$$x(l) = \frac{r-p}{l(l+2n)} + \frac{v_0 + 2ns_0}{l+2n} + \frac{ls_0}{l+2n} \qquad \textbf{(13.2.8)}$$

Using pairs 1, 13, 10, and 12 from Table 1.1, we invert equation (13.2.8) from the Laplace domain into the time domain and obtain

the solution of differential equation of motion (13.2.1) with the initial conditions of motion (13.2.2):

$$x = \frac{r-p}{2n}[t + \frac{1}{2n}(e^{-2nt} - 1)] + \frac{v_0 + 2ns_0}{2n}(1 - e^{-2nt}) + s_0 e^{-2nt} \quad (13.2.9)$$

Performing some algebraic actions with equation (13.2.9), we write:

$$x = \frac{1}{2n}[(r-p)t + \left(\frac{r-p}{2n} - v_0\right)e^{-2nt} - \frac{r-p}{2n} + v_0 + 2ns_0] \quad (13.2.10)$$

The first derivative of equation (13.2.10) represents the velocity of the system:

$$\frac{dx}{dt} = \frac{r-p}{2n}(1 - e^{-2nt}) + v_0 e^{-2nt} \quad (13.2.11)$$

Taking the second derivative of equation (13.2.10), we determine the acceleration:

$$\frac{d^2x}{dt^2} = (r - p - 2nv_0)e^{-2nt} \quad (13.2.12)$$

In this case, the velocity approaches its maximum value when the acceleration approaches zero. Because $r - p - 2nv_0 > 0$, therefore, according to equation (13.2.12), we may write:

$$e^{-2nT} \to 0 \quad (13.2.13)$$

where T is the time that the process lasts. It should be noted that T tends to infinity.

Combining expression (13.2.13) with equation (13.2.11) and substituting notations (13.2.4), (13.2.5), and (13.2.6), we determine the maximum value of the velocity v_{max}:

$$v_{max} \to \frac{R - P}{C} \quad (13.2.14)$$

In this case, the acceleration has the maximum value at the beginning of the process when $t = 0$. Hence, equating in equation (13.2.12) the time to zero, we determine the maximum value of the acceleration a_{max}:

$$a_{max} = r - p - 2nv_0 \qquad (13.2.15)$$

For public transportation systems, the maximum value of the acceleration according to equation (13.2.12) should comply with the norms of public health and safety.

In general, the force applied to the system in any given time equals the sum of the force of inertia and the resisting forces. In this case, the acceleration according to equation (13.2.15) approaches zero when the velocity according to expression (13.2.14) approaches its maximum value. Therefore, in this case, the value of the maximum force applied to the system equals the sum of the maximum value of the damping force and the constant resisting force.

Multiplying expression (13.2.14) by C, we determine the maximum value of the damping force F_d:

$$F_d = R - P \qquad (13.2.16)$$

Adding the force P to equation (13.2.16), we calculate the maximum value of the force R_{max} applied to the system:

$$R_{max} = R \qquad (13.2.17)$$

The stress calculations of the system are based on the maximum value of the force according to equation (13.2.17). Multiplying equation (13.2.17) by expression (13.2.14), we calculate the power N of the energy source:

$$N = \frac{R(R - P)}{C} \qquad (13.2.18)$$

13.2.2 Initial Displacement Equals Zero

The initial conditions of motion are:

$$\text{for} \quad t = 0 \quad x = 0; \quad \frac{dx}{dt} = v_0 \qquad (13.2.19)$$

Solving differential equation of motion (13.2.1) with the initial conditions of motion (13.2.19), we obtain:

$$x = \frac{1}{2n}[(r-p)t + \left(\frac{r-p}{2n} - v_0\right)e^{-2nt} - \frac{r-p}{2n} + v_0]$$ **(13.2.20)**

The rest of the parameters and their analysis are the same as in the previous case.

13.2.3 Initial Velocity Equals Zero

The initial conditions of motion are:

$$\text{for} \quad t = 0 \quad x = s_0; \quad \frac{dx}{dt} = 0$$ **(13.2.21)**

The solution of differential equation of motion (13.2.1) with the initial conditions of motion (13.2.21) reads:

$$x = \frac{1}{2n}[(r-p)t + \frac{r-p}{2n}e^{-2nt} - \frac{r-p}{2n} + 2ns_0]$$ **(13.2.22)**

The first and the second derivatives of equation (13.2.22) represent the velocity and acceleration respectively:

$$\frac{dx}{dt} = \frac{r-p}{2n}(1 - e^{-2nt})$$ **(13.2.23)**

$$\frac{d^2x}{dt^2} = (r-p)e^{-2nt}$$ **(13.2.24)**

Equation (13.2.23) shows that the maximum value of the velocity is the same as in the previous cases and does not depend on the initial velocity.

According to equation (13.2.24), the maximum value of the acceleration a_{max} occurs at $t = 0$. Hence, we have:

$$a_{max} = r - p$$ **(13.2.25)**

Therefore, for public transportation systems, the value of the acceleration according to equation (13.2.25) should comply with the norms of public health and safety.

The rest of the parameters and their analysis are the same as in the previous cases.

13.2.4 Both the Initial Displacement and Velocity Equal Zero

The initial conditions of motion are:

$$\text{for} \quad t = 0 \quad x = 0; \quad \frac{dx}{dt} = 0 \tag{13.2.26}$$

Solving differential equation of motion (13.2.1) with the initial conditions of motion (13.2.26), we obtain:

$$x = \frac{1}{2n}[(r-p)t + \frac{r-p}{2n}e^{-2nt} - \frac{r-p}{2n}] \tag{13.2.27}$$

The rest of the parameters and their analysis are the same as in the previous cases.

13.3 Harmonic Force $A \sin(\omega_1 t + \lambda)$

This section describes engineering systems subjected to the action of the force of inertia, the damping force, and the constant resisting force as the resisting forces (Row 13 of Guiding Table 2.1) and the harmonic force (Column 3) as the active force. The current problem could be related to some vibratory systems intended for the interaction with viscoelastoplastic media that exert damping and constant resisting forces as the reaction to their plastic deformation (more information related the deformation of these media is presented in section 1.2). This problem could be also related to the interaction of a vibratory system with a specific hydraulic link. Considerations related to the behavior of a constant resisting force applied to a vibratory system are discussed in section 1.2.

The system is moving in the horizontal direction. We want to determine the basic parameters of motion. Figure 13.3 shows the model of a system subjected to the action of a harmonic force, a damping force, and a constant resisting force.

Based on the considerations above and on the model in Figure 13.3, we can assemble the left and right sides of the differential

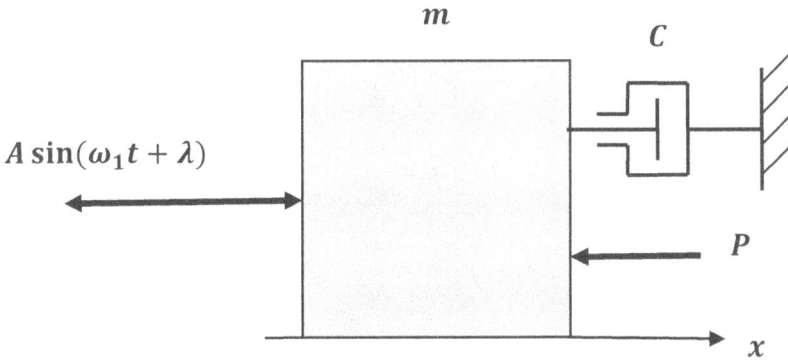

Figure 13.3 Model of a system subjected to a harmonic force, damping force, and a constant resisting force

equation of motion. The left side consists of the force of inertia, the damping force, and the constant resisting force, while the right side consists of the harmonic force. Therefore, the differential equation of motion reads:

$$m\frac{d^2x}{dt^2} + C\frac{dx}{dt} + P = A\sin(\omega_1 t + \lambda) \tag{13.3.1}$$

Differential equation of motion (13.3.1) has different solutions for various initial conditions of motion. These solutions and their analyses are presented below.

13.3.1 General Initial Conditions

The general initial conditions of motion are:

$$\text{for} \quad t = 0 \quad x = s_0; \quad \frac{dx}{dt} = v_0 \tag{13.3.2}$$

where s_0 and v_0 are the initial displacement and initial velocity respectively.

Transforming the sinusoidal function in equation (13.3.1) and dividing the latter by m, we have:

$$\frac{d^2x}{dt^2} + 2n\frac{dx}{dt} + p = a\sin\omega_1 t\cos\lambda + a\cos\omega_1 t\sin\lambda \tag{13.3.3}$$

where n is the damping factor and:

$$2n = \frac{C}{m} \tag{13.3.4}$$

$$p = \frac{P}{m} \tag{13.3.5}$$

$$a = \frac{A}{m} \tag{13.3.6}$$

Using Laplace Transform Pairs 3, 4, 5, 6, and 7 from Table 1.1, we convert differential equation of motion (13.3.3) with the initial conditions of motion (13.3.2) from the time domain into the Laplace domain. The resulting algebraic equation of motion in the Laplace domain reads:

$$l^2 x(l) - lv_0 - l^2 s_0 + 2nlx(l) - 2nls_0 + p = \frac{a\omega_1 l}{l^2 + \omega_1^2}\cos\lambda + \frac{al^2}{l^2 + \omega_1^2}\sin\lambda \tag{13.3.7}$$

Applying some conventional procedures to equation (13.3.7), we write:

$$x(l)l(l + 2n) = l(2ns_0 + v_0) + s_0 l^2 - p + \frac{a\omega_1 l}{l^2 + \omega_1^2}\cos\lambda + \frac{al^2}{l^2 + \omega_1^2}\sin\lambda \tag{13.3.8}$$

Solving equation (13.3.8) for the Laplace domain displacement $x(l)$, we obtain:

$$x(l) = \frac{2ns_0 + v_0}{l + 2n} + \frac{ls_0}{l + 2n} - \frac{p}{l(l + 2n)}$$

$$+ \frac{a\omega_1}{(l + 2n)(l^2 + \omega_1^2)}\cos\lambda$$

$$+ \frac{la}{(l + 2n)(l^2 + \omega_1^2)}\sin\lambda \tag{13.3.9}$$

Based on pairs 1, 10, 12, 13, 19, and 20 from Table 1.1, we invert equation (13.3.9) from the Laplace domain into the time domain and obtain the solution of differential equation of motion (13.3.1) with the initial conditions of motion (13.3.2):

$$x = \frac{2ns_0 + v_0}{2n}(1 - e^{-2nt}) + s_0 e^{-2nt} - \frac{p}{2n}\left[t + \frac{1}{2n}(e^{-2nt} - 1)\right]$$

$$+ \frac{2na(1 - \cos\omega_1 t)\cos\lambda}{\omega_1(\omega_1^2 + 4n^2)} - \frac{a\sin\omega_1 t\cos\lambda}{\omega_1^2 + 4n^2}$$

$$+ \frac{a\omega_1(1 - e^{-2nt})\cos\lambda}{2n(\omega_1^2 + 4n^2)} + \frac{a(1 - \cos\omega_1 t)\sin\lambda}{\omega_1^2 + 4n^2}$$

$$+ \frac{2na\sin\omega_1 t\sin\lambda}{\omega_1(\omega_1^2 + 4n^2)} - \frac{a(1 - e^{-2nt})\sin\lambda}{\omega_1^2 + 4n^2} \qquad \textbf{(13.3.10)}$$

Using basic algebra, we transform equation (13.3.10) to the following shape:

$$x = s_0 + \frac{v_0}{2n}(1 - e^{-2nt}) - \frac{p}{2n}\left[t + \frac{1}{2n}(e^{-2nt} - 1)\right]$$

$$+ \frac{2na(1 - \cos\omega_1 t)\cos\lambda}{\omega_1(\omega_1^2 + 4n^2)} - \frac{a\sin\omega_1 t\cos\lambda}{\omega_1^2 + 4n^2}$$

$$+ \frac{a\omega_1(1 - e^{-2nt})\cos\lambda}{2n(\omega_1^2 + 4n^2)} + \frac{a(1 - \cos\omega_1 t)\sin\lambda}{\omega_1^2 + 4n^2}$$

$$+ \frac{2na\sin\omega_1 t\sin\lambda}{\omega_1(\omega_1^2 + 4n^2)} - \frac{a(1 - e^{-2nt})\sin\lambda}{\omega_1^2 + 4n^2} \qquad \textbf{(13.3.11)}$$

Taking the first derivative of equation (13.3.11), we determine the velocity of the system:

$$\frac{dx}{dt} = v_0 e^{-2nt} - \frac{p}{2n}(1 - e^{-2nt}) + \frac{2na\sin\omega_1 t\cos\lambda}{\omega_1^2 + 4n^2}$$

$$- \frac{a\omega_1\cos\omega_1 t\cos\lambda}{\omega_1^2 + 4n^2} + \frac{a\omega_1 e^{-2nt}\cos\lambda}{\omega_1^2 + 4n^2}$$

$$+ \frac{a\omega_1\sin\omega_1 t\sin\lambda}{\omega_1^2 + 4n^2} + \frac{2na\cos\omega_1 t\sin\lambda}{\omega_1^2 + 4n^2} - \frac{2nae^{-2nt}\sin\lambda}{\omega_1^2 + 4n^2}$$

$$\qquad \textbf{(13.3.12)}$$

The second derivative of equation (13.3.11) represents the acceleration of the system:

$$\frac{d^2x}{dt^2} = -2nv_0e^{-2nt} - pe^{-2nt}$$

$$+ \frac{2na\omega_1 \cos\omega_1 t \cos\lambda}{\omega_1^2 + 4n^2} + \frac{a\omega_1^2 \sin\omega_1 t \cos\lambda}{\omega_1^2 + 4n^2}$$

$$- \frac{2na\omega_1 e^{2nt} \cos\lambda}{\omega_1^2 + 4n^2} + \frac{a\omega_1^2 \cos\omega_1 t \sin\lambda}{\omega_1^2 + 4n^2}$$

$$- \frac{2na\omega_1 \sin\omega_1 t \sin\lambda}{\omega_1^2 + 4n^2} + \frac{4n^2 ae^{-2nt} \sin\lambda}{\omega_1^2 + 4n^2} \qquad \textbf{(13.3.13)}$$

13.3.2 Initial Displacement Equals Zero

The initial conditions of motion are:

$$\text{for} \quad t = 0 \quad x = 0; \quad \frac{dx}{dt} = v_0 \qquad \textbf{(13.3.14)}$$

The solution of differential equation of motion (13.3.1) with the initial conditions of motion (13.3.14) reads:

$$x = \frac{v_0}{2n}(1 - e^{-2nt}) - \frac{p}{2n}\left[t + \frac{1}{2n}(e^{-2nt} - 1)\right]$$

$$+ \frac{2na(1 - \cos\omega_1 t)\cos\lambda}{\omega_1(\omega_1^2 + 4n^2)} - \frac{a\sin\omega_1 t \cos\lambda}{\omega_1^2 + 4n^2}$$

$$+ \frac{a\omega_1(1 - e^{-2nt})\cos\lambda}{2n(\omega_1^2 + 4n^2)} + \frac{a(1 - \cos\omega_1 t)\sin\lambda}{\omega_1^2 + 4n^2}$$

$$+ \frac{2na\sin\omega_1 t \sin\lambda}{\omega_1(\omega_1^2 + 4n^2)} - \frac{a(1 - e^{-2nt})\sin\lambda}{\omega_1^2 + 4n^2} \qquad \textbf{(13.3.15)}$$

The equations for the velocity and the acceleration are the same as in the previous case.

13.3.3 Initial Velocity Equals Zero

The initial conditions of motion are:

$$\text{for} \quad t = 0 \quad x = s_0; \quad \frac{dx}{dt} = 0 \qquad \textbf{(13.3.16)}$$

Solving differential equation of motion (13.3.1) with the initial conditions of motion (13.3.16), we obtain:

$$x = s_0 - \frac{p}{2n}\left[t + \frac{1}{2n}(e^{-2nt} - 1)\right] + \frac{2na(1 - \cos\omega_1 t)\cos\lambda}{\omega_1(\omega_1^2 + 4n^2)}$$

$$- \frac{a\sin\omega_1 t\cos\lambda}{\omega_1^2 + 4n^2} + \frac{a\omega_1(1 - e^{-2nt})\cos\lambda}{2n(\omega_1^2 + 4n^2)} + \frac{a(1 - \cos\omega_1 t)\sin\lambda}{\omega_1^2 + 4n^2}$$

$$+ \frac{2na\sin\omega_1 t\sin\lambda}{\omega_1(\omega_1^2 + 4n^2)} - \frac{a(1 - e^{-2nt})\sin\lambda}{\omega_1^2 + 4n^2} \qquad (13.3.17)$$

Taking the first and the second derivatives of equation (13.3.17), we determine the velocity and the acceleration of the system respectively:

$$\frac{dx}{dt} = \frac{p}{2n}(e^{-2nt} - 1) + \frac{2na\sin\omega_1 t\cos\lambda}{\omega_1^2 + 4n^2}$$

$$- \frac{a\omega_1\cos\omega_1 t\cos\lambda}{\omega_1^2 + 4n^2} + \frac{a\omega_1 e^{-2nt}\cos\lambda}{\omega_1^2 + 4n^2} + \frac{a\omega_1\sin\omega_1 t\sin\lambda}{\omega_1^2 + 4n^2}$$

$$+ \frac{2na\cos\omega_1 t\sin\lambda}{\omega_1^2 + 4n^2} - \frac{2nae^{-2nt}\sin\lambda}{\omega_1^2 + 4n^2} \qquad (13.3.18)$$

$$\frac{d^2x}{dt^2} = - pe^{-2nt}$$

$$+ \frac{2na\omega_1\cos\omega_1 t\cos\lambda}{\omega_1^2 + 4n^2} + \frac{a\omega_1^2\sin\omega_1 t\cos\lambda}{\omega_1^2 + 4n^2}$$

$$- \frac{2na\omega_1 e^{-2nt}\cos\lambda}{\omega_1^2 + 4n^2} + \frac{a\omega_1^2\cos\omega_1 t\sin\lambda}{\omega_1^2 + 4n^2}$$

$$- \frac{2na\omega_1\sin\omega_1 t\sin\lambda}{\omega_1^2 + 4n^2} + \frac{4n^2 ae^{-2nt}\sin\lambda}{\omega_1^2 + 4n^2} \qquad (13.3.19)$$

13.3.4 Both the Initial Displacement and Velocity Equal Zero
The initial conditions of motion are:

$$\text{for} \quad t = 0 \quad x = 0; \quad \frac{dx}{dt} = 0 \qquad (13.3.20)$$

The solution of differential equation of motion (13.3.1) with the initial conditions of motion (13.3.20) reads:

$$x = \frac{p}{2n}\left[\frac{1}{2n}(1-e^{-2nt})-t\right]$$
$$+\frac{2na(1-\cos\omega_1 t)\cos\lambda}{\omega_1(\omega_1^2+4n^2)}-\frac{a\sin\omega_1 t\cos\lambda}{\omega_1^2+4n^2}$$
$$+\frac{a\omega_1(1-e^{-2nt})\cos\lambda}{2n(\omega_1^2+4n^2)}+\frac{a(1-\cos\omega_1 t)\sin\lambda}{\omega_1^2+4n^2}$$
$$+\frac{2na\sin\omega_1 t\sin\lambda}{\omega_1(\omega_1^2+4n^2)}-\frac{a(1-e^{-2nt})\sin\lambda}{\omega_1^2+4n^2} \qquad \textbf{(13.3.21)}$$

The equations for the velocity and acceleration are the same as for the previous case.

13.4 Time-Dependent Force $Q\left(\rho+\frac{\mu t}{\tau}\right)$

The intersection of Row 13 and Column 4 in Guiding Table 2.1 indicates that the engineering systems described in this section are subjected to the action of the force of inertia, damping force, and constant resisting force as the resisting forces, and the time-dependent force as the active force.

The current problem could be related to the acceleration of an upward moving system during its initial phase of motion. In this case, the damping force represents the air resistance, and the force of gravity is the constant resisting force, while the active force is increasing during a predetermined interval of time. Alternatively, the current problem could be associated with the deformation of certain viscoelastoplastic media that exert damping and constant resisting forces as a reaction to their plastic deformation (additional related information is presented in section 1.2). This problem could be also related to the interaction of a system with a specific hydraulic link.

The system is moving in the horizontal direction. We want to determine the basic parameters of motion, the values of the velocity and acceleration at the end of the predetermined interval of time.

Figure 13.4 shows the model of a system subjected to the action of a time-dependent force, a damping force, and a constant resisting force.

Based on the considerations above and the model in Figure 13.4, we compose the left and right sides of the differential equation of motion. The left side consists of the force of inertia, the damping force, and the constant resisting force, while the right side includes the time-dependent force.

Hence, the differential equation of motion reads:

$$m\frac{d^2x}{dt^2} + C\frac{dx}{dt} + P = Q(\rho + \frac{\mu t}{\tau}) \qquad \text{(13.4.1)}$$

Differential equation of motion (13.4.1) has different solutions for various initial conditions of motion. These solutions and their analyses are presented below.

13.4.1 General Initial Conditions

The general initial conditions of motion are:

$$\text{for} \quad t = 0 \quad x = s_0; \quad \frac{dx}{dt} = v_0 \qquad \text{(13.4.2)}$$

where s_0 and v_0 are the initial displacement and initial velocity respectively.

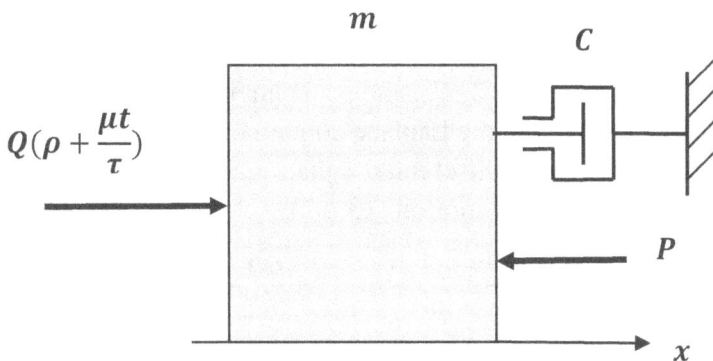

Figure 13.4 Model of a system subjected to a time-dependent force, a damping force, and a constant resisting force

Dividing equation (13.4.1) by m, we obtain:

$$\frac{d^2x}{dt^2} + 2n\frac{dx}{dt} + p = q(\rho + \frac{\mu t}{\tau}) \qquad (13.4.3)$$

where n is the damping factor and:

$$2n = \frac{C}{m} \qquad (13.4.4)$$

$$p = \frac{P}{m} \qquad (13.4.5)$$

$$q = \frac{Q}{m} \qquad (13.4.6)$$

Using Laplace Transform Pairs 3, 4, 5, and 2 from Table 1.1, we convert differential equation of motion (13.4.3) with the initial conditions of motion (13.4.2) from the time domain into the Laplace domain, and obtain the corresponding algebraic equation of motion in the Laplace domain:

$$l^2 x(l) - lv_0 - l^2 s_0 + 2nlx(l) - 2nls_0 + p = q\rho + \frac{q\mu}{\tau l} \qquad (13.4.7)$$

The solution of equation (13.4.7) for the Laplace domain displacement $x(l)$ reads:

$$x(l) = \frac{v_0 + 2ns_0}{l + 2n} + \frac{ls_0}{l + 2n} + \frac{q\rho - p}{l(l + 2n)} + \frac{q\mu}{\tau l^2(l + 2n)} \qquad (13.4.8)$$

Based on pairs 1, 10, 12, 13, and 31 from Table 1.1, we invert the equation (13.4.8) from the Laplace domain into the time domain and obtain the solution of differential equation of motion (13.4.1) with the initial conditions of motion (13.4.2):

$$x = \frac{1}{2n}(v_0 + 2ns_0)(1 - e^{-2nt}) + s_0 e^{-2nt} + \frac{q\rho - p}{2n}\left[t + \frac{1}{2n}(e^{-2nt} - 1)\right]$$

$$+ \frac{q\mu}{4n\tau}[t^2 - \frac{1}{n}t - \frac{1}{2n^2}(e^{-2nt} - 1)] \qquad (13.4.9)$$

Applying some algebraic procedures to equation (13.4.9), we have:

$$x = \frac{1}{2n}\{v_0 + 2ns_0 - v_0 e^{-2nt} + \frac{q\mu}{2\tau}t^2$$

$$+ \left(q\rho - p - \frac{q\mu}{2n\tau}\right)[t + \frac{1}{2n}(e^{-2nt} - 1)]\} \qquad \textbf{(13.4.10)}$$

The first derivative of the equation (13.4.10) represents the velocity of the system:

$$\frac{dx}{dt} = v_0 e^{-2nt} + \frac{q\mu}{2n\tau}t + \frac{1}{2n}\left(q\rho - p - \frac{q\mu}{2n\tau}\right)(1 - e^{-2nt}) \qquad \textbf{(13.4.11)}$$

Taking the second derivative of equation (13.4.10), we determine the acceleration:

$$\frac{d^2x}{dt^2} = -2nv_0 e^{-2nt} + \frac{q\mu}{2n\tau} + (q\rho - p - \frac{q\mu}{2n\tau})e^{-2nt} \qquad \textbf{(13.4.12)}$$

Substituting the time τ into equations (13.4.11), (13.4.12), we calculate the velocity v and acceleration a at end of the predetermined interval of time respectively:

$$v = v_0 e^{-2n\tau} + \frac{q\mu}{2n\tau}\tau + \frac{1}{2n}\left(q\rho - p - \frac{q\mu}{2n\tau}\right)(1 - e^{-2n\tau}) \qquad \textbf{(13.4.13)}$$

$$a = -2nv_0 e^{-2n\tau} + \frac{q\mu}{2n\tau} + (q\rho - p - \frac{q\mu}{2n\tau})e^{-2n\tau} \qquad \textbf{(13.4.14)}$$

13.4.2 Initial Displacement Equals Zero

The initial conditions of motion are:

$$\text{for} \quad t = 0 \quad x = 0; \quad \frac{dx}{dt} = v_0 \qquad \textbf{(13.4.15)}$$

The solution of differential equation of motion (13.4.1) with the initial conditions of motion (13.4.15) reads:

$$x = \frac{1}{2n}\{v_0 - v_0 e^{-2nt} + \frac{q\mu}{2\tau}t^2 + \left(q\rho - p - \frac{q\mu}{2n\tau}\right)[t + \frac{1}{2n}(e^{-2nt} - 1)]\}$$

$$\textbf{(13.4.16)}$$

The rest of the parameters are the same as in the previous case.

13.4.3 Initial Velocity Equals Zero

The initial conditions of motion are:

$$\text{for} \quad t = 0 \quad x = s_0; \quad \frac{dx}{dt} = 0 \qquad (13.4.17)$$

Solving differential equation of motion (13.4.1) with the initial conditions of motion (13.4.17), we have:

$$x = \frac{1}{2n}\{2ns_0 - v_0 e^{-2nt} + \frac{q\mu}{2\tau}t^2 + \left(q\rho - p - \frac{q\mu}{2n\tau}\right)[t + \frac{1}{2n}(e^{-2nt} - 1)]\}$$

$$(13.4.18)$$

Taking the first and the second derivatives of equation (13.4.18), we determine the velocity and acceleration of the system respectively:

$$\frac{dx}{dt} = \frac{q\mu}{2n\tau}t + \frac{1}{2n}\left(q\rho - p - \frac{q\mu}{2n\tau}\right)(1 - e^{-2nt}) \qquad (13.4.19)$$

$$\frac{d^2x}{dt^2} = \frac{q\mu}{2n\tau} + (q\rho - p - \frac{q\mu}{2n\tau})e^{-2nt} \qquad (13.4.20)$$

Substituting the time τ into equations (13.4.19) and (13.4.20), we calculate the values of the velocity v and the acceleration a at the end of the predetermined interval of time respectively:

$$\frac{dx}{dt} = \frac{q\mu}{2n\tau}\tau + \frac{1}{2n}\left(q\rho - p - \frac{q\mu}{2n\tau}\right)(1 - e^{-2n\tau}) \qquad (13.4.21)$$

$$\frac{d^2x}{dt^2} = \frac{q\mu}{2n\tau} + (q\rho - p - \frac{q\mu}{2n\tau})e^{-2n\tau} \qquad (13.4.22)$$

13.4.4 Both the Initial Displacement and Velocity Equal Zero

The initial conditions of motion are:

$$\text{for} \quad t = 0 \quad x = 0; \quad \frac{dx}{dt} = 0 \qquad (13.4.23)$$

The solution of the differential equation of motion (13.4.1) with the initial conditions of motion (13.4.23) reads:

$$x = \frac{1}{2n}\{\frac{q\mu}{2\tau}t^2 + \left(q\rho - p - \frac{q\mu}{2n\tau}\right)[t + \frac{1}{2n}(e^{-2nt} - 1)]\} \qquad (13.4.24)$$

The rest of the parameters are the same as for the previous case.

13.5 Constant Force R and Harmonic Force $A\sin(\omega_1 t + \lambda)$

This section describes engineering systems experiencing the action of the force of inertia, the damping force, and the constant resisting force as the resisting forces (Row 13 in Guiding Table 2.1) and the constant active force and the harmonic force (Column 5). The current problem could be related to the working processes of vibratory systems interacting with certain viscoelastoplastic media. The corresponding information related to the reaction of these media to their plastic deformation and regarding the behavior of a constant resisting force applied to a vibratory system is presented in section 1.2. The current problem could also reflect the interaction of a vibratory system with a specific hydraulic link.

The system is moving in the horizontal direction. We want to determine the basic parameters of motion. Figure 13.5 shows the model of a system subjected to the action of a constant active force, a harmonic force, a damping force, and a constant resisting force.

The considerations above and the model shown in Figure 13.5 let us assemble the left and right sides of the differential equation of motion of the system. The left side includes the force of inertia, the damping force, and the constant resisting force, while the right side consists of the constant active force and the harmonic force. Therefore, the differential equation of motion reads:

$$m\frac{d^2x}{dt^2} + C\frac{dx}{dt} + P = R + A\sin(\omega_1 t + \lambda) \qquad (13.5.1)$$

Differential equation of motion (13.5.1) has different solutions for various initial conditions of motion. These solutions and their analyses are presented below.

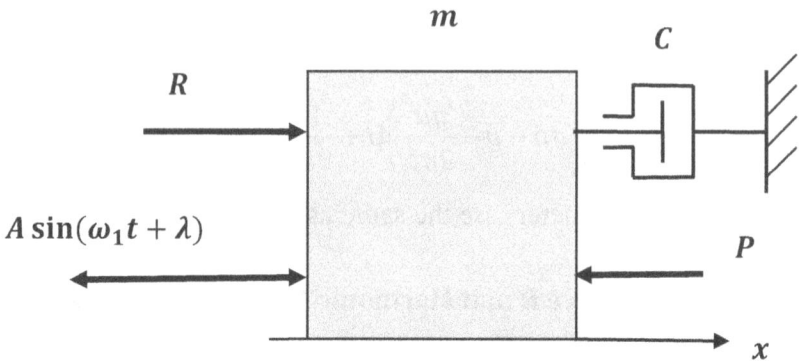

Figure 13.5 Model of a system subjected to a constant active force, a harmonic force, a damping force, and a constant resisting force

13.5.1 General Initial Conditions

The general initial conditions of motion are:

$$\text{for} \quad t = 0 \quad x = s_0; \quad \frac{dx}{dt} = v_0 \qquad (13.5.2)$$

Transforming the sinusoidal function in equation (13.5.1) and dividing the latter by m we obtain:

$$\frac{d^2x}{dt^2} + 2n\frac{dx}{dt} + p = r + a\sin\omega_1 t \cos\lambda + a\cos\omega_1 t \sin\lambda \qquad (13.5.3)$$

where n is the damping factor and:

$$2n = \frac{C}{m} \qquad (13.5.4)$$

$$p = \frac{P}{m} \qquad (13.5.5)$$

$$r = \frac{R}{m} \qquad (13.5.6)$$

$$a = \frac{A}{m} \qquad (13.5.7)$$

Based on Laplace Transform Pairs 3, 4, 5, 6, and 7 from Table 1.1, we convert differential equation of motion (13.5.3) with the initial conditions of motion (13.5.2) from the time domain into the Laplace domain, and obtain the resulting algebraic equation of motion in the Laplace domain:

$$l^2 x(l) - lv_0 - l^2 s_0 + 2nlx(l) - 2nls_0 + p = r + \frac{a\omega_1 l \cos\lambda}{l^2 + \omega_1^2} + \frac{al^2 \sin\lambda}{l^2 + \omega_1^2}$$

$$(13.5.8)$$

Applying basic algebra to equation (13.5.8), we have:

$$x(l)l(l + 2n) = r - p + l(v_0 + 2ns_0) + l^2 s_0 + \frac{a\omega_1 l \cos\lambda}{l^2 + \omega_1^2} + \frac{al^2 \sin\lambda}{l^2 + \omega_1^2}$$

$$(13.5.9)$$

The solution of equation (13.5.9) for the displacement $x(l)$ in Laplace domain reads:

$$x(l) = \frac{r - p}{l(l + 2n)} + \frac{v_0 + 2ns_0}{l + 2n} + \frac{ls_0}{l + 2n}$$

$$+ \frac{a\omega_1 \cos\lambda}{(l + 2n)(l^2 + \omega_1^2)} + \frac{al \sin\lambda}{(l + 2n)(l^2 + \omega_1^2)} \quad (13.5.10)$$

Based on pairs 1, 13, 10, 12, 19, and 20 from Table 1.1, we invert equation (13.5.10) from the Laplace domain into the time domain and obtain the solution of differential equation of motion (13.5.1) with the initial conditions of motion (13.5.2):

$$x = \frac{r - p}{2n}[t + \frac{1}{2n}(e^{-2nt} - 1)] + \frac{v_0 + 2ns_0}{2n}(1 - e^{-2nt}) + s_0 e^{-2nt}$$

$$+ \frac{2na(1 - \cos\omega_1 t)\cos\lambda}{\omega_1(\omega_1^2 + 4n^2)} - \frac{a\sin\omega_1 t \cos\lambda}{\omega_1^2 + 4n^2}$$

$$+ \frac{a\omega_1(1 - e^{-2nt})\cos\lambda}{2n(\omega_1^2 + 4n^2)} + \frac{a(1 - \cos\omega_1 t)\sin\lambda}{\omega_1^2 + 4n^2}$$

$$+ \frac{2na\sin\omega_1 t \sin\lambda}{\omega_1(\omega_1^2 + 4n^2)} - \frac{a(1 - e^{-2nt})\sin\lambda}{\omega_1^2 + 4n^2} \quad (13.5.11)$$

Using algebraic actions, we transform equation (13.5.11) to the following shape:

$$x = s_0 + \frac{v_0}{2n}(1 - e^{-2nt}) + \frac{r-p}{2n}[t + \frac{1}{2n}(e^{-2nt} - 1)]$$

$$+ \frac{2na(1 - \cos\omega_1 t)\cos\lambda}{\omega_1(\omega_1^2 + 4n^2)} - \frac{a\sin\omega_1 t\cos\lambda}{\omega_1^2 + 4n^2}$$

$$+ \frac{a\omega_1(1 - e^{-2nt})\cos\lambda}{2n(\omega_1^2 + 4n^2)} + \frac{a(1 - \cos\omega_1 t)\sin\lambda}{\omega_1^2 + 4n^2}$$

$$+ \frac{2na\sin\omega_1 t\sin\lambda}{\omega_1(\omega_1^2 + 4n^2)} - \frac{a(1 - e^{-2nt})\sin\lambda}{\omega_1^2 + 4n^2} \qquad \textbf{(13.5.12)}$$

Taking the first derivative of equation (13.5.12), we determine the velocity of the system:

$$\frac{dx}{dt} = v_0 e^{-2nt} + \frac{r-p}{2n}(1 - e^{-2nt}) + \frac{2na\sin\omega_1 t\cos\lambda}{\omega_1^2 + 4n^2}$$

$$- \frac{a\omega_1\cos\omega_1 t\cos\lambda}{\omega_1^2 + 4n^2} + \frac{a\omega_1 e^{-2nt}\cos\lambda}{\omega_1^2 + 4n^2}$$

$$+ \frac{a\omega_1\sin\omega_1 t\sin\lambda}{\omega_1^2 + 4n^2} + \frac{2na\cos\omega_1 t\sin\lambda}{\omega_1^2 + 4n^2}$$

$$- \frac{2nae^{-2nt}\sin\lambda}{\omega_1^2 + 4n^2} \qquad \textbf{(13.5.13)}$$

The second derivative of equation (13.5.12) represents the acceleration:

$$\frac{d^2x}{dt^2} = (r - p - 2nv_0)e^{-2nt}$$

$$+ \frac{2na\omega_1\cos\omega_1 t\cos\lambda}{\omega_1^2 + 4n^2} + \frac{a\omega_1^2\sin\omega_1 t\cos\lambda}{\omega_1^2 + 4n^2} - \frac{2na\omega_1 e^{-2nt}\cos\lambda}{\omega_1^2 + 4n^2}$$

$$+ \frac{a\omega_1^2\cos\omega_1 t\sin\lambda}{\omega_1^2 + 4n^2} - \frac{2na\omega_1\sin\omega_1 t\sin\lambda}{\omega_1^2 + 4n^2} + \frac{4n^2 ae^{-2nt}\sin\lambda}{\omega_1^2 + 4n^2}$$

$$\textbf{(13.5.14)}$$

13.5.2 Initial Displacement Equals Zero

The initial conditions of motion are:

$$\text{for} \quad t = 0 \quad x = 0; \quad \frac{dx}{dt} = v_0 \qquad (13.5.15)$$

Solving differential equation of motion (13.5.1) with the initial conditions of motion (13.5.15), we have:

$$x = \frac{v_0}{2n}(1 - e^{-2nt}) + \frac{r - p}{2n}[t + \frac{1}{2n}(e^{-2nt} - 1)]$$
$$+ \frac{2na(1 - \cos\omega_1 t)\cos\lambda}{\omega_1(\omega_1^2 + 4n^2)} - \frac{a\sin\omega_1 t\cos\lambda}{\omega_1^2 + 4n^2}$$
$$+ \frac{a\omega_1(1 - e^{-2nt})\cos\lambda}{2n(\omega_1^2 + 4n^2)} + \frac{a(1 - \cos\omega_1 t)\sin\lambda}{\omega_1^2 + 4n^2}$$
$$+ \frac{2na\sin\omega_1 t\sin\lambda}{\omega_1(\omega_1^2 + 4n^2)} - \frac{a(1 - e^{-2nt})\sin\lambda}{\omega_1^2 + 4n^2} \qquad (13.5.16)$$

The equations for the velocity and acceleration are the same as in the previous case.

13.5.3 Initial Velocity Equals Zero

The initial conditions of motion are:

$$\text{for} \quad t = 0 \quad x = s_0; \quad \frac{dx}{dt} = 0 \qquad (13.5.17)$$

The solution of differential equation of motion (13.5.1) with the initial conditions of motion (13.5.17) reads:

$$x = s_0 + \frac{r - p}{2n}[t + \frac{1}{2n}(e^{-2nt} - 1)] + \frac{2na(1 - \cos\omega_1 t)\cos\lambda}{\omega_1(\omega_1^2 + 4n^2)}$$
$$- \frac{a\sin\omega_1 t\cos\lambda}{\omega_1^2 + 4n^2} + \frac{a\omega_1(1 - e^{-2nt})\cos\lambda}{2n(\omega_1^2 + 4n^2)} + \frac{a(1 - \cos\omega_1 t)\sin\lambda}{\omega_1^2 + 4n^2}$$
$$+ \frac{2na\sin\omega_1 t\sin\lambda}{\omega_1(\omega_1^2 + 4n^2)} - \frac{a(1 - e^{-2nt})\sin\lambda}{\omega_1^2 + 4n^2} \qquad (13.5.18)$$

Taking the first derivative of equation (13.5.18), we determine the velocity of the system:

$$\frac{dx}{dt} = \frac{r-p}{2n}(1-e^{-2nt}) + \frac{2na\sin\omega_1 t\cos\lambda}{\omega_1^2+4n^2}$$

$$-\frac{a\omega_1\cos\omega_1 t\cos\lambda}{\omega_1^2+4n^2} + \frac{a\omega_1 e^{-2nt}\cos\lambda}{\omega_1^2+4n^2} + \frac{a\omega_1\sin\omega_1 t\sin\lambda}{\omega_1^2+4n^2}$$

$$+\frac{2na\cos\omega_1 t\sin\lambda}{\omega_1^2+4n^2} - \frac{2nae^{-2nt}\sin\lambda}{\omega_1^2+4n^2} \qquad (13.5.19)$$

The second derivative of equation (13.5.18) represents the acceleration:

$$\frac{d^2x}{dt^2} = (r-p)e^{-2nt} + \frac{2na\omega_1\cos\omega_1 t\cos\lambda}{\omega_1^2+4n^2}$$

$$+\frac{a\omega_1^2\sin\omega_1 t\cos\lambda}{\omega_1^2+4n^2} - \frac{2na\omega_1 e^{-2nt}\cos\lambda}{\omega_1^2+4n^2} + \frac{a\omega_1^2\cos\omega_1 t\sin\lambda}{\omega_1^2+4n^2}$$

$$-\frac{2na\omega_1\sin\omega_1 t\sin\lambda}{\omega_1^2+4n^2} + \frac{4n^2ae^{-2nt}\sin\lambda}{\omega_1^2+4n^2} \qquad (13.5.20)$$

13.5.4 Both the Initial Displacement and Velocity Equal Zero
The initial conditions of motion are:

$$\text{for}\quad t=0\quad x=0;\quad \frac{dx}{dt}=0 \qquad (13.5.21)$$

Solving differential equation of motion (13.5.1) with the initial conditions of motion (13.5.21), we obtain:

$$x = \frac{r-p}{2n}[t + \frac{1}{2n}(e^{-2nt}-1)] + \frac{2na(1-\cos\omega_1 t)\cos\lambda}{\omega_1(\omega_1^2+4n^2)}$$

$$-\frac{a\sin\omega_1 t\cos\lambda}{\omega_1^2+4n^2} + \frac{a\omega_1(1-e^{-2nt})\cos\lambda}{2n(\omega_1^2+4n^2)} + \frac{a(1-\cos\omega_1 t)\sin\lambda}{\omega_1^2+4n^2}$$

$$+\frac{2na\sin\omega_1 t\sin\lambda}{\omega_1(\omega_1^2+4n^2)} - \frac{a(1-e^{-2nt})\sin\lambda}{\omega_1^2+4n^2} \qquad (13.5.22)$$

The equations for the velocity and acceleration are the same as in the previous case.

13.6 Harmonic Force $A \sin(\omega_1 t + \lambda)$ and Time-Dependent Force $Q\left(\rho + \frac{\mu t}{\tau}\right)$

The intersection of Row 13 in Guiding Table 2.1 with Column 6 indicates that this section describes engineering systems subjected to the action of the force of inertia, the damping force, and constant resisting force as the resisting forces and to the action of a harmonic force and a time-dependent force as the active forces.

The current problem could be associated with the working process of a vibratory system interacting with a viscoelastoplastic medium. At the beginning of the process, along with the harmonic force, the system utilizes a time-dependent force that is acting a predetermined interval of time. The corresponding information related to the reaction of a viscoelastoplastic medium to its plastic deformation and to the behavior of a constant resisting force applied to a vibratory system is presented in section 1.2.

The system is moving in the horizontal direction. We want to determine the basic parameters of motion. Figure 13.6 shows the model of a system subjected to the action of a harmonic force, a time-dependent force, a damping force, and a constant resisting force.

Based on the considerations above and the model in Figure 13.6, we can compose the left and right sides of the differential equation of motion. The left side consists of the force of inertia, the damping force, and the constant resisting force. The right side of the equation includes the harmonic force and the time-dependent force. Hence, the differential equation of motion reads:

$$m\frac{d^2x}{dt^2} + C\frac{dx}{dt} + P = A\sin(\omega_1 t + \lambda) + Q(\rho + \frac{\mu t}{\tau}) \quad \textbf{(13.6.1)}$$

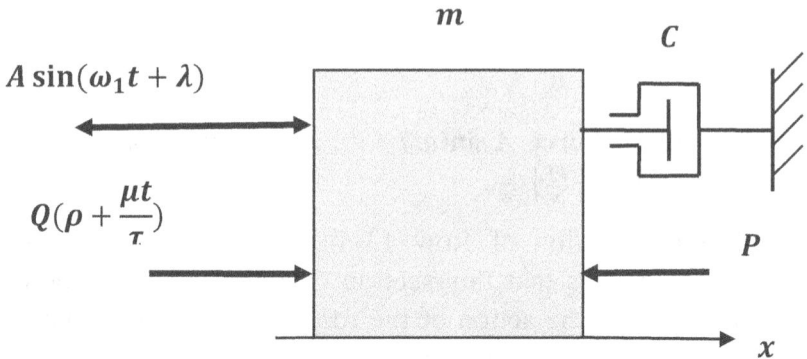

$A \sin(\omega_1 t + \lambda)$

$Q(\rho + \dfrac{\mu t}{\tau})$

m

C

P

x

Figure 13.6 Model of a system subjected to the action of a harmonic force, a time-dependent force, a damping force, and a constant resisting force

Differential equation of motion (13.6.1) has different solutions for various initial conditions of motion. These solutions and their analyses are presented below.

13.6.1 General Initial Conditions

The general initial conditions of motion are:

$$\text{for} \quad t = 0 \quad x = s_0; \quad \frac{dx}{dt} = v_0 \qquad (13.6.2)$$

where s_0 and v_0 are the initial displacement and initial velocity respectively.

Transforming equation (13.6.1), we may write:

$$m\frac{d^2x}{dt^2} + C\frac{dx}{dt} + P = Q\rho + Q\frac{\mu t}{\tau} + A\sin\omega_1 t \cos\lambda + A\cos\omega_1 t \sin\lambda$$

$$(13.6.3)$$

Dividing equation (13.6.3) by m, we have:

$$\frac{d^2x}{dt^2} + 2n\frac{dx}{dt} + p = q\rho + q\frac{\mu t}{\tau} + a\sin\omega_1 t \cos\lambda + a\cos\omega_1 t \sin\lambda$$

$$(13.6.4)$$

where n is the damping factor and:

$$2n = \frac{C}{m} \tag{13.6.5}$$

$$p = \frac{P}{m} \tag{13.6.6}$$

$$q = \frac{Q}{m} \tag{13.6.7}$$

$$a = \frac{A}{m} \tag{13.6.8}$$

Using Laplace Transform Pairs 3, 4, 5, 6, and 7 from Table 1.1, we convert differential equation of motion (13.6.4) with the initial conditions of motion (13.6.2) from the time domain into the Laplace domain. The resulting algebraic equation of motion in the Laplace domain reads:

$$l^2 x(l) - l v_0 - l^2 s_0 + 2nlx(l) - 2nls_0 + p$$

$$= q\rho + \frac{q\mu}{\tau l} + \frac{a\omega_1 l}{l^2 + \omega_1^2}\cos\lambda + \frac{al^2}{l^2 + \omega_1^2}\sin\lambda \tag{13.6.9}$$

Based on some algebraic procedures with the equation (13.6.9), we may write:

$$x(l)l(l + 2n) = l(v_0 + 2ns_0) + l^2 s_0 + q\rho - p + \frac{q\mu}{\tau l}$$

$$+ \frac{a\omega_1 l}{l^2 + \omega_1^2}\cos\lambda + \frac{al^2}{l^2 + \omega_1^2}\sin\lambda \tag{13.6.10}$$

Solving equation (13.6.10) for the Laplace domain displacement $x(l)$, we have:

$$x(l) = \frac{v_0 + 2ns_0}{l + 2n} + \frac{ls_0}{l + 2n} + \frac{q\rho - p}{l(l + 2n)} + \frac{q\mu}{l^2 \tau(l + 2n)}$$

$$+ \frac{a\omega_1 \cos\lambda}{(l + 2n)(l^2 + \omega_1^2)} + \frac{al\sin\lambda}{(l + 2n)(l^2 + \omega_1^2)} \tag{13.6.11}$$

Based on pairs 1, 10, 12, 13, 31, 19, and 20 from Table 1.1, we invert equation (13.6.11) from the Laplace domain into the time domain and obtain the solution of differential equation of motion (13.6.1) with the initial conditions of motion (13.6.2):

$$
x = \frac{v_0 + 2ns_0}{2n}(1 - e^{-2nt}) + s_0 e^{-2nt} + \frac{qp - p}{2n}\left[t + \frac{1}{2n}(e^{-2nt} - 1) \right]
$$

$$
+ \frac{q\mu}{4n\tau}\left[t^2 - \frac{1}{n}t - \frac{1}{2n^2}(e^{-2nt} - 1) \right] + \frac{2na(1 - \cos\omega_1 t)\cos\lambda}{\omega_1(\omega_1^2 + 4n^2)}
$$

$$
- \frac{a\sin\omega_1 t \cos\lambda}{\omega_1^2 + 4n^2} + \frac{a\omega_1(1 - e^{-2nt})\cos\lambda}{2n(\omega_1^2 + 4n^2)} + \frac{a(1 - \cos\omega_1 t)\sin\lambda}{\omega_1^2 + 4n^2}
$$

$$
+ \frac{2na\sin\omega_1 t \sin\lambda}{\omega_1(\omega_1^2 + 4n^2)} - \frac{a(1 - e^{-2nt})\sin\lambda}{\omega_1^2 + 4n^2} \tag{13.6.12}
$$

Taking the first derivative of equation (13.6.12), we determine the velocity of the system:

$$
\frac{dx}{dt} = (v_0 - \frac{qp - p}{2n} + \frac{q\mu}{4n^2\tau})e^{-2nt} + \frac{qp - p}{2n} + \frac{q\mu}{2n\tau}(t - \frac{1}{2n})
$$

$$
+ \frac{2na\sin\omega_1 t \cos\lambda}{\omega_1^2 + 4n^2} - \frac{a\omega_1\cos\omega_1 t \cos\lambda}{\omega_1^2 + 4n^2} + \frac{a\omega_1 e^{-2nt}\cos\lambda}{\omega_1^2 + 4n^2}
$$

$$
+ \frac{a\omega_1\sin\omega_1 t \sin\lambda}{\omega_1^2 + 4n^2} + \frac{2na\cos\omega_1 t \sin\lambda}{\omega_1^2 + 4n^2} - \frac{2nae^{-2nt}\sin\lambda}{\omega_1^2 + 4n^2} \tag{13.6.13}
$$

The second derivative of equation (13.6.12) represents the acceleration:

$$
\frac{d^2x}{dt^2} = \left(-2nv_0 + qp - p - \frac{q\mu}{2n\tau} \right)e^{-2nt} + \frac{q\mu}{2n\tau}
$$

$$
+ \frac{2na\omega_1\cos\omega_1 t \cos\lambda}{\omega_1^2 + 4n^2} + \frac{a\omega_1^2\sin\omega_1 t \cos\lambda}{\omega_1^2 + 4n^2}
$$

$$
- \frac{2na\omega_1 e^{-2nt}\cos\lambda}{\omega_1^2 + 4n^2} + \frac{a\omega_1^2\cos\omega_1 t \sin\lambda}{\omega_1^2 + 4n^2}
$$

$$
- \frac{2na\omega_1\sin\omega_1 t \sin\lambda}{\omega_1^2 + 4n^2} + \frac{4n^2 ae^{-2nt}\sin\lambda}{\omega_1^2 + 4n^2} \tag{13.6.14}
$$

13.6.2 Initial Displacement Equals Zero

The initial conditions of motion are:

$$\text{for} \quad t = 0 \quad x = 0; \quad \frac{dx}{dt} = v_0 \qquad (13.6.15)$$

The solution of differential equation of motion (13.6.1) with the initial conditions of motion (13.6.15) reads:

$$x = \frac{v_0 + 2ns_0}{2n}(1 - e^{-2nt}) + \frac{q\rho - p}{2n}\left[t + \frac{1}{2n}(e^{-2nt} - 1)\right]$$

$$+ \frac{q\mu}{4n\tau}\left[t^2 - \frac{1}{n}t - \frac{1}{2n^2}(e^{-2nt} - 1)\right] + \frac{2na(1 - \cos\omega_1 t)\cos\lambda}{\omega_1(\omega_1^2 + 4n^2)}$$

$$- \frac{a\sin\omega_1 t\cos\lambda}{\omega_1^2 + 4n^2} + \frac{a\omega_1(1 - e^{-2nt})\cos\lambda}{2n(\omega_1^2 + 4n^2)} + \frac{a(1 - \cos\omega_1 t)\sin\lambda}{\omega_1^2 + 4n^2}$$

$$+ \frac{2na\sin\omega_1 t\sin\lambda}{\omega_1(\omega_1^2 + 4n^2)} - \frac{a(1 - e^{-2nt})\sin\lambda}{\omega_1^2 + 4n^2} \qquad (13.6.16)$$

The rest of the parameters are the same as in the previous case.

13.6.3 Initial Velocity Equals Zero

The initial conditions of motion are:

$$\text{for} \quad t = 0 \quad x = s_0; \quad \frac{dx}{dt} = 0 \qquad (13.6.17)$$

Solving differential equation of motion (13.6.1) with the initial conditions of motion (13.6.17), we obtain:

$$x = s_0 + \frac{q\rho - p}{2n}\left[t + \frac{1}{2n}(e^{-2nt} - 1)\right] + \frac{q\mu}{4n\tau}\left[t^2 - \frac{1}{n}t - \frac{1}{2n^2}(e^{-2nt} - 1)\right]$$

$$+ \frac{2na(1 - \cos\omega_1 t)\cos\lambda}{\omega_1(\omega_1^2 + 4n^2)} - \frac{a\sin\omega_1 t\cos\lambda}{\omega_1^2 + 4n^2}$$

$$+ \frac{a\omega_1(1 - e^{-2nt})\cos\lambda}{2n(\omega_1^2 + 4n^2)} + \frac{a(1 - \cos\omega_1 t)\sin\lambda}{\omega_1^2 + 4n^2}$$

$$+ \frac{2na\sin\omega_1 t\sin\lambda}{\omega_1(\omega_1^2 + 4n^2)} - \frac{a(1 - e^{-2nt})\sin\lambda}{\omega_1^2 + 4n^2} \qquad (13.6.18)$$

The first derivative of equation (13.6.18) represents the velocity of the system:

$$\frac{dx}{dt} = \frac{1}{2n}\left(q\rho - p - \frac{q\mu}{2n\tau}\right)(1 - e^{-2nt}) + \frac{q\mu}{2n\tau}t + \frac{2na\sin\omega_1 t\cos\lambda}{\omega_1^2 + 4n^2}$$

$$-\frac{a\omega_1\cos\omega_1 t\cos\lambda}{\omega_1^2 + 4n^2} + \frac{a\omega_1 e^{-2nt}\cos\lambda}{\omega_1^2 + 4n^2} + \frac{a\omega_1\sin\omega_1 t\sin\lambda}{\omega_1^2 + 4n^2}$$

$$+\frac{2na\cos\omega_1 t\sin\lambda}{\omega_1^2 + 4n^2} - \frac{2nae^{-2nt}\sin\lambda}{\omega_1^2 + 4n^2} \qquad (13.6.19)$$

Taking the second derivative of the equation (13.6.18), we determine the acceleration:

$$\frac{d^2x}{dt^2} = \left(q\rho - p - \frac{q\mu}{2n\tau}\right)e^{-2nt} + \frac{q\mu}{2n\tau} + \frac{2na\omega_1\cos\omega_1 t\cos\lambda}{\omega_1^2 + 4n^2}$$

$$+\frac{a\omega_1^2\sin\omega_1 t\cos\lambda}{\omega_1^2 + 4n^2} - \frac{2na\omega_1 e^{-2nt}\cos\lambda}{\omega_1^2 + 4n^2} + \frac{a\omega_1^2\cos\omega_1 t\sin\lambda}{\omega_1^2 + 4n^2}$$

$$-\frac{2na\omega_1\sin\omega_1 t\sin\lambda}{\omega_1^2 + 4n^2} + \frac{4n^2 ae^{-2nt}\sin\lambda}{\omega_1^2 + 4n^2} \qquad (13.6.20)$$

13.6.4 Both the Initial Displacement and Velocity Equal Zero

The initial conditions of motion are:

$$\text{for} \quad t = 0 \quad x = 0; \quad \frac{dx}{dt} = 0 \qquad (13.6.21)$$

The solution of differential equation of motion (13.6.1) with the initial conditions of motion (13.6.21) reads:

$$x = \frac{q\rho - p}{2n}\left[t + \frac{1}{2n}(e^{-2nt} - 1)\right] + \frac{q\mu}{4n\tau}\left[t^2 - \frac{1}{n}t - \frac{1}{2n^2}(e^{-2nt} - 1)\right]$$

$$+\frac{2na(1 - \cos\omega_1 t)\cos\lambda}{\omega_1(\omega_1^2 + 4n^2)} - \frac{a\sin\omega_1 t\cos\lambda}{\omega_1^2 + 4n^2} + \frac{a\omega_1(1 - e^{-2nt})\cos\lambda}{2n(\omega_1^2 + 4n^2)}$$

$$+\frac{a(1 - \cos\omega_1 t)\sin\lambda}{\omega_1^2 + 4n^2} + \frac{2na\sin\omega_1 t\sin\lambda}{\omega_1(\omega_1^2 + 4n^2)} - \frac{a(1 - e^{-2nt})\sin\lambda}{\omega_1^2 + 4n^2} \qquad (13.6.22)$$

The equations for the velocity and the acceleration are the same as in the previous case.

14

DAMPING, CONSTANT RESISTANCE, AND FRICTION

Engineering systems subjected to the force of inertia, the damping force, the constant resisting force, and the friction force as the resisting forces are described in this chapter. These resisting forces are marked with the plus signs in Row 14 of Guiding Table 2.1. The numbers of this chapter's six sections are shown in the same row, corresponding to Columns 1 through 6, which name the active forces applied to these systems.

The left sides of the differential equations of motion for the engineering systems throughout this chapter are identical and comprise the resisting forces mentioned above. The right sides of these equations are different, corresponding to the active forces applied to the systems.

The problems described in this chapter could be related to the working processes of engineering systems interacting with visco-elastoplastic media or with specific hydraulic links.

14.1 Active Force Equals Zero

The intersection of Row 14 and Column 1 in the Guiding Table 2.1 indicates that the engineering systems described in this section are subjected to the force of inertia, the damping force, the constant resisting force, and the friction force as the resisting forces, while the active force equals zero. The considerations related to the motion of a system in the absence of active forces are presented in section 1.3.

The current problem could be related to the motion of a system that is intended to interact with a viscoelastoplastic medium that exerts the damping force, the constant resisting force, and the friction force as the reaction to its plastic deformation. The considerations regarding the deformation of a viscoelastoplastic medium are presented in section 1.2.

The system is moving in the horizontal direction. We want to determine the basic parameters of motion and the maximum values of the displacement, the acceleration, and the force applied to the system. Figure 14.1 shows the model of a system subjected to the action of a damping force, a constant resisting force, and a friction force.

Based on the considerations above and the model in Figure 14.1, we can assemble the left and right sides of the differential equation of motion. The left side comprises the force of inertia, the damping force, the constant resisting force, and the friction force, while the right side equals zero. Hence, the differential equation of motion reads:

$$m\frac{d^2x}{dt^2} + C\frac{dx}{dt} + P + F = 0 \qquad (14.1.1)$$

The differential equation of motion (14.1.1) has different solutions for various initial conditions of motion. These solutions and their analyses are presented below.

14.1.1 General Initial Conditions

The general initial conditions of motion are:

$$\text{for} \quad t = 0 \quad x = s_0; \quad \frac{dx}{dt} = v_0 \qquad (14.1.2)$$

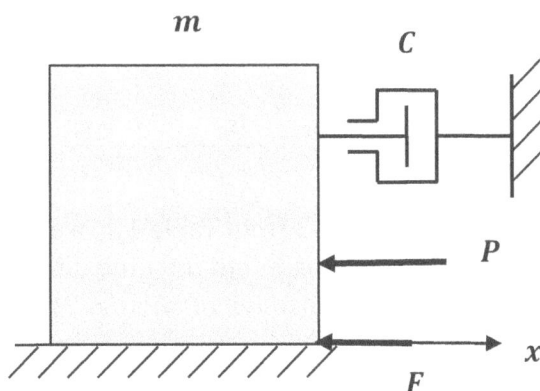

Figure 14.1 Model of a system subjected to a damping force, a constant resisting force, and a friction force

where s_0 and v_0 are the initial displacement and initial velocity respectively.

Dividing equation (14.1.1) by m we may write:

$$\frac{d^2x}{dt^2} + 2n\frac{dx}{dt} + p + f = 0 \qquad (14.1.3)$$

where n is the damping factor and:

$$2n = \frac{C}{m} \qquad (14.1.4)$$

$$p = \frac{P}{m} \qquad (14.1.5)$$

$$f = \frac{F}{m} \qquad (14.1.6)$$

Using Laplace Transform Pairs 3, 4, and 5 from Table 1.1, we convert differential equation of motion (14.1.3) with the initial conditions of motion (14.1.2) from the time domain into the Laplace domain and obtain the corresponding algebraic equation of motion in the Laplace domain:

$$l^2x(l) - lv_0 - l^2s_0 + 2nlx(l) - 2nls_0 + p + f = 0 \qquad (14.1.7)$$

Solving equation (14.1.7) for the displacement $x(l)$ in the Laplace domain, we have:

$$x(l) = \frac{v_0 + 2ns_0}{l + 2n} + \frac{ls_0}{l + 2n} - \frac{p + f}{l(l + 2n)} \quad (14.1.8)$$

Based on pairs 1, 10, 12, and 13 from Table 1.1, we invert equation (14.1.8) from the Laplace domain into the time domain and obtain the solution of differential equation of motion (14.1.1) with the initial conditions of motion (14.1.2):

$$x = \frac{v_0 + 2ns_0}{2n}(1 - e^{-2nt}) + s_0 e^{-2nt} - \frac{p + f}{2n}[t + \frac{1}{2n}(e^{-2nt} - 1)] \quad (14.1.9)$$

Applying algebraic procedures to equation (14.1.9), we have:

$$x = \frac{1}{2n}[2ns_0 + v_0 + \frac{p + f}{2n} - (p + f)t - \left(v_0 + \frac{p + f}{2n}\right)e^{-2nt}] \quad (14.1.10)$$

Taking the first and second derivatives of equation (14.1.10), we determine the velocity and acceleration of the system respectively:

$$\frac{dx}{dt} = (v_0 + \frac{p + f}{2n})e^{-2nt} - \frac{p + f}{2n} \quad (14.1.11)$$

$$\frac{d^2x}{dt^2} = -2n(v_0 + \frac{p + f}{2n})e^{-2nt} \quad (14.1.12)$$

Obviously, we are considering a deceleration process at the end of which the velocity becomes equals to zero.

Equating the velocity according to equation (14.1.11) to zero, we have:

$$(v_0 + \frac{p + f}{2n})e^{-2nT} - \frac{p + f}{2n} = 0 \quad (14.1.13)$$

where T is the time that the process lasts.

Based on equation (14.1.13), we may write:

$$e^{-2nT} = \frac{p + f}{2nv_0 + p + f} \quad (14.1.14)$$

From equation (14.1.14), we determine the time T:

$$T = \frac{1}{2n} \ln \frac{2nv_0 + p + f}{p + f} \qquad (14.1.15)$$

Combining equations (14.1.14) and (14.1.15) with equation (14.1.10), we calculate the maximum displacement s_{max} of the system:

$$s_{max} = \frac{1}{2n}(2ns_0 + v_0 - \frac{p + f}{2n} \ln \frac{2nv_0 + p + f}{p + f}) \qquad (14.1.16)$$

In this case, the maximum value of the force applied to the system equals to the maximum value of the force of inertia that is associated with the maximum value of the acceleration (actually, deceleration) at the beginning of the process of motion when $t = 0$. Thus, from equation (14.1.12), we obtain the maximum absolute value of the acceleration a_{max}:

$$a_{max} = |2nv_0 + p + f| \qquad (14.1.17)$$

Multiplying equation (14.1.17) by m and substituting notations (14.1.4), (14.1.5), and (14.1.6), we determine the maximum force R_{max} applied to the system:

$$R_{max} = Cv_0 + P + F \qquad (14.1.18)$$

The value of the force according to equation (14.1.18) should be used for stress calculations of the system.

14.1.2 Initial Displacement Equals Zero

The initial conditions of motion are:

$$\text{for} \quad t = 0 \quad x = 0; \quad \frac{dx}{dt} = v_0 \qquad (14.1.19)$$

The solution of differential equation of motion (14.1.1) with the initial conditions of motion (14.1.2) reads:

$$x = \frac{1}{2n}[v_0 + \frac{p + f}{2n} - (p + f)t - \left(v_0 + \frac{p + f}{2n}\right)e^{-2nt}] \qquad (14.1.20)$$

The rest of the parameters and their analysis are the same as for the previous case.

14.2 Constant Force R

The engineering systems described in this section are subjected to the action of the force of inertia, the damping force, the constant resisting force, and the friction force as the resisting forces (Row 14 in Guiding Table 2.1) and the constant active force (Column 2). The current problems could be associated with the working processes of systems intended for interaction with a viscoelastoplastic medium that exerts a damping force, a constant resisting force, and a friction force as the reaction to its deformation (more details related to the deformation of a viscoelastoplastic medium are presented in section 1.2). These problems could be also related to the working processes of some pneumatically operated machines that are characterized by the decreasing of the air pressure force applied to a certain component (piston) during the acceleration of the latter. This pressure force represents an active force that decreases proportionally to the increasing of the velocity of this component. More related considerations are discussed in section 1.3, where equation (1.3.1) reflects the mentioned above decreasing of the air pressure force. This equation shows that a corresponding damping force in the left side of the differential equation of motion can account for the above-mentioned decrease of the active force.

The system moves on a horizontal frictional surface. We want to determine the basic parameters of motion, the maximum values of the velocity and the acceleration of the system, the maximum force applied to the system, and the power of the energy source. Figure 14.2 shows the model of a system subjected to the action of a constant active force, a damping force, a constant resisting force, and a friction force.

Based on the considerations above and the model in Figure 14.2, we compose the left and right sides of the differential equation of motion of this system. The left side consists of the force of inertia, the damping force, the constant resisting force, and the friction force, while the right side includes a constant active force. Therefore, the differential equation of motion reads:

$$m\frac{d^2x}{dt^2} + C\frac{dx}{dt} + P + F = R \qquad (14.2.1)$$

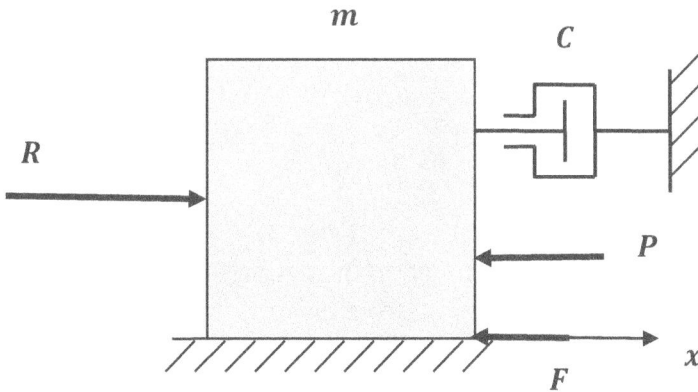

Figure 14.2 Model of a system subjected to a constant active force, a damping force, a constant resisting force, and a friction force

Differential equation of motion (14.2.1) has different solutions for various initial conditions of motion. These solutions and their analyses are presented below.

14.2.1 General Initial Conditions

The general initial conditions of motion are:

$$\text{for} \quad t = 0 \quad x = s_0; \quad \frac{dx}{dt} = v_0 \qquad (14.2.2)$$

where s_0 and v_0 are the initial displacement and initial velocity respectively.

Dividing equation (14.2.1) by m, we have:

$$\frac{d^2x}{dt^2} + 2n\frac{dx}{dt} + p + f = r \qquad (14.2.3)$$

where n is the damping factor and:

$$2n = \frac{C}{m} \qquad (14.2.4)$$

$$p = \frac{P}{m} \qquad (14.2.5)$$

$$f = \frac{F}{m} \tag{14.2.6}$$

$$r = \frac{R}{m} \tag{14.2.7}$$

Using Laplace Transform Pairs 3, 4 and 5 from Table 1.1, we convert differential equation of motion (14.2.3) with the initial conditions of motion (14.2.2) from the time domain into the Laplace domain. The resulting algebraic equation of motion in the Laplace domain reads:

$$l^2 x(l) - lv_0 - l^2 s_0 + 2nlx(l) - 2nls_0 + p + f = r \tag{14.2.8}$$

Solving equation (12.2.7) for the displacement $x(l)$ in Laplace domain, we may write:

$$x(l) = \frac{r - p - f}{l(l + 2n)} + \frac{v_0 + 2ns_0}{l + 2n} + \frac{ls_0}{l + 2n} \tag{14.2.9}$$

Using pairs 1, 13, 10, and 12 from Table 1.1, we invert equation (14.2.9) from the Laplace domain into the time domain and obtain the solution of differential equation of motion (14.2.1) with the initial conditions of motion (14.2.2):

$$x = \frac{r - p - f}{2n}[t + \frac{1}{2n}(e^{-2nt} - 1)] + \frac{v_0 + 2ns_0}{2n}(1 - e^{-2nt}) + s_0 e^{-2nt} \tag{14.2.10}$$

Applying algebraic procedures to equation (14.2.10), we have:

$$x = \frac{1}{2n}[(r - p - f)t + \left(\frac{r - p - f}{2n} - v_0\right)e^{-2nt} - \frac{r - p - f}{2n} + v_0 + 2ns_0] \tag{14.2.11}$$

Taking the first derivative of equation (14.2.11), we determine the velocity of the system:

$$\frac{dx}{dt} = \frac{r - p - f}{2n}(1 - e^{-2nt}) + v_0 e^{-2nt} \tag{14.2.12}$$

The second derivative of equation (14.2.11) represents the acceleration:

$$\frac{d^2x}{dt^2} = (r - p - f - 2nv_0)e^{-2nt} \qquad (14.2.13)$$

In this case, when the acceleration approaches zero, the velocity approaches its maximum value. Because $(r - p - f - 2nv_0) > 0$, then according to equation (14.2.123) we may write:

$$e^{-2nT} \rightarrow 0 \qquad (14.2.14)$$

where T is the time that the acceleration process lasts. Actually, T tends to the infinity.

Combining expression (14.2.14) with equation (14.2.12) and substituting notations (14.2.4), (14.2.5), (14.2.6), and (14.2.7), we determine the maximum value of the velocity v_{max}:

$$v_{max} \rightarrow \frac{R - P - F}{C} \qquad (14.2.15)$$

In order to calculate the maximum force applied to the system, we need first to determine the maximum value of the acceleration, which in this case occurs at the beginning of the process when $t = 0$. Therefore, based on equation (14.2.13), we determine the maximum value of the acceleration a_{max}:

$$a_{max} = r - p - f - 2nv_0 \qquad (14.2.16)$$

Multiplying equation (14.2.16) by m, we determine the maximum value of the force of inertia force F_i:

$$F_i = m(r - p - f - 2nv_0 \qquad (14.2.17)$$

Adding to the force of inertia according to equation (14.2.17) the resisting forces Cv_0, P, and F, and also substituting notations (14.2.4), (14.2.5), (14.2.6), and (14.2.7), we determine the maximum value of the force R_{max} applied to the system:

$$R_{max} = R \qquad (14.2.18)$$

We can obtain the same result in the case when the acceleration approaches zero and, therefore, the force of inertia approaches zero. For any given time, the maximum force applied to the system equals the sum of the force of inertia and the resisting forces. Hence, for this case, the force applied to the system equals the sum of the resisting forces. Thus, the damping force equals the product of multiplying the velocity according to equation (14.2.15) by the damping coefficient C. Adding to this damping force the forces P and F, we obtain the maximum value of the force applied to the system. This force has the same value as in equation (14.2.18).

The stress analysis of the system should be based on the maximum value of the force according to equation (14.2.18). Multiplying the maximum force according to equation (14.2.18) by the maximum velocity according to expression (14.2.15), we determine the power N of the energy source :

$$N = \frac{R(R-P-F)}{C} \qquad (14.2.19)$$

14.2.2 Initial Displacement Equals Zero

The initial conditions of motion are:

$$\text{for} \quad t = 0 \quad x = 0; \quad \frac{dx}{dt} = v_0 \qquad (14.2.20)$$

The solution of differential equation of motion (14.2.1) with the initial conditions of motion (14.2.20) reads:

$$x = \frac{1}{2n}[(r-p-f)t + \left(\frac{r-p-f}{2n} - v_0\right)e^{-2nt} - \frac{r-p-f}{2n} + v_0]$$

$$(14.2.21)$$

The rest of the parameters and their analysis is the same as in the previous case.

14.2.3 Initial Velocity Equals Zero

The initial conditions of motion are:

$$\text{for} \quad t = 0 \quad x = s_0; \quad \frac{dx}{dt} = 0 \qquad (14.2.22)$$

Solving differential equation of motion (14.2.1) with the initial conditions of motion (14.2.22), we obtain:

$$x = \frac{1}{2n}[(r-p-f)t + \frac{r-p-f}{2n}e^{-2nt} - \frac{r-p-f}{2n} + 2ns_0] \qquad \textbf{(14.2.23)}$$

The first and the second derivatives of equation (14.2.23) represent the velocity and the acceleration respectively:

$$\frac{dx}{dt} = \frac{r-p-f}{2n}\left(1-e^{-2nt}\right) \qquad \textbf{(14.2.24)}$$

$$\frac{d^2x}{dt^2} = (r-p-f)e^{-2nt} \qquad \textbf{(14.2.25)}$$

The rest of the parameters and their analysis are the same as in the previous case.

14.2.4 Both the Initial Displacement and Velocity Equal Zero

The initial conditions of motion are:

$$\text{for} \quad t=0 \quad x=0; \quad \frac{dx}{dt}=0 \qquad \textbf{(14.2.26)}$$

The solution of differential equation of motion (14.2.1) with the initial conditions of motion (14.2.26) reads:

$$x = \frac{1}{2n}[(r-p-f)t + \frac{r-p-f}{2n}e^{-2nt} - \frac{r-p-f}{2n}] \qquad \textbf{(14.2.27)}$$

The rest of the parameters and their analysis is the same as in the previous case.

14.3 Harmonic Force $A \sin(\omega_1 t + \lambda)$

The intersection of Row 14 with Column 3 in Guiding Table 2.1 indicates that this section describes engineering systems subjected to the force of inertia, the damping force, the constant resisting force, and the friction force as the resisting forces and the harmonic force as the active force. The current problem could be associated with working processes of certain vibratory machines intended for the interaction with viscoelastoplastic media that exert damping forces, constant resisting, and friction forces as the reaction to their plastic deformation. The considerations related to the deformation of viscoelastoplastic media as well as to the behavior of a constant resisting force and also a friction force applied to a vibratory system are discussed in section 1.2.

The system is moving in the horizontal direction. We want to determine the basic parameters of motion. Figure 14.3 shows the model of a system subjected to the action of a harmonic force, a damping force, a constant resisting force, and a friction force.

The considerations above and the model in the Figure 14.3 let us compose the left and right sides of the differential equation of motion. The left side consists of the force of inertia, the damping force, the constant resisting force, and the friction force, while the right side includes the harmonic force. Hence, the differential equation of motion reads:

$$m\frac{d^2x}{dt^2} + C\frac{dx}{dt} + P + F = A\sin(\omega_1 t + \lambda) \qquad (14.3.1)$$

Differential equation of motion (14.3.1) has different solutions for various initial conditions of motion. These solutions and their analyses are presented below.

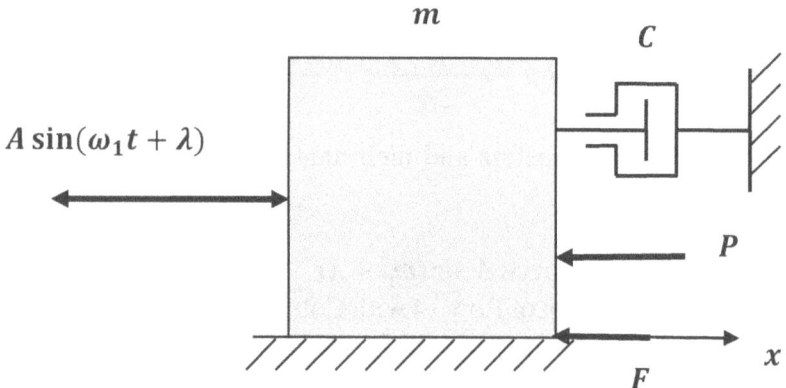

Figure 14.3 Model of a system subjected to a harmonic force, a damping force, a constant resisting force, and a friction force

14.3.1 General Initial Conditions

The general initial conditions of motion are:

$$\text{for} \quad t = 0 \quad x = s_0; \quad \frac{dx}{dt} = v_0 \quad (14.3.2)$$

where s_0 and v_0 are the initial displacement and initial velocity respectively.

Transforming the sinusoidal function of equation (14.3.1) and dividing the latter by m, we may write:

$$\frac{d^2 x}{dt^2} + 2n\frac{dx}{dt} + p + f = a\sin\omega_1 t \cos\lambda + a\cos\omega_1 t \sin\lambda \quad (14.3.3)$$

where n is the damping factor and:

$$2n = \frac{C}{m} \quad (14.3.4)$$

$$p = \frac{P}{m} \quad (14.3.5)$$

$$f = \frac{F}{m} \quad (14.3.6)$$

$$a = \frac{A}{m} \quad (14.3.7)$$

Using Laplace Transform Pairs 3, 4, 5, 6, and 7 from Table 1.1, we convert differential equation of motion (14.3.3) with the initial conditions of motion (14.3.2) from the time domain into the Laplace domain. The resulting algebraic equation of motion in the Laplace domain reads:

$$l^2 x(l) - l v_0 - l^2 s_0 + 2nlx(l) - 2nls_0 + p + f$$
$$= \frac{a\omega_1 l}{l^2 + \omega_1^2}\cos\lambda + \frac{al^2}{l^2 + \omega_1^2}\sin\lambda \quad (14.3.8)$$

Applying some basic algebra to equation (14.3.8), we have:

$$x(l)l(l+2n) = l(2ns_0 + v_0) + s_0 l^2 - p - f + \frac{a\omega_1 l}{l^2 + \omega_1^2}\cos\lambda$$

$$+ \frac{al^2}{l^2 + \omega_1^2}\sin\lambda \qquad \textbf{(14.3.9)}$$

Solving equation (14.3.9) for the Laplace domain displacement $x(l)$, we have:

$$x(l) = \frac{2ns_0 + v_0}{l+2n} + \frac{ls_0}{l+2n} - \frac{p+f}{l(l+2n)} + \frac{a\omega_1}{(l+2n)(l^2 + \omega_1^2)}\cos\lambda$$

$$+ \frac{la}{(l+2n)(l^2 + \omega_1^2)}\sin\lambda \qquad \textbf{(14.3.10)}$$

Applying pairs 1, 10, 12, 13, 19, and 20 from Table 1.1, we invert equation (14.3.10) from the Laplace domain into the time domain and obtain the solution of differential equation of motion (14.3.1) with the initial conditions of motion (14.3.2):

$$x = \frac{2ns_0 + v_0}{2n}(1 - e^{-2nt}) + s_0 e^{-2nt} - \frac{p+f}{2n}\left[t + \frac{1}{2n}(e^{-2nt} - 1)\right]$$

$$+ \frac{2na(1 - \cos\omega_1 t)\cos\lambda}{\omega_1(\omega_1^2 + 4n^2)} - \frac{a\sin\omega_1 t\cos\lambda}{\omega_1^2 + 4n^2} + \frac{a\omega_1(1 - e^{-2nt})\cos\lambda}{2n(\omega_1^2 + 4n^2)}$$

$$+ \frac{a(1 - \cos\omega_1 t)\sin\lambda}{\omega_1^2 + 4n^2} + \frac{2na\sin\omega_1 t\sin\lambda}{\omega_1(\omega_1^2 + 4n^2)} - \frac{a(1 - e^{-2nt})\sin\lambda}{\omega_1^2 + 4n^2}$$

$$\textbf{(14.3.11)}$$

Applying the appropriate algebraic actions to equation (14.3.11), we write:

$$x = s_0 + \frac{v_0}{2n}(1 - e^{-2nt}) - \frac{p+f}{2n}\left[t + \frac{1}{2n}(e^{-2nt} - 1)\right]$$

$$+ \frac{2na(1 - \cos\omega_1 t)\cos\lambda}{\omega_1(\omega_1^2 + 4n^2)} - \frac{a\sin\omega_1 t\cos\lambda}{\omega_1^2 + 4n^2} + \frac{a\omega_1(1 - e^{-2nt})\cos\lambda}{2n(\omega_1^2 + 4n^2)}$$

$$+ \frac{a(1 - \cos\omega_1 t)\sin\lambda}{\omega_1^2 + 4n^2} + \frac{2na\sin\omega_1 t\sin\lambda}{\omega_1(\omega_1^2 + 4n^2)} - \frac{a(1 - e^{-2nt})\sin\lambda}{\omega_1^2 + 4n^2}$$

$$\textbf{(14.3.12)}$$

The first derivative of equation (14.3.12) represents the velocity of the system:

$$\frac{dx}{dt} = v_0 e^{-2nt} - \frac{p+f}{2n}(1-e^{-2nt}) + \frac{2na \sin \omega_1 t \cos \lambda}{\omega_1^2 + 4n^2}$$

$$- \frac{a\omega_1 \cos \omega_1 t \cos \lambda}{\omega_1^2 + 4n^2} + \frac{a\omega_1 e^{-2nt} \cos \lambda}{\omega_1^2 + 4n^2} + \frac{a\omega_1 \sin \omega_1 t \sin \lambda}{\omega_1^2 + 4n^2}$$

$$+ \frac{2na \cos \omega_1 t \sin \lambda}{\omega_1^2 + 4n^2} - \frac{2nae^{-2nt} \sin \lambda}{\omega_1^2 + 4n^2} \qquad (14.3.13)$$

Taking the second derivative of equation (14.3.12), we determine the acceleration of the system:

$$\frac{d^2x}{dt^2} = -2nv_0 e^{-2nt} - (p+f)e^{-2nt} + \frac{2na\omega_1 \cos \omega_1 t \cos \lambda}{\omega_1^2 + 4n^2}$$

$$+ \frac{a\omega_1^2 \sin \omega_1 t \cos \lambda}{\omega_1^2 + 4n^2} - \frac{2na\omega_1 e^{-2nt} \cos \lambda}{\omega_1^2 + 4n^2} + \frac{a\omega_1^2 \cos \omega_1 t \sin \lambda}{\omega_1^2 + 4n^2}$$

$$- \frac{2na\omega_1 \sin \omega_1 t \sin \lambda}{\omega_1^2 + 4n^2} + \frac{4n^2 a e^{-2nt} \sin \lambda}{\omega_1^2 + 4n^2} \qquad (14.3.14)$$

14.3.2 Initial Displacement Equals Zero

The initial conditions of motion are:

$$\text{for} \quad t = 0 \quad x = 0; \quad \frac{dx}{dt} = v_0 \qquad (14.3.15)$$

Solving differential equation of motion (14.3.1) with the initial conditions of motion (14.3.15), we obtain:

$$x = \frac{v_0}{2n}(1-e^{-2nt}) - \frac{p+f}{2n}\left[t + \frac{1}{2n}(e^{-2nt}-1)\right] + \frac{2na(1-\cos \omega_1 t)\cos \lambda}{\omega_1(\omega_1^2 + 4n^2)}$$

$$- \frac{a \sin \omega_1 t \cos \lambda}{\omega_1^2 + 4n^2} + \frac{a\omega_1(1-e^{-2nt})\cos \lambda}{2n(\omega_1^2 + 4n^2)} + \frac{a(1-\cos \omega_1 t)\sin \lambda}{\omega_1^2 + 4n^2}$$

$$+ \frac{2na \sin \omega_1 t \sin \lambda}{\omega_1(\omega_1^2 + 4n^2)} - \frac{a(1-e^{-2nt})\sin \lambda}{\omega_1^2 + 4n^2} \qquad (14.3.16)$$

The equations for the velocity and the acceleration are the same as in the previous case.

14.3.3 Initial Velocity Equals Zero

The initial conditions of motion are:

$$\text{for} \quad t = 0 \quad x = s_0; \quad \frac{dx}{dt} = 0 \tag{14.3.17}$$

The solution of differential equation of motion (14.3.1) with the initial conditions of motion (14.3.17) reads:

$$
x = s_0 - \frac{p+f}{2n}\left[t + \frac{1}{2n}(e^{-2nt} - 1)\right] + \frac{2na(1 - \cos\omega_1 t)\cos\lambda}{\omega_1(\omega_1^2 + 4n^2)}
$$

$$
- \frac{a\sin\omega_1 t\cos\lambda}{\omega_1^2 + 4n^2} + \frac{a\omega_1(1 - e^{-2nt})\cos\lambda}{2n(\omega_1^2 + 4n^2)} + \frac{a(1 - \cos\omega_1 t)\sin\lambda}{\omega_1^2 + 4n^2}
$$

$$
+ \frac{2na\sin\omega_1 t\sin\lambda}{\omega_1(\omega_1^2 + 4n^2)} - \frac{a(1 - e^{-2nt})\sin\lambda}{\omega_1^2 + 4n^2} \tag{14.3.18}
$$

Taking the first and second derivatives of equation (14.3.18), we determine the velocity and the acceleration of the system respectively:

$$
\frac{dx}{dt} = \frac{p+f}{2n}(e^{-2nt} - 1) + \frac{2na\sin\omega_1 t\cos\lambda}{\omega_1^2 + 4n^2} - \frac{a\omega_1\cos\omega_1 t\cos\lambda}{\omega_1^2 + 4n^2}
$$

$$
+ \frac{a\omega_1 e^{-2nt}\cos\lambda}{\omega_1^2 + 4n^2} + \frac{a\omega_1\sin\omega_1 t\sin\lambda}{\omega_1^2 + 4n^2} + \frac{2na\cos\omega_1 t\sin\lambda}{\omega_1^2 + 4n^2}
$$

$$
- \frac{2nae^{-2nt}\sin\lambda}{\omega_1^2 + 4n^2} \tag{14.3.19}
$$

$$
\frac{d^2x}{dt^2} = -(p+f)e^{-2nt} + \frac{2na\omega_1\cos\omega_1 t\cos\lambda}{\omega_1^2 + 4n^2} + \frac{a\omega_1^2\sin\omega_1 t\cos\lambda}{\omega_1^2 + 4n^2}
$$

$$
- \frac{2na\omega_1 e^{-2nt}\cos\lambda}{\omega_1^2 + 4n^2} + \frac{a\omega_1^2\cos\omega_1 t\sin\lambda}{\omega_1^2 + 4n^2} - \frac{2na\omega_1\sin\omega_1 t\sin\lambda}{\omega_1^2 + 4n^2}
$$

$$
+ \frac{4n^2 ae^{-2nt}\sin\lambda}{\omega_1^2 + 4n^2} \tag{14.3.20}
$$

14.3.4 Both the Initial Displacement and Velocity Equal Zero

The initial conditions of motion are:

$$\text{for} \quad t = 0 \quad x = 0; \quad \frac{dx}{dt} = 0 \qquad (14.3.21)$$

Solving differential equation of motion (14.3.1) with the initial conditions of motion (14.3.21), we obtain:

$$x = \frac{p+f}{2n}\left[\frac{1}{2n}(1-e^{-2nt})-t\right]+\frac{2na(1-\cos\omega_1 t)\cos\lambda}{\omega_1(\omega_1^2+4n^2)}$$

$$-\frac{a\sin\omega_1 t\cos\lambda}{\omega_1^2+4n^2}+\frac{a\omega_1(1-e^{-2nt})\cos\lambda}{2n(\omega_1^2+4n^2)}+\frac{a(1-\cos\omega_1 t)\sin\lambda}{\omega_1^2+4n^2}$$

$$+\frac{2na\sin\omega_1 t\sin\lambda}{\omega_1(\omega_1^2+4n^2)}-\frac{a(1-e^{-2nt})\sin\lambda}{\omega_1^2+4n^2} \qquad (14.3.22)$$

The equations for the velocity and acceleration are the same as for the previous case.

14.4 Time-Dependent Force $Q\left(\rho+\frac{\mu t}{\tau}\right)$

According to Guiding Table 2.1, this section describes engineering systems subjected to the action of the force of inertia, the damping force, the constant resisting force, and the friction force as the resisting forces (Row 14) and to the action of the time-dependent force as the active force (Column 4).

The current problem could be related to the working process of a system intended to interact with a viscoelastoplastic medium that exerts a damping force, a constant resisting force, and a friction force during the phase of its plastic deformation (more related considerations regarding the viscoelastoplastic medium are presented in section 1.2). The time-dependent force is acting for a limited interval of time during the initial phase of the working process.

The system is moving in the horizontal direction. We want to determine the basic parameters of motion, the values of the velocity and the acceleration at the end of the predetermined interval of time.

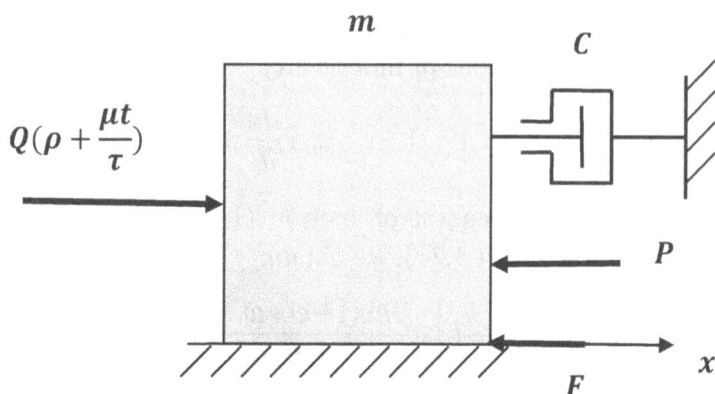

Figure 14.4 Model of a system subjected to a time-dependent force, a damping force, a constant resisting force, and a friction force

Figure 14.4 shows the model of a system subjected to the action of a time-dependent force, a damping force, a constant resisting force, and a friction force.

The considerations above and the model in Figure 14.4 let us assemble the left and right sides of the differential equation of motion. The left side consists of the force of inertia, the damping force, the constant resisting force, and the friction force, while the right side includes the time-dependent force. Therefore, the differential equation of motion reads:

$$m\frac{d^2x}{dt^2} + C\frac{dx}{dt} + P + F = Q(\rho + \frac{\mu t}{\tau}) \qquad (14.4.1)$$

Differential equation of motion (14.4.1) has different solutions for various initial conditions of motion. These solutions and their analyses are presented below.

14.4.1 General Initial Conditions

The general initial conditions of motion are:

$$\text{for} \quad t = 0 \quad x = s_0; \quad \frac{dx}{dt} = v_0 \qquad (14.4.2)$$

where s_0 and v_0 are the initial displacement and initial velocity respectively.

Dividing equation (14.4.1) by m, we have:

$$\frac{d^2x}{dt^2} + 2n\frac{dx}{dt} + p + f = q(\rho + \frac{\mu t}{\tau}) \qquad (14.4.3)$$

where n is the damping factor and:

$$2n = \frac{C}{m} \qquad (14.4.4)$$

$$p = \frac{P}{m} \qquad (14.4.5)$$

$$f = \frac{F}{m} \qquad (14.4.6)$$

$$q = \frac{Q}{m} \qquad (14.4.7)$$

Using Laplace Transform Pairs 3, 4, 5, and 2 from Table 1.1, we convert differential equation of motion (14.4.3) with the initial conditions of motion (14.4.2) from the time domain into the Laplace domain, and obtain the corresponding algebraic equation of motion in the Laplace domain:

$$l^2x(l) - lv_0 - l^2s_0 + 2nlx(l) - 2nls_0 + p + f = q\rho + \frac{q\mu}{\tau l} \qquad (14.4.8)$$

The solution of equation (14.4.8) for the Laplace domain displacement $x(l)$ reads:

$$x(l) = \frac{v_0 + 2ns_0}{l + 2n} + \frac{ls_0}{l + 2n} + \frac{q\rho - p - f}{l(l + 2n)} + \frac{q\mu}{\tau l^2(l + 2n)} \qquad (14.4.9)$$

Applying pairs 1, 10, 12, 13, and 31 from Table 1.1, we invert equation (14.4.9) from the Laplace domain into the time domain and obtain the solution of differential equation of motion (14.4.1) with the initial conditions of motion (14.4.2):

$$x = \frac{1}{2n}(v_0 + 2ns_0 - v_0e^{-2nt}) + \frac{q\rho - p - f}{2n}\left[t + \frac{1}{2n}(e^{-2nt} - 1)\right]$$

$$+ \frac{q\mu}{4n\tau}[t^2 - \frac{1}{n}t - \frac{1}{2n^2}(e^{-2nt} - 1)] \qquad (14.4.10)$$

Based on some algebraic procedures with equation (14.4.10), we write:

$$x = \frac{1}{2n}\{v_0 + 2ns_0 - v_0 e^{-2nt} + \frac{q\mu}{2\tau}t^2$$

$$+ \left(q\rho - p - f - \frac{q\mu}{2n\tau}\right)[t + \frac{1}{2n}(e^{-2nt} - 1)]\} \qquad (14.4.11)$$

Taking the first derivative of equation (14.4.11), we determine the velocity of the system:

$$\frac{dx}{dt} = v_0 e^{-2nt} + \frac{q\mu}{2n\tau}t + \frac{1}{2n}\left(q\rho - p - f - \frac{q\mu}{2n\tau}\right)(1 - e^{-2nt}) \qquad (14.4.12)$$

The second derivative of equation (14.4.11) represents the acceleration:

$$\frac{d^2 x}{dt^2} = -2nv_0 e^{-2nt} + \frac{q\mu}{2n\tau} + (q\rho - p - f - \frac{\mu}{2n\tau})e^{-2nt} \qquad (14.4.13)$$

Substituting the time τ into equations (14.4.12), (14.4.13), we determine the velocity v and acceleration a at the end of the predetermined interval of time:

$$v = v_0 e^{-2n\tau} + \frac{q\mu}{2n\tau}\tau + \frac{1}{2n}\left(q\rho - p - f - \frac{q\mu}{2n\tau}\right)(1 - e^{-2n\tau}) \qquad (14.4.14)$$

$$a = -2nv_0 e^{-2n\tau} + \frac{q\mu}{2n\tau} + (q\rho - p - f - \frac{q\mu}{2n\tau})e^{-2n\tau} \qquad (14.4.15)$$

14.4.2 Initial Displacement Equals Zero
The initial conditions of motion are:

$$\text{for} \quad t = 0 \quad x = 0; \quad \frac{dx}{dt} = v_0 \qquad (14.4.16)$$

Solving differential equation of motion (14.4.1) with the initial conditions of motion (14.4.16), we obtain:

$$x = \frac{1}{2n}\{v_0 - v_0 e^{-2nt} + \frac{q\mu}{2\tau}t^2 + \left(q\rho - p - f - \frac{q\mu}{2n\tau}\right)[t + \frac{1}{2n}(e^{-2nt} - 1)]\}$$

$$(14.4.17)$$

The rest of the parameters are the same as in the previous case.

14.4.3 Initial Velocity Equals Zero

The initial conditions of motion are:

$$\text{for} \quad t = 0 \quad x = s_0; \quad \frac{dx}{dt} = 0 \qquad \textbf{(14.4.18)}$$

The solution of differential equation of motion (14.4.1) with the initial conditions of motion (14.4.18) reads:

$$x = \frac{1}{2n}\{2ns_0 + \frac{q\mu}{2\tau}t^2 + \left(q\rho - p - f - \frac{q\mu}{2n\tau}\right)[t + \frac{1}{2n}(e^{-2nt} - 1)]\}$$

$$\textbf{(14.4.19)}$$

Taking the first and second derivatives of equation (14.4.19), we determine the velocity and acceleration of the system respectively:

$$\frac{dx}{dt} = \frac{q\mu}{2n\tau}t + \frac{1}{2n}\left(q\rho - p - f - \frac{q\mu}{2n\tau}\right)(1 - e^{-2nt}) \quad \textbf{(14.4.20)}$$

$$\frac{d^2x}{dt^2} = \frac{q\mu}{2n\tau} + (q\rho - p - f - \frac{q\mu}{2n\tau})e^{-2nt} \qquad \textbf{(14.4.21)}$$

Substituting the time τ into equations (14.4.20) and (14.4.21), we determine the values of the velocity v and the acceleration a at the end of the predetermined interval of time:

$$\frac{dx}{dt} = \frac{q\mu}{2n\tau}\tau + \frac{1}{2n}\left(q\rho - p - f - \frac{q\mu}{2n\tau}\right)(1 - e^{-2n\tau}) \quad \textbf{(14.4.22)}$$

$$\frac{d^2x}{dt^2} = \frac{q\mu}{2n\tau} + (q\rho - p - f - \frac{q\mu}{2n\tau})e^{-2n\tau} \qquad \textbf{(14.4.23)}$$

14.4.4 Both the Initial Displacement and Velocity Equal Zero

The initial conditions of motion are:

$$\text{for} \quad t = 0 \quad x = 0; \quad \frac{dx}{dt} = 0 \qquad \textbf{(14.4.24)}$$

The solution of differential equation of motion equation (14.4.1) with the initial conditions of motion (14.4.24) reads:

$$x = \frac{1}{2n}\{\frac{q\mu}{2\tau}t^2 + \left(q\rho - p - f - \frac{q\mu}{2n\tau}\right)[t + \frac{1}{2n}(e^{-2nt} - 1)]\} \qquad (14.4.25)$$

The rest of the parameters are the same as for the previous case.

14.5 Constant Force R and Harmonic Force $A \sin(\omega_1 t + \lambda)$

Guiding Table 2.1 indicates that the engineering systems discussed in this section are subjected to the action of the force of inertia, the damping force, the constant resisting force, and the friction force as the resisting forces (Row 14), and to the constant active force and the harmonic force as the active forces (Column 5).

The current problem could be related to the working process of a vibratory system intended for the interaction with certain viscoelastoplastic media that exert damping forces, constant resisting forces, and friction forces as the reaction to their deformation. Additional considerations related to the deformation of viscoelastoplastic media and to the behavior of a constant resisting force as well as a friction force applied to a vibratory system are discussed in section 1.2.

The system is moving in the horizontal direction. We want to determine the basic parameters of motion. Figure 14.5 shows the model of a system subjected to the action of a constant active force, a harmonic force, a damping force, a constant resisting force, and a friction force.

Accounting for the considerations above and the model shown in Figure 14.5, we can compose the left and right sides of the differential equation of motion of the system. The left side consists of the force of inertia, the damping force, the constant resisting force, and the friction force. The right side of this equation includes the constant active force and the harmonic force. Hence, the differential equation of motion reads:

$$m\frac{d^2x}{dt^2} + C\frac{dx}{dt} + P + F = R + A\sin(\omega_1 t + \lambda) \qquad (14.5.1)$$

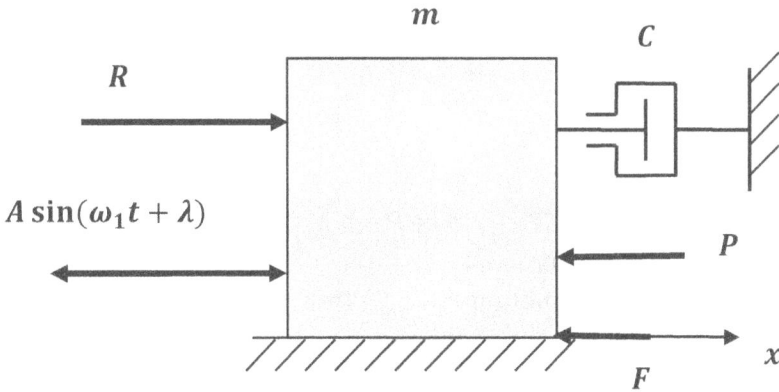

Figure 14.5 Model of a system subjected to a constant active force, a harmonic force, a damping force, a constant resisting force, and a friction force

Differential equation of motion (14.5.1) has different solutions for various initial conditions of motion. These solutions and their analyses are presented below.

14.5.1 General Initial Conditions

The general initial conditions of motion are:

$$\text{for} \quad t = 0 \quad x = s_0; \quad \frac{dx}{dt} = v_0 \tag{14.5.2}$$

Transforming the sinusoidal function in equation (14.5.1) and dividing the latter by m, we have:

$$\frac{d^2x}{dt^2} + 2n\frac{dx}{dt} + p + f = r + a\sin\omega_1 t \cos\lambda + a\cos\omega_1 t \sin\lambda \tag{14.5.3}$$

where n is the damping factor and:

$$2n = \frac{C}{m} \tag{14.5.4}$$

$$p = \frac{P}{m} \tag{14.5.5}$$

$$f = \frac{F}{m} \tag{14.5.6}$$

$$r = \frac{R}{m} \tag{14.5.7}$$

$$a = \frac{A}{m} \tag{14.5.8}$$

Using Laplace Transform Pairs 3, 4, 5, 6, and 7 from Table 1.1, we convert differential equation of motion (14.5.3) with the initial conditions of motion (14.5.2) from the time domain into the Laplace domain and obtain the resulting algebraic equation of motion in the Laplace domain:

$$l^2 x(l) - lv_0 - l^2 s_0 + 2nlx(l) - 2nls_0 + p + f = r + \frac{a\omega_1 l \cos\lambda}{l^2 + \omega_1^2} + \frac{al^2 \sin\lambda}{l^2 + \omega_1^2} \tag{14.5.9}$$

Applying some basic algebra to equation (14.5.9), we may write:

$$x(l)l(l+2n) = r - p - f + l(v_0 + 2ns_0) + l^2 s_0 + \frac{a\omega_1 l \cos\lambda}{l^2 + \omega_1^2} + \frac{al^2 \sin\lambda}{l^2 + \omega_1^2} \tag{14.5.10}$$

The solution of equation (14.5.10 for the displacement $x(l)$ in Laplace domain reads:

$$x(l) = \frac{r - p - f}{l(l+2n)} + \frac{v_0 + 2ns_0}{l+2n} + \frac{ls_0}{l+2n} + \frac{a\omega_1 \cos\lambda}{(l+2n)(l^2 + \omega_1^2)}$$
$$+ \frac{al \sin\lambda}{(l+2n)(l^2 + \omega_1^2)} \tag{14.5.11}$$

Applying pairs 1, 13, 10, 12, 19, and 20 from Table 1.1, we invert equation (14.5.11) from the Laplace domain into the time domain and obtain the solution of differential equation of motion (14.5.1) with the initial conditions of motion (14.5.2):

$$x = \frac{r - p - f}{2n}[t + \frac{1}{2n}(e^{-2nt} - 1)] + \frac{v_0 + 2ns_0}{2n}(1 - e^{-2nt}) + s_0 e^{-2nt}$$
$$+ \frac{2na(1 - \cos\omega_1 t)\cos\lambda}{\omega_1(\omega_1^2 + 4n^2)} - \frac{a \sin\omega_1 t \cos\lambda}{\omega_1^2 + 4n^2} + \frac{a\omega_1(1 - e^{-2nt})\cos\lambda}{2n(\omega_1^2 + 4n^2)}$$

$$+\frac{a(1-\cos\omega_1 t)\sin\lambda}{\omega_1^2+4n^2}+\frac{2na\sin\omega_1 t\sin\lambda}{\omega_1(\omega_1^2+4n^2)}-\frac{a(1-e^{-2nt})\sin\lambda}{\omega_1^2+4n^2}$$

$$(14.5.12)$$

Based on some algebraic procedures, we transform equation (14.5.12) to the following shape:

$$x=s_0+\frac{v_0}{2n}(1-e^{-2nt})+\frac{r-p-f}{2n}\left[t+\frac{1}{2n}(e^{-2nt}-1)\right]$$

$$+\frac{2na(1-\cos\omega_1 t)\cos\lambda}{\omega_1(\omega_1^2+4n^2)}-\frac{a\sin\omega_1 t\cos\lambda}{\omega_1^2+4n^2}+\frac{a\omega_1(1-e^{-2nt})\cos\lambda}{2n(\omega_1^2+4n^2)}$$

$$+\frac{a(1-\cos\omega_1 t)\sin\lambda}{\omega_1^2+4n^2}+\frac{2na\sin\omega_1 t\sin\lambda}{\omega_1(\omega_1^2+4n^2)}-\frac{a(1-e^{-2nt})\sin\lambda}{\omega_1^2+4n^2}$$

$$(14.5.13)$$

The first derivative of equation (14.5.13) represents the velocity of the system:

$$\frac{dx}{dt}=v_0 e^{-2nt}+\frac{r-p-f}{2n}(1-e^{-2nt})+\frac{2na\sin\omega_1 t\cos\lambda}{\omega_1^2+4n^2}$$

$$-\frac{a\omega_1\cos\omega_1 t\cos\lambda}{\omega_1^2+4n^2}+\frac{a\omega_1 e^{-2nt}\cos\lambda}{\omega_1^2+4n^2}+\frac{a\omega_1\sin\omega_1 t\sin\lambda}{\omega_1^2+4n^2}$$

$$+\frac{2na\cos\omega_1 t\sin\lambda}{\omega_1^2+4n^2}-\frac{2nae^{-2nt}\sin\lambda}{\omega_1^2+4n^2}\qquad(14.5.14)$$

Taking the second derivative of equation (14.5.13), we determine the acceleration:

$$\frac{d^2x}{dt^2}=(r-p-f-2nv_0)e^{-2nt}+\frac{2na\omega_1\cos\omega_1 t\cos\lambda}{\omega_1^2+4n^2}$$

$$+\frac{a\omega_1^2\sin\omega_1 t\cos\lambda}{\omega_1^2+4n^2}-\frac{2na\omega_1 e^{-2nt}\cos\lambda}{\omega_1^2+4n^2}+\frac{a\omega_1^2\cos\omega_1 t\sin\lambda}{\omega_1^2+4n^2}$$

$$-\frac{2na\omega_1\sin\omega_1 t\sin\lambda}{\omega_1^2+4n^2}+\frac{4n^2 ae^{-2nt}\sin\lambda}{\omega_1^2+4n^2}$$

$$(14.5.15)$$

14.5.2 Initial Displacement Equals Zero

The initial conditions of motion are:

$$\text{for} \quad t = 0 \quad x = 0; \quad \frac{dx}{dt} = v_0 \qquad (14.5.16)$$

The solution of differential equation of motion (14.5.1) with the initial conditions of motion (14.5.16) reads:

$$
\begin{aligned}
x = {} & \frac{v_0}{2n}(1 - e^{-2nt}) + \frac{r - p - f}{2n}[t + \frac{1}{2n}(e^{-2nt} - 1)] \\
& + \frac{2na(1 - \cos\omega_1 t)\cos\lambda}{\omega_1(\omega_1^2 + 4n^2)} - \frac{a\sin\omega_1 t\cos\lambda}{\omega_1^2 + 4n^2} \\
& + \frac{a\omega_1(1 - e^{-2nt})\cos\lambda}{2n(\omega_1^2 + 4n^2)} + \frac{a(1 - \cos\omega_1 t)\sin\lambda}{\omega_1^2 + 4n^2} \\
& + \frac{2na\sin\omega_1 t\sin\lambda}{\omega_1(\omega_1^2 + 4n^2)} - \frac{a(1 - e^{-2nt})\sin\lambda}{\omega_1^2 + 4n^2}
\end{aligned} \qquad (14.5.17)
$$

The equations for the velocity and acceleration are the same as in the previous case.

14.5.3 Initial Velocity Equals Zero

The initial conditions of motion are:

$$\text{for} \quad t = 0 \quad x = s_0; \quad \frac{dx}{dt} = 0 \qquad (14.5.18)$$

Solving differential equation of motion (14.5.1) with the initial conditions of motion (14.5.18), we obtain:

$$
\begin{aligned}
x = {} & s_0 + \frac{r - p - f}{2n}[t + \frac{1}{2n}(e^{-2nt} - 1)] + \frac{2na(1 - \cos\omega_1 t)\cos\lambda}{\omega_1(\omega_1^2 + 4n^2)} \\
& - \frac{a\sin\omega_1 t\cos\lambda}{\omega_1^2 + 4n^2} + \frac{a\omega_1(1 - e^{-2nt})\cos\lambda}{2n(\omega_1^2 + 4n^2)} + \frac{a(1 - \cos\omega_1 t)\sin\lambda}{\omega_1^2 + 4n^2} \\
& + \frac{2na\sin\omega_1 t\sin\lambda}{\omega_1(\omega_1^2 + 4n^2)} - \frac{a(1 - e^{-2nt})\sin\lambda}{\omega_1^2 + 4n^2}
\end{aligned} \qquad (14.5.19)
$$

The first derivative of equation (14.5.19) represents the velocity of the system:

$$\frac{dx}{dt} = \frac{r-p-f}{2n}(1-e^{-2nt}) + \frac{2na\sin\omega_1 t\cos\lambda}{\omega_1^2 + 4n^2} - \frac{a\omega_1\cos\omega_1 t\cos\lambda}{\omega_1^2 + 4n^2}$$

$$+ \frac{a\omega_1 e^{-2nt}\cos\lambda}{\omega_1^2 + 4n^2} + \frac{a\omega_1\sin\omega_1 t\sin\lambda}{\omega_1^2 + 4n^2} + \frac{2na\cos\omega_1 t\sin\lambda}{\omega_1^2 + 4n^2}$$

$$- \frac{2nae^{-2nt}\sin\lambda}{\omega_1^2 + 4n^2} \tag{14.5.20}$$

Taking the second derivative of equation (14.5.19), we determine the acceleration:

$$\frac{d^2x}{dt^2} = (r-p-f)e^{-2nt} + \frac{2na\omega_1\cos\omega_1 t\cos\lambda}{\omega_1^2 + 4n^2} + \frac{a\omega_1^2\sin\omega_1 t\cos\lambda}{\omega_1^2 + 4n^2}$$

$$- \frac{2na\omega_1 e^{-2nt}\cos\lambda}{\omega_1^2 + 4n^2} + \frac{a\omega_1^2\cos\omega_1 t\sin\lambda}{\omega_1^2 + 4n^2} - \frac{2na\omega_1\sin\omega_1 t\sin\lambda}{\omega_1^2 + 4n^2}$$

$$+ \frac{4n^2 ae^{-2nt}\sin\lambda}{\omega_1^2 + 4n^2} \tag{14.5.21}$$

14.5.4 Both the Displacement and Velocity Equal Zero

The initial conditions of motion are:

$$\text{for}\quad t=0\quad x=0;\quad \frac{dx}{dt}=0 \tag{14.5.22}$$

The solution of differential equation motion (14.5.1) with the initial conditions of motion (14.5.22) reads:

$$x = \frac{r-p-f}{2n}[t + \frac{1}{2n}(e^{-2nt}-1)] + \frac{2na(1-\cos\omega_1 t)\cos\lambda}{\omega_1(\omega_1^2 + 4n^2)}$$

$$- \frac{a\sin\omega_1 t\cos\lambda}{\omega_1^2 + 4n^2} + \frac{a\omega_1(1-e^{-2nt})\cos\lambda}{2n(\omega_1^2 + 4n^2)} + \frac{a(1-\cos\omega_1 t)\sin\lambda}{\omega_1^2 + 4n^2}$$

$$+ \frac{2na\sin\omega_1 t\sin\lambda}{\omega_1(\omega_1^2 + 4n^2)} - \frac{a(1-e^{-2nt})\sin\lambda}{\omega_1^2 + 4n^2} \tag{14.5.23}$$

The equations for the velocity and acceleration are the same as in the previous case.

14.6   Harmonic Force $A \sin(\omega_1 t + \lambda)$

and Time-Dependent Force $Q\left(\rho + \frac{\mu t}{\tau}\right)$

According to Guiding Table 2.1, the engineering systems discussed in this section are subjected to the force of inertia, the damping force, the constant resisting force, and the friction force as the resisting forces (Row 14) and to the harmonic force and time-dependent force as the active forces (Column 6).

The current problem could be related to vibratory systems intended for the interaction with viscoelastoplastic media. At the beginning of the working processes, these systems may utilize, along with the harmonic force, a time-dependent force that acts during a predetermined interval of time. Note that the viscoelastoplastic media exert damping forces, constant resisting forces, and friction forces as a reaction to their plastic deformation. Detailed considerations related to the deformation of viscoelastoplastic media as well as to the behavior of a constant resisting force and a friction force applied to a vibratory system are presented in section 1.2.

The system is moving in the horizontal direction. We want to determine the basic parameters of motion. Figure 14.6 shows the model of a system subjected to the action of a harmonic force, a time-dependent force, a damping force, a constant resisting force, and a friction force.

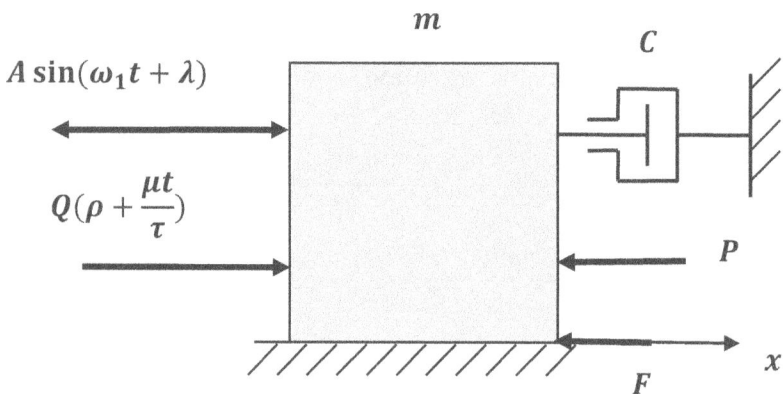

Figure 14.6 Model of a system subjected to a harmonic force, a time-dependent force, a damping force, a constant resisting force, and a friction force

The considerations above and the model in Figure 14.6 let us assemble the left and right sides of the differential equation of motion. The left side consists of the force of inertia, the damping force, the constant resisting force, and the friction force. The right side includes the harmonic force and the time-dependent force. Therefore, the differential equation of motion reads:

$$m\frac{d^2x}{dt^2}+C\frac{dx}{dt}+P+F = A\sin(\omega_1 t+\lambda)+Q(\rho+\frac{\mu t}{\tau}) \quad (14.6.1)$$

Differential equation of motion (14.6.1) has different solutions for various initial conditions of motion. These solutions and their analyses are presented below.

14.6.1 General Initial Conditions
The general initial conditions of motion are:

$$\text{for} \quad t=0 \quad x=s_0; \quad \frac{dx}{dt}=v_0 \quad (14.6.2)$$

where s_0 and v_0 are the initial displacement and initial velocity respectively.

Transforming equation (14.6.1), we may write:

$$m\frac{d^2x}{dt^2}+C\frac{dx}{dt}+P+F = Q\rho+Q\frac{\mu t}{\tau}+A\sin\omega_1 t\cos\lambda+A\cos\omega_1 t\sin\lambda$$

$$(14.6.3)$$

Dividing equation (14.6.3) by m, we have:

$$\frac{d^2x}{dt^2}+2n\frac{dx}{dt}+p+f = q\rho+q\frac{\mu t}{\tau}+a\sin\omega_1 t\cos\lambda+a\cos\omega_1 t\sin\lambda$$

$$(14.6.4)$$

where n is the damping factor and:

$$2n=\frac{C}{m} \quad (14.6.5)$$

$$p=\frac{P}{m} \quad (14.6.6)$$

$$f = \frac{F}{m} \tag{14.6.7}$$

$$q = \frac{Q}{m} \tag{14.6.8}$$

$$a = \frac{A}{m} \tag{14.6.9}$$

Using Laplace Transform Pairs 3, 4, 5, 6, and 7 from Table 1.1, we convert differential equation of motion (14.6.4) with the initial conditions of motion (14.6.2) from the time domain into the Laplace domain. The resulting algebraic equation of motion in the Laplace domain reads:

$$l^2 x(l) - lv_0 - l^2 s_0 + 2nlx(l) - 2nls_0 + p + f = q\rho + \frac{q\mu}{\tau l}$$

$$+ \frac{a\omega_1 l}{l^2 + \omega_1^2}\cos\lambda + \frac{al^2}{l^2 + \omega_1^2}\sin\lambda \tag{14.6.10}$$

Applying some algebraic procedures to equation (14.6.10), we have:

$$x(l)l(l + 2n) = l(v_0 + 2ns_0) + l^2 s_0 + q\rho - p - f$$

$$+ \frac{q\mu}{\tau l} + \frac{a\omega_1 l}{l^2 + \omega_1^2}\cos\lambda + \frac{al^2}{l^2 + \omega_1^2}\sin\lambda \tag{14.6.11}$$

Solving equation (14.6.11) for the Laplace domain displacement $x(l)$, we may write:

$$x(l) = \frac{v_0 + 2ns_0}{l + 2n} + \frac{ls_0}{l + 2n} + \frac{q\rho - p - f}{l(l + 2n)} + \frac{q\mu}{l^2 \tau(l + 2n)}$$

$$+ \frac{a\omega_1 \cos\lambda}{(l + 2n)(l^2 + \omega_1^2)} + \frac{al \sin\lambda}{(l + 2n)(l^2 + \omega_1^2)} \tag{14.6.12}$$

Applying pairs 1, 10, 12, 13, 31, 19, and 20 from Table 1.1, we invert equation (14.6.12) from the Laplace domain into the time domain and obtain the solution of differential equation of motion (14.6.1) with the initial conditions of motion (14.6.2):

$$x = \frac{v_0 + 2ns_0}{2n}(1 - e^{-2nt}) + s_0 e^{-2nt} + \frac{q\rho - p - f}{2n}\left[t + \frac{1}{2n}(e^{-2nt} - 1)\right]$$

$$+ \frac{q\mu}{4n\tau}\left[t^2 - \frac{1}{n}t - \frac{1}{2n^2}(e^{-2nt} - 1)\right] + \frac{2na(1 - \cos\omega_1 t)\cos\lambda}{\omega_1(\omega_1^2 + 4n^2)}$$

$$- \frac{a\sin\omega_1 t \cos\lambda}{\omega_1^2 + 4n^2} + \frac{a\omega_1(1 - e^{-2nt})\cos\lambda}{2n(\omega_1^2 + 4n^2)} + \frac{a(1 - \cos\omega_1 t)\sin\lambda}{\omega_1^2 + 4n^2}$$

$$+ \frac{2na\sin\omega_1 t \sin\lambda}{\omega_1(\omega_1^2 + 4n^2)} - \frac{a(1 - e^{-2nt})\sin\lambda}{\omega_1^2 + 4n^2} \hspace{2cm} \textbf{(14.6.13)}$$

The first derivative of equation (14.6.13) represents the velocity of the system:

$$\frac{dx}{dt} = (v_0 - \frac{q\rho - p - f}{2n} + \frac{q\mu}{4n^2\tau})e^{-2nt} + \frac{q\rho - p - f}{2n} + \frac{q\mu}{2n\tau}(t - \frac{1}{2n})$$

$$+ \frac{2na\sin\omega_1 t \cos\lambda}{\omega_1^2 + 4n^2} - \frac{a\omega_1\cos\omega_1 t \cos\lambda}{\omega_1^2 + 4n^2} + \frac{a\omega_1 e^{-2nt}\cos\lambda}{\omega_1^2 + 4n^2}$$

$$+ \frac{a\omega_1\sin\omega_1 t \sin\lambda}{\omega_1^2 + 4n^2} + \frac{2na\cos\omega_1 t \sin\lambda}{\omega_1^2 + 4n^2} - \frac{2nae^{-2nt}\sin\lambda}{\omega_1^2 + 4n^2}$$

$$\hspace{8cm} \textbf{(14.6.14)}$$

Taking the second derivative of equation (14.6.13), we determine the acceleration:

$$\frac{d^2x}{dt^2} = \left(-2nv_0 + q\rho - p - f - \frac{q\mu}{2n\tau}\right)e^{-2nt} + \frac{q\mu}{2n\tau} + \frac{2na\omega_1\cos\omega_1 t \cos\lambda}{\omega_1^2 + 4n^2}$$

$$+ \frac{a\omega_1^2\sin\omega_1 t \cos\lambda}{\omega_1^2 + 4n^2} - \frac{2na\omega_1 e^{-2nt}\cos\lambda}{\omega_1^2 + 4n^2} + \frac{a\omega_1^2\cos\omega_1 t \sin\lambda}{\omega_1^2 + 4n^2}$$

$$- \frac{2na\omega_1\sin\omega_1 t \sin\lambda}{\omega_1^2 + 4n^2} + \frac{4n^2ae^{-2nt}\sin\lambda}{\omega_1^2 + 4n^2} \hspace{2cm} \textbf{(14.6.15)}$$

14.6.2 Initial Displacement Equals Zero

The initial conditions of motion are:

$$\text{for} \quad t = 0 \quad x = 0; \quad \frac{dx}{dt} = v_0 \hspace{2cm} \textbf{(14.6.16)}$$

Solving differential equation of motion (14.6.1) with the initial conditions of motion (14.6.16), we obtain:

$$
x = \frac{v_0 + 2ns_0}{2n}(1 - e^{-2nt}) + \frac{q\rho - p - f}{2n}\left[t + \frac{1}{2n}(e^{-2nt} - 1)\right]
$$

$$
+ \frac{q\mu}{4n\tau}\left[t^2 - \frac{1}{n}t - \frac{1}{2n^2}(e^{-2nt} - 1)\right] + \frac{2na(1 - \cos\omega_1 t)\cos\lambda}{\omega_1(\omega_1^2 + 4n^2)}
$$

$$
- \frac{a\sin\omega_1 t\cos\lambda}{\omega_1^2 + 4n^2} + \frac{a\omega_1(1 - e^{-2nt})\cos\lambda}{2n(\omega_1^2 + 4n^2)} + \frac{a(1 - \cos\omega_1 t)\sin\lambda}{\omega_1^2 + 4n^2}
$$

$$
+ \frac{2na\sin\omega_1 t\sin\lambda}{\omega_1(\omega_1^2 + 4n^2)} - \frac{a(1 - e^{-2nt})\sin\lambda}{\omega_1^2 + 4n^2} \qquad (14.6.17)
$$

The rest of the parameters are the same as in the previous case.

14.6.3 Initial Velocity Equals Zero

The initial conditions of motion are:

$$
\text{for} \quad t = 0 \quad x = s_0; \quad \frac{dx}{dt} = 0 \qquad (14.6.18)
$$

The solution of differential equation of motion (14.6.1) with the initial conditions of motion (14.6.18) reads:

$$
x = s_0 + \frac{q\rho - p - f}{2n}\left[t + \frac{1}{2n}(e^{-2nt} - 1)\right]
$$

$$
+ \frac{q\mu}{4n\tau}\left[t^2 - \frac{1}{n}t - \frac{1}{2n^2}(e^{-2nt} - 1)\right] + \frac{2na(1 - \cos\omega_1 t)\cos\lambda}{\omega_1(\omega_1^2 + 4n^2)}
$$

$$
- \frac{a\sin\omega_1 t\cos\lambda}{\omega_1^2 + 4n^2} + \frac{a\omega_1(1 - e^{-2nt})\cos\lambda}{2n(\omega_1^2 + 4n^2)} + \frac{a(1 - \cos\omega_1 t)\sin\lambda}{\omega_1^2 + 4n^2}
$$

$$
+ \frac{2na\sin\omega_1 t\sin\lambda}{\omega_1(\omega_1^2 + 4n^2)} - \frac{a(1 - e^{-2nt})\sin\lambda}{\omega_1^2 + 4n^2} \qquad (14.6.19)
$$

Taking the first derivative of equation (14.6.19), we determine the velocity of the system:

$$
\frac{dx}{dt} = \frac{1}{2n}\left(q\rho - p - f - \frac{q\mu}{2n\tau}\right)(1 - e^{-2nt}) + \frac{q\mu}{2n\tau}t + \frac{2na\sin\omega_1 t\cos\lambda}{\omega_1^2 + 4n^2}
$$

$$-\frac{a\omega_1 \cos\omega_1 t \cos\lambda}{\omega_1^2 + 4n^2} + \frac{a\omega_1 e^{-2nt} \cos\lambda}{\omega_1^2 + 4n^2} + \frac{a\omega_1 \sin\omega_1 t \sin\lambda}{\omega_1^2 + 4n^2}$$

$$+\frac{2na \cos\omega_1 t \sin\lambda}{\omega_1^2 + 4n^2} - \frac{2nae^{-2nt} \sin\lambda}{\omega_1^2 + 4n^2} \tag{14.6.20}$$

The second derivative of equation (14.6.19) represents the acceleration:

$$\frac{d^2x}{dt^2} = \left(q\rho - p - f - \frac{q\mu}{2n\tau}\right)e^{-2nt} + \frac{q\mu}{2n\tau} + \frac{2na\omega_1 \cos\omega_1 t \cos\lambda}{\omega_1^2 + 4n^2}$$

$$+\frac{a\omega_1^2 \sin\omega_1 t \cos\lambda}{\omega_1^2 + 4n^2} - \frac{2na\omega_1 e^{-2nt} \cos\lambda}{\omega_1^2 + 4n^2} + \frac{a\omega_1^2 \cos\omega_1 t \sin\lambda}{\omega_1^2 + 4n^2}$$

$$-\frac{2na\omega_1 \sin\omega_1 t \sin\lambda}{\omega_1^2 + 4n^2} + \frac{4n^2 ae^{-2nt} \sin\lambda}{\omega_1^2 + 4n^2} \tag{14.6.21}$$

14.6.4 Both the Initial Displacement and Velocity Equal Zero
The initial conditions of motion are:

$$\text{for} \quad t = 0 \ \ x = 0; \quad \frac{dx}{dt} = 0 \tag{14.6.22}$$

Solving differential equation of motion (14.6.21) with the initial conditions of motion (14.6.22), we obtain:

$$x = \frac{q\rho - p - f}{2n}\left[t + \frac{1}{2n}(e^{-2nt} - 1)\right] + \frac{q\mu}{4n\tau}\left[t^2 - \frac{1}{n}t - \frac{1}{2n^2}(e^{-2nt} - 1)\right]$$

$$+\frac{2na(1 - \cos\omega_1 t)\cos\lambda}{\omega_1(\omega_1^2 + 4n^2)} - \frac{a\sin\omega_1 t \cos\lambda}{\omega_1^2 + 4n^2} + \frac{a\omega_1(1 - e^{-2nt})\cos\lambda}{2n(\omega_1^2 + 4n^2)}$$

$$+\frac{a(1 - \cos\omega_1 t)\sin\lambda}{\omega_1^2 + 4n^2} + \frac{2na\sin\omega_1 t \sin\lambda}{\omega_1(\omega_1^2 + 4n^2)} - \frac{a(1 - e^{-2nt})\sin\lambda}{\omega_1^2 + 4n^2} \tag{14.6.23}$$

The equations for the velocity and the acceleration are the same as in the previous case.

15

DAMPING AND STIFFNESS

This chapter discusses engineering systems subjected to the action of the force of inertia, the damping force, and the stiffness force. These resisting forces are marked with the plus sign in Row 15 of Guiding Table 2.1. The active forces applied to these systems are indicated by the intersection of this row with Columns 1 through 6. In this chapter, the left sides of the differential equations of motion of the systems in all sections are identical and consist of the sum of the force of inertia, the damping force, and the stiffness force. The right sides of these equations consist of the corresponding active forces applied to these systems. The section titles reflect the characteristics of the active forces that are associated with the problems.

The problems considered in this chapter could be associated with the working processes of engineering systems interacting with viscoelastoplastic media.

15.1 Active Force Equals Zero

The intersection of Row 15 of Guiding Table 2.1 with Column 1 indicates that this section describes engineering systems subjected to the action of the force of inertia, the damping force, and the stiffness force as the resisting forces, while the active force equals zero.

The considerations related to the motion of a system in the absence of active forces are described in section 1.3. The current problem could be related to a system performing damped vibrations. This problem could also be associated with the interaction of a system with a viscoelastoplastic medium that exerts damping and stiffness resisting forces as the reaction to its deformation (see section 1.2).

The motion occurs in the horizontal direction. We want to determine the basic parameters of motion. Figure 15.1 shows the model of a system subjected to the action of a damping force and a stiffness force.

The considerations above and the model in Figure 15.1 let us compose the left and right sides of the differential equation of motion. The left side consists of the force of inertia, the damping force, and the stiffness force, while the right side of the equation equals zero. Therefore, the differential equation of motion reads:

$$m\frac{d^2x}{dt^2} + C\frac{dx}{dt} + Kx = 0 \qquad (15.1.1)$$

Differential equation of motion (15.1.1) has different solutions for various initial conditions of motion. These solutions are presented below.

15.1.1 General Initial Conditions

The general initial conditions of motion are:

$$\text{for} \quad t = 0 \quad x = s_0; \quad \frac{dx}{dt} = v_0 \qquad (15.1.2)$$

where s_0 and v_0 are the initial displacement and initial velocity respectively.

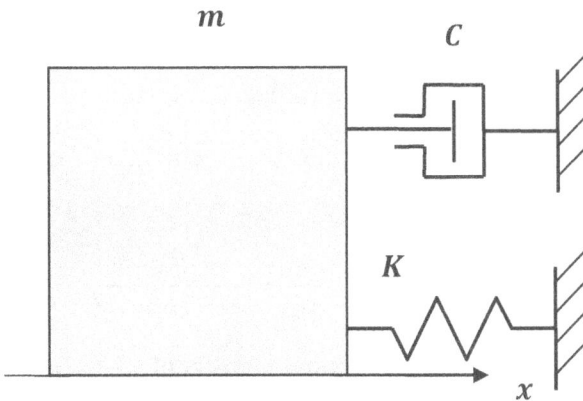

Figure 15.1 Model of a system subjected to a damping force and a stiffness force

Dividing equation (15.1.1) by m, we may write:

$$\frac{d^2x}{dt^2} + 2n\frac{dx}{dt} + \omega_0^2\, x = 0 \qquad (15.1.3)$$

where n is the damping factor and ω_0 is the natural frequency, while:

$$2n = \frac{C}{m} \qquad (15.1.4)$$

$$\omega_0^2 = \frac{K}{m} \qquad (15.1.5)$$

Using Laplace Transform Pairs 3, 4, 5, and 1 from Table 1.1, we convert differential equation of motion (15.1.3) with the initial conditions of motion (15.1.2) from the time domain into the Laplace domain and obtain the corresponding algebraic equation of motion in the Laplace domain:

$$l^2x(l) - lv_0 - l^2s_0 + 2nlx(l) - 2nls_0 + \omega_0^2\, x(l) = 0 \qquad (15.1.6)$$

Applying basic algebra to equation (15.1.6), we have:

$$x(l)(l^2 + 2nl + \omega_0^2) = l(v_0 + 2ns_0) + l^2 s_0 \qquad \textbf{(15.1.7)}$$

Solving equation (15.1.7) for the displacement $x(l)$ in the Laplace domain, we obtain:

$$x(l) = \frac{l(v_0 + 2ns_0)}{l^2 + 2nl + \omega_0^2} + \frac{l^2 s_0}{l^2 + 2nl + \omega_0^2} \qquad \textbf{(15.1.8)}$$

Transforming the denominators in equation (15.1.8), we may write:

$$l^2 + 2nl + \omega_0^2 + n^2 - n^2 = (l+n)^2 + \omega^2 \qquad \textbf{(15.1.9)}$$

where

$$\omega^2 = \omega_0^2 - n^2 \qquad \textbf{(15.1.10)}$$

while ω^2 could be positive, equal to zero, or negative.

All these three cases are considered below.

a. Case $\omega^2 > 0$

Adjusting the denominators in equation (15.1.8) according to equation (15.1.9), we may write:

$$x(l) = \frac{l(v_0 + 2ns_0)}{(l+n)^2 + \omega^2} + \frac{l^2 s_0}{(l+n)^2 + \omega^2} \qquad \textbf{(15.1.11)}$$

Applying pairs 1, 24, and 27 from Table 1.1, we invert equation (15.1.11) from the Laplace domain into the time domain and obtain the solution of differential equation of motion (15.1.1) with the initial conditions of motion (15.1.2):

$$x = \frac{v_0 + 2ns_0}{\omega} e^{-nt} \sin \omega t + s_0 e^{-nt} (\cos \omega t - \frac{n}{\omega} \sin \omega t) \qquad \textbf{(15.1.12)}$$

Applying conventional algebra to equation (15.1.12), we have:

$$x = \frac{e^{-nt}}{\omega} [(v_0 + s_0 n) \sin \omega t + s_0 \omega \cos \omega t] \qquad \textbf{(15.1.13)}$$

Taking the first and second derivatives of equation (15.1.13), we determine the velocity and acceleration of the system respectively:

$$\frac{dx}{dt} = \frac{e^{-nt}}{\omega}\{v_0\omega\cos\omega t - [v_0n + s_0(n^2 + \omega^2)]\sin\omega t\} \quad \textbf{(15.1.14)}$$

$$\frac{d^2x}{dt^2} = e^{-nt}\{\frac{1}{\omega}[s_0n(\omega^2 + n^2) - v_0(\omega^2 - n^2)]\sin\omega t$$

$$-[2v_0n + s_0(\omega^2 + n^2)]\cos\omega t\} \quad \textbf{(15.1.15)}$$

The system is in vibratory motion.

b. Case $\omega^2 = 0$

In equation (15.1.11), equating ω^2 to zero, we obtain:

$$x(l) = \frac{l(v_0 + 2ns_0)}{(l+n)^2} + \frac{l^2s_0}{(l+n)^2} \quad \textbf{(15.1.16)}$$

Based on pairs 1, 37, and 38 from Table 1.1, we invert equation (15.1.16) from the Laplace domain into the time domain and obtain for this case the solution of differential equation (15.1.1) with the initial conditions of motion (15.1.2):

$$x = (v_0 + 2ns_0)te^{-nt} + s_0(1 - nt)e^{-nt} \quad \textbf{(15.1.17)}$$

After transforming equation (15.1.16), we may write:

$$x = e^{-nt}[t(v_0 + s_0n) + s_0] \quad \textbf{(15.1.18)}$$

Taking the first and second derivatives of equation (15.1.18), we determine the velocity and the acceleration respectively:

$$\frac{dx}{dt} = e^{-nt}[v_0 - nt(v_0 + s_0n)] \quad \textbf{(15.1.19)}$$

$$\frac{d^2x}{dt^2} = e^{-nt}[tn^2(v_0 + s_0n) - 2nv_0 - s_0n^2] \quad \textbf{(15.1.20)}$$

The system is in rectilinear motion.

c. Case $\omega^2 < 0$

Taking in equation (15.1.11) ω^2 with the negative sign, we may write:

$$x(l) = \frac{l(v_0 + 2ns_0)}{(l+n)^2 - \omega^2} + \frac{l^2 s_0}{(l+n)^2 - \omega^2} \tag{15.1.21}$$

Using pairs 1, 25, and 28 from Table 1.1, we invert equation (15.1.21) from the Laplace domain into the time domain and obtain the solution of differential equation of motion (15.1.1) with the initial conditions (15.1.2):

$$x = \frac{v_0 + 2ns_0}{\omega} e^{-nt} \sinh \omega t + \frac{s_0}{\omega} e^{-nt} (\omega \cosh \omega t - n \sinh \omega t) \tag{15.1.22}$$

Applying some basic algebra to equation (15.1.22), we obtain:

$$x = \frac{1}{\omega} e^{-nt} [(v_0 + s_0 n) \sinh \omega t + s_0 \omega \cosh \omega t] \tag{15.1.23}$$

Taking the first and second derivatives of equation (15.1.22), we determine the velocity and acceleration respectively:

$$\frac{dx}{dt} = \frac{1}{\omega} e^{-nt} \{ v_0 \omega \cosh \omega t - [v_0 n - s_0 (n^2 - \omega^2)] \sinh \omega t \} \tag{15.1.24}$$

$$\frac{d^2 x}{dt^2} = e^{-nt} \{ \frac{1}{\omega} [v_0 (n^2 + \omega^2) - s_0 n (n^2 - \omega^2)] \sinh \omega t$$

$$- [2 v_0 n - s_0 (n^2 - \omega^2)] \cosh \omega t \} \tag{15.1.25}$$

The system is in rectilinear motion.

15.1.2 Initial Displacement Equals Zero

The initial conditions of motion are:

$$\text{for} \quad t = 0 \quad x = 0; \quad \frac{dx}{dt} = v_0 \tag{15.1.26}$$

a. Case $\omega^2 > 0$

The solution of differential equation of motion (15.1.1) with the initial conditions of motion (15.1.26) reads:

$$x = \frac{v_0}{\omega} e^{-nt} \sin \omega t \qquad (15.1.27)$$

Taking the first and second derivatives of equation (15.1.27), we determine the velocity and the acceleration respectively:

$$\frac{dx}{dt} = \frac{v_0}{\omega} e^{-nt} (\omega \cos \omega t - n \sin \omega t) \qquad (15.1.28)$$

$$\frac{d^2 x}{dt^2} = \frac{v_0}{\omega} e^{-nt} [(n^2 - \omega^2) \sin \omega t - 2n\omega \cos \omega t] \qquad (15.1.29)$$

b. Case $\omega^2 = 0$

Solving differential equation of motion (15.1.1) with the initial conditions of motion (15.1.26), we obtain:

$$x = v_0 t e^{-nt} \qquad (15.1.30)$$

Taking the first and second derivatives of equation (15.1.30), we determine the velocity and the acceleration respectively:

$$\frac{dx}{dt} = v_0 (1 - nt) e^{-nt} \qquad (15.1.31)$$

$$\frac{d^2 x}{dt^2} = v_0 n(nt - 2) e^{-nt} \qquad (15.1.32)$$

c. Case $\omega^2 < 0$

The solution of differential equation of motion (15.1.1) with the initial conditions of motion (15.1.26) reads:

$$x = \frac{1}{\omega} v_0 e^{-nt} \sinh \omega t \qquad (15.1.33)$$

Taking the first and second derivatives of equation (15.1.33), we determine the velocity and acceleration respectively:

$$\frac{dx}{dt} = \frac{1}{\omega} v_0 e^{-nt}(\omega \cosh \omega t - n \sinh \omega t) \qquad (15.1.34)$$

$$\frac{d^2x}{dt^2} = \frac{1}{\omega} v_0 e^{-nt}[(n^2 + \omega^2)\sinh \omega t - 2n\omega \cosh \omega t] \qquad (15.1.35)$$

15.1.3 Initial Velocity Equals Zero

In order to consider the motion of the system into the positive direction, as shown in Figure 15.1, the initial displacement should be taken with the negative sign. Therefore, the initial conditions of motion are:

$$\text{for} \quad t = 0 \quad x = -s_0; \quad \frac{dx}{dt} = 0 \qquad (15.1.36)$$

a. Case $\omega^2 > 0$

The solution of differential equation of motion (15.1.1) with the initial conditions of motion (15.1.36) reads:

$$x = -\frac{1}{\omega} s_0 e^{-nt}(n \sin \omega t - \omega \cos \omega t) \qquad (15.1.37)$$

Taking the first and second derivatives of equation (15.1.37), we determine the velocity and the acceleration respectively:

$$\frac{dx}{dt} = \frac{1}{\omega} s_0 e^{-nt}(n^2 - \omega^2)\sin \omega t \qquad (15.1.38)$$

$$\frac{d^2x}{dt^2} = \frac{1}{\omega} s_0 e^{-nt}(n^2 + \omega^2)(\omega \cos \omega t - n \sin \omega t) \qquad (15.1.39)$$

b. Case $\omega^2 = 0$

Solving differential equation of motion (15.1.1) with the initial conditions of motion (15.1.36), we obtain:

$$x = -s_0 e^{-nt}(nt + 1) \qquad (15.1.40)$$

Taking the first and second derivatives of equation (15.1.40), we determine the velocity and the acceleration respectively:

$$\frac{dx}{dt} = s_0 n^2 t e^{-nt} \qquad (15.1.41)$$

$$\frac{d^2x}{dt^2} = -s_0 n^2 e^{-nt}(nt-1) \qquad (15.1.42)$$

c. Case $\omega^2 < 0$

The solution of differential equation of motion (15.1.1) with initial conditions of motion (15.1.36) reads:

$$x = -\frac{1}{\omega} s_0 e^{-nt}(\omega \cosh \omega t + n \sinh \omega t) \qquad (15.1.43)$$

Taking the first and second derivatives of equation (15.1.43), we determine the velocity and the acceleration respectively:

$$\frac{dx}{dt} = \frac{1}{\omega} s_0 e^{-nt}(n^2 - \omega^2)\sinh \omega t] \qquad (15.1.44)$$

$$\frac{d^2x}{dt^2} = -s_0 e^{-nt}[(n^2 - \omega^2)(\cosh \omega t + \frac{n}{\omega}\sinh \omega t] \qquad (15.1.45)$$

15.2 Constant Force R

According to Guiding Table 2.1, this section describes engineering systems subjected to the action of the force of inertia, the damping force, and the stiffness force as the resisting forces (Row 15) and to the constant active force (Column 2). The current problem could be associated with the working process of systems intended for interaction with viscoelastoplastic media that exert damping and stiffness forces as a reaction to their deformation. Additional considerations related to the deformation of viscoelastoplastic media are discussed in section 1.2.

The system moves in the horizontal direction. We want to determine the law of motion of this engineering system. Figure 15.2

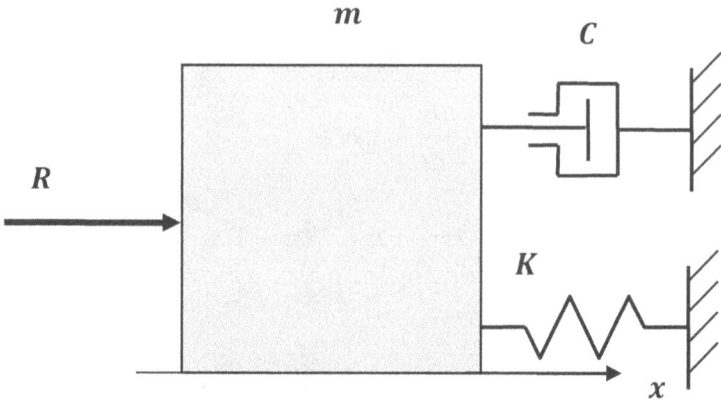

Figure 15.2 Model of a system subjected to a constant active force, a damping force, and a stiffness force

shows the model of a system subjected to the action of a constant active force, a damping force, and a stiffness force.

Based on the considerations above and the model in Figure 15.2, we assemble the left and right sides of the differential equation of motion of this system. The left side consists of the force of inertia, the damping force, and the stiffness force, while the right side includes the constant active force. Therefore, the differential equation of motion reads:

$$m\frac{d^2x}{dt^2} + C\frac{dx}{dt} + Kx = R \qquad (15.2.1)$$

Differential equation of motion (15.2.1) has different solutions for various initial conditions of motion. These solutions are presented below.

15.2.1 General Initial Conditions

The general initial conditions of motion are:

$$\text{for} \quad t = 0 \quad x = s_0; \quad \frac{dx}{dt} = v_0 \qquad (15.2.2)$$

where s_0 and v_0 are the initial displacement and initial velocity respectively.

Dividing equation (15.2.1) by m, we may write:

$$\frac{d^2x}{dt^2} + 2n\frac{dx}{dt} + \omega_0^2 x = r \qquad (15.2.3)$$

where n is the damping factor and ω_0 is the natural frequency, while:

$$2n = \frac{C}{m} \qquad (15.2.4)$$

$$\omega_0^2 = \frac{K}{m} \qquad (15.2.5)$$

$$r = \frac{R}{m} \qquad (15.2.6)$$

Based on Laplace Transform Pairs 3,4,5, and 1 from Table 1.1, we convert differential equation of motion (15.2.3) with the initial conditions of motion (15.2.2) from the time domain into the Laplace domain and obtain the corresponding algebraic equation of motion in the Laplace domain:

$$l^2 x(l) - lv_0 - l^2 s_0 + 2nlx(l) - 2nls_0 + \omega_0^2 x(l) = r \qquad (15.2.7)$$

Applying basic algebra to equation (15.1.7), we have:

$$x(l)(l^2 + 2nl + \omega_0^2) = r + l(v_0 + 2ns_0) + l^2 s_0 \qquad (15.2.8)$$

Solving equation (15.2.8) for the displacement $x(l)$ in the Laplace domain, we obtain:

$$x(l) = \frac{r}{l^2 + 2nl + \omega_0^2} + \frac{l(v_0 + 2ns_0)}{l^2 + 2nl + \omega_0^2} + \frac{l^2 s_0}{l^2 + 2nl + \omega_0^2} \qquad (15.2.9)$$

Applying basic algebra to the denominators in equation (15.2.9), we may write:

$$l^2 + 2nl + \omega_0^2 + n^2 - n^2 = (l+n)^2 + \omega^2 \qquad (15.2.10)$$

where:

$$\omega^2 = \omega_0^2 - n^2 \qquad (15.2.11)$$

while ω^2 could be positive, equal to zero, or negative.

All these three cases are considered below.

a. Case $\omega^2 > 0$

Adjusting the denominators in equation (15.2.9) according to equation (15.2.1), we obtain:

$$x(l) = \frac{r}{(l+n)^2 + \omega^2} + \frac{l(v_0 + 2ns_0)}{(l+n)^2 + \omega^2} + \frac{l^2 s_0}{(l+n)^2 + \omega^2} \tag{15.2.12}$$

Applying pairs 1, 22, 24, and 27 from Table 1.1, we invert equation (15.2.12) from the Laplace domain into the time domain and obtain the solution of differential equation of motion (15.2.1) with the initial conditions of motion (15.2.2):

$$x = \frac{r}{\omega^2 + n^2}[1 - e^{-nt}(\cos\omega t + \frac{n}{\omega}\sin\omega t)] + \frac{1}{\omega}(v_0 + 2ns_0)e^{-nt}\sin\omega t$$

$$+ s_0 e^{-nt}(\cos\omega t - \frac{n}{\omega}\sin\omega t) \tag{15.2.13}$$

b. Case $\omega^2 = 0$

Equating ω^2 in equation (15.2.12) to zero, we have:

$$x(l) = \frac{r}{(l+n)^2} + \frac{l(v_0 + 2ns_0)}{(l+n)^2} + \frac{l^2 s_0}{(l+n)^2} \tag{15.2.14}$$

Using pairs 1, 36, 37, and 38 from Table 1.1, we invert equation (15.2.14) from the Laplace domain into the time domain and obtain the solution of differential equation of motion (15.2.1) with the initial conditions of motion (15.2.2):

$$x = \frac{r}{n^2}[1 - e^{-nt}(1 + nt)] + (v_0 + 2ns_0)te^{-nt} + s_0 e^{-nt}(1 - nt) \tag{15.2.15}$$

c. Case $\omega^2 < 0$

Taking ω^2 in equation (15.2.11) with the negative sign, we may write:

$$x(l) = \frac{r}{(l+n)^2 - \omega^2} + \frac{l(v_0 + 2ns_0)}{(l+n)^2 - \omega^2} + \frac{l^2 s_0}{(l+n)^2 - \omega^2} \tag{15.2.16}$$

Using pairs 1, 23, 25, and 28 from Table 1.1, we invert equation (15.2.16) from the Laplace domain into the time domain and obtain the solution of differential equation of motion (15.2.1) with the initial conditions of motion (15.2.2):

$$x = \frac{r}{n^2 - \omega^2}[1 - e^{-nt}(\cosh \omega t + \frac{n}{\omega} \sinh \omega t)] + \frac{v_0 + 2ns_0}{\omega} e^{-nt} \sinh \omega t$$

$$+ s_0 e^{-nt}(\cosh \omega t - \frac{n}{\omega} \sinh \omega t) \qquad (15.2.17)$$

15.2.2 Initial Displacement Equals Zero
The initial conditions of motion are:

$$\text{for} \quad t = 0 \quad x = 0; \quad \frac{dx}{dt} = v_0 \qquad (15.2.18)$$

The solutions of differential equation of motion (15.2.1) with the initial conditions of motion (15.2.18) for the following cases are presented below.

a. Case $\omega^2 > 0$

$$x = \frac{r}{\omega^2 + n^2}[1 - e^{-nt}(\cos \omega t + \frac{n}{\omega} \sin \omega t)] + \frac{1}{\omega} v_0 e^{-nt} \sin \omega t \quad (15.2.19)$$

b. Case $\omega^2 = 0$

$$x = \frac{r}{n^2}\left[1 - e^{-nt}(1 + nt)\right] + v_0 t e^{-nt} \qquad (15.2.20)$$

c. Case $\omega^2 < 0$

$$x = \frac{r}{n^2 - \omega^2}[1 - e^{-nt}(\cosh \omega t + \frac{n}{\omega} \sinh \omega t)] + \frac{v_0}{\omega} e^{-nt} \sinh \omega t \quad (15.2.21)$$

15.2.3 Initial Velocity Equals Zero
The initial conditions of motion are:

$$\text{for} \quad t = 0 \quad x = s_0; \quad \frac{dx}{dt} = 0 \qquad (15.2.22)$$

The solutions of differential equation of motion (15.2.1) with the initial conditions of motion (15.2.22) for the following cases are presented below.

a. **Case $\omega^2 > 0$**

$$x = \frac{r}{\omega^2 + n^2}[1 - e^{-nt}(\cos\omega t + \frac{n}{\omega}\sin\omega t)] + \frac{1}{\omega}2ns_0 e^{-nt}\sin\omega t$$

$$+ s_0 e^{-nt}(\cos\omega t - \frac{n}{\omega}\sin\omega t) \qquad (15.2.23)$$

b. **Case $\omega^2 = 0$**

$$x = \frac{r}{n^2}\left[1 - e^{-nt}(1 + nt)\right] + 2ns_0 te^{-nt} + s_0 e^{-nt}(1 - nt) \qquad (15.2.24)$$

c. **Case $\omega^2 < 0$**

$$x = \frac{r}{n^2 - \omega^2}[1 - e^{-nt}(\cosh\omega t + \frac{n}{\omega}\sinh\omega t)] + \frac{2ns_0}{\omega}e^{-nt}\sinh\omega t$$

$$+ s_0 e^{-nt}(\cosh\omega t - \frac{n}{\omega}\sinh\omega t) \qquad (15.2.25)$$

15.2.4 Both the Displacement and Velocity Equal Zero
The initial conditions of motion are:

$$\text{for} \quad t = 0 \quad x = 0; \quad \frac{dx}{dt} = 0 \qquad (15.2.26)$$

The solutions of differential equation of motion (15.2.1) with the initial conditions of motion (15.2.26) for the following cases are presented below.

a. **Case $\omega^2 > 0$**

$$x = \frac{r}{n^2 - \omega^2}[1 - e^{-nt}(\cos\omega t + \frac{n}{\omega}\sin\omega t)] \qquad (15.2.27)$$

b. Case $\omega^2 = 0$

$$x = \frac{r}{n^2}\left[1 - e^{-nt}(1 + nt)\right] \qquad (15.2.28)$$

c. Case $\omega^2 < 0$

$$x = \frac{r}{n^2 - \omega^2}[1 - e^{-nt}(\cosh \omega t + \frac{n}{\omega}\sinh \omega t)] \qquad (15.2.29)$$

15.3 Harmonic Force $A \sin(\omega_1 t + \lambda)$

This section describes engineering systems subjected to the force of inertia, the damping force, and the stiffness force as the resisting forces (Row 15 of Guiding Table 2.1) and a the harmonic force as the active force (Column3). The current problem could be associated with vibratory machines intended for the deformation of viscoelastioplastic media. These media exert damping and stiffness forces as the reaction to their deformation (see section 1.2).

The system is moving in the horizontal direction. We want to determine the law of motion of the system. Figure 15.3 shows the model of a system subjected to the action of a harmonic force, a damping force, and a stiffness force.

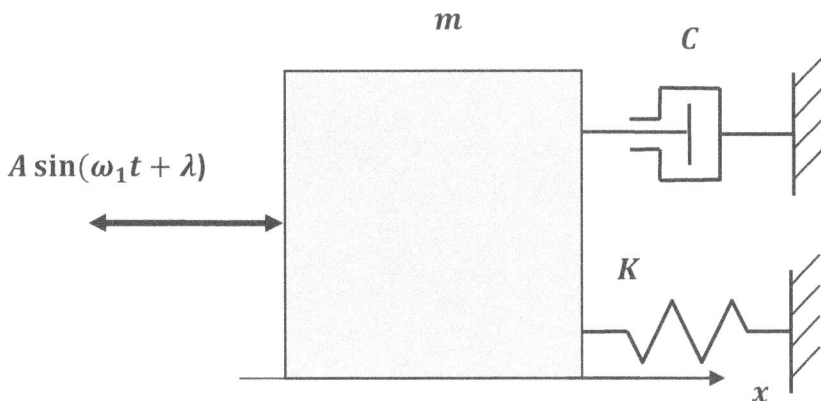

Figure 15.3 Model of a system subjected to a harmonic force, a damping force, and a stiffness force

Based on the considerations above and the model in Figure 15.3, we can compose the left and right sides of the differential equation of motion. The left side consists of the force of inertia, the damping force, and the stiffness force, while the right side includes the harmonic force. Therefore, the differential equation of motion reads:

$$m\frac{d^2x}{dt^2} + C\frac{dx}{dt} + Kx = A\sin(\omega_1 t + \lambda) \qquad (15.3.1)$$

Differential equation of motion (15.3.1) has different solutions for various initial conditions of motion. These solutions are presented below.

15.3.1 General Initial Conditions

The general initial conditions of motion are:

$$\text{for} \quad t = 0 \quad x = s_0; \quad \frac{dx}{dt} = v_0 \qquad (15.3.2)$$

where s_0 and v_0 are the initial displacement and initial velocity respectively.

Transforming the sinusoidal function of equation (15.3.1) and dividing the latter by m, we obtain:

$$\frac{d^2x}{dt^2} + 2n\frac{dx}{dt} + \omega_0^2 x = a\sin\omega_1 t\cos\lambda + a\cos\omega_1 t\sin\lambda \qquad (15.3.3)$$

where n is the damping factor and ω_0 is the natural frequency, while:

$$2n = \frac{C}{m} \qquad (15.3.4)$$

$$\omega_0^2 = \frac{K}{m} \qquad (15.3.5)$$

$$a = \frac{A}{m} \qquad (15.3.6)$$

Using Laplace Transform Pairs 3, 4, 5, 6, and 7 from Table 1.1, we convert differential equation of motion (15.3.3) with the initial conditions of motion (15.3.2) from the time domain into the Laplace domain. The resulting algebraic equation of motion in the Laplace domain reads:

$$l^2 x(l) - l v_0 - l^2 s_0 + 2nl x(l) - 2nl s_0 + \omega_0^2 x(l)$$

$$= \frac{a\omega_1 l}{l^2 + \omega_1^2} \cos \lambda + \frac{al^2}{l^2 + \omega_1^2} \sin \lambda \qquad (15.3.7)$$

Solving equation (15.3.7) for the displacement $x(l)$ in the Laplace domain, we obtain:

$$x(l) = \frac{l(v_0 + 2ns_0)}{l^2 + 2nl + \omega_0^2} + \frac{l^2 s_0}{l^2 + 2nl + \omega_0^2} + \frac{a\omega_1 l}{(l^2 + \omega_1^2)(l^2 + 2nl + \omega_0^2)} \cos \lambda$$

$$+ \frac{al^2}{(l^2 + \omega_1^2)(l^2 + 2nl + \omega_0^2)} \sin \lambda \qquad (15.3.8)$$

Transforming the denominators in equation (15.3.8), we may write:

$$l^2 + 2nl + \omega_0^2 + n^2 - n^2 = (l+n)^2 + \omega^2 \qquad (15.3.9)$$

where:

$$\omega^2 = \omega_0^2 - n^2 \qquad (15.3.10)$$

while ω^2 could be positive, equal to zero, or negative.
   All these three cases are considered below.

a. Case $\omega^2 > 0$
   Based on the equation (15.3.9), we adjust the denominators in the equation (15.3.8) and we may write:

$$x(l) = \frac{l(v_0 + 2ns_0)}{(l+n)^2 + \omega^2} + \frac{l^2 s_0}{(l+n)^2 + \omega^2} + \frac{a\omega_1 l \cos \lambda}{(l^2 + \omega_1^2)[(l+n)^2 + \omega^2]}$$

$$+ \frac{al^2 \sin \lambda}{(l^2 + \omega_1^2)[(l+n)^2 + \omega^2]} \qquad (15.3.11)$$

Using Laplace Transform Pairs 1, 24, 27, 42, and 43 from Table 1.1, we invert equation (15.3.11) from the Laplace domain into the time domain and obtain the solution of differential equation of motion (15.3.1) with the initial conditions of motion (15.3.2):

$$x = \frac{1}{\omega}(v_0 + 2ns_0)e^{-nt}\sin\omega t + s_0 e^{-nt}(\cos\omega t - \frac{n}{\omega}\sin\omega t)$$

$$+\frac{a\omega_1\cos\lambda}{(\omega_1^2 - \omega_0^2)^2 + 4n^2\omega_1^2}\{2n[e^{-nt}(\cos\omega t + \frac{n}{\omega}\sin\omega t) - \cos\omega_1 t]$$

$$-(\omega_1^2 - \omega_0^2)(\frac{1}{\omega_1}\sin\omega_1 t - \frac{1}{\omega}e^{-nt}\sin\omega t)\} + \frac{a\sin\lambda}{(\omega_1^2 - \omega_0^2)^2 + 4n^2\omega_1^2}$$

$$\times\{(\omega_1^2 - \omega_0^2)[e^{-nt}(\cos\omega t + \frac{n}{\omega}\sin\omega t) - \cos\omega_1 t]$$

$$+2n\omega_1(\sin\omega_1 t - \frac{\omega_1}{\omega}e^{-nt}\sin\omega t)\} \qquad (15.3.12)$$

b.　Case $\omega^2 = 0$

In equation (15.3.11), equating ω^2 to zero, we have:

$$x(l) = \frac{l(v_0 + 2ns_0)}{(l+n)^2} + \frac{l^2 s_0}{(l+n)^2} + \frac{a\omega_1 l}{(l^2 + \omega_1^2)(l+n)^2}\cos\lambda$$

$$+\frac{al^2}{(l^2 + \omega_1^2)(l+n)^2}\sin\lambda \qquad (15.3.13)$$

Applying pairs 1, 37, 38, 46, and 47 from Table 1.1, we invert equation (15.3.13) from the Laplace domain into the time domain and obtain the solution of differential equation of motion (15.3.1) with the initial conditions of motion (15.3.2):

$$x = (v_0 + 2ns_0)te^{-nt} + s_0(1 - nt)e^{-nt}$$

$$+\frac{a\omega_1\cos\lambda}{4n^2\omega_1^2 + (\omega_1^2 - n^2)^2}\{2n\left[e^{-nt}(1 + nt) - \cos\omega_1 t\right]$$

$$-(\omega_1^2 - n^2)(\frac{1}{\omega_1}\sin\omega_1 t - te^{-nt})\} + \frac{a\sin\lambda}{4n^2\omega_1^2 + (\omega_1^2 - n^2)^2}$$

$$\times\{2n\omega_1(\sin\omega_1 t - \omega_1 te^{-nt}) - (\omega_1^2 - n^2)[e^{-nt}(1+nt) - \cos\omega_1 t]\}$$

$$(15.3.14)$$

c. Case $\omega^2 < 0$

The equation for the displacement in the Laplace domain in this case reads:

$$x(l) = \frac{l(v_0 + 2ns_0)}{(l+n)^2 - \omega^2} + \frac{l^2 s_0}{(l+n)^2 - \omega^2} + \frac{a\omega_1 l}{(l^2 + \omega_1^2)[(l+n)^2 - \omega^2]}\cos\lambda$$

$$+ \frac{al^2}{(l^2 + \omega_1^2)[(l+n)^2 - \omega^2]}\sin\lambda \qquad (15.3.15)$$

Using pairs 1, 25, 28, 48, and 49 from Table 1.1, we invert equation (15.3.15) from the Laplace domain into the time domain and obtain the solution of differential equation of motion (15.3.1) with the initial conditions of motion (15.3.2):

$$x = \frac{v_0 + 2ns_0}{\omega}e^{-nt}\sinh\omega t + \frac{s_0}{\omega}e^{-nt}(\omega\cosh\omega t - n\sinh\omega t)$$

$$+ \frac{a\omega_1\cos\lambda}{4n^2\omega_1^2 + (\omega_1^2 - \omega_0^2)^2}\{e^{-nt}[2n\cosh\omega t + \frac{1}{\omega}(\omega_1^2 + \omega^2)\sinh\omega t]$$

$$-2n\cos\omega_1 t - \frac{1}{\omega_1}(\omega_1^2 - \omega_0^2)\sin\omega_1 t\} + \frac{(\omega_1^2 - \omega_0^2)a\sin\lambda}{4n^2\omega_1^2 + (\omega_1^2 - \omega_0^2)^2}$$

$$\times\{\frac{\omega_1^2}{\omega^2}(1 - \cos\omega_1 t) - \frac{\omega_1^2}{n^2 - \omega^2}[1 - e^{-nt}(\cosh\omega t - \frac{n}{\omega}\sinh\omega t)]$$

$$+ \frac{2n\omega_1}{\omega_1^2 - \omega_0^2}(\sin\omega_1 t - \frac{\omega_1}{\omega}e^{-nt}\sinh\omega t)\} \qquad (15.3.16)$$

15.3.2 Initial Displacement Equals Zero

The initial conditions of motion are:

$$\text{for} \quad t = 0 \quad x = 0; \quad \frac{dx}{dt} = v_0 \qquad (15.3.17)$$

The solutions of differential equation of motion (15.3.1) with the initial conditions of motion (15.3.17) for the following cases are presented below.

a. Case $\omega^2 > 0$

$$x = \frac{1}{\omega} v_0 e^{-nt} \sin \omega t + \frac{a\omega_1 \cos \lambda}{(\omega_1^2 - \omega_0^2)^2 + 4n^2\omega_1^2}$$

$$\times \{2n[e^{-nt}(\cos \omega t + \frac{n}{\omega}\sin \omega t) - \cos \omega_1 t]$$

$$-(\omega_1^2 - \omega_0^2)(\frac{1}{\omega_1}\sin \omega_1 t - \frac{1}{\omega}e^{-nt}\sin \omega t)\}$$

$$+\frac{a \sin \lambda}{(\omega_1^2 - \omega_0^2)^2 + 4n^2\omega_1^2}\{(\omega_1^2 - \omega_0^2)[e^{-nt}(\cos \omega t + \frac{n}{\omega}\sin \omega t) - \cos \omega_1 t]$$

$$+2n\omega_1(\sin \omega_1 t - \frac{\omega_1}{\omega}e^{-nt}\sin \omega t)\} \qquad (15.3.18)$$

b. Case $\omega^2 = 0$

$$x = v_0 t e^{-nt} + \frac{a\omega_1 \cos \lambda}{4n^2\omega_1^2 + (\omega_1^2 - n^2)^2}\{2n[e^{-nt}(1 + nt) - \cos \omega_1 t]$$

$$-(\omega_1^2 - n^2)(\frac{1}{\omega_1}\sin \omega_1 t - te^{-nt})\} + \frac{a \sin \lambda}{4n^2\omega_1^2 + (\omega_1^2 - n^2)^2}$$

$$\times \{2n\omega_1(\sin \omega_1 t - \omega_1 te^{-nt}) - (\omega_1^2 - n^2)[e^{-nt}(1 + nt) - \cos \omega_1 t]\}$$

$$(15.3.19)$$

c. Case $\omega^2 < 0$

$$x = \frac{v_0}{\omega} e^{-nt} \sinh \omega t$$

$$+ \frac{a\omega_1 \cos \lambda}{4n^2\omega_1^2 + (\omega_1^2 - \omega_0^2)^2} \{e^{-nt}[2n\cosh\omega t + \frac{1}{\omega}(\omega_1^2 + \omega^2)\sinh\omega t]$$

$$- 2n\cos\omega_1 t - \frac{1}{\omega_1}(\omega_1^2 - \omega_0^2)\sin\omega_1 t\}$$

$$+ \frac{(\omega_1^2 - \omega_0^2)a\sin\lambda}{4n^2\omega_1^2 + (\omega_1^2 - \omega_0^2)^2} \{\frac{\omega_1^2}{\omega^2}(1 - \cos\omega_1 t)$$

$$- \frac{\omega_1^2}{n^2 - \omega^2}[1 - e^{-nt}(\cosh\omega t - \frac{n}{\omega}\sinh\omega t)]$$

$$+ \frac{2n\omega_1}{\omega_1^2 - \omega_0^2}(\sin\omega_1 t - \frac{\omega_1}{\omega}e^{-nt}\sinh\omega t)\}$$ **(15.3.20)**

15.3.3 Initial Velocity Equals Zero

The initial conditions of motion are:

$$\text{for} \quad t = 0 \quad x = s_0; \quad \frac{dx}{dt} = 0 \qquad \textbf{(15.3.21)}$$

The solutions of differential equation of motion (15.3.1) with the initial conditions of motion (15.3.21) for the following cases are presented below.

a. Case $\omega^2 > 0$

$$x = \frac{1}{\omega} 2ns_0 e^{-nt} \sin\omega t + s_0 e^{-nt}(\cos\omega t - \frac{n}{\omega}\sin\omega t)$$

$$+ \frac{a\omega_1 \cos\lambda}{(\omega_1^2 - \omega_0^2)^2 + 4n^2\omega_1^2} \{2n[e^{-nt}(\cos\omega t + \frac{n}{\omega}\sin\omega t) - \cos\omega_1 t]$$

$$- (\omega_1^2 - \omega_0^2)(\frac{1}{\omega_1}\sin\omega_1 t - \frac{1}{\omega}e^{-nt}\sin\omega t)\}$$

$$+\frac{a\sin\lambda}{(\omega_1^2-\omega_0^2)^2+4n^2\omega_1^2}\{(\omega_1^2-\omega_0^2)[e^{-nt}(\cos\omega t+\frac{n}{\omega}\sin\omega t)-\cos\omega_1 t]$$

$$+2n\omega_1(\sin\omega_1 t-\frac{\omega_1}{\omega}e^{-nt}\sin\omega t)\} \tag{15.3.22}$$

b. Case $\omega^2 = 0$

$$x=2ns_0 te^{-nt}+s_0(1-nt)e^{-nt}+\frac{a\omega_1\cos\lambda}{4n^2\omega_1^2+(\omega_1^2-n^2)^2}$$

$$\times\{2n\left[e^{-nt}(1+nt)-\cos\omega_1 t\right]-(\omega_1^2-n^2)(\frac{1}{\omega_1}\sin\omega_1 t-te^{-nt})\}$$

$$+\frac{a\sin\lambda}{4n^2\omega_1^2+(\omega_1^2-n^2)^2}\{2n\omega_1(\sin\omega_1 t-\omega_1 te^{-nt})$$

$$-(\omega_1^2-n^2)[e^{-nt}(1+nt)-\cos\omega_1 t]\} \tag{15.3.23}$$

c. Case $\omega^2 < 0$

$$x=\frac{2ns_0}{\omega}\sinh\omega t+\frac{s_0}{\omega}e^{-nt}(\omega\cosh\omega t-n\sinh\omega t)$$

$$+\frac{a\omega_1\cos\lambda}{4n^2\omega_1^2+(\omega_1^2-\omega_0^2)^2}\{e^{-nt}[2n\cosh\omega t$$

$$+\frac{1}{\omega}(\omega_1^2+\omega^2)\sinh\omega t]-2n\cos\omega_1 t-\frac{1}{\omega_1}(\omega_1^2-\omega_0^2)\sin\omega_1 t\}$$

$$+\frac{(\omega_1^2-\omega_0^2)a\sin\lambda}{4n^2\omega_1^2+(\omega_1^2-\omega_0^2)^2}\{\frac{\omega_1^2}{\omega^2}(1-\cos\omega_1 t)$$

$$-\frac{\omega_1^2}{n^2-\omega^2}[1-e^{-nt}(\cosh\omega t-\frac{n}{\omega}\sinh\omega t)]$$

$$+\frac{2n\omega_1}{\omega_1^2-\omega_0^2}(\sin\omega_1 t-\frac{\omega_1}{\omega}e^{-nt}\sinh\omega t)\} \tag{15.3.24}$$

15.3.4 Both the Initial Displacement and Velocity Equal Zero

The initial conditions of motion are:

$$\text{for} \quad t = 0 \quad x = 0; \quad \frac{dx}{dt} = 0 \qquad (15.3.25)$$

The solutions of differential equation of motion (15.3.1) with the initial conditions of motion (15.3.25) for the following cases are presented below.

a. Case $\omega^2 > 0$

$$x = \frac{a\omega_1 \cos\lambda}{(\omega_1^2 - \omega_0^2)^2 + 4n^2\omega_1^2} \{2n[e^{-nt}(\cos\omega t + \frac{n}{\omega}\sin\omega t) - \cos\omega_1 t]$$

$$-(\omega_1^2 - \omega_0^2)(\frac{1}{\omega_1}\sin\omega_1 t - \frac{1}{\omega}e^{-nt}\sin\omega t)\}$$

$$+\frac{a\sin\lambda}{(\omega_1^2 - \omega_0^2)^2 + 4n^2\omega_1^2} \{(\omega_1^2 - \omega_0^2)[e^{-nt}(\cos\omega t + \frac{n}{\omega}\sin\omega t)$$

$$-\cos\omega_1 t] + 2n\omega_1(\sin\omega_1 t - \frac{\omega_1}{\omega}e^{-nt}\sin\omega t)\} \qquad (15.3.26)$$

b. Case $\omega^2 = 0$

$$x = \frac{a\omega_1 \cos\lambda}{4n^2\omega_1^2 + (\omega_1^2 - n^2)^2} \{2n[e^{-nt}(1+nt) - \cos\omega_1 t]$$

$$-(\omega_1^2 - n^2)(\frac{1}{\omega_1}\sin\omega_1 t - te^{-nt})\}$$

$$+\frac{a\sin\lambda}{4n^2\omega_1^2 + (\omega_1^2 - n^2)^2} \{2n\omega_1(\sin\omega_1 t - \omega_1 te^{-nt})$$

$$-(\omega_1^2 - n^2)[e^{-nt}(1+nt) - \cos\omega_1 t]\} \qquad (15.3.27)$$

c. Case $\omega^2 < 0$

$$x = \frac{a\omega_1 \cos\lambda}{4n^2\omega_1^2 + (\omega_1^2 - \omega_0^2)^2} \{e^{-nt}[2n\cosh\omega t + \frac{1}{\omega}(\omega_1^2 + \omega^2)\sinh\omega t]$$

$$-2n\cos\omega_1 t - \frac{1}{\omega_1}(\omega_1^2 - \omega_0^2)\sin\omega_1 t\}$$

$$+\frac{(\omega_1^2 - \omega_0^2)a\sin\lambda}{4n^2\omega_1^2 + (\omega_1^2 - \omega_0^2)^2}\{\frac{\omega_1^2}{\omega^2}(1 - \cos\omega_1 t)$$

$$-\frac{\omega_1^2}{n^2 - \omega^2}[1 - e^{-nt}(\cosh\omega t - \frac{n}{\omega}\sinh\omega t)]$$

$$+\frac{2n\omega_1}{\omega_1^2 - \omega_0^2}(\sin\omega_1 t - \frac{\omega_1}{\omega}e^{-nt}\sinh\omega t)\} \qquad (15.3.28)$$

15.4 Time-Dependent Force $Q\left(\rho + \frac{\mu t}{\tau}\right)$

This section describes engineering systems subjected to the force of inertia, the damping force, and the stiffness force as the resisting forces (Row 15 of Guiding Table 2.1) and the time-dependent force as the active force (Column 4). The current problem could be associated with the working processes of engineering systems interacting with viscoelastoplastic media that exert damping forces and stiffness forces as the reaction to their deformation (see section 1.2). Sometimes during the initial phase of the working process, the system is subjected to a time-dependent force that is lasting a predetermined time.

The system is moving in the horizontal direction. We want to determine the system's law of motion. Figure 15.4 shows the model of a system subjected to the action of a time-dependent force, a damping force, and a stiffness force.

Based on the considerations above and the model in Figure 15.4, we can compose the left and right sides of the differential equation of motion. The left side includes the force of inertia, the damping force and the stiffness force, while the right side of this equation consists of the time-dependent force. Hence, the differential equation of motion reads:

$$m\frac{d^2x}{dt^2} + C\frac{dx}{dt} + Kx = Q(\rho + \frac{\mu t}{\tau}) \qquad (15.4.1)$$

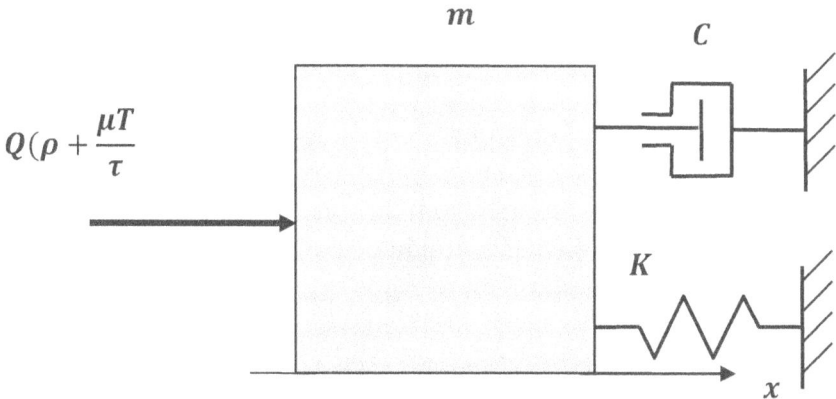

Figure 15.4 Model of a system subjected to a time-dependent force, a damping force, and a stiffness force

Differential equation of motion (15.4.1) has different solutions for various initial conditions of motion. These solutions are presented below.

15.4.1 General Initial Conditions

The general initial conditions of motion are:

$$\text{for}\quad t=0\quad x=s_0;\quad \frac{dx}{dt}=v_0 \tag{15.4.2}$$

where s_0 and v_0 are the initial displacement and initial velocity respectively.

Dividing equation (15.4.1) by m, we obtain:

$$\frac{d^2x}{dt^2}+2n\frac{dx}{dt}+\omega_0^2 x = q(\rho+\frac{\mu t}{\tau}) \tag{15.4.3}$$

where n is the damping factor and ω_0 is the natural frequency, while:

$$2n=\frac{C}{m} \tag{15.4.4}$$

$$\omega_0^2=\frac{K}{m} \tag{15.4.5}$$

$$q = \frac{Q}{m} \qquad \textbf{(15.4.6)}$$

Using Laplace Transform Pairs 3, 4, 5, and 2 from Table 1.1, we convert differential equation of motion (15.4.3) with the initial conditions of motion (15.4.2) from the time domain into the Laplace domain, and obtain the corresponding algebraic equation of motion in the Laplace domain:

$$l^2 x(l) - l v_0 - l^2 s_0 + 2nl x(l) - 2nl s_0 + \omega_0^2 x(l) = q\rho + \frac{q\mu}{\tau l} \qquad \textbf{(15.4.7)}$$

Solving equation (15.4.7) for the displacement $x(l)$ in the Laplace domain, we have:

$$x(l) = \frac{l(v_0 + 2ns_0)}{l^2 + 2nl + \omega_0^2} + \frac{l^2 s_0}{l^2 + 2nl + \omega_0^2}$$

$$+ \frac{q\rho}{l^2 + 2nl + \omega_0^2} + \frac{q\mu}{\tau l(l^2 + 2nl + \omega_0^2)} \qquad \textbf{(15.4.8)}$$

Transforming the denominators in equation (15.4.8), we may write:

$$l^2 + 2nl + \omega_0^2 + n^2 - n^2 = (l+n)^2 + \omega^2 \qquad \textbf{(15.4.9)}$$

where:

$$\omega^2 = \omega_0^2 - n^2 \qquad \textbf{(15.4.10)}$$

while ω^2 could be positive, equal to zero, or negative.
All these three cases are considered below.

a. Case $\omega^2 > 0$

Adjusting the denominators in equation (15.4.8) according to equation (15.4.9), we may write:

$$x(l) = \frac{l(v_0 + 2ns_0)}{(l+n)^2 + \omega^2} + \frac{l^2 s_0}{(l+n)^2 + \omega^2}$$

$$+ \frac{q\rho}{(l+n)^2 + \omega^2} + \frac{q\mu}{l\tau[(l+n)^2 + \omega^2)]} \qquad \textbf{(15.4.11)}$$

Using pairs 1, 24, 27, 22 and 39 from Table 1.1, we invert equation (15.4.11) from the Laplace domain into the time domain and obtain the solution of differential equation of motion (15.4.1) with the initial conditions of motion (15.4.2):

$$x = \frac{v_0 + 2ns_0}{\omega}e^{-nt}\sin\omega t + s_0(\cos\omega t - \frac{n}{\omega}\sin\omega t)e^{-nt}$$

$$+\frac{q\rho}{\omega^2 + n^2}[1 - e^{-nt}(\cos\omega t + \frac{n}{\omega}\sin\omega t)]$$

$$+\frac{q\mu}{\omega\tau(\omega^2 + n^2)^2}\{(\omega^2 + n^2)\omega t - 2n\omega$$

$$-e^{-nt}[(\omega^2 - n^2)\sin\omega t - 2n\omega\cos\omega t]\} \qquad (15.4.12)$$

b. Case $\omega^2 = 0$

Equating ω^2 in equation (15.4.11) to zero, we obtain:

$$x(l) = \frac{l(v_0 + 2ns_0)}{(l+n)^2} + \frac{l^2 s_0}{(l+n)^2} + \frac{q\rho}{(l+n)^2} + \frac{q\mu}{l\tau(l+n)^2} \qquad (15.4.13)$$

Based on pairs 1, 37, 38, 36, 15 from Table 1.1, we invert equation (15.4.13) from the Laplace domain into the time domain and obtain the solution of differential equation of motion (15.4.1) with the initial conditions of motion (15.4.2):

$$x = (v_0 + 2ns_0)te^{-nt} + s_0(1 - nt)e^{-nt} + \frac{q\rho}{n^2}[1 - e^{-nt}(1 + nt)]$$

$$+\frac{q\mu}{\tau n^2}[t - \frac{2}{n} + e^{-nt}(\frac{2}{n} + t)] \qquad (15.4.14)$$

c. Case $\omega^2 < 0$

Taking ω^2 in equation (15.4.11) with the negative sign, we may write:

$$x(l) = \frac{l(v_0 + 2ns_0)}{(l+n)^2 - \omega^2} + \frac{l^2 s_0}{(l+n)^2 - \omega^2}$$

$$+\frac{q\rho}{(l+n)^2-\omega^2}+\frac{q\mu}{l\tau[(l+n)^2-\omega^2]} \tag{15.4.15}$$

Applying pairs 1, 25, 28, 23, and 40 from Table 1.1, we invert equation (15.4.15) from the Laplace domain into the time domain and obtain the solution of differential equation of motion (15.4.1) with the initial conditions of motion (15.4.2):

$$x=\frac{v_0+2ns_0}{\omega}e^{-nt}\sinh\omega t+\frac{s_0}{\omega}e^{-nt}(\omega\cosh\omega t-n\sinh\omega t)$$

$$+\frac{q\rho}{n^2-\omega^2}[1-e^{-nt}(\cosh\omega t+\frac{n}{\omega}\sinh\omega t)]$$

$$+\frac{q\mu}{\omega\tau(n^2-\omega^2)^2}[(n^2-\omega^2)(\omega t+e^{-nt}\sinh\omega t)$$

$$-2n\omega(1-\cosh\omega t)] \tag{15.4.16}$$

15.4.2 Initial Displacement Equals Zero
The initial conditions of motion are:

$$\text{for}\quad t=0\quad x=0;\quad \frac{dx}{dt}=v_0 \tag{15.4.17}$$

The solutions of differential equation of motion (15.4.1) with the initial conditions of motion (15.4.17) for the following cases are presented below.

a. Case $\omega^2>0$

$$x=\frac{v_0}{\omega}e^{-nt}\sin\omega t+\frac{q\rho}{\omega^2+n^2}[1-e^{-nt}(\cos\omega t+\frac{n}{\omega}\sin\omega t)]$$

$$+\frac{q\mu}{\omega\tau(\omega^2+n^2)^2}\{(\omega^2+n^2)\omega t-2n\omega$$

$$-e^{-nt}[(\omega^2-n^2)\sin\omega t-2n\omega\cos\omega t]\} \tag{15.4.18}$$

b. Case $\omega^2 = 0$

$$x = v_0 t e^{-nt} + \frac{q\rho}{n^2}[1 - e^{-nt}(1+nt)] + \frac{q\mu}{\tau n^2}[t - \frac{2}{n} + e^{-nt}(\frac{2}{n} + t)]$$

(15.4.19)

c. Case $\omega^2 < 0$

$$x = \frac{v_0}{\omega}e^{-nt}\sinh\omega t + \frac{q\rho}{n^2 - \omega^2}[1 - e^{-nt}(\cosh\omega t + \frac{n}{\omega}\sinh\omega t]$$

$$+ \frac{q\mu}{\omega\tau(n^2 - \omega^2)^2}[(n^2 - \omega^2)(\omega t + e^{-nt}\sinh\omega t)$$

$$-2n\omega(1 - \cosh\omega t)]$$

(15.4.20)

15.4.3 Initial Velocity Equals Zero

The initial conditions of motion are:

$$\text{for} \quad t = 0 \quad x = s_0; \quad \frac{dx}{dt} = 0$$

(15.4.21)

The solutions of differential equation of motion (15.4.1) with the initial conditions of motion (15.4.21) for the following cases are presented below.

a. Case $\omega^2 > 0$

$$x = \frac{2ns_0}{\omega}e^{-nt}\sin\omega t + s_0(\cos\omega t - \frac{n}{\omega}\sin\omega t)e^{-nt}$$

$$+ \frac{q\rho}{\omega^2 + n^2}[1 - e^{-nt}(\cos\omega t + \frac{n}{\omega}\sin\omega t)]$$

$$+ \frac{q\mu}{\omega\tau(\omega^2 + n^2)^2}\{(\omega^2 + n^2)\omega t - 2n\omega$$

$$- e^{-nt}[(\omega^2 - n^2)\sin\omega t - 2n\omega\cos\omega t]\}$$

(15.4.22)

b. Case $\omega^2 = 0$

$$x = 2ns_0te^{-nt} + s_0(1-nt)e^{-nt} + \frac{q\rho}{n^2}\left[1-e^{-nt}(1+nt)\right]$$

$$+\frac{q\mu}{\tau n^2}[t-\frac{2}{n}+e^{-nt}\left(\frac{2}{n}+t\right)]] \qquad (15.4.23)$$

c. Case $\omega^2 < 0$

$$x = \frac{2ns_0}{\omega}e^{-nt}\sinh\omega t + \frac{s_0}{\omega}e^{-nt}(\omega\cosh\omega t - n\sinh\omega t)$$

$$+\frac{q\rho}{n^2-\omega^2}[1-e^{-nt}(\cosh\omega t + \frac{n}{\omega}\sinh\omega t)]$$

$$+\frac{q\mu}{\omega\tau(n^2-\omega^2)^2}[(n^2-\omega^2)(\omega t + e^{-nt}\sinh\omega t)$$

$$-2n\omega(1-\cosh\omega t)] \qquad (15.4.24)$$

15.4.4 Both the Displacement and Velocity Equal Zero

The initial conditions of motion are:

$$\text{for} \quad t = 0 \quad x = 0; \quad \frac{dx}{dt} = 0 \qquad (15.4.25)$$

The solutions of differential equation of motion (15.4.1) with initial conditions of motion (15.4.25) for the following cases are presented below.

a. Case $\omega^2 > 0$

$$x = \frac{q\rho}{\omega^2+n^2}[1-e^{-nt}(\cos\omega t + \frac{n}{\omega}\sin\omega t)]$$

$$+\frac{q\mu}{\omega\tau(\omega^2+n^2)^2}\{(\omega^2+n^2)\omega t - 2n\omega$$

$$-e^{-nt}[(\omega^2-n^2)\sin\omega t - 2n\omega\cos\omega t]\} \qquad (15.4.26)$$

b. Case $\omega^2 = 0$

$$x = \frac{q\rho}{n^2}\left[1 - e^{-nt}(1+nt)\right] + \frac{q\mu}{n^2}\left[t - \frac{2}{n} + e^{-nt}\left(\frac{2}{n} + t\right)\right] \qquad (15.4.27)$$

c. Case $\omega^2 < 0$

$$x = \frac{q\rho}{n^2 - \omega^2}\left[1 - e^{-nt}(\cosh\omega t + \frac{n}{\omega}\sinh\omega t)\right]$$

$$+ \frac{q\mu}{\omega\tau(n^2 - \omega^2)^2}\left[(n^2 - \omega^2)(\omega t + e^{-nt}\sinh\omega t)\right.$$

$$-2n\omega(1 - \cosh\omega t)] \qquad (15.4.28)$$

15.5 Constant Force R and Harmonic Force $A\sin(\omega_1 t + \lambda)$

The intersection of Row 15 and Column 5 of Guiding Table 2.1 indicates that this section describes engineering systems subjected to the force of inertia, the damping force, and the stiffness force as the resisting forces and to the constant active force and the harmonic force as the active forces. The problem could be associated with the working processes of vibratory systems intended for the deformation of viscoelastoplastic media that exert the damping and stiffness forces representing their reaction to the deformation. Additional considerations related to the deformation of viscoelastoplastic media are presented in section 1.2.

The system is moving in the horizontal direction. We want to determine the law of motion of the system. Figure 15.5 shows the model of a system subjected to the action of a constant active force, harmonic force, damping force, and stiffness force.

The considerations above and the model shown in Figure 15.5 let us compose the left and right sides of the differential equation of motion of the system. The left side consists of the force of inertia, the damping force, and the stiffness force. The right side of this

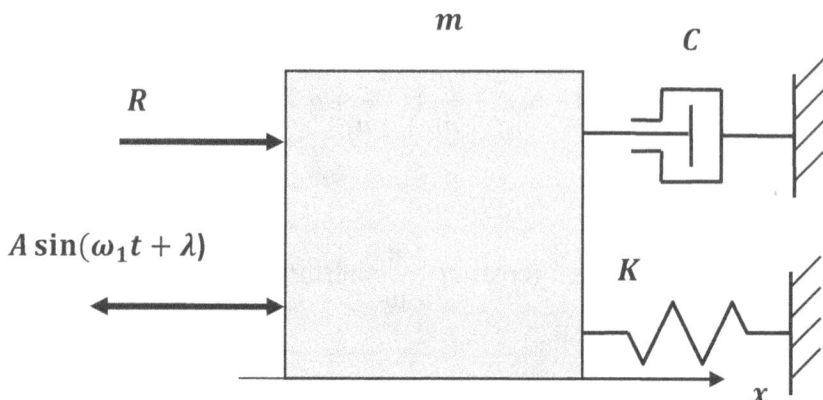

Figure 15.5 Model of a system subjected to a constant active force, a harmonic force, a damping force, and a stiffness force

equation includes the constant active force and the harmonic force. Therefore, the differential equation of motion reads:

$$m\frac{d^2x}{dt^2} + C\frac{dx}{dt} + Kx = R + A\sin(\omega_1 t + \lambda) \qquad (15.5.1)$$

Differential equation of motion (15.5.1) has different solutions for various initial conditions of motion. These solutions are presented below.

15.5.1 General Initial Conditions

The general initial conditions of motion are:

$$\text{for} \quad t = 0 \quad x = s_0; \quad \frac{dx}{dt} = v_0 \qquad (15.5.2)$$

Transforming the sinusoidal function in equation (15.5.1) and dividing the latter by m, we obtain:

$$\frac{d^2x}{dt^2} + 2n\frac{dx}{dt} + \omega_0^2 x = r + a\sin\omega_1 t\cos\lambda + a\cos\omega_1 t\sin\lambda \qquad (15.5.3)$$

where n is the damping factor and ω_0 in the natural frequency, while:

$$2n = \frac{C}{m} \tag{15.5.4}$$

$$\omega_0^2 = \frac{K}{m} \tag{15.5.5}$$

$$r = \frac{R}{m} \tag{15.5.6}$$

$$a = \frac{A}{m} \tag{15.5.7}$$

Based on Laplace Transform Pairs 3, 4, 1, 5, 6, and 7 from Table 1.1, we convert differential equation of motion (15.5.3) with the initial conditions of motion (15.5.2) from the time domain into the Laplace domain, and obtain the corresponding algebraic equation of motion in the Laplace domain:

$$l^2 x(l) - l v_0 - l^2 s_0 + 2nlx(l) - 2nls_0 + \omega_0^2 x(l)$$

$$= r + \frac{a\omega_1 l \cos \lambda}{l^2 + \omega_1^2} + \frac{al^2 \sin \lambda}{l^2 + \omega_1^2} \tag{15.5.8}$$

Applying basic algebra to equation (15.5.8), we may write:

$$x(l)(l^2 + 2n + \omega_0^2)$$

$$= r + l(v_0 + 2ns_0) + l^2 s_0 + \frac{a\omega_1 l \cos \lambda}{l^2 + \omega_1^2} + \frac{al^2 \sin \lambda}{l^2 + \omega_1^2} \tag{15.5.9}$$

The solution of equation (15.5.9) for the displacement $x(l)$ in Laplace domain reads:

$$x(l) = \frac{r}{l^2 + 2nl + \omega_0^2} + \frac{l(v_0 + 2ns_0)}{l^2 + 2nl + \omega_0^2} + \frac{l^2 s_0}{l^2 + 2nl + \omega_0^2}$$

$$+ \frac{a\omega_1 l \cos \lambda}{(l^2 + \omega_1^2)(l^2 + 2nl + \omega_0^2)} + \frac{al^2 \sin \lambda}{(l^2 + \omega_1^2)(l^2 + 2nl + \omega_0^2)} \tag{15.5.10}$$

Transforming the denominators in equation (15.5.10), we may write:

$$l^2 + 2nl + \omega_0^2 + n^2 - n^2 = (l+n)^2 + \omega^2 \qquad (15.5.11)$$

where:

$$\omega^2 = \omega_0^2 - n^2 \qquad (15.5.12)$$

while ω^2 could be positive, equal to zero, or negative.
All these three cases are considered below.

a. Case $\omega^2 > 0$

Combining equations (15.3.8) and (15.3.9), we write:

$$x(l) = \frac{r}{(l+n)^2 + \omega^2} + \frac{l(v_0 + 2ns_0)}{(l+n)^2 + \omega^2} + \frac{l^2 s_0}{(l+n)^2 + \omega^2}$$

$$+ \frac{a\omega_1 l \cos \lambda}{(l^2 + \omega_1^2)[(l+n)^2 + \omega^2]} + \frac{al^2 \sin \lambda}{(l^2 + \omega_1^2)[(l+n)^2 + \omega^2]} \qquad (15.5.13)$$

Using pairs 1, 22, 24, 27, 42, and 43 from Table 1.1, we invert equation (15.5.13) from the Laplace domain into the time domain and obtain the solution of differential equation of motion (15.5.1) with the initial conditions of motion (15.2.2):

$$x = \frac{r}{\omega_0^2}[1 - e^{-nt}(\cos \omega t + \frac{n}{\omega} \sin \omega t)] + \frac{1}{\omega}(v_0 + 2ns_0)e^{-nt} \sin \omega t$$

$$+ s_0 e^{-nt}(\cos \omega t - \frac{n}{\omega} \sin \omega t)$$

$$+ \frac{a\omega_1 \cos \lambda}{(\omega_1^2 - \omega_0^2)^2 + 4n^2\omega_1^2}\{2n[e^{-nt}(\cos \omega t + \frac{n}{\omega} \sin \omega t) - \cos \omega_1 t]$$

$$- (\omega_1^2 - \omega_0^2)(\frac{1}{\omega_1} \sin \omega_1 t - \frac{1}{\omega} e^{-nt} \sin \omega t)\}$$

$$+ \frac{a \sin \lambda}{(\omega_1^2 - \omega_0^2)^2 + 4n^2\omega_1^2}\{(\omega_1^2 - \omega_0^2)[e^{-nt}(\cos \omega t + \frac{n}{\omega} \sin \omega t)$$

$$-\cos\omega_1 t] + 2n\omega_1 (\sin\omega_1 t - \frac{\omega_1}{\omega} e^{-nt} \sin\omega t)\} \qquad (15.5.14)$$

b. Case $\omega^2 = 0$

Equating ω^2 to zero in the equation (15.5.13), we have:

$$x(l) = \frac{r}{(l+n)^2} + \frac{l(v_0 + 2ns_0)}{(l+n)^2} + \frac{l^2 s_0}{(l+n)^2}$$

$$+ \frac{a\omega_1 l \cos\lambda}{(l^2 + \omega_1^2)(l+n)^2} + \frac{al^2 \sin\lambda}{(l^2 + \omega_1^2)(l+n)^2} \qquad (15.5.15)$$

Using pairs 1, 36, 37, 38, 46, and 47 from Table 1.1, we invert equation (15.5.15) from the Laplace domain into the time domain and obtain the solution of differential equation of motion (15.5.1) with the initial conditions of motion (15.5.2):

$$x = \frac{r}{n^2}\left[1 - e^{-nt}(1+nt)\right] + (v_0 + 2ns_0)te^{-nt}$$

$$+ s_0(1-nt)e^{-nt} + \frac{a\omega_1 \cos\lambda}{4n^2\omega_1^2 + (\omega_1^2 - n^2)^2}\left\{2n\left[e^{-nt}(1+nt) - \cos\omega_1 t\right]\right.$$

$$- (\omega_1^2 - n^2)(\frac{1}{\omega_1}\sin\omega_1 t - te^{-nt})\}$$

$$+ \frac{a\sin\lambda}{4n^2\omega_1^2 + (\omega_1^2 - n^2)^2}\left\{2n\omega_1(\sin\omega_1 t - \omega_1 te^{-nt})\right.$$

$$- (\omega_1^2 - n^2)[e^{-nt}(1+nt) - \cos\omega_1 t]\} \qquad (15.5.16)$$

c. Case $\omega^2 < 0$

In equation (15.5.13), taking ω^2 with the negative sign, we may write:

$$x(l) = \frac{r}{(l+n)^2 - \omega^2} + \frac{l(v_0 + 2ns_0)}{(l+n)^2 - \omega^2} + \frac{l^2 s_0}{(l+n)^2 - \omega^2}$$

$$+\frac{a\omega_1 l \cos \lambda}{(l^2 + \omega_1^2)[(l+n)^2 - \omega^2]} + \frac{a l^2 \sin \lambda}{(l^2 + \omega_1^2)[(l+n)^2 - \omega^2]} \quad \textbf{(15.5.17)}$$

Applying pairs 1, 23, 25, 28, 48, and 49 from Table 1.1, we invert equation (15.5.17) from the Laplace domain into the time domain and obtain the solution of differential equation of motion (15.5.1) with the initial conditions of motion (15.2.2):

$$x = \frac{r}{n^2 - \omega^2}[1 - e^{-nt}(\cosh \omega t + \frac{n}{\omega}\sinh \omega t)] + \frac{v_0 + 2ns_0}{\omega}e^{-nt}\sinh \omega t$$

$$+\frac{s_0}{\omega}e^{-nt}(\omega \cosh \omega t - n \sinh \omega t)$$

$$+\frac{a\omega_1 \cos \lambda}{4n^2\omega_1^2 + (\omega_1^2 - \omega_0^2)^2}\{e^{-nt}[2n\cosh \omega t + \frac{1}{\omega}(\omega_1^2 + \omega^2)\sinh \omega t]$$

$$-2n\cos \omega_1 t - \frac{1}{\omega_1}(\omega_1^2 - \omega_0^2)\sin \omega_1 t\}$$

$$+\frac{(\omega_1^2 - \omega_0^2)a\sin \lambda}{4n^2\omega_1^2 + (\omega_1^2 - \omega_0^2)^2}\{\frac{\omega_1^2}{\omega^2}(1 - \cos \omega_1 t)$$

$$-\frac{\omega_1^2}{n^2 - \omega^2}[1 - e^{-nt}(\cosh \omega t - \frac{n}{\omega}\sinh \omega t)]$$

$$+\frac{2n\omega_1}{\omega_1^2 - \omega_0^2}(\sin \omega_1 t - \frac{\omega_1}{\omega}e^{-nt}\sinh \omega t)\} \quad \textbf{(15.5.18)}$$

15.5.2 Initial Displacement Equals Zero

$$\text{For} \quad t = 0 \quad x = 0; \quad \frac{dx}{dt} = v_0 \quad \textbf{(15.5.19)}$$

The solutions of differential equation of motion (15.5.1) with the initial conditions of motion (15.5.19) for the following cases are presented below.

a. Case $\omega^2 > 0$

$$x = \frac{r}{\omega_0^2}[1 - e^{-nt}(\cos\omega t + \frac{n}{\omega}\sin\omega t)] + \frac{1}{\omega}v_0 e^{-nt}\sin\omega t$$

$$+\frac{a\omega_1\cos\lambda}{(\omega_1^2 - \omega_0^2)^2 + 4n^2\omega_1^2}\{2n[e^{-nt}(\cos\omega t + \frac{n}{\omega}\sin\omega t) - \cos\omega_1 t]$$

$$-(\omega_1^2 - \omega_0^2)(\frac{1}{\omega_1}\sin\omega_1 t - \frac{1}{\omega}e^{-nt}\sin\omega t)\} + \frac{a\sin\lambda}{(\omega_1^2 - \omega_0^2)^2 + 4n^2\omega_1^2}$$

$$\times\{(\omega_1^2 - \omega_0^2)[e^{-nt}(\cos\omega t + \frac{n}{\omega}\sin\omega t) - \cos\omega_1 t]$$

$$+2n\omega_1(\sin\omega_1 t - \frac{\omega_1}{\omega}e^{-nt}\sin\omega t)\} \qquad\qquad (15.5.20)$$

b. Case $\omega^2 = 0$

$$x = \frac{r}{n^2}\left[1 - e^{-nt}(1 + nt)\right] + v_0 t e^{-nt}$$

$$+\frac{a\omega_1\cos\lambda}{4n^2\omega_1^2 + (\omega_1^2 - n^2)^2}\{2n\left[e^{-nt}(1 + nt) - \cos\omega_1 t\right]$$

$$-(\omega_1^2 - n^2)(\frac{1}{\omega_1}\sin\omega_1 t - te^{-nt})\}$$

$$+\frac{a\sin\lambda}{4n^2\omega_1^2 + (\omega_1^2 - n^2)^2}\{2n\omega_1(\sin\omega_1 t - \omega_1 te^{-nt})$$

$$-(\omega_1^2 - n^2)[e^{-nt}(1 + nt) - \cos\omega_1 t]\} \qquad\qquad (15.5.21)$$

c. Case $\omega^2 < 0$

$$x = \frac{r}{n^2 - \omega^2}[1 - e^{-nt}(\cosh\omega t + \frac{n}{\omega}\sinh\omega t)] + \frac{v_0}{\omega}e^{-nt}\sinh\omega t$$

$$+\frac{a\omega_1\cos\lambda}{4n^2\omega_1^2 + (\omega_1^2 - \omega_0^2)^2}\{e^{-nt}[2n\cosh\omega t + \frac{1}{\omega}(\omega_1^2 + \omega^2)\sinh\omega t]$$

$$-2n\cos\omega_1 t - \frac{1}{\omega_1}(\omega_1^2 - \omega_0^2)\sin\omega_1 t\} + \frac{(\omega_1^2 - \omega_0^2)a\sin\lambda}{4n^2\omega_1^2 + (\omega_1^2 - \omega_0^2)^2}$$

$$\times\{\frac{\omega_1^2}{\omega^2}(1-\cos\omega_1 t) - \frac{\omega_1^2}{n^2 - \omega^2}[1 - e^{-nt}(\cosh\omega t - \frac{n}{\omega}\sinh\omega t)]$$

$$+\frac{2n\omega_1}{\omega_1^2 - \omega_0^2}(\sin\omega_1 t - \frac{\omega_1}{\omega}e^{-nt}\sinh\omega t)\}\qquad(15.5.22)$$

15.5.3 Initial Velocity Equals Zero

The initial conditions of motion are:

$$\text{for}\quad t = 0\quad x = s_0;\quad \frac{dx}{dt} = 0\qquad(15.5.23)$$

The solutions of differential equation of motion (15.5.1) with the initial conditions of motion (15.5.23) for the following cases are presented below.

a.   Case $\omega^2 > 0$

$$x = \frac{r}{\omega_0^2}[1 - e^{-nt}(\cos\omega t + \frac{n}{\omega}\sin\omega t)] + \frac{1}{\omega}2ns_0 e^{-nt}\sin\omega t$$

$$+s_0 e^{-nt}(\cos\omega t - \frac{n}{\omega}\sin\omega t)$$

$$+\frac{a\omega_1\cos\lambda}{(\omega_1^2 - \omega_0^2)^2 + 4n^2\omega_1^2}\{2n[e^{-nt}(\cos\omega t + \frac{n}{\omega}\sin\omega t) - \cos\omega_1 t]$$

$$-(\omega_1^2 - \omega_0^2)(\frac{1}{\omega_1}\sin\omega_1 t - \frac{1}{\omega}e^{-nt}\sin\omega t)\} + \frac{a\sin\lambda}{(\omega_1^2 - \omega_0^2)^2 + 4n^2\omega_1^2}$$

$$\times\{(\omega_1^2 - \omega_0^2)[e^{-nt}(\cos\omega t + \frac{n}{\omega}\sin\omega t) - \cos\omega_1 t]$$

$$+2n\omega_1(\sin\omega_1 t - \frac{\omega_1}{\omega}e^{-nt}\sin\omega t)\}\qquad(15.5.24)$$

b. Case $\omega^2 = 0$

$$x = \frac{r}{n^2}\left[1 - e^{-nt}(1+nt)\right] + 2ns_0 t e^{-nt}$$

$$+ s_0(1-nt)e^{-nt} + \frac{a\omega_1 \cos\lambda}{4n^2\omega_1^2 + (\omega_1^2 - n^2)^2}$$

$$\times\{2n\left[e^{-nt}(1+nt) - \cos\omega_1 t\right] - (\omega_1^2 - n^2)(\frac{1}{\omega_1}\sin\omega_1 t - te^{-nt})\}$$

$$+ \frac{a\sin\lambda}{4n^2\omega_1^2 + (\omega_1^2 - n^2)^2}\{2n\omega_1(\sin\omega_1 t - \omega_1 te^{-nt})$$

$$-(\omega_1^2 - n^2)[e^{-nt}(1+nt) - \cos\omega_1 t]\} \tag{15.5.25}$$

c. Case $\omega^2 < 0$

$$x = \frac{r}{n^2 - \omega^2}[1 - e^{-nt}(\cosh\omega t + \frac{n}{\omega}\sinh\omega t)] + \frac{2ns_0}{\omega}e^{-nt}\sinh\omega t$$

$$+ \frac{s_0}{\omega}e^{-nt}(\omega\cosh\omega t - n\sinh\omega t)$$

$$+ \frac{a\omega_1 \cos\lambda}{4n^2\omega_1^2 + (\omega_1^2 - \omega_0^2)^2}\{e^{-nt}[2n\cosh\omega t + \frac{1}{\omega_1}(\omega_1^2 + \omega^2)\sinh\omega t]$$

$$-2n\cos\omega_1 t - \frac{1}{\omega_1}(\omega_1^2 - \omega_0^2)\sin\omega_1 t\}$$

$$+ \frac{(\omega_1^2 - \omega_0^2)a\sin\lambda}{4n^2\omega_1^2 + (\omega_1^2 - \omega_0^2)^2}\{\frac{\omega_1^2}{\omega^2}(1 - \cos\omega_1 t)$$

$$- \frac{\omega_1^2}{n^2 - \omega^2}[1 - e^{-nt}(\cosh\omega t - \frac{n}{\omega}\sinh\omega t)]$$

$$+ \frac{2n\omega_1}{\omega_1^2 - \omega_0^2}(\sin\omega_1 t - \frac{\omega_1}{\omega}e^{-nt}\sinh\omega t)\} \tag{15.5.26}$$

15.5.4 Both the Displacement and Velocity Equal Zero

The initial conditions of motion are:

$$\text{for} \quad t = 0 \quad x = 0; \quad \frac{dx}{dt} = 0 \tag{15.5.27}$$

The solutions of differential equation of motion (15.5.1) with the initial conditions of motion (15.5.27) for the following cases are presented below.

a. Case $\omega^2 > 0$

$$x = \frac{r}{\omega_0^2}[1 - e^{-nt}(\cos \omega t + \frac{n}{\omega}\sin \omega t)]$$

$$+ \frac{a\omega_1 \cos \lambda}{(\omega_1^2 - \omega_0^2)^2 + 4n^2\omega_1^2}\{2n[e^{-nt}(\cos \omega t + \frac{n}{\omega}\sin \omega t) - \cos \omega_1 t]$$

$$- (\omega_1^2 - \omega_0^2)(\frac{1}{\omega_1}\sin \omega_1 t - \frac{1}{\omega}e^{-nt}\sin \omega t)\} + \frac{a \sin \lambda}{(\omega_1^2 - \omega_0^2)^2 + 4n^2\omega_1^2}$$

$$\times \{(\omega_1^2 - \omega_0^2)[e^{-nt}(\cos \omega t + \frac{n}{\omega}\sin \omega t) - \cos \omega_1 t]$$

$$+ 2n\omega_1(\sin \omega_1 t - \frac{\omega_1}{\omega}e^{-nt}\sin \omega t)\} \tag{15.5.28}$$

b. Case $\omega^2 = 0$

$$x = \frac{r}{n^2}\left[1 - e^{-nt}(1 + nt)\right] + \frac{a\omega_1 \cos \lambda}{4n^2\omega_1^2 + (\omega_1^2 - n^2)^2}\{2n[e^{-nt}(1 + nt)$$

$$- \cos \omega_1 t] - (\omega_1^2 - n^2)(\frac{1}{\omega_1}\sin \omega_1 t - te^{-nt})\}$$

$$+ \frac{a \sin \lambda}{4n^2\omega_1^2 + (\omega_1^2 - n^2)^2}\{2n\omega_1(\sin \omega_1 t - \omega_1 te^{-nt})$$

$$- (\omega_1^2 - n^2)[e^{-nt}(1 + nt) - \cos \omega_1 t]\} \tag{15.5.29}$$

c. Case $\omega^2 < 0$

$$x = \frac{r}{n^2 - \omega^2}[1 - e^{-nt}(\cosh \omega t + \frac{n}{\omega}\sinh \omega t)]$$

$$+ \frac{a\omega_1 \cos \lambda}{4n^2\omega_1^2 + (\omega_1^2 - \omega_0^2)^2}\{e^{-nt}[2n\cosh \omega t + \frac{1}{\omega}(\omega_1^2 + \omega^2)\sinh \omega t]$$

$$-2n\cos\omega_1 t - \frac{1}{\omega_1}(\omega_1^2 - \omega_0^2)\sin \omega_1 t\}$$

$$+ \frac{(\omega_1^2 - \omega_0^2)a\sin \lambda}{4n^2\omega_1^2 + (\omega_1^2 - \omega_0^2)^2}\{\frac{\omega_1^2}{\omega^2}(1 - \cos \omega_1 t)$$

$$- \frac{\omega_1^2}{n^2 - \omega^2}[1 - e^{-nt}(\cosh \omega t - \frac{n}{\omega}\sinh \omega t)]$$

$$+ \frac{2n\omega_1}{\omega_1^2 - \omega_0^2}(\sin \omega_1 t - \frac{\omega_1}{\omega}e^{-nt}\sinh \omega t)\} \qquad (15.5.30)$$

15.6 Harmonic Force $A \sin(\omega_1 t + \lambda)$ and Time-Dependent Force $Q\left(\rho + \frac{\mu t}{\tau}\right)$

The intersection of Row 15 and Column 6 in Guiding Table 2.1 indicates that this section describes engineering systems subjected to the force of inertia, the damping force, and stiffness force as the resisting forces, and to the harmonic force and time-dependent force as the active forces. The current problem could be associated with the working process of a vibratory system interacting with a visco-elastoplastic medium. A time-dependent force is sometimes applied during the initial phase of the working process. This force is acting along with the harmonic force for just a predetermined interval of time. Additional considerations related to the deformation of visco-elastoplastic media are presented in section 1.2.

The system is moving in the horizontal direction. We want to determine the law of motion of the system. Figure 15.6 shows the model of a system subjected to the action of a harmonic force, a time-dependent force, a damping force, and a stiffness force.

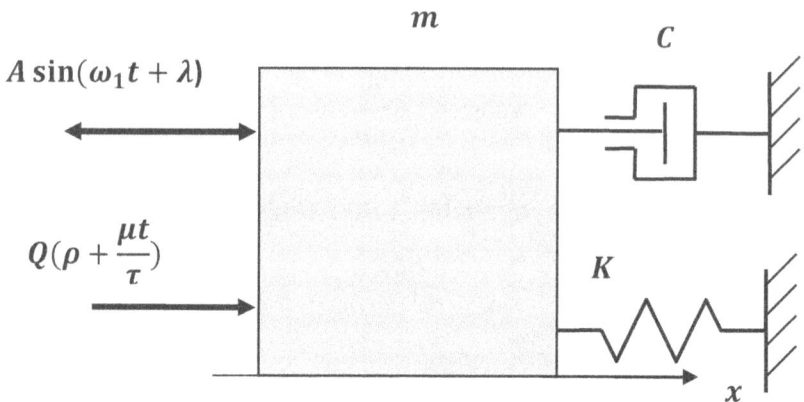

Figure 15.6 Model of a system subjected to a harmonic force, a time-dependent force, a damping force, and a stiffness force

Based on the considerations above and the model in Figure 15.6, we can assemble the left and right sides of the differential equation of motion. The left side includes the force of inertia, the damping force, and the stiffness force. The right side consists of the harmonic force and the time-dependent force. Hence, the differential equation of motion reads:

$$m\frac{d^2x}{dt^2}+C\frac{dx}{dt}+Kx = A\sin(\omega_1 t+\lambda)+Q(\rho+\frac{\mu t}{\tau}) \quad (15.6.1)$$

The differential equation of motion (15.6.1) has different solutions for various initial conditions of motion. These solutions are presented below.

15.6.1 General Initial Conditions

The general initial conditions of motion are:

$$\text{for} \quad t=0 \quad x=s_0; \quad \frac{dx}{dt}=v_0 \quad (15.6.2)$$

where s_0 and v_0 are the initial displacement and initial velocity respectively.

Transforming equation (15.6.1), we may write:

$$m\frac{d^2x}{dt^2} + C\frac{dx}{dt} + Kx = A\sin\omega_1 t\cos\lambda + A\cos\omega_1 t\sin\lambda + Q\rho + Q\frac{\mu t}{\tau}$$

$$(15.6.3)$$

Dividing equation (15.6.3) by m, we have:

$$\frac{d^2x}{dt^2} + 2n\frac{dx}{dt} + \omega_0^2 x = a\sin\omega_1 t\cos\lambda$$

$$+ a\cos\omega_1 t\sin\lambda + q\rho + \frac{q\mu}{\tau l} \qquad (15.6.4)$$

where n is the damping factor and ω_0 is the natural frequency, while:

$$2n = \frac{C}{m} \qquad (15.6.5)$$

$$\omega_0^2 = \frac{K}{m} \qquad (15.6.6)$$

$$a = \frac{A}{m} \qquad (15.6.7)$$

$$q = \frac{Q}{m} \qquad (15.6.8)$$

Using Laplace Transform Pairs 3, 4, 5, 6, 7, and 2 from Table 1.1, we convert differential equation of motion (15.6.4) with the initial conditions of motion (15.6.2) from the time domain into the Laplace domain. The resulting algebraic equation of motion in the Laplace domain reads:

$$l^2 x(l) - lv_0 - l^2 s_0 + 2nlx(l) - 2nls_0 + \omega_0^2 x(l)$$

$$= \frac{a\omega_1 l\cos\lambda}{l^2 + \omega_1^2} + \frac{al^2\sin\lambda}{l^2 + \omega_1^2} + q\rho + \frac{q\mu}{l\tau} \qquad (15.6.9)$$

Solving equation (15.6.9) for the displacement $x(l)$ in the Laplace domain, we have:

$$x(l) = \frac{l(v_0 + 2ns_0)}{l^2 + 2nl + \omega_0^2} + \frac{l^2 s_0}{l^2 + 2nl + \omega_0^2}$$

$$+ \frac{a\omega_1 l \cos \lambda}{(l^2 + \omega_1^2)(l^2 + 2nl + \omega_0^2)} + \frac{al^2 \sin \lambda}{(l^2 + \omega_1^2)(l^2 + 2nl + \omega_0^2)}$$

$$+ \frac{q\rho}{l^2 + 2nl + \omega_0^2} + \frac{q\mu}{\tau l(l^2 + 2nl + \omega_0^2)} \qquad \textbf{(15.6.10)}$$

Applying some algebra to the denominators in equation (15.6.10), we may write:

$$l^2 + 2nl + \omega_0^2 + n^2 - n^2 = (l+n)^2 + \omega^2 \qquad \textbf{(15.6.11)}$$

where:

$$\omega^2 = \omega_0^2 - n^2 \qquad \textbf{(15.6.12)}$$

while ω^2 could be positive, equal to zero, or negative.
   All these three cases are considered below.

a. Case $\omega^2 > 0$

   Transforming the denominators in equation (15.6.10) according to equation (15.6.11), we have:

$$x(l) = \frac{l(v_0 + 2ns_0)}{(l+n)^2 + \omega^2} + \frac{l^2 s_0}{(l+n)^2 + \omega^2}$$

$$+ \frac{a\omega_1 l \cos \lambda}{(l^2 + \omega_1^2)[(l+n)^2 + \omega^2]} + \frac{al^2 \sin \lambda}{(l^2 + \omega_1^2)[(l+n)^2 + \omega^2]}$$

$$+ \frac{q\rho}{(l+n)^2 + \omega^2} + \frac{q\mu}{\tau l[(l+n)^2 + \omega^2]} \qquad \textbf{(15.6.13)}$$

Using pairs 1, 24, 27, 42, 43, 22, and 39 from Table 1.1, we invert equation (15.6.13) from the Laplace domain into the time domain

and obtain the solution of differential equation of motion (15.6.1) with the initial conditions of motion (15.6.2):

$$x = \frac{1}{\omega}(v_0 + 2ns_0)e^{-nt}\sin\omega t + s_0 e^{-nt}(\cos\omega t - \frac{n}{\omega}\sin\omega t)$$

$$+\frac{a\omega_1\cos\lambda}{(\omega_1^2 - \omega_0^2)^2 + 4n^2\omega_1^2}\{2n[e^{-nt}(\cos\omega t + \frac{n}{\omega}\sin\omega t) - \cos\omega_1 t]$$

$$-(\omega_1^2 - \omega_0^2)(\frac{1}{\omega_1}\sin\omega_1 t - \frac{1}{\omega}e^{-nt}\sin\omega t)\} + \frac{a\sin\lambda}{(\omega_1^2 - \omega_0^2)^2 + 4n^2\omega_1^2}$$

$$\times\{(\omega_1^2 - \omega_0^2)[e^{-nt}(\cos\omega t + \frac{n}{\omega}\sin\omega t) - \cos\omega_1 t]$$

$$+2n\omega_1(\sin\omega_1 t - \frac{\omega_1}{\omega}e^{-nt}\sin\omega t)\}$$

$$+\frac{q\rho}{\omega_0^2}[1 - e^{-nt}(\cos\omega t + \frac{n}{\omega}\sin\omega t)]$$

$$+\frac{q\mu}{\omega\tau(\omega^2 + n^2)^2}\{(\omega^2 + n^2)\omega t - 2n\omega$$

$$-e^{-nt}[(\omega^2 - n^2)\sin\omega t - 2n\omega\cos\omega t]\} \qquad (15.6.14)$$

b. Case $\omega^2 = 0$

Equating ω^2 to zero in equation (15.6.13), we obtain:

$$x(l) = \frac{l(v_0 + 2ns_0)}{(l+n)^2} + \frac{l^2 s_0}{(l+n)^2} + \frac{a\omega_1 l}{(l^2 + \omega_1^2)(l+n)^2}\cos\lambda$$

$$+\frac{al^2}{(l^2 + \omega_1^2)(l+n)^2}\sin\lambda + \frac{q\rho}{(l+n)^2} + \frac{q\mu}{l\tau(l+n)^2} \qquad (15.6.15)$$

Applying pairs 1, 37, 38, 46, 47, 36, and 15 from Table 1.1, we invert equation (15.6.15) from the Laplace domain into the time domain and obtain the solution of differential equation of motion (15.6.1) with the initial conditions of motion (15.6.2):

$$x = (v_0 + 2ns_0)te^{-nt} + s_0(1 - nt)e^{-nt}$$

$$+ \frac{a\omega_1 \cos\lambda}{4n^2\omega_1^2 + (\omega_1^2 - n^2)^2} \{2n(1 - \cos\omega_1 t) - \frac{1}{\omega_1}(\omega_1^2 - n^2)\sin\omega_1 t$$

$$-2n\left[1 - e^{-nt}(1 + nt)\right] + (\omega_1^2 - n^2)te^{-nt}\}$$

$$+ \frac{a\sin\lambda}{4n^2\omega_1^2 + (\omega_1^2 - n^2)^2} \{2n\omega_1 \sin\omega_1 t$$

$$-(\omega_1^2 - n^2)(1 - \cos\omega_1 t) - (\omega_1^2 - n^2)[1 - e^{-nt}(1 + nt)] - 2n\omega_1^2 te^{-nt}\}$$

$$+ \frac{q\rho}{n^2}[1 - e^{-nt}(1 + nt)] + \frac{q\mu}{\tau n^2}[t - \frac{2}{n} + e^{-nt}\left(\frac{2}{n} + t\right)] \qquad \textbf{(15.6.16)}$$

c. Case $\omega^2 < 0$

Taking ω^2 with the negative sign in equation (15.6.13), we may write:

$$x(l) = \frac{l(v_0 + 2ns_0)}{(l+n)^2 - \omega^2} + \frac{l^2 s_0}{(l+n)^2 - \omega^2} + \frac{a\omega_1 l}{(l^2 + \omega_1^2)[(l+n)^2 - \omega^2]}\cos\lambda$$

$$+ \frac{al^2}{(l^2 + \omega_1^2)[(l+n)^2 - \omega^2]}\sin\lambda + \frac{q\rho}{(l+n)^2 - \omega^2} + \frac{q\mu}{l\tau(l+n)^2 - \omega^2}$$

$$\textbf{(15.6.17)}$$

Using pairs 1, 25, 28, 48, 49, 23, and 40 from Table 1.1, we invert equation (15.6.17) from the Laplace domain into the time domain and obtain the solution of differential equation of motion (15.6.1) with the initial conditions of motion (15.6.2):

$$x = \frac{v_0 + 2ns_0}{\omega}\sinh\omega t + \frac{s_0}{\omega}e^{-nt}(\omega\cosh\omega t - n\sinh\omega t)$$

$$+ \frac{a\omega_1 \cos\lambda}{4n^2\omega_1^2 + (\omega_1^2 - \omega_0^2)^2} \{e^{-nt}[2n\cosh\omega t + \frac{1}{\omega}(\omega_1^2 + \omega^2)\sinh\omega t]$$

$$-2n\cos\omega_1 t - \frac{1}{\omega_1}(\omega_1^2 - \omega_0^2)\sin\omega_1 t\}$$

$$+\frac{(\omega_1^2 - \omega_0^2)a\sin\lambda}{4n^2\omega_1^2 + (\omega_1^2 - \omega_0^2)^2}\{\frac{\omega_1^2}{\omega^2}(1 - \cos\omega_1 t)$$

$$-\frac{\omega_1^2}{n^2 - \omega^2}[1 - e^{-nt}(\cosh\omega t - \frac{n}{\omega}\sinh\omega t)]$$

$$+\frac{2n\omega_1}{\omega_1^2 - \omega_0^2}(\sin\omega_1 t - \frac{\omega_1}{\omega}e^{-nt}\sinh\omega t)\}$$

$$+\frac{q\rho}{n^2 - \omega^2}[1 - e^{-nt}(\cosh\omega t + \frac{n}{\omega}\sinh\omega t)]$$

$$+\frac{q\mu}{\omega\tau(n^2 - \omega^2)^2}[(n^2 - \omega^2)(\omega t + e^{-nt}\sinh\omega t) - 2n\omega(1 - \cosh\omega t)]$$

$$(15.6.18)$$

15.6.2 Initial Displacement Equals Zero

The initial conditions of motion are:

$$\text{for}\quad t = 0\quad x = 0;\quad \frac{dx}{dt} = v_0 \qquad (15.6.19)$$

The solutions of differential equation of motion (15.6.1) with the initial conditions of motion (15.6.19) for the following cases are presented below.

a. Case $\omega^2 > 0$

$$x = \frac{1}{\omega}v_0 e^{-nt}\sin\omega t + \frac{a\omega_1\cos\lambda}{(\omega_1^2 - \omega_0^2)^2 + 4n^2\omega_1^2}\{2n[e^{-nt}(\cos\omega t$$

$$+\frac{n}{\omega}\sin\omega t) - \cos\omega_1 t] - (\omega_1^2 - \omega_0^2)(\frac{1}{\omega_1}\sin\omega_1 t - \frac{1}{\omega}e^{-nt}\sin\omega t)\}$$

$$+\frac{a\sin\lambda}{(\omega_1^2 - \omega_0^2)^2 + 4n^2\omega_1^2}\{(\omega_1^2 - \omega_0^2)[e^{-nt}(\cos\omega t + \frac{n}{\omega}\sin\omega t) - \cos\omega_1 t]$$

$$+2n\omega_1(\sin\omega_1 t-\frac{\omega_1}{\omega}e^{-nt}\sin\omega t)\}$$

$$+\frac{q\rho}{n^2-\omega^2}[1-e^{-nt}(\cos\omega t+\frac{n}{\omega}\sin\omega t)]$$

$$+\frac{q\mu}{\omega\tau(\omega^2+n^2)^2}\{(\omega^2+n^2)\omega t$$

$$-2n\omega-e^{-nt}[(\omega^2-n^2)\sin\omega t-2n\omega\cos\omega t]\}\qquad\textbf{(15.6.20)}$$

b. Case $\omega^2=0$

$$x=v_0te^{-nt}+\frac{a\omega_1\cos\lambda}{4n^2\omega_1^2+(\omega_1^2-n^2)^2}\left\{2n\left[e^{-nt}(1+nt)-\cos\omega_1 t\right]\right.$$

$$-(\omega_1^2-n^2)(\frac{1}{\omega_1}\sin\omega_1 t-te^{-nt})\}$$

$$+\frac{a\sin\lambda}{4n^2\omega_1^2+(\omega_1^2-n^2)^2}\{2n\omega_1(\sin\omega_1 t-\omega_1 te^{-nt})$$

$$-(\omega_1^2-n^2)[e^{-nt}(1+nt)-\cos\omega_1 t]\}$$

$$+\frac{q\rho}{n^2}[1-e^{-nt}(1+nt)]+\frac{q\mu}{\tau n^2}[t-\frac{2}{n}+e^{-nt}\left(\frac{2}{n}+t\right)]\qquad\textbf{(15.6.21)}$$

c. Case $\omega^2<0$

$$x=\frac{v_0}{\omega}\sinh\omega t$$

$$+\frac{a\omega_1\cos\lambda}{4n^2\omega_1^2+(\omega_1^2-\omega_0^2)^2}\{e^{-nt}[2n\cosh\omega t+\frac{1}{\omega}(\omega_1^2+\omega^2)\sinh\omega t]$$

$$-2n\cos\omega_1 t-\frac{1}{\omega_1}(\omega_1^2-\omega_0^2)\sin\omega_1 t\}$$

$$+\frac{(\omega_1^2-\omega_0^2)a\sin\lambda}{4n^2\omega_1^2+(\omega_1^2-\omega_0^2)^2}\{\frac{\omega_1^2}{\omega^2}(1-\cos\omega_1 t)$$

$$-\frac{\omega_1^2}{n^2-\omega^2}[1-e^{-nt}(\cosh\omega t-\frac{n}{\omega}\sinh\omega t)]$$

$$+\frac{2n\omega_1}{\omega_1^2-\omega_0^2}(\sin\omega_1 t-\frac{\omega_1}{\omega}e^{-nt}\sinh\omega t)\}$$

$$+\frac{q\rho}{n^2-\omega^2}[1-e^{-nt}(\cosh\omega t+\frac{n}{\omega}\sinh\omega t)]+\frac{q\mu}{\omega\tau(n^2-\omega^2)^2}$$

$$\times[(n^2-\omega^2)(\omega t+e^{-nt}\sinh\omega t)-2n\omega(1-\cosh\omega t)] \quad \textbf{(15.6.22)}$$

15.6.3 Initial Velocity Equals Zero
The initial conditions of motion are:

$$\text{for}\quad t=0\quad x=s_0;\quad \frac{dx}{dt}=0 \quad \textbf{(15.6.23)}$$

The solutions of differential equation of motion (15.6.1) with the initial conditions of motion (15.6.23) for the following cases are presented below.

a. Case $\omega^2>0$

$$x=\frac{1}{\omega}2ns_0e^{-nt}\sin\omega t+s_0e^{-nt}(\cos\omega t-\frac{n}{\omega}\sin\omega t)$$

$$+\frac{a\omega_1\cos\lambda}{(\omega_1^2-\omega_0^2)^2+4n^2\omega_1^2}\{2n[e^{-nt}(\cos\omega t+\frac{n}{\omega}\sin\omega t)-\cos\omega_1 t]$$

$$-(\omega_1^2-\omega_0^2)(\frac{1}{\omega_1}\sin\omega_1 t-\frac{1}{\omega}e^{-nt}\sin\omega t)\}$$

$$+\frac{a\sin\lambda}{(\omega_1^2-\omega_0^2)^2+4n^2\omega_1^2}\{(\omega_1^2-\omega_0^2)[e^{-nt}(\cos\omega t$$

$$+\frac{n}{\omega}\sin\omega t)-\cos\omega_1 t]+2n\omega_1(\sin\omega_1 t-\frac{\omega_1}{\omega}e^{-nt}\sin\omega t)\}$$

$$+\frac{q\rho}{\omega_0^2}[1-e^{-nt}(\cos\omega t+\frac{n}{\omega}\sin\omega t)]+\frac{q\mu}{\omega\tau(\omega^2+n^2)^2}$$

$$\times\{(\omega^2+n^2)\omega t-2n\omega-e^{-nt}[\omega^2-n^2]\sin\omega t-2n\omega\cos\omega t]\}$$

$$\textbf{(15.6.24)}$$

b. Case $\omega^2 = 0$

$$x = 2ns_0te^{-nt} + s_0(1-nt)e^{-nt}$$

$$+\frac{a\omega_1\cos\lambda}{4n^2\omega_1^2 + (\omega_1^2 - n^2)^2}\{2n\left[e^{-nt}(1+nt) - \cos\omega_1 t\right]$$

$$-(\omega_1^2 - n^2)(\frac{1}{\omega_1}\sin\omega_1 t - te^{-nt})\}$$

$$+\frac{a\sin\lambda}{4n^2\omega_1^2 + (\omega_1^2 - n^2)^2}\{2n\omega_1(\sin\omega_1 t - \omega_1 te^{-nt})$$

$$-(\omega_1^2 - n^2)[e^{-nt}(1+nt) - \cos\omega_1 t]\}$$

$$+\frac{q\rho}{n^2}[1 - e^{-nt}(1+nt)] + \frac{q\mu}{\tau n^2}[t - \frac{2}{n} + e^{-nt}\left(\frac{2}{n} + t\right)] \quad \textbf{(15.6.25)}$$

c. Case $\omega^2 < 0$

$$x = \frac{2ns_0}{\omega}\sinh\omega t + \frac{s_0}{\omega}e^{-nt}(\omega\cosh\omega t - n\sinh\omega t)$$

$$+\frac{a\omega_1\cos\lambda}{4n^2\omega_1^2 + (\omega_1^2 - \omega_0^2)^2}\{e^{-nt}[2n\cosh\omega t + \frac{1}{\omega}(\omega_1^2 + \omega^2)\sinh\omega t]$$

$$-2n\cos\omega_1 t - \frac{1}{\omega_1}(\omega_1^2 - \omega_0^2)\sin\omega_1 t\} + \frac{(\omega_1^2 - \omega_0^2)a\sin\lambda}{4n^2\omega_1^2 + (\omega_1^2 - \omega_0^2)^2}$$

$$\times\{\frac{\omega_1^2}{\omega^2}(1 - \cos\omega_1 t) - \frac{\omega_1^2}{n^2 - \omega^2}[1 - e^{-nt}(\cosh\omega t - \frac{n}{\omega}\sinh\omega t)]$$

$$+\frac{2n\omega_1}{\omega_1^2 - \omega_0^2}(\sin\omega_1 t - \frac{\omega_1}{\omega}e^{-nt}\sinh\omega t)\}$$

$$+\frac{q\rho}{n^2 - \omega^2}[1 - e^{-nt}(\cosh\omega t + \frac{n}{\omega}\sinh\omega t)]$$

$$+\frac{q\mu}{\omega\tau(n^2 - \omega^2)^2}[(n^2 - \omega^2)(\omega t + e^{-nt}\sinh\omega t) - 2n\omega(1 - \cosh\omega t)]$$

$$\textbf{(15.6.26)}$$

15.6.4 Both the Initial Displacement and Velocity Equals Zero

The initial conditions of motion are:

$$\text{for} \quad t = 0 \quad x = 0; \quad \frac{dx}{dt} = 0 \qquad (15.6.27)$$

The solutions of differential equation of motion (15.6.1) with the initial conditions of motion (15.6.27) for the following cases are presented below.

a. **Case $\omega^2 > 0$**

$$x = \frac{a\omega_1 \cos \lambda}{(\omega_1^2 - \omega_0^2)^2 + 4n^2\omega_1^2} \{2n[e^{-nt}(\cos \omega t + \frac{n}{\omega}\sin \omega t) - \cos \omega_1 t]$$

$$- (\omega_1^2 - \omega_0^2)(\frac{1}{\omega_1}\sin \omega_1 t - \frac{1}{\omega}e^{-nt}\sin \omega t)\} + \frac{a\sin \lambda}{(\omega_1^2 - \omega_0^2)^2 + 4n^2\omega_1^2}$$

$$\times \{(\omega_1^2 - \omega_0^2)[e^{-nt}(\cos \omega t + \frac{n}{\omega}\sin \omega t) - \cos \omega_1 t]$$

$$+ 2n\omega_1(\sin \omega_1 t - \frac{\omega_1}{\omega}e^{-nt}\sin \omega t)\}$$

$$+ \frac{q\rho}{\omega_0^2}[1 - e^{-nt}(\cos \omega t + \frac{n}{\omega}\sin \omega t)] + \frac{q\mu}{\omega\tau(\omega^2 + n^2)^2}$$

$$\times \{(\omega^2 + n^2)\omega t - 2n\omega - e^{-nt}[(\omega^2 - n^2)\sin \omega t - 2n\omega \cos \omega t]\}$$

$$(15.6.28)$$

b. **Case $\omega^2 = 0$**

$$x = \frac{a\omega_1 \cos \lambda}{4n^2\omega_1^2 + (\omega_1^2 - n^2)^2} \{2n\left[e^{-nt}(1 + nt) - \cos \omega_1 t\right]$$

$$- (\omega_1^2 - n^2)(\frac{1}{\omega_1}\sin \omega_1 t - te^{-nt})\}$$

$$+ \frac{a\sin \lambda}{4n^2\omega_1^2 + (\omega_1^2 - n^2)^2} \{2n\omega_1(\sin \omega_1 t - \omega_1 te^{-nt})$$

$$- (\omega_1^2 - n^2)[e^{-nt}(1 + nt) - \cos \omega_1 t]\}$$

$$+ \frac{q\rho}{n^2}[1 - e^{-nt}(1 + nt)] + \frac{q\mu}{\tau n^2}[t - \frac{2}{n} + e^{-nt}\left(\frac{2}{n} + t\right)] \quad (15.6.29)$$

c. Case $\omega^2 < 0$

$$x = \frac{a\omega_1 \cos\lambda}{4n^2\omega_1^2 + (\omega_1^2 - \omega_0^2)^2} \{e^{-nt}[2n\cosh\omega t + \frac{1}{\omega}(\omega_1^2 + \omega^2)\sinh\omega t]$$

$$-2n\cos\omega_1 t - \frac{1}{\omega_1}(\omega_1^2 - \omega_0^2)\sin\omega_1 t\} + \frac{(\omega_1^2 - \omega_0^2)a\sin\lambda}{4n^2\omega_1^2 + (\omega_1^2 - \omega_0^2)^2}$$

$$\times\{\frac{\omega_1^2}{\omega^2}(1 - \cos\omega_1 t) - \frac{\omega_1^2}{n^2 - \omega^2}[1 - e^{-nt}(\cosh\omega t - \frac{n}{\omega}\sinh\omega t)]$$

$$+ \frac{2n\omega_1}{\omega_1^2 - \omega_0^2}(\sin\omega_1 t - \frac{\omega_1}{\omega}e^{-nt}\sinh\omega t)\}$$

$$+ \frac{q\rho}{n^2 - \omega^2}[1 - e^{-nt}(\cosh\omega t + \frac{n}{\omega}\sinh\omega t)]$$

$$+ \frac{q\mu}{\omega\tau(n^2 - \omega^2)^2}[(n^2 - \omega^2)(\omega t + e^{-nt}\sinh\omega t) - 2n\omega(1 - \cosh\omega t)]$$

$$\textbf{(15.6.30)}$$

16

DAMPING, STIFFNESS, AND FRICTION

This chapter discusses the problems of engineering systems subjected to the force of inertia, the damping force, the stiffness force, and the friction force as the resisting forces — these are the forces marked with the plus sign in Row 16 of Guiding Table 2.1. The active forces applied to the systems are shown on the intersection of this row with Columns 1 through 6. The titles of this chapter's sections reflect the characteristics of the active forces applied to the systems. Throughout this chapter, the left sides of the differential equations of motion are identical and consist of the force of inertia, the damping force, the stiffness force, and the friction force. The right sides of these equations differ from each other and consist of the corresponding active forces applied to the systems.

The problems described in this chapter could reflect the working processes of engineering systems interacting with viscoelasto-plastic media.

16.1 Active Force Equals Zero

This section describes engineering systems subjected to the force of inertia, the damping force, the stiffness force, and the friction force as the resisting forces (Row 16 of Guiding Table 2.1) in the absence of active forces (Column 1). The considerations related to the motion of a system in the absence of active forces are discussed in section 1.3. According to the characteristics of the resisting forces, the current problem could be related to a vibratory system or to a system interacting with viscoelastoplastic media that in certain cases exert the damping, the stiffness, and the friction forces as the reaction to their deformation. Additional information related to the deformation of these types of media and the considerations regarding the behavior of a friction force applied to a vibratory system are presented in section 1.2.

The system is moving in the horizontal direction. We want to determine the law of motion of the system. Figure 16.1 shows the model of a system subjected to the action of a damping force, a stiffness force, and a friction force.

Based on the considerations above and on the model in Figure 16.1, we can assemble the left and right sides of the differential equation of motion. The left side consists of the force of inertia,

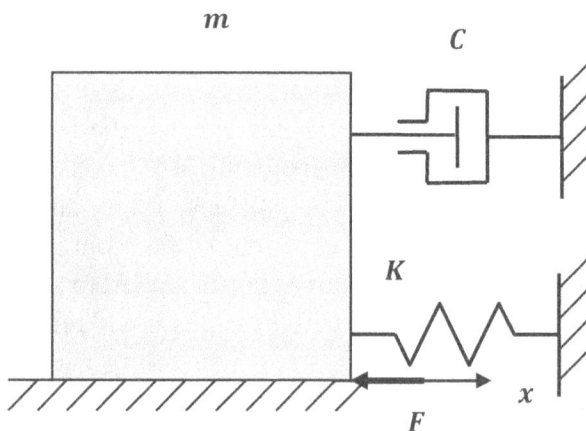

Figure 16.1 Model of a system subjected to a damping force, a stiffness force, and a friction force

the damping force, the stiffness force, and the friction force, while the right side equals zero. Hence, the differential equation of motion reads:

$$m\frac{d^2x}{dt^2}+C\frac{dx}{dt}+Kx+F=0 \qquad (16.1.1)$$

Differential equation of motion (16.1.1) has different solutions for various initial conditions of motion. These solutions are presented below.

16.1.1 General Initial Conditions

The general initial conditions of motion are:

$$\text{for} \quad t=0 \quad x=s_0; \quad \frac{dx}{dt}=v_0 \qquad (16.1.2)$$

where s_0 and v_0 are the initial displacement and initial velocity respectively.

Dividing equation (16.1.1) by m, we may write:

$$\frac{d^2x}{dt^2}+2n\frac{dx}{dt}+\omega_0^2 x+f=0 \qquad (16.1.3)$$

where n is the damping factor and ω_0 is the natural frequency, while:

$$2n=\frac{C}{m} \qquad (16.1.4)$$

$$\omega_0^2=\frac{K}{m} \qquad (16.1.5)$$

$$f=\frac{F}{m} \qquad (16.1.6)$$

Based on Laplace Transform Pairs 3, 4, 5, and 1 from Table 1.1, we convert differential equation of motion (16.1.3) with the initial conditions of motion (16.1.2) from the time domain into the Laplace domain and obtain the corresponding algebraic equation of motion in the Laplace domain:

$$l^2 x(l) - lv_0 - l^2 s_0 + 2nlx(l) - 2nls_0 + \omega_0^2 x(l) + f = 0 \quad \textbf{(16.1.7)}$$

Applying basic algebra to equation (16.1.7), we may write:

$$x(l)(l^2 + 2nl + \omega_0^2) = l(v_0 + 2ns_0) + l^2 s_0 - f \quad \textbf{(16.1.8)}$$

The solution of equation (16.1.8) for the displacement $x(l)$ in the Laplace domain reads:

$$x(l) = \frac{l(v_0 + 2ns_0)}{l^2 + 2nl + \omega_0^2} + \frac{l^2 s_0}{l^2 + 2nl + \omega_0^2} - \frac{f}{l^2 + 2nl + \omega_0^2} \quad \textbf{(16.1.9)}$$

Applying algebraic procedures with the denominators of equation (16.1.9), we may write:

$$l^2 + 2nl + \omega_0^2 + n^2 - n^2 = (l+n)^2 + \omega^2 \quad \textbf{(16.1.10)}$$

where:

$$\omega^2 = \omega_0^2 - n^2 \quad \textbf{(16.1.11)}$$

while ω^2 could be positive, equal to zero, or negative.
    All these three cases are considered below.

a. Case $\omega^2 > 0$

    Transforming the denominators in equation (16.1.9) according to equation (16.1.10), we have:

$$x(l) = \frac{l(v_0 + 2ns_0)}{(l+n)^2 + \omega^2} + \frac{l^2 s_0}{(l+n)^2 + \omega^2} - \frac{f}{(l+n)^2 + \omega^2} \quad \textbf{(16.1.12)}$$

Using pairs 1, 24, 27, and 22 from Table 1.1, we invert equation (16.1.12) from the Laplace domain into the time domain and obtain the solution of differential equation of motion (16.1.1) with the initial conditions of motion (16.1.2):

$$x = \frac{v_0 + 2ns_0}{\omega} e^{-nt} \sin \omega t + s_0 e^{-nt} (\cos \omega t - \frac{n}{\omega} \sin \omega t)$$

$$-\frac{f}{\omega^2+n^2}[1-e^{-nt}(\cos\omega t+\frac{n}{\omega}\sin\omega t)] \qquad (16.1.13)$$

b. Case $\omega^2 = 0$

Equating in equation (16.1.12) ω^2 to zero, we obtain:

$$x(l) = \frac{l(v_0+2ns_0)}{(l+n)^2}+\frac{l^2 s_0}{(l+n)^2}-\frac{f}{(l+n)^2} \qquad (16.1.14)$$

Using the pairs 1, 37, 38, and 36 from Table 1.1, we invert equation (16.1.14) from the Laplace domain into the time domain and obtain the solution of differential equation (16.1.1) with the initial conditions of motion (16.1.2):

$$x = (v_0+2ns_0)te^{-nt}+s_0(1-nt)e^{-nt}-\frac{f}{n^2}[1-e^{-nt}(1+nt)] \quad (16.1.15)$$

c. Case $\omega^2 < 0$

In equation (16.1.12), taking ω^2 with the negative sign, we may write:

$$x(l) = \frac{l(v_0+2ns_0)}{(l+n)^2-\omega^2}+\frac{l^2 s_0}{(l+n)^2-\omega^2}-\frac{f}{(l+n)^2-\omega^2} \qquad (16.1.16)$$

Applying pairs 1, 25, 28, and 23 from Table 1.1, we invert equation (16.1.16) from the Laplace domain into the time domain and obtain the solution of differential equation of motion (16.1.1) with the initial conditions of motion (16.1.2):

$$x = \frac{v_0+2ns_0}{\omega}e^{-nt}\sinh\omega t+\frac{s_0}{\omega}e^{-nt}(\omega\cosh\omega t-n\sinh\omega t)$$

$$-\frac{f}{n^2-\omega^2}[1-e^{-nt}(\omega\cosh\omega t+\frac{n}{\omega}\sinh\omega t)] \qquad (16.1.17)$$

16.1.2 Initial Displacement Equals Zero

The initial conditions of motion are:

$$\text{for} \quad t = 0 \quad x = 0; \quad \frac{dx}{dt} = v_0 \qquad \textbf{(16.1.18)}$$

The solutions of differential equation of motion (16.1.1) with the initial conditions of motion (16.1.18) for the following cases are presented below.

a. **Case $\omega^2 > 0$**

$$x = \frac{v_0}{\omega} e^{-nt} \sin \omega t - \frac{f}{\omega^2 + n^2} [1 - e^{-nt} (\cos \omega t + \frac{n}{\omega} \sin \omega t)] \qquad \textbf{(16.1.19)}$$

b. **Case $\omega^2 = 0$**

$$x = v_0 t e^{-nt} - \frac{f}{n^2} [1 - e^{-nt} (1 + nt)] \qquad \textbf{(16.1.20)}$$

c. **Case $\omega^2 < 0$**

$$x = \frac{1}{\omega} v_0 e^{-nt} \sinh \omega t$$

$$- \frac{f}{n^2 - \omega^2} [1 - e^{-nt} (\omega \cosh \omega t + \frac{n}{\omega} \sinh \omega t)] \qquad \textbf{(16.1.21)}$$

16.1.3 Initial Velocity Equals Zero

According to Figure 16.1, the motion of the system should occur in the positive direction. This becomes possible if the initial displacement is taken with the negative sign.

Hence, the initial conditions of motion are:

$$\text{for} \quad t = 0 \quad x = -s_0; \quad \frac{dx}{dt} = 0 \qquad \textbf{(16.1.22)}$$

In addition, it should be emphasized that for the current case it is assumed that $|Ks_0| > |F|$.

The solutions of differential equation of motion (16.1.1) with the initial conditions of motion (16.1.22) for the following cases are presented below.

a. Case $\omega^2 > 0$

$$x = -\frac{2ns_0}{\omega}e^{-nt}\sin\omega t - s_0 e^{-nt}(\cos\omega t - \frac{n}{\omega}\sin\omega t)$$

$$-\frac{f}{\omega^2 + n^2}[1 - e^{-nt}(\cos\omega t + \frac{n}{\omega}\sin\omega t)] \qquad (16.1.23)$$

b. Case $\omega^2 = 0$

$$x = -2ns_0 te^{-nt} - s_0(1 - nt)e^{-nt} - \frac{f}{n^2}[1 - e^{-nt}(1 + nt)] \qquad (16.1.24)$$

c. Case $\omega^2 < 0$

$$x = -\frac{2ns_0}{\omega}e^{-nt}\sinh\omega t - \frac{s_0}{\omega}e^{-nt}(\omega\cosh\omega t - n\sinh\omega t)$$

$$-\frac{f}{n^2 - \omega^2}[1 - e^{-nt}(\omega\cosh\omega t + \frac{n}{\omega}\sinh\omega t)] \qquad (16.1.25)$$

16.2 Constant Force R

According to Guiding Table 2.1, this section describes engineering systems subjected to the action of the force of inertia, the damping force, the stiffness force, and the friction force as the resisting forces (Row 16) and constant active force (Column 2).

The current problem could be associated with the working processes of engineering systems interacting with viscoelastoplastic media that exert damping, stiffness, and friction forces as the reaction to deformation. Additional considerations regarding the deformation of viscoelastoplastic media as well as the behavior of a friction force applied to a vibratory system are presented in section 1.2.

The system moves in the horizontal direction. We want to determine the law of motion of this engineering system. Figure 16.2

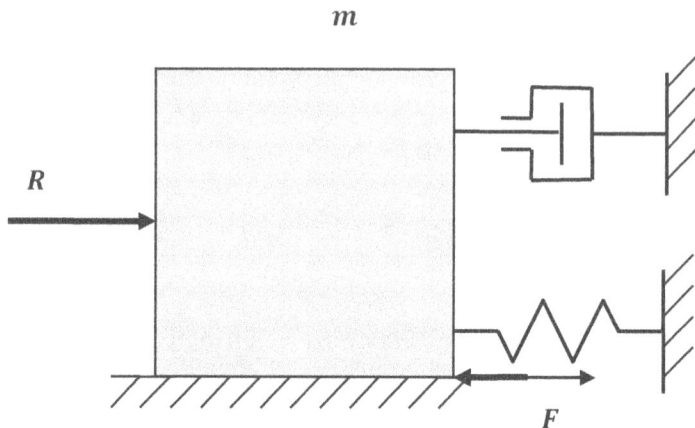

Figure 16.2 Model of a system subjected to a constant active force, a damping force, a stiffness force, and a friction force

shows the model of a system subjected to the action of a constant active force, a damping force, a stiffness force, and a friction force.

The considerations above and the model in Figure 16.2 allow us to compose the left and right sides of the differential equation of motion of this system. The left side includes the force of inertia, the damping force, the stiffness force, and the friction force. The right side consists of a constant active force. Hence, the differential equation of motion reads:

$$m\frac{d^2x}{dt^2} + C\frac{dx}{dt} + Kx + F = R \qquad (16.2.1)$$

Differential equation of motion (16.2.1) has different solutions for various initial conditions of motion. These solutions are presented below.

16.2.1 General Initial Conditions

The general initial conditions of motion are:

$$\text{for} \quad t = 0 \quad x = s_0; \quad \frac{dx}{dt} = v_0 \qquad (16.2.2)$$

where s_0 and v_0 are the initial displacement and initial velocity respectively.

Dividing equation (16.2.1) by m, we may write:

$$\frac{d^2x}{dt^2} + 2n\frac{dx}{dt} + \omega_0^2 x + f = r \qquad (16.2.3)$$

where n is the damping factor and ω_0 is the natural frequency, while:

$$2n = \frac{C}{m} \qquad (16.2.4)$$

$$\omega_0^2 = \frac{K}{m} \qquad (16.2.5)$$

$$f = \frac{F}{m} \qquad (16.2.6)$$

$$r = \frac{R}{m} \qquad (16.2.7)$$

Based on Laplace Transform Pairs 3, 4, 5, and 1 from Table 1.1, we convert differential equation of motion (16.2.3) with the initial conditions of motion (16.2.2) from the time domain into the Laplace domain and obtain the corresponding algebraic equation of motion in the Laplace domain:

$$l^2 x(l) - lv_0 - l^2 s_0 + 2nlx(l) - 2nls_0 + \omega_0^2 x(l) + f = r \qquad (16.2.8)$$

Applying basic algebra to equation (16.2.8), we write:

$$x(l)(l^2 + 2nl + \omega_0^2) = r - f + l(v_0 + 2ns_0) + l^2 s_0 \qquad (16.2.9)$$

The solution of equation (16.2.9) for the displacement $x(l)$ in the Laplace domain reads:

$$x(l) = \frac{r - f}{l^2 + 2nl + \omega_0^2} + \frac{l(v_0 + 2ns_0)}{l^2 + 2nl + \omega_0^2} + \frac{l^2 s_0}{l^2 + 2nl + \omega_0^2} \qquad (16.2.10)$$

Transforming the denominators of equation (16.2.10), we write:

$$l^2 + 2nl + \omega_0^2 + n^2 - n^2 = (l + n)^2 + \omega^2 \qquad (16.2.11)$$

where:

$$\omega^2 = \omega_0^2 - n^2 \qquad (16.2.12)$$

while ω^2 could be positive, equal to zero, or negative.

All these three cases are considered below.

a. Case $\omega^2 > 0$

Using equation (16.2.11), we adjust the denominators in equation (16.2.10) and we may write:

$$x(l) = \frac{r - f}{(l+n)^2 + \omega^2} + \frac{l(v_0 + 2ns_0)}{(l+n)^2 + \omega^2} + \frac{l^2 s_0}{(l+n)^2 + \omega^2} \qquad (16.2.13)$$

Based on pairs 1, 22, 24, and 27 from Table 1.1, we invert equation (16.2.13) from the Laplace domain into the time domain and obtain the solution of differential equation of motion (16.2.1) with the initial conditions of motion (16.2.2):

$$x = \frac{r - f}{\omega^2 + n^2}[1 - e^{-nt}(\cos \omega t + \frac{n}{\omega} \sin \omega t)] + \frac{1}{\omega}(v_0 + 2ns_0)e^{-nt} \sin \omega t$$

$$+ s_0 e^{-nt}(\cos \omega t - \frac{n}{\omega} \sin \omega t) \qquad (16.2.14)$$

b. Case $\omega^2 = 0$

Equating ω^2 to zero in equation (16.2.13), we have:

$$x(l) = \frac{r - f}{(l+n)^2} + \frac{l(v_0 + 2ns_0)}{(l+n)^2} + \frac{l^2 s_0}{(l+n)^2} \qquad (16.2.15)$$

Applying pairs 1, 36, 37, and 38 from Table 1.1, we invert equation (16.2.15) from the Laplace domain into the time domain and obtain the solution of differential equation of motion (16.2.1) with the initial conditions of motion (16.2.2):

$$x = \frac{r - f}{n^2}[1 - e^{-nt}(1 + nt)] + (v_0 + 2ns_0)te^{-nt} + s_0 e^{-nt}(1 - nt) \ (16.2.16)$$

c. Case $\omega^2 < 0$

Taking ω^2 with the negative sign in equation (16.2.13), we may write:

$$x(l) = \frac{r - f}{(l+n)^2 - \omega^2} + \frac{l(v_0 + 2ns_0)}{(l+n)^2 - \omega^2} + \frac{l^2 s_0}{(l+n)^2 - \omega^2} \qquad \textbf{(16.2.17)}$$

Using pairs 1, 23, 25, and 28 from Table 1.1, we invert equation (16.2.17) from the Laplace domain into the time domain and obtain the solution of differential equation of motion (16.2.1) with the initial conditions of motion (16.2.2):

$$x = \frac{r - f}{n^2 - \omega^2}[1 - e^{-nt}(\cosh \omega t + \frac{n}{\omega}\sinh \omega t)] + \frac{v_0 + 2ns_0}{\omega}e^{-nt}\sinh \omega t$$

$$+ \frac{s_0}{\omega}e^{-nt}(\omega \cosh \omega t - n\sinh \omega t) \qquad \textbf{(16.2.18)}$$

16.2.2 Initial Displacement Equals Zero

The initial conditions of motion are:

$$\text{for} \quad t = 0 \quad x = 0; \quad \frac{dx}{dt} = v_0 \qquad \textbf{(16.2.19)}$$

The solutions of differential equation of motion (16.2.1) with the initial conditions of motion (16.2.19) for the following cases are presented below.

a. Case $\omega^2 > 0$

$$x = \frac{r - f}{\omega^2 + n^2}[1 - e^{-nt}(\cos \omega t + \frac{n}{\omega}\sin \omega t)] + \frac{1}{\omega}v_0 e^{-nt}\sin \omega t \quad \textbf{(16.2.20)}$$

b. Case $\omega^2 = 0$

$$x = \frac{r - f}{n^2}[1 - e^{-nt}(1 + nt)] + v_0 t e^{-nt} \qquad \textbf{(16.2.21)}$$

c. Case $\omega^2 < 0$

$$x = \frac{r-f}{n^2-\omega^2}[1-e^{-nt}(\cosh\omega t + \frac{n}{\omega}\sinh\omega t)]$$

$$+\frac{v_0}{\omega}e^{-nt}\sinh\omega t \qquad\qquad (16.2.22)$$

16.2.3 Initial Velocity Equals Zero

The initial conditions of motion are:

$$\text{for}\quad t=0 \quad x=s_0; \quad \frac{dx}{dt}=0 \qquad (16.2.23)$$

The solutions of differential equation of motion (16.2.1) with the initial conditions of motion (16.2.23) for the following cases are presented below.

a. Case $\omega^2 > 0$

$$x = \frac{r-f}{\omega^2+n^2}[1-e^{-nt}(\cos\omega t + \frac{n}{\omega}\sin\omega t)]$$

$$+\frac{1}{\omega}2ns_0e^{-nt}\sin\omega t + s_0e^{-nt}(\cos\omega t - \frac{n}{\omega}\sin\omega t) \quad (16.2.24)$$

b. Case $\omega^2 = 0$

$$x = \frac{r-f}{n^2}[1-e^{-nt}(1+nt)]+2ns_0te^{-nt}+s_0e^{-nt}(1-nt) \qquad (16.2.25)$$

c. Case $\omega^2 < 0$

$$x = \frac{r-f}{n^2-\omega^2}[1-e^{-nt}(\cosh\omega t + \frac{n}{\omega}\sinh\omega t)]$$

$$+\frac{2ns_0}{\omega}e^{-nt}\sinh\omega t + \frac{s_0}{\omega}e^{-nt}(\omega\cosh\omega t - n\sinh\omega t) \quad (16.2.26)$$

16.2.4 Both the Initial Displacement and Velocity Equal Zero

The initial conditions of motion are:

$$\text{for} \quad t = 0 \quad x = 0; \quad \frac{dx}{dt} = 0 \quad \quad (16.2.27)$$

The solutions of differential equation of motion (16.2.1) with the initial conditions of motion (16.2.27) for the following cases are presented below.

a. Case $\omega^2 > 0$

$$x = \frac{r - f}{\omega^2 + n^2}[1 - e^{-nt}(\cos\omega t + \frac{n}{\omega}\sin\omega t)] \quad \quad (16.2.28)$$

b. Case $\omega^2 = 0$

$$x = \frac{r - f}{n^2}[1 - e^{-nt}(1 + nt)] \quad \quad (16.2.29)$$

c. Case $\omega^2 < 0$

$$x = \frac{r - f}{n^2 - \omega^2}[1 - e^{-nt}(\cosh\omega t + \frac{n}{\omega}\sinh\omega t)] \quad \quad (16.2.30)$$

16.3 Harmonic Force $A\sin(\omega_1 t + \lambda)$

The intersection of Row 16 and Column 3 in Guiding Table 2.1 indicates that this section describes engineering systems subjected to the action of the force of inertia, the damping force, the stiffness force, and the friction force as the resisting forces, and to the harmonic force as the active force.

The current problem could be associated with the working process of vibratory systems intended for the deformation of visco-elastoplastic media that exert damping, stiffness, and friction forces as the reaction to their deformation. Additional considerations related to the deformation of viscoelastoplastic media as well to the behavior of a friction force applied to a vibratory system are discussed in section 1.2.

The system is moving in the horizontal direction. We want to determine the law of motion of the system. Figure 16.3 represents the model of a system subjected to the action of a harmonic force, a damping force, a stiffness force, and a friction force.

The considerations above and the model in Figure 16.3 let us assemble the left and right sides of the differential equation of motion. The left side consists of the force of inertia, the damping force, the stiffness force, and the friction force, while the right side includes the harmonic force. Therefore, the differential equation of motion reads:

$$m\frac{d^2x}{dt^2} + C\frac{dx}{dt} + Kx + F = A\sin(\omega_1 t + \lambda) \qquad \textbf{(16.3.1)}$$

Differential equation of motion (16.3.1) has different solutions for various initial conditions of motion. These solutions are presented below.

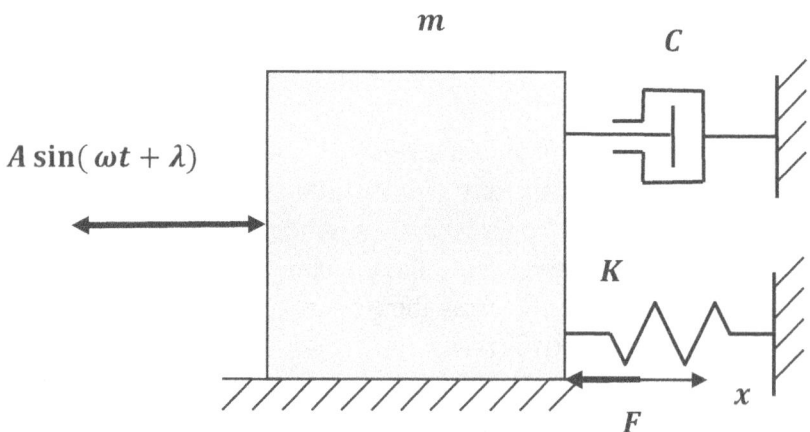

Figure 16.3 Model of a system subjected to a harmonic force, a damping force, a stiffness force

16.3.1 General Initial Conditions

The general initial conditions of motion are:

$$\text{for} \quad t = 0 \quad x = s_0; \quad \frac{dx}{dt} = v_0 \tag{16.3.2}$$

where s_0 and v_0 are the initial displacement and initial velocity respectively.

Transforming the sinusoidal function in equation (16.3.1) and dividing the latter by m, we have:

$$\frac{d^2x}{dt^2} + 2n\frac{dx}{dt} + \omega_0^2 x + f = a\sin\omega_1 t\cos\lambda + a\cos\omega_1 t\sin\lambda \tag{16.3.3}$$

where n is the damping factor and ω_0 is the natural frequency, while:

$$2n = \frac{C}{m} \tag{16.3.4}$$

$$\omega_0^2 = \frac{K}{m} \tag{16.3.5}$$

$$f = \frac{F}{m} \tag{16.3.6}$$

$$a = \frac{A}{m} \tag{16.3.7}$$

Based on Laplace Transform Pairs 3, 4, 5, 6, and 7 from Table 1.1, we convert differential equation of motion (16.3.3) with the initial conditions of motion (16.3.2) from the time domain into the Laplace domain and obtain the corresponding algebraic equation of motion in the Laplace domain:

$$l^2 x(l) - lv_0 - l^2 s_0 + 2nlx(l) - 2nls_0 + \omega_0^2 x(l) + f$$

$$= \frac{a\omega_1 l}{l^2 + \omega_1^2}\cos\lambda + \frac{al^2}{l^2 + \omega_1^2}\sin\lambda \tag{16.3.8}$$

Solving equation (16.3.8) for the displacement $x(l)$ in the Laplace domain, we have:

$$x(l) = \frac{l(v_0 + 2ns_0)}{l^2 + 2nl + \omega_0^2} + \frac{l^2 s_0}{l^2 + 2nl + \omega_0^2} + \frac{a\omega_1 l}{(l^2 + \omega_1^2)(l^2 + 2nl + \omega_0^2)} \cos \lambda$$

$$+ \frac{al^2}{(l^2 + \omega_1^2)(l^2 + 2nl + \omega_0^2)} \sin \lambda - \frac{f}{l^2 + 2nl + \omega_0^2}$$

$$(16.3.9)$$

Applying basic algebra to the denominators in equation (16.3.9), we may write:

$$l^2 + 2nl + \omega_0^2 + n^2 - n^2 = (l+n)^2 + \omega^2 \qquad (16.3.10)$$

where:

$$\omega^2 = \omega_0^2 - n^2 \qquad (16.3.11)$$

while ω^2 could be positive, equal to zero, or negative.
All these three cases are considered below.

a. Case $\omega^2 > 0$

Transforming the denominators in equations (16.3.9) based on the equation (16.3.10), we write:

$$x(l) = \frac{l(v_0 + 2ns_0)}{(l+n)^2 + \omega^2} + \frac{l^2 s_0}{(l+n)^2 + \omega^2} + \frac{a\omega_1 l \cos \lambda}{(l^2 + \omega_1^2)[(l+n)^2 + \omega^2]}$$

$$+ \frac{al^2 \sin \lambda}{(l^2 + \omega_1^2)[(l+n)^2 + \omega^2]} - \frac{f}{(l+n)^2 + \omega^2} \qquad (16.3.12)$$

Using Laplace Transform Pairs 1, 24, 27, 42, 43, and 22 from Table 1.1, we invert equation (16.3.12) from the Laplace domain into the time domain and obtain the solution of differential equation of motion (16.3.1) with the initial conditions of motion (16.3.2):

$$x = \frac{1}{\omega}(v_0 + 2ns_0)e^{-nt}\sin\omega t + s_0 e^{-nt}(\cos\omega t - \frac{n}{\omega}\sin\omega t)$$

$$+\frac{a\omega_1\cos\lambda}{\left(\omega_1^2 - \omega_0^2\right)^2 + 4n^2\omega_1^2}\{2n[e^{-nt}(\cos\omega t + \frac{n}{\omega}\sin\omega t) - \cos\omega_1 t]$$

$$-(\omega_1^2 - \omega_0^2)(\frac{1}{\omega_1}\sin\omega_1 t - \frac{1}{\omega}e^{-nt}\sin\omega t)\}$$

$$+\frac{a\sin\lambda}{\left(\omega_1^2 - \omega_0^2\right)^2 + 4n^2\omega_1^2}\{(\omega_1^2 - \omega_0^2)[e^{-nt}(\cos\omega t + \frac{n}{\omega}\sin\omega t) - \cos\omega_1 t]$$

$$+2n\omega_1(\sin\omega_1 t - \frac{\omega_1}{\omega}e^{-nt}\sin\omega t)\}$$

$$-\frac{f}{\omega^2 + n^2}[1 - e^{-nt}(\cos\omega t + \frac{n}{\omega}\sin\omega t)] \qquad\qquad (16.3.13)$$

b. **Case $\omega^2 = 0$**

Equating ω^2 to zero in equation (16.3.12), we obtain:

$$x(l) = \frac{l(v_0 + 2ns_0)}{(l+n)^2} + \frac{l^2 s_0}{(l+n)^2} + \frac{a\omega_1 l}{(l^2 + \omega_1^2)(l+n)^2}\cos\lambda$$

$$+\frac{al^2}{(l^2 + \omega_1^2)(l+n)^2}\sin\lambda - \frac{f}{(l+n)^2} \qquad\qquad (16.3.14)$$

Applying pairs 1, 37, 38, 46, 47, and 36 from Table 1.1, we invert equation (16.3.14) from the Laplace domain into the time domain and obtain the solution of differential equation of motion (16.3.1) with the initial conditions of motion (16.3.2):

$$x = (v_0 + 2ns_0)te^{-nt} + s_0(1 - nt)e^{-nt}$$

$$+\frac{a\omega_1\cos\lambda}{4n^2\omega_1^2 + (\omega_1^2 - n^2)^2}\{2n[e^{-nt}(1+nt) - \cos\omega_1 t]$$

$$-(\omega_1^2 - n^2)(\frac{1}{\omega_1}\sin\omega_1 t - te^{-nt}))\}$$

$$+\frac{a\sin\lambda}{4n^2\omega_1^2 + (\omega_1^2 - n^2)^2}\{2n\omega_1(\sin\omega_1 t - \omega_1 te^{-nt})$$

$$-(\omega_1^2 - n^2)[e^{-nt}(1+nt) - \cos\omega_1 t]\} - \frac{f}{n^2}[1 - e^{-nt}(1+nt)]$$

$$(16.3.15)$$

c. Case $\omega^2 < 0$

Taking in equation (16.3.12) ω^2 with the negative sign, we may write:

$$x(l) = \frac{l(v_0 + 2ns_0)}{(l+n)^2 - \omega^2} + \frac{l^2 s_0}{(l+n)^2 - \omega^2} + \frac{a\omega_1 l}{(l^2 + \omega_1^2)[(l+n)^2 - \omega^2]}\cos\lambda$$

$$+\frac{al^2}{(l^2 + \omega_1^2)[(l+n)^2 - \omega^2]}\sin\lambda - \frac{f}{(l+n)^2 - \omega^2} \qquad (16.3.16)$$

Using pairs $1, 25, 28, 48, 49,$ and 23 from Table 1.1, we invert equation (16.3.16) from the Laplace domain into the time domain and obtain the solution of differential equation of motion (16.3.1) with the initial conditions of motion (16.3.2):

$$x = \frac{v_0 + 2ns_0}{\omega}e^{-nt}\sinh\omega t + \frac{s_0}{\omega}e^{-nt}(\omega\cosh\omega t - n\sinh\omega t)$$

$$+\frac{a\omega_1\cos\lambda}{4n^2\omega_1^2 + (\omega_1^2 - \omega_0^2)^2}\{e^{-nt}[2n\cosh\omega t + \frac{1}{\omega}(\omega_1^2 + \omega^2)\sinh\omega t]$$

$$- 2n\cos\omega_1 t - \frac{1}{\omega_1}(\omega_1^2 - \omega_0^2)\sin\omega_1 t\}$$

$$+\frac{(\omega_1^2 - \omega_0^2)a\sin\lambda}{4n^2\omega_1^2 + (\omega_1^2 - \omega_0^2)^2}\{\frac{\omega_1^2}{\omega^2}(1 - \cos\omega_1 t)$$

$$-\frac{\omega_1^2}{n^2-\omega^2}[1-e^{-nt}(\cos h\,\omega t-\frac{n}{\omega}\sin h\,\omega t)]$$

$$+\frac{2n\omega_1}{\omega_1^2-\omega_0^2}(\sin\omega_1 t-\frac{n}{\omega}e^{-nt}\sin h\,\omega t)\}$$

$$-\frac{f}{n^2-\omega^2}[1-e^{-nt}(\cos h\,\omega t+\frac{n}{\omega}\sin h\,\omega t)] \qquad (16.3.17)$$

16.3.2 Initial Displacement Equals Zero

The initial conditions of motion are:

$$\text{for} \quad t=0 \quad x=0; \quad \frac{dx}{dt}=v_0 \qquad (16.3.18)$$

The solutions of differential equation of motion (16.3.1) with the initial conditions of motion (16.3.18) for the following cases are presented below.

a. Case $\omega^2 > 0$

$$x = \frac{1}{\omega}v_0 e^{-nt}\sin\omega t$$

$$+\frac{a\omega_1\cos\lambda}{(\omega_1^2-\omega_0^2)^2+4n^2\omega_1^2}\{2n[e^{-nt}(\cos\omega t+\frac{n}{\omega}\sin\omega t)-\cos\omega_1 t]$$

$$-(\omega_1^2-\omega_0^2)(\frac{1}{\omega_1}\sin\omega_1 t-\frac{1}{\omega}e^{-nt}\sin\omega t)\}$$

$$+\frac{a\sin\lambda}{(\omega_1^2-\omega_0^2)^2+4n^2\omega_1^2}\{(\omega_1^2-\omega_0^2)[e^{-nt}(\cos\omega t+\frac{n}{\omega}\sin\omega t)-\cos\omega_1 t]$$

$$+2n\omega_1(\sin\omega_1 t-\frac{\omega_1}{\omega}e^{-nt}\sin\omega t)\}$$

$$-\frac{f}{\omega^2+n^2}[1-e^{-nt}(\cos\omega t+\frac{n}{\omega}\sin\omega t)] \qquad (16.3.19)$$

b. Case $\omega^2 = 0$

$$x = v_0 t e^{-nt} + \frac{a\omega_1 \cos\lambda}{4n^2\omega_1^2 + (\omega_1^2 - n^2)^2}\{2n[e^{-nt}(1+nt) - \cos\omega_1 t]$$

$$-(\omega_1^2 - n^2)(\frac{1}{\omega_1}\sin\omega_1 t - te^{-nt})\}$$

$$+\frac{a\sin\lambda}{4n^2\omega_1^2 + (\omega_1^2 - n^2)^2}\{2n\omega_1(\sin\omega_1 t - \omega_1 te^{-nt})$$

$$-(\omega_1^2 - n^2)[e^{-nt}(1+nt) - \cos\omega_1 t] - \frac{f}{n^2}[1 - e^{-nt}(1+nt)]$$

$$(16.3.20)$$

c. Case $\omega^2 < 0$

$$x = \frac{v_0}{\omega}e^{-nt}\sinh\omega t$$

$$+\frac{a\omega_1\cos\lambda}{4n^2\omega_1^2 + (\omega_1^2 - \omega_0^2)^2}\{e^{-nt}[2n\cosh\omega t + \frac{1}{\omega}(\omega_1^2 + \omega^2)\sinh\omega t]$$

$$-2n\cos\omega_1 t - \frac{1}{\omega_1}(\omega_1^2 - \omega_0^2)\sin\omega_1 t\}$$

$$+\frac{(\omega_1^2 - \omega_0^2)a\sin\lambda}{4n^2\omega_1^2 + (\omega_1^2 - \omega_0^2)^2}\{\frac{\omega_1^2}{\omega^2}(1 - \cos\omega_1 t)$$

$$-\frac{\omega_1^2}{n^2 - \omega^2}[1 - e^{-nt}(\cosh\omega t - \frac{n}{\omega}\sinh\omega t)]$$

$$-\frac{\omega_1^2}{n^2 - \omega^2}[1 - e^{-nt}(\cosh\omega t - \frac{n}{\omega}\sinh\omega t)]$$

$$+\frac{2n\omega_1}{\omega_1^2 - \omega_0^2}(\sin\omega_1 t - \frac{n}{\omega}e^{-nt}\sinh\omega t)\}$$

$$-\frac{f}{n^2 - \omega^2}[1 - e^{-nt}(\cosh\omega t + \frac{n}{\omega}\sinh\omega t)] \qquad (16.3.21)$$

16.3.3 Initial Velocity Equals Zero

The initial conditions of motion are:

$$\text{for} \quad t = 0 \quad x = s_0; \quad \frac{dx}{dt} = 0 \qquad (16.3.22)$$

The solutions of differential equation of motion (16.3.1) with the initial conditions of motion (16.3.22) for the following cases are presented below.

a. Case $\omega^2 > 0$

$$x = \frac{1}{\omega} 2 n s_0 e^{-nt} \sin \omega t + s_0 e^{-nt} (\cos \omega t - \frac{n}{\omega} \sin \omega t)$$

$$+ \frac{a\omega_1 \cos \lambda}{(\omega_1^2 - \omega_0^2)^2 + 4n^2\omega_1^2} \{ 2n[e^{-nt}(\cos \omega t + \frac{n}{\omega} \sin \omega t) - \cos \omega_1 t]$$

$$- (\omega_1^2 - \omega_0^2)(\frac{1}{\omega_1} \sin \omega_1 t - \frac{1}{\omega} e^{-nt} \sin \omega t) \}$$

$$+ \frac{a \sin \lambda}{(\omega_1^2 - \omega_0^2)^2 + 4n^2\omega_1^2} \{ (\omega_1^2 - \omega_0^2)[e^{-nt}(\cos \omega t + \frac{n}{\omega} \sin \omega t) - \cos \omega_1 t]$$

$$+ 2n\omega_1(\sin \omega_1 t - \frac{\omega_1}{\omega} e^{-nt} \sin \omega t) \}$$

$$- \frac{f}{\omega^2 + n^2} [1 - e^{-nt}(\cos \omega t + \frac{n}{\omega} \sin \omega t)] \qquad (16.3.23)$$

b. Case $\omega^2 = 0$

$$x = 2 n s_0 t e^{-nt} + s_0 (1 - nt) e^{-nt}$$

$$+ \frac{a\omega_1 \cos \lambda}{4n^2\omega_1^2 + (\omega_1^2 - n^2)^2} \{ 2n[e^{-nt}(1 + nt) - \cos \omega_1 t]$$

$$- (\omega_1^2 - n^2)(\frac{1}{\omega_1} \sin \omega_1 t - te^{-nt}) \}$$

$$+\frac{a\sin\lambda}{4n^2\omega_1^2+(\omega_1^2-n^2)^2}\{2n\omega_1(\sin\omega_1 t-\omega_1 te^{-nt})$$

$$-(\omega_1^2-n^2)[e^{-nt}(1+nt)-\cos\omega_1 t]\}-\frac{f}{n^2}[1-e^{-nt}(1+nt)]$$

$$(16.3.24)$$

c. Case $\omega^2<0$

$$x=\frac{2ns_0}{\omega}e^{-nt}\sinh\omega t+\frac{s_0}{\omega}e^{-nt}(\omega\cosh\omega t-n\sinh\omega t)$$

$$+\frac{a\omega_1\cos\lambda}{4n^2\omega_1^2+(\omega_1^2-\omega_0^2)^2}\{e^{-nt}[2n\cosh\omega t+\frac{1}{\omega}(\omega_1^2+\omega^2)\sinh\omega t]$$

$$-2n\cos\omega_1 t-\frac{1}{\omega_1}(\omega_1^2-\omega_0^2)\sin\omega_1 t\}$$

$$+\frac{(\omega_1^2-\omega_0^2)a\sin\lambda}{4n^2\omega_1^2+(\omega_1^2-\omega_0^2)^2}\{\frac{\omega_1^2}{\omega^2}(1-\cos\omega_1 t)$$

$$-\frac{\omega_1^2}{n^2-\omega^2}[1-e^{-nt}(\cosh\omega t-\frac{n}{\omega}\sinh\omega t)]$$

$$+\frac{2n\omega_1}{\omega_1^2-\omega_0^2}(\sin\omega_1 t-\frac{n}{\omega}e^{-nt}\sinh\omega t)\}$$

$$+\frac{f}{n^2-\omega^2}[1-e^{-nt}(\cosh\omega t+\tfrac{n}{\omega}\sinh\omega t)]\qquad(16.3.25)$$

16.3.4 Both the Initial Displacement and Velocity Equal Zero
The initial conditions of motion are:

$$\text{for}\quad t=0\quad x=0;\quad \frac{dx}{dt}=0\qquad(16.3.26)$$

The solutions of differential equation of motion (16.3.1) with the initial conditions of motion (16.3.26) for the following cases are presented below.

a. Case $\omega^2 > 0$

$$x = \frac{a\omega_1 \cos \lambda}{(\omega_1^2 - \omega_0^2)^2 + 4n^2\omega_1^2}\{2n[e^{-nt}(\cos \omega t + \frac{n}{\omega}\sin \omega t) - \cos \omega_1 t]$$

$$-(\omega_1^2 - \omega_0^2)(\frac{1}{\omega_1}\sin \omega_1 t - \frac{1}{\omega}e^{-nt}\sin \omega t)\}$$

$$+\frac{a\sin \lambda}{4n^2\omega_1^2 + (\omega_1^2 - \omega_0^2)^2}\{(\omega_1^2 - \omega_0^2)[e^{-nt}(\cos \omega t + \frac{n}{t}\sin \omega t)$$

$$-\cos \omega_1 t] + 2n\omega_1(\sin \omega_1 t - \frac{\omega_1}{\omega}e^{-nt}\sin \omega t)\}$$

$$-\frac{f}{\omega^2 + n^2}[1 - e^{-nt}(\cos \omega t + \frac{n}{\omega}\sin \omega t)] \qquad \text{(16.3.27)}$$

b. Case $\omega^2 = 0$

$$x = \frac{a\omega_1 \cos \lambda}{4n^2\omega_1^2 + (\omega_1^2 - n^2)^2}\{2n[e^{-nt}(1 + nt) - \cos \omega_1 t]$$

$$-(\omega_1^2 - n^2)\frac{1}{\omega_1}\sin \omega_1 t - te^{-nt})\}$$

$$+\frac{a\sin \lambda}{4n^2\omega_1^2 + (\omega_1^2 - n^2)^2}\{2n\omega_1(\sin \omega_1 t - \omega_1 te^{-nt})$$

$$-(\omega_1^2 - n^2)[e^{-nt}(1 + nt) - \cos \omega_1 t]\}$$

$$-\frac{f}{n^2}[1 - e^{-nt}(1 + nt)]$$

$$\text{(16.3.28)}$$

c. Case $\omega^2 < 0$

$$x = \frac{a\omega_1 \cos \lambda}{4n^2\omega_1^2 + (\omega_1^2 - \omega_0^2)^2} \{e^{-nt}[2n\cosh\omega t + \frac{1}{\omega}(\omega_1^2 + \omega^2)\sinh\omega t]$$

$$-2n\cos\omega_1 t - \frac{1}{\omega_1}(\omega_1^2 - \omega_0^2)\sin\omega_1 t\}$$

$$+\frac{(\omega_1^2 - \omega_0^2)a\sin\lambda}{4n^2\omega_1^2 + (\omega_1^2 - \omega_0^2)^2}\{\frac{\omega_1^2}{\omega^2}(1 - \cos\omega_1 t)$$

$$-\frac{\omega_1^2}{n^2 - \omega^2}[1 - e^{-nt}(\cosh\omega t - \frac{n}{\omega}\sinh\omega t)]$$

$$+\frac{2n\omega_1}{\omega_1^2 - \omega_0^2}(\sin\omega_1 t - \frac{n}{\omega}e^{-nt}\sinh\omega t)\}$$

$$-\frac{f}{n^2 - \omega^2}[1 - e^{-nt}(\cosh\omega t + \frac{n}{\omega}\sinh\omega t)] \qquad (16.3.29)$$

16.4 Time-Dependent Force $Q\left(\rho + \frac{\mu t}{\tau}\right)$

This section describes engineering systems subjected to the force of inertia, the damping force, the stiffness force, and the friction force as resisting forces (Row 16 in Guiding Table 2.1) and the time-dependent force as the active force (Column 4).

The current problem could be associated with the working processes of engineering systems interacting with viscoelastoplastic media. In certain cases, a viscoelastoplastic medium exerts a damping force, a stiffness force, and a friction force during its deformation. More information related to the deformation of viscoelastoplastic media is presented in section 1.2. Sometimes during the initial phase of the working process, the system is subjected to a time-dependent force that acts a predetermined interval of time.

The system is moving in the horizontal direction. We want to determine the law of motion of the system. Figure 16.4 shows the model of a system subjected to the action of a time-dependent force, a damping force, a stiffness force, and a friction force.

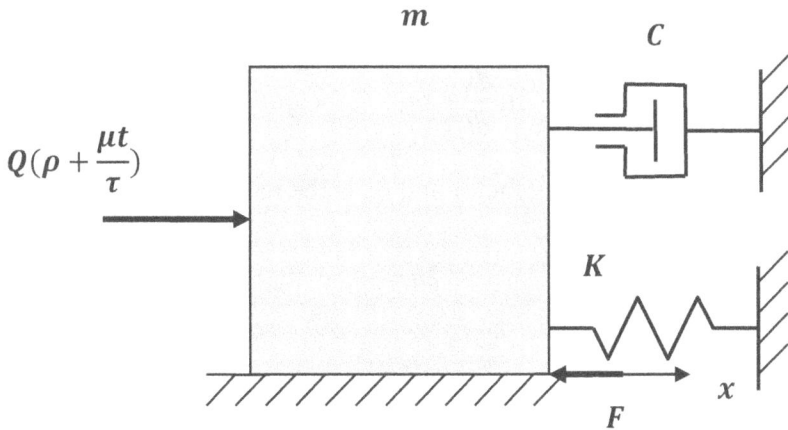

Figure 16.4 Model of a system subjected to a time-dependent force, a damping force, a stiffness force, and a friction force

The considerations above and the model in Figure 16.4 let us compose the left and right sides of the differential equation of motion. The left side consists of the force of inertia, the damping force, the stiffness force, and the friction force, while the right side includes the time-dependent force. Therefore, the differential equation of motion reads:

$$m\frac{d^2x}{dt^2} + C\frac{dx}{dt} + Kx + F = Q(\rho + \frac{\mu t}{\tau}) \qquad (16.4.1)$$

Differential equation of motion (16.4.1) has different solutions for various initial conditions of motion. These solutions are presented below.

16.4.1 General Initial Conditions
The general initial conditions of motion are:

$$\text{for} \quad t = 0 \quad x = s_0; \quad \frac{dx}{dt} = v_0 \qquad (16.4.2)$$

where s_0 and v_0 are the initial displacement and initial velocity respectively.

Dividing equation (16.4.1) by m, we have:

$$\frac{d^2x}{dt^2} + 2n\frac{dx}{dt} + \omega_0^2 x + f = q(\rho + \frac{\mu t}{\tau}) \qquad \textbf{(16.4.3)}$$

where n is the damping factor and ω_0 is the natural frequency, while:

$$2n = \frac{C}{m} \qquad \textbf{(16.4.4)}$$

$$\omega_0^2 = \frac{K}{m} \qquad \textbf{(16.4.5)}$$

$$f = \frac{F}{m} \qquad \textbf{(16.4.6)}$$

$$q = \frac{Q}{m} \qquad \textbf{(16.4.7)}$$

Based on Laplace Transform Pairs 3, 4, 5, and 2 from Table 1.1, we convert differential equation of motion (16.4.3) with the initial conditions of motion (16.4.2) from the time domain into the Laplace domain and obtain the resulting algebraic equation of motion in the Laplace domain:

$$l^2 x(l) - lv_0 - l^2 s_0 + 2nlx(l) - 2nls_0 + \omega_0^2 x(l) - f = q\rho + \frac{q\mu}{\tau l} \quad \textbf{(16.4.8)}$$

The solution of equation (16.4.8) for the displacement $x(l)$ in the Laplace domain reads:

$$x(l) = \frac{l(v_0 + 2ns_0)}{l^2 + 2nl + \omega_0^2} + \frac{l^2 s_0}{l^2 + 2nl + \omega_0^2}$$

$$+ \frac{q\rho - f}{l^2 + 2nl + \omega_0^2} + \frac{q\mu}{l\tau(l^2 + 2nl + \omega_0^2)} \qquad \textbf{(16.4.9)}$$

Applying basic algebra to the denominators of equation (16.4.9), we may write:

$$l^2 + 2nl + \omega_0^2 + n^2 - n^2 = (l + n)^2 + \omega^2 \qquad \textbf{(16.4.10)}$$

where:
$$\omega^2 = \omega_0^2 - n^2 \qquad (16.4.11)$$

while ω^2 could be positive, equal to zero, or negative.

All these three cases are considered below.

a. Case $\omega^2 > 0$

Based on equation (16.4.10), we adjust the denominators in equation (16.4.9), and we have:

$$x(l) = \frac{l(v_0 + 2ns_0)}{(l+n)^2 + \omega^2} + \frac{l^2 s_0}{(l+n)^2 + \omega^2}$$

$$+ \frac{q\rho - f}{(l+n)^2 + \omega^2} + \frac{q\mu}{l\tau[(l+n)^2 + \omega^2)]} \qquad (16.4.12)$$

Using pairs 1, 24, 27, 22 and 39 from Table 1.1, we invert equation (16.4.12) from the Laplace domain into the time domain and obtain the solution of differential equation of motion (16.4.1) with the initial conditions of motion (16.4.2):

$$x = \frac{v_0 + 2ns_0}{\omega} e^{-nt} \sin \omega t + s_0 (\cos \omega t - \frac{n}{\omega} \sin \omega t) e^{-nt}$$

$$+ \frac{q\rho - f}{\omega^2 + n^2} [1 - e^{-nt} (\cos \omega t + \frac{n}{\omega} \sin \omega t)]$$

$$+ \frac{q\mu}{\omega\tau(\omega^2 + n^2)^2} \{ (\omega^2 + n^2)\omega t - 2n\omega$$

$$- e^{-nt} [(\omega^2 - n^2) \sin \omega t - 2n\omega \cos \omega t] \} \qquad (16.4.13)$$

b. Case $\omega^2 = 0$

In equation (16.4.12), equating ω^2 to zero, we write:

$$x(l) = \frac{l(v_0 + 2ns_0)}{(l+n)^2} + \frac{l^2 s_0}{(l+n)^2} + \frac{q\rho - f}{(l+n)^2} + \frac{q\mu}{l\tau(l+n)^2} \qquad (16.4.14)$$

Based on pairs 1, 37, 38, 36, and 15 from Table 1.1, we invert equation (16.4.14) from the Laplace domain into the time domain and obtain the solution of differential equation of motion (16.4.1) with the initial conditions of motion (16.4.2):

$$x = (v_0 + 2ns_0)te^{-nt} + s_0(1-nt)e^{-nt} + \frac{q\rho - f}{n^2}[1 - e^{-nt}(1+nt)]$$

$$+ \frac{q\mu}{\tau n^2}[t - \frac{2}{n} + e^{-nt}\left(\frac{2}{n}+t\right)] \tag{16.4.15}$$

c. Case $\omega^2 < 0$

In equation (16.4.12), taking ω^2 with the negative sign, we may write:

$$x(l) = \frac{l(v_0 + 2ns_0)}{(l+n)^2 - \omega^2} + \frac{l^2 s_0}{(l+n)^2 - \omega^2}$$

$$+ \frac{q\rho - f}{(l+n)^2 - \omega^2} + \frac{q\mu}{l\tau[(l+n)^2 - \omega^2]} \tag{16.4.16}$$

Applying pairs 1, 25, 28, 23, and 40 from Table 1.1, we invert equation (16.4.16) from the Laplace domain into the time domain and obtain the solution of differential equation of motion (16.4.1) with the initial conditions of motion (16.2.2):

$$x = \frac{v_0 + 2ns_0}{\omega}e^{-nt}\sin h\omega + \frac{s_0}{\omega}e^{-nt}(\omega\cos h\omega t - n\sin h\omega t)$$

$$+ \frac{q\rho - f}{n^2 - \omega^2}[1 - e^{-nt}(\cos h\omega t + \frac{n}{\omega}\sin h\omega t)]$$

$$+ \frac{q\mu}{\omega\tau(n^2 - \omega^2)^2}[(n^2 - \omega^2)(\omega t + e^{-nt}\sin h\omega t) - 2n\omega(1 - \cos h\omega t)]$$

$$\tag{16.4.17}$$

16.4.2 Initial Displacement Equals Zero

The initial conditions of motion are:

$$\text{for} \quad t = 0 \quad x = 0; \quad \frac{dx}{dt} = v_0 \tag{16.4.18}$$

The solutions of differential equation of motion (16.4.1) with the initial conditions of motion (16.4.18) for the following cases are presented below.

a. Case $\omega^2 > 0$

$$x = \frac{v_0}{\omega} e^{-nt} \sin \omega t + \frac{q\rho - f}{\omega^2 + n^2} [1 - e^{-nt} (\cos \omega t + \frac{n}{\omega} \sin \omega t)]$$

$$+ \frac{q\mu}{\omega \tau (\omega^2 + n^2)^2} \{ (\omega^2 + n^2) \omega t - 2n\omega$$

$$- e^{-nt} [(\omega^2 - n^2) \sin \omega t - 2n\omega \cos \omega t] \}$$ (16.4.19)

b. Case $\omega^2 = 0$

$$x = v_0 t e^{-nt} + \frac{q\rho - f}{n^2} [1 - e^{-nt} (1 + nt)] + \frac{q\mu}{\tau n^2} [t - \frac{2}{n} + e^{-nt} (\frac{2}{n} + t)]$$

(16.4.20)

c. Case $\omega^2 < 0$

$$x = \frac{v_0}{\omega} e^{-nt} \sin h\omega t + + \frac{q\rho - f}{n^2 - \omega^2} [1 - e^{-nt} (\cos h\omega t + \frac{n}{\omega} \sin h\omega t)]$$

$$+ \frac{q\mu}{\omega \tau (n^2 - \omega^2)^2} [(n^2 - \omega^2)(\omega t + e^{-nt} \sinh \omega t) - 2n\omega (1 - \cosh \omega t)]$$

(16.4.21)

16.4.3 Initial Velocity Equals Zero
The initial conditions of motion are:

$$\text{for} \quad t = 0 \quad x = s_0; \quad \frac{dx}{dt} = 0$$ (16.4.22)

The solutions of differential equation of motion (16.4.1) with the initial conditions of motion (16.4.22) for the following cases are presented below.

a. Case $\omega^2 > 0$

$$x = \frac{2ns_0}{\omega}e^{-nt}\sin\omega t + s_0(\cos\omega t - \frac{n}{\omega}\sin\omega t)e^{-nt}$$

$$+\frac{qp-f}{\omega^2+n^2}[1-e^{-nt}(\cos\omega t + \frac{n}{\omega}\sin\omega t)]$$

$$+\frac{q\mu}{\omega\tau(\omega^2+n^2)^2}\{(\omega^2+n^2)\omega t$$

$$-2n\omega - e^{-nt}[(\omega^2-n^2)\sin\omega t - 2n\omega\cos\omega t]\} \qquad \textbf{(16.4.23)}$$

b. Case $\omega^2 = 0$

$$x = 2ns_0te^{-nt} + s_0(1-nt)e^{-nt} + \frac{qp-f}{\tau n^2}[1-e^{-nt}(1+nt)]$$

$$+\frac{q\mu}{\tau n^2}[t - \frac{2}{n} + e^{-nt}\left(\frac{2}{n}+t\right)] \qquad \textbf{(16.4.24)}$$

c. Case $\omega^2 < 0$

$$x = \frac{2ns_0}{\omega}e^{-nt}\sin h\omega t + \frac{s_0}{\omega}e^{-nt}(\omega\cos h\omega t - n\sin h\omega t)$$

$$+\frac{qp-f}{n^2-\omega^2}[1-e^{-nt}(\cos h\omega t + \frac{n}{\omega}\sin h\omega t)]$$

$$+\frac{q\mu}{\omega\tau(n^2-\omega^2)^2}[(n^2-\omega^2)(\omega t + e^{-nt}\sin h\omega t)$$

$$-2n\omega(1-\cos h\omega t)]$$

$$\textbf{(16.4.25)}$$

16.4.4 Both the Initial Displacement and Velocity Equal Zero

The initial conditions of motion are:

$$\text{for} \quad t = 0 \quad x = 0; \quad \frac{dx}{dt} = 0 \qquad (16.4.26)$$

The solutions of differential equation of motion (16.4.1) with the initial conditions of motion (16.4.26) for the following cases are presented below.

a. Case $\omega^2 > 0$

$$x = \frac{q\rho - f}{\omega^2 + n^2}[1 - e^{-nt}(\cos \omega t + \frac{n}{\omega}\sin \omega t)]$$

$$+ \frac{q\mu}{\omega \tau(\omega^2 + n^2)^2}\{(\omega^2 + n^2)\omega t - 2n\omega$$

$$- e^{-nt}[(\omega^2 - n^2)\sin \omega t - 2n\omega \cos \omega t]\} \qquad (16.4.27)$$

b. Case $\omega^2 = 0$

$$x = \frac{q\rho - f}{n^2}[1 - e^{-nt}(1 + nt)] + \frac{q\mu}{\tau n^2}[t - \frac{2}{n} + e^{-nt}\left(\frac{2}{n} + t\right)] \ (16.4.28)$$

c. Case $\omega^2 < 0$

$$x = + \frac{q\rho - f}{n^2 - \omega^2}[1 - e^{-nt}(\cos h\omega t + \frac{n}{\omega}\sin h\omega t)]$$

$$+ \frac{q\mu}{\omega \tau(n^2 - \omega^2)^2}[(n^2 - \omega^2)(\omega t + e^{-nt}\sin h\omega t) - 2n\omega(1 - \cos h\omega t)]$$

$$(16.4.29)$$

16.5 Constant Force R and Harmonic Force $A\sin(\omega_1 t + \lambda)$

The intersection of Row 16 and Column 5 in Guiding Table 2.1 indicates that this section describes engineering systems subjected to the force of inertia, the damping force, the stiffness force, and the friction force as the resisting forces and the constant active force and harmonic force as the active forces. The current problem could be associated with the working process of a vibratory system intended for the interaction with a viscoelastoplastic medium that exerts the damping, the stiffness, and the friction forces as a reaction to its deformation. Additional considerations related to the deformation of a viscoelastoplastic medium as well as to the behavior of a friction force applied to a vibratory system are presented in section 1.2.

The system is moving in the horizontal direction. We want to determine the law of motion of the system. Figure 16.5 shows the model of a system subjected to the action of a constant active force, a harmonic force, a damping force, a stiffness force, and a friction force.

Accounting for the considerations above and the model shown in Figure 16.5, we may compose the left and right sides of the differential equation of motion of the system. The left side consists of the force of inertia, the damping force, the stiffness force, and the friction force. The right side includes the constant active force and the harmonic force. Hence, the differential equation of motion reads:

$$m\frac{d^2x}{dt^2} + C\frac{dx}{dt} + Kx + F = R + A\sin(\omega_1 t + \lambda) \qquad \textbf{(16.5.1)}$$

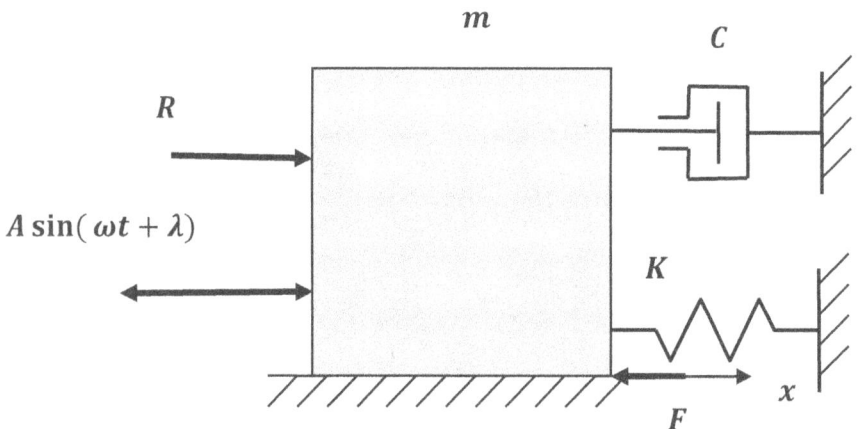

Figure 16.5 Model of a system subjected to a constant active force, a harmonic force, a damping force, a stiffness force, and a friction force

Differential equation of motion (16.5.1) has different solutions for various initial conditions of motion. These solutions are presented below.

16.5.1 General Initial Conditions

The general initial conditions of motion are:

$$\text{for} \quad t = 0 \quad x = s_0; \quad \frac{dx}{dt} = v_0 \qquad (16.5.2)$$

Transforming the sinusoidal function in equation (16.5.1) and dividing the latter by m, we obtain:

$$\frac{d^2x}{dt^2} + 2n\frac{dx}{dt} + \omega_0^2 x + f = r + a\sin\omega_1 t\cos\lambda + a\cos\omega_1 t\sin\lambda \quad (16.5.3)$$

where n is the damping factor and ω_0 in the natural frequency, while:

$$2n = \frac{C}{m} \qquad (16.5.4)$$

$$\omega_0^2 = \frac{K}{m} \qquad (16.5.5)$$

$$f = \frac{F}{m} \qquad (16.5.6)$$

$$r = \frac{R}{m} \qquad (16.5.7)$$

$$a = \frac{A}{m} \qquad (16.5.8)$$

Using Laplace Transform Pairs 3, 4, 1, 5, 6, and 7 from Table 1.1, we convert differential equation of motion (16.5.3) with the initial conditions of motion (16.5.2) from the time domain into the Laplace domain and obtain the resulting algebraic equation of motion in the Laplace domain:

$$l^2x(l) - lv_0 - l^2s_0 + 2nlx(l) - 2nls_0 + \omega_0^2\,x(l) + f$$

$$= r + \frac{a\omega_1 l \cos \lambda}{l^2 + \omega_1^2} + \frac{al^2 \sin \lambda}{l^2 + \omega_1^2} \qquad (16.5.9)$$

Applying algebraic procedures to equation (16.5.9), we have:

$$x(l)(l^2 + 2nl + \omega_0^2) = r - f + l(v_0 + 2ns_0) + l^2 s_0$$

$$+ \frac{a\omega_1 l \cos \lambda}{l^2 + \omega_1^2} + \frac{al^2 \sin \lambda}{l^2 + \omega_1^2} \qquad (16.5.10)$$

Solving equation (16.5.10) for the displacement $x(l)$ in the Laplace domain, we write:

$$x(l) = \frac{r - f}{l^2 + 2nl + \omega_0^2} + \frac{l(v_0 + 2ns_0)}{l^2 + 2nl + \omega_0^2} + \frac{l^2 s_0}{l^2 + 2nl + \omega_0^2}$$

$$+ \frac{a\omega_1 l \cos \lambda}{(l^2 + \omega_1^2)(l^2 + 2nl + \omega_0^2)} + \frac{al^2 \sin \lambda}{(l^2 + \omega_1^2)(l^2 + 2nl + \omega_0^2)} \qquad (16.5.11)$$

Applying basic algebra to the denominators in equation (16.5.11), we have:

$$l^2 + 2nl + \omega_0^2 + n^2 - n^2 = (l + n)^2 + \omega^2 \qquad (16.5.12)$$

where:

$$\omega^2 = \omega_0^2 - n^2 \qquad (16.5.13)$$

while ω^2 could be positive, equal to zero, or negative.
All these three cases are considered below.

a. Case $\omega^2 > 0$

Adjusting the denominators in equation (16.5.11) according to equation (16.5.12), we write:

$$x(l) = \frac{r - f}{(l + n)^2 + \omega^2} + \frac{l(v_0 + 2ns_0)}{(l + n)^2 + \omega^2} + \frac{l^2 s_0}{(l + n)^2 + \omega^2}$$

$$+\frac{a\omega_1 l\cos\lambda}{(l^2+\omega_1^2)[(l+n)^2+\omega^2]}+\frac{al^2\sin\lambda}{(l^2+\omega_1^2)[(l+n)^2+\omega^2]}\quad\textbf{(16.5.14)}$$

Using pairs $1, 22, 24, 27, 42$, and 43 from Table 1.1, we invert equation (16.5.14) from the Laplace domain into the time domain and obtain the solution of differential equation of motion (16.5.1) with the initial conditions of motion (16.2.2):

$$x=\frac{r-f}{\omega_0^2}[1-e^{-nt}(\cos\omega t+\frac{n}{\omega}\sin\omega t)]$$

$$+\frac{1}{\omega}(v_0+2ns_0)e^{-nt}\sin\omega t+s_0 e^{-nt}(\cos\omega t-\frac{n}{\omega}\sin\omega t)$$

$$+\frac{a\omega_1\cos\lambda}{(\omega_1^2-\omega_0^2)^2+4n^2\omega_1^2}\{2n[e^{-nt}(\cos\omega t+\frac{n}{\omega}\sin\omega t)-\cos\omega_1 t]$$

$$-(\omega_1^2-\omega_0^2)(\frac{1}{\omega_1}\sin\omega_1 t-\frac{1}{\omega}e^{-nt}\sin\omega t)\}$$

$$+\frac{a\sin\lambda}{(\omega_1^2-\omega_0^2)^2+4n^2\omega_1^2}\{(\omega_1^2-\omega_0^2)[e^{-nt}(\cos\omega t+\frac{n}{\omega}\sin\omega t)$$

$$-\cos\omega_1 t]+2n\omega_1(\sin\omega_1 t-\frac{\omega_1}{\omega}e^{-nt}\sin\omega t)\}\quad\textbf{(16.5.15)}$$

b. Case $\omega^2=0$

In equation (16.5.14), equating ω^2 to zero, we obtain:

$$x(l)=\frac{r-f}{(l+n)^2}+\frac{l(v_0+2ns_0)}{(l+n)^2}+\frac{l^2 s_0}{(l+n)^2}+\frac{a\omega_1 l\cos\lambda}{(l^2+\omega_1^2)(l+n)^2}$$

$$+\frac{al^2\sin\lambda}{(l^2+\omega_1^2)(l+n)^2}\quad\textbf{(16.5.16)}$$

Applying pairs $1, 36, 37, 38, 46$, and 47 from Table 1.1, we invert equation (16.5.16) from the Laplace domain into the time domain

and obtain the solution of differential equation of motion (16.5.1) with the initial conditions of motion (16.5.2):

$$x = \frac{r-f}{n^2}[1 - e^{-nt}(1+nt)] + (v_0 + 2ns_0)te^{-nt}$$

$$+ s_0(1-nt)e^{-nt} + \frac{a\omega_1 \cos\lambda}{4n^2\omega_1^2 + (\omega_1^2 - n^2)^2}\{2n[e^{-nt}(1+nt) - \cos\omega_1 t]$$

$$- (\omega_1^2 - n^2)(\frac{1}{\omega_1}\sin\omega_1 t - te^{-nt})\}$$

$$+ \frac{a\sin\lambda}{4n^2\omega_1^2 + (\omega_1^2 - n^2)^2}\{2n\omega_1(\sin\omega_1 t - \omega_1 te^{-nt})$$

$$- (\omega_1^2 - n^2)[e^{-nt}(1+nt) - \cos\omega_1 t]\} \qquad (16.5.17)$$

c. Case $\omega^2 < 0$

In equation (16.5.14), taking ω^2 with the negative sign, we may write:

$$x(l) = \frac{r-f}{(l+n)^2 - \omega^2} + \frac{l(v_0 + 2ns_0)}{(l+n)^2 - \omega^2} + \frac{l^2 s_0}{(l+n)^2 - \omega^2}$$

$$+ \frac{a\omega_1 l \cos\lambda}{(l^2 + \omega_1^2)[(l+n)^2 - \omega^2]} + \frac{al^2 \sin\lambda}{(l^2 + \omega_1^2)[(l+n)^2 - \omega^2]} \quad (16.5.18)$$

Based on pairs 1, 23, 25, 28, 48, and 49 from Table 1.1, we invert equation (16.5.18) from the Laplace domain into the time domain and obtain the solution of differential equation of motion (16.5.1) with the initial conditions of motion (16.2.2):

$$x = \frac{r-f}{n^2 - \omega_2}[1 - e^{-nt}(\cos h\omega t + \frac{n}{\omega}\sin h\omega t)]$$

$$+ \frac{v_0 + 2ns_0}{\omega}e^{-nt}\sin h\omega t + \frac{s_0}{\omega}e^{-nt}(\omega\cos h\omega t - n\sin h\omega t)$$

$$+ \frac{a\omega_1 \cos\lambda}{4n^2\omega_1^2 + (\omega_1^2 - \omega_0^2)^2}\{2n[e^{-nt}(\cos h\omega t + \frac{n}{\omega}\sin h\omega t)$$

$$-\cos\omega_1 t] - (\omega_1^2 - \omega_0^2)(\frac{1}{\omega_1}\sin\omega_1 t - \frac{1}{\omega}e^{-nt}\sin h\omega t)\}$$

$$+\frac{(\omega_1^2 - \omega_0^2)a\sin\lambda}{4n^2\omega_1^2 + (\omega_1^2 - \omega_0^2)^2}\{\frac{\omega_1^2}{\omega^2}(1 - \cos\omega_1 t)$$

$$-\frac{\omega_1^2}{n^2 - \omega^2}[1 - e^{-nt}(\cosh\omega t - \frac{n}{\omega}\sinh\omega t)]$$

$$+\frac{2n\omega_1}{\omega_1^2 - \omega_0^2}(\sin\omega_1 t - \frac{\omega_1}{\omega}e^{-nt}\sinh\omega t)\} \qquad \textbf{(16.5.19)}$$

16.5.2 Initial Displacement Equals Zero

The initial conditions of motion are:

$$\text{for} \quad t = 0 \quad x = 0; \quad \frac{dx}{dt} = v_0 \qquad \textbf{(16.5.20)}$$

The solutions of differential equation of motion (16.5.1) with the initial conditions of motion (16.5.20) for the following cases are presented below.

a.   Case $\omega^2 > 0$

$$x = \frac{r - f}{\omega_0^2}[1 - e^{-nt}(\cos\omega t + \frac{n}{\omega}\sin\omega t)] + \frac{1}{\omega}v_0 e^{-nt}\sin\omega t$$

$$+\frac{a\omega_1\cos\lambda}{(\omega_1^2 - \omega_0^2)^2 + 4n^2\omega_1^2}\{2n[e^{-nt}(\cos\omega t + \frac{n}{\omega}\sin\omega t) - \cos\omega_1 t]$$

$$-(\omega_1^2 - \omega_0^2)(\frac{1}{\omega_1}\sin\omega_1 t - \frac{1}{\omega}e^{-nt}\sin\omega t)\}$$

$$+\frac{a\sin\lambda}{(\omega_1^2 - \omega_0^2)^2 + 4n^2\omega_1^2}\{(\omega_1^2 - \omega_0^2)[e^{-nt}(\cos\omega t + \frac{n}{\omega}\sin\omega t)$$

$$-\cos\omega_1 t] + 2n\omega_1(\sin\omega_1 t - \frac{\omega_1}{\omega}e^{-nt}\sin\omega t)\} \qquad \textbf{(16.5.21)}$$

b. Case $\omega^2 = 0$

$$x = \frac{r-f}{n^2}[1 - e^{-nt}(1+nt)] + v_0 t e^{-nt}$$

$$+ \frac{a\omega_1 \cos\lambda}{4n^2\omega_1^2 + (\omega_1^2 - n^2)^2}\{2n[e^{-nt}(1+nt) - \cos\omega_1 t]$$

$$-(\omega_1^2 - n^2)(\frac{1}{\omega_1}\sin\omega_1 t - te^{-nt})\}$$

$$+ \frac{a\sin\lambda}{4n^2\omega_1^2 + (\omega_1^2 - n^2)^2}\{2n\omega_1(\sin\omega_1 t - \omega_1 t e^{-nt})$$

$$-(\omega_1^2 - n^2)[e^{-nt}(1+nt) - \cos\omega_1 t]\} \qquad \textbf{(16.5.22)}$$

c. Case $\omega^2 < 0$

$$x = \frac{r-f}{n^2 - \omega_2}[1 - e^{-nt}(\cos h\omega t + \frac{n}{\omega}\sin h\omega t)] + \frac{v_0}{\omega}e^{-nt}\sin h\omega t$$

$$+ \frac{a\omega_1 \cos\lambda}{4n^2\omega_1^2 + (\omega_1^2 - \omega_0^2)^2}\{e^{-nt}[2n\cos h\omega t + \frac{1}{\omega}(\omega_1^2 + \omega^2)\sin h\omega t]$$

$$-2n\cos\omega_1 t - \frac{1}{\omega_1}(\omega_1^2 - \omega_0^2)\sin\omega_1 t\}$$

$$+ \frac{(\omega_1^2 - \omega_0^2)a\sin\lambda}{4n^2\omega_1^2 + (\omega_1^2 - \omega_0^2)^2}\{\frac{\omega_1^2}{\omega^2}(1 - \cos\omega_1 t)$$

$$-\frac{\omega_1^2}{n^2 - \omega^2}[1 - e^{-nt}(\cos h\omega t - \frac{n}{\omega}\sin h\omega t)]$$

$$+ \frac{2n\omega_1}{\omega_1^2 - \omega_0^2}(\sin\omega_1 t - \frac{\omega_1}{\omega}e^{-nt}\sin h\omega t)\} \qquad \textbf{(16.5.23)}$$

16.5.3 Initial Velocity Equals Zero

The initial conditions of motion are:

$$\text{for} \quad t = 0 \quad x = s_0; \quad \frac{dx}{dt} = 0 \qquad \textbf{(16.5.24)}$$

The solutions of differential equation of motion (16.5.1) with the initial conditions of motion (16.5.24) for the following cases are presented below.

a. Case $\omega^2 > 0$

$$x = \frac{r-f}{\omega_0^2}[1 - e^{-nt}(\cos\omega t + \frac{n}{\omega}\sin\omega t)] + \frac{1}{\omega}2ns_0e^{-nt}\sin\omega t$$

$$+ s_0e^{-nt}(\cos\omega t - \frac{n}{\omega}\sin\omega t)$$

$$+\frac{a\omega_1\cos\lambda}{(\omega_1^2 - \omega_0^2)^2 + 4n^2\omega_1^2}\{2n[e^{-nt}(\cos\omega t + \frac{n}{\omega}\sin\omega t) - \cos\omega_1 t]$$

$$- (\omega_1^2 - \omega_0^2)(\frac{1}{\omega_1}\sin\omega_1 t - \frac{1}{\omega}e^{-nt}\sin\omega t)\}$$

$$+\frac{a\sin\lambda}{(\omega_1^2 - \omega_0^2)^2 + 4n^2\omega_1^2}\{(\omega_1^2 - \omega_0^2)[e^{-nt}(\cos\omega t + \frac{n}{\omega}\sin\omega t)$$

$$- \cos\omega_1 t] + 2n\omega_1(\sin\omega_1 t - \frac{\omega_1}{\omega}e^{-nt}\sin\omega t)\} \qquad (16.5.25)$$

b. Case $\omega^2 = 0$

$$x = \frac{r-f}{n^2}[1 - e^{-nt}(1 + nt)] + 2ns_0te^{-nt}$$

$$+ s_0(1 - nt)e^{-nt} + \frac{a\omega_1\cos\lambda}{4n^2\omega_1^2 + (\omega_1^2 - n^2)^2}\{2n[e^{-nt}(1 + nt)$$

$$- \cos\omega_1 t] - (\omega_1^2 - n^2)(\frac{1}{\omega_1}\sin\omega_1 t - te^{-nt})\}$$

$$+\frac{a\sin\lambda}{4n^2\omega_1^2 + (\omega_1^2 - n^2)^2}\{2n\omega_1(\sin\omega_1 t - \omega_1 te^{-nt})$$

$$-(\omega_1^2 - n^2)[e^{-nt}(1 + nt) - \cos\omega_1 t]\} \qquad (16.5.26)$$

c. Case $\omega^2 < 0$

$$x = \frac{r-f}{n^2 - \omega_2}[1 - e^{-nt}(\cos h\omega t + \frac{n}{\omega}\sin h\omega t)] + \frac{2ns_0}{\omega}e^{-nt}\sin h\omega t$$

$$+\frac{s_0}{\omega}e^{-nt}(\omega\cos h\omega t - n\sin h\omega t)$$

$$+\frac{a\omega_1 \cos\lambda}{4n^2\omega_1^2 + (\omega_1^2 - \omega_0^2)^2}\{e^{-nt}[2n\cos h\omega t$$

$$+\frac{1}{\omega}(\omega_1^2 + \omega^2)\sin h\omega t] - 2n\cos\omega_1 t - \frac{1}{\omega_1}(\omega_1^2 - \omega_0^2)\sin\omega_1 t\}$$

$$+\frac{(\omega_1^2 - \omega_0^2)a\sin\lambda}{4n^2\omega_1^2 + (\omega_1^2 - \omega_0^2)^2}\{\frac{\omega_1^2}{\omega^2}(1 - \cos\omega_1 t)$$

$$-\frac{\omega_1^2}{n^2 - \omega^2}[1 - e^{-nt}(\cosh\omega t - \frac{n}{\omega}\sinh\omega t)]$$

$$+\frac{2n\omega_1}{\omega_1^2 - \omega_0^2}(\sin\omega_1 t - \frac{\omega_1}{\omega}e^{-nt}\sin h\omega t)\} \qquad (16.5.27)$$

16.5.4 Both the Displacement and Velocity Equal Zero

The initial conditions of motion are:

$$\text{for } t = 0 \quad x = 0; \quad \frac{dx}{dt} = 0 \qquad (16.5.28)$$

The solutions of differential equation of motion (16.5.1) with the initial conditions of motion (16.5.28) for the following cases are presented below.

a. Case $\omega^2 > 0$

$$x = \frac{r-f}{\omega_0^2}[1 - e^{-nt}(\cos\omega t + \frac{n}{\omega}\sin\omega t)]$$

$$+\frac{a\omega_1 \cos\lambda}{(\omega_1^2 - \omega_0^2)^2 + 4n^2\omega_1^2}\{2n[e^{-nt}(\cos\omega t + \frac{n}{\omega}\sin\omega t) - \cos\omega_1 t]$$

$$-(\omega_1^2 - \omega_0^2)(\frac{1}{\omega_1}\sin\omega_1 t - \frac{1}{\omega}e^{-nt}\sin\omega t)\}$$

$$+\frac{a\sin\lambda}{(\omega_1^2-\omega_0^2)^2+4n^2\omega_1^2}\{(\omega_1^2-\omega_0^2)[e^{-nt}(\cos\omega t+\frac{n}{\omega}\sin\omega t)$$

$$-\cos\omega_1 t]+2n\omega_1(\sin\omega_1 t-\frac{\omega_1}{\omega}e^{-nt}\sin\omega t)\} \qquad (16.5.29)$$

b. Case $\omega^2 = 0$

$$x = \frac{r-f}{n^2}[1-e^{-nt}(1+nt)]$$

$$+\frac{a\omega_1\cos\lambda}{4n^2\omega_1^2+(\omega_1^2-n^2)^2}\{2n[e^{-nt}(1+nt)-\cos\omega_1 t]$$

$$-(\omega_1^2-n^2)(\frac{1}{\omega_1}\sin\omega_1 t-te^{-nt})\}$$

$$+\frac{a\sin\lambda}{4n^2\omega_1^2+(\omega_1^2-n^2)^2}\{2n\omega_1(\sin\omega_1-te^{-nt})$$

$$-(\omega_1^2-n^2)[e^{-nt}(1+nt)-\cos\omega_1 t]\} \qquad (16.5.30)$$

c. Case $\omega^2 < 0$

$$x = \frac{r-f}{n^2-\omega_2}[1-e^{-nt}(\cos h\omega t+\frac{n}{\omega}\sin h\omega t)]$$

$$+\frac{a\omega_1\cos\lambda}{4n^2\omega_1^2+(\omega_1^2-\omega_0^2)^2}\{e^{-nt}[2n\cos h\omega t$$

$$+\frac{1}{\omega}(\omega_1^2+\omega^2)\sin h\omega t]-2n\cos\omega_1 t-\frac{1}{\omega_1}(\omega_1^2-\omega_0^2)\sin\omega_1 t\}$$

$$+\frac{(\omega_1^2-\omega_0^2)a\sin\lambda}{4n^2\omega_1^2+(\omega_1^2-\omega_0^2)^2}\{\frac{\omega_1^2}{\omega^2}(1-\cos\omega_1 t)$$

$$-\frac{\omega_1^2}{n^2-\omega^2}[1-e^{-nt}(\cos h\omega t-\frac{n}{\omega}\sin h\omega t)]$$

$$+\frac{2n\omega_1}{\omega_1^2-\omega_0^2}(\sin \omega_1 t-\frac{\omega_1}{\omega}e^{-nt}\sin h\omega t)\}\qquad(16.5.31)$$

16.6 Harmonic Force $A\sin(\omega_1 t+\lambda)$ and Time-Dependent Force $Q\left(\rho+\frac{\mu t}{\tau}\right)$

Guiding Table 2.1 indicates that this section describes the engineering systems subjected to the force of inertia, the damping force, the stiffness force, and the friction force as the resisting forces (Row 16 in Guiding Table 2.1) and to the harmonic force and time-dependent force as the active forces (Column 6).

The current problem could be associated with a vibratory system that the working process of which is characterized by the initial phase utilizing a time-dependent force that is acting for a predetermined interval of time. The system could be intended for the interaction with viscoelastoplastic media that may exert damping, stiffness, and friction forces as the reaction to deformation. Additional considerations related to the deformation of viscoelastoplastic media as well as to the behavior of a friction force applied to a vibratory system are presented in section 1.2.

The system is moving in the horizontal direction. We want to determine the law of motion of the system. Figure 16.6 shows the model of a system subjected to the action of a harmonic force, a time-dependent force, a damping force, a stiffness force, and a friction force.

The considerations above and the model in Figure 16.6 let us compose the left and right sides of the differential equation of motion. The left side consists of the force of inertia, the damping force, the stiffness force, and the friction force. The right side of the equation includes the harmonic force and the time-dependent force. Therefore, the differential equation of motion reads:

$$m\frac{d^2x}{dt^2}+C\frac{dx}{dt}+Kx+F=A\sin(\omega_1 t+\lambda)+Q(\rho+\frac{\mu t}{\tau})\qquad(16.6.1)$$

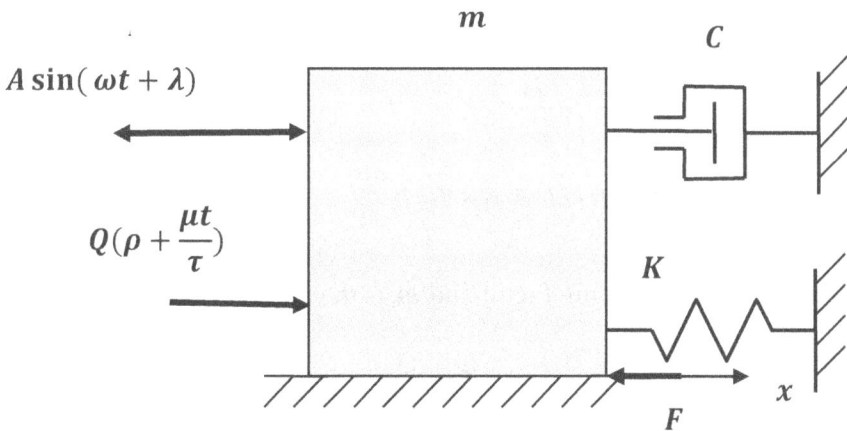

Figure 16.6 Model of a system subjected to a harmonic force, a time-dependent force, a damping force, a stiffness force, and a friction force

The differential equation of motion (16.6.1) has different solutions for various initial conditions of motion. These solutions are presented below.

16.6.1 General Initial Conditions

The general initial conditions of motion are:

$$\text{for} \quad t = 0 \quad x = s_0; \quad \frac{dx}{dt} = v_0 \qquad (16.6.2)$$

where s_0 and v_0 are the initial displacement and initial velocity respectively.

Transforming equation (16.6.1), we have:

$$m\frac{d^2x}{dt^2} + C\frac{dx}{dt} + Kx + F = A\sin\omega_1 t\cos\lambda$$

$$+A\cos\omega_1 t\sin\lambda + Qp + Q\frac{\mu t}{\tau} \qquad (16.6.3)$$

Dividing equation (16.6.3) by m, we may write:

$$\frac{d^2x}{dt^2} + 2n\frac{dx}{dt} + \omega_0^2\, x + f = a\sin\omega_1 t\cos\lambda$$

$$+a\cos\omega_1 t\sin\lambda + q\rho + \frac{q\mu}{\tau l} \qquad (16.6.4)$$

where n is the damping factor and ω_0 is the natural frequency, while:

$$2n = \frac{C}{m} \qquad (16.6.5)$$

$$\omega_0^2 = \frac{K}{m} \qquad (16.6.6)$$

$$f = \frac{F}{m} \qquad (16.6.7)$$

$$a = \frac{A}{m} \qquad (16\,.6.8)$$

$$q = \frac{Q}{m} \qquad (16.6.9)$$

Based on Laplace Transform Pairs 3, 4, 5, 6, 7, and 2 from Table 1.1, we convert the differential equation of motion (16.6.4) with the initial conditions of motion (16.6.2) from the time domain into the Laplace domain. The resulting algebraic equation of motion in the Laplace domain reads:

$$l^2 x(l) - l v_0 - l^2 s_0 + 2nlx(l) - 2nls_0 + \omega_0^2\, x(l) + f$$

$$= \frac{a\omega_1 l\cos\lambda}{l^2 + \omega_1^2} + \frac{al^2\sin\lambda}{l^2 + \omega_1^2} + q\rho + \frac{q\mu}{l\tau} \qquad (16.6.10)$$

Solving equation (16.6.10) for the displacement $x(l)$ in the Laplace domain, we write:

$$x(l) = \frac{l(v_0 + 2ns_0)}{l^2 + 2nl + \omega_0^2} + \frac{l^2 s_0}{l^2 + 2nl + \omega_0^2}$$

$$+ \frac{a\omega_1 l \cos\lambda}{(l^2 + \omega_1^2)(l^2 + 2nl + \omega_0^2)} + \frac{al^2 \sin\lambda}{(l^2 + \omega_1^2)(l^2 + 2nl + \omega_0^2)}$$

$$+ \frac{q\rho - f}{l^2 + 2nl + \omega_0^2} + \frac{q\mu}{l\tau(l^2 + 2nl + \omega_0^2)} \qquad \textbf{(16.6.11)}$$

Applying basic algebra to the denominators in equation (16.6.11), we have:

$$l^2 + 2nl + \omega_0^2 + n^2 - n^2 = (l+n)^2 + \omega^2 \qquad \textbf{(16.6.12)}$$

where:

$$\omega^2 = \omega_0^2 - n^2 \qquad \textbf{(16.6.13)}$$

while ω^2 could be positive, equal to zero, or negative.
   All these three cases are considered below.

a. Case $\omega^2 > 0$

Based on equation (16.6.12), we adjust the denominators in equation (16.6.11) and we may write:

$$x(l) = \frac{l(v_0 + 2ns_0)}{(l+n)^2 + \omega^2} + \frac{l^2 s_0}{(l+n)^2 + \omega^2} + \frac{a\omega_1 l \cos\lambda}{(l^2 + \omega_1^2)[(l+n)^2 + \omega^2]}$$

$$+ \frac{al^2 \sin\lambda}{(l^2 + \omega_1^2)[(l+n)^2 + \omega^2]} + \frac{q\rho - f}{(l+n)^2 + \omega^2} + \frac{q\mu}{l\tau[(l+n)^2 + \omega^2]}$$

$$\textbf{(16.6.14)}$$

Applying pairs $1, 24, 27, 42, 43, 22$, and 39 from Table 1.1, we invert the equation (16.6.14) from the Laplace domain into the time domain and obtain the solution of differential equation of motion (16.6.1) with the initial conditions of motion (16.6.2):

$$x = \frac{1}{\omega}(v_0 + 2ns_0)e^{-nt}\sin\omega t + s_0 e^{-nt}(\cos\omega t - \frac{n}{\omega}\sin\omega t)$$

$$+\frac{a\omega_1 \cos\lambda}{(\omega_1^2 - \omega_0^2)^2 + 4n^2\omega_1^2}\{2n[e^{-nt}(\cos\omega t + \frac{n}{\omega}\sin\omega t) - \cos\omega_1 t]$$

$$-(\omega_1^2 - \omega_0^2)(\frac{1}{\omega_1}\sin\omega_1 t - \frac{1}{\omega}e^{-nt}\sin\omega t)\}$$

$$+\frac{a\sin\lambda}{(\omega_1^2 - \omega_0^2)^2 + 4n^2\omega_1^2}\{(\omega_1^2 - \omega_0^2)[e^{-nt}(\cos\omega t + \frac{n}{\omega}\sin\omega t)$$

$$-\cos\omega_1 t] + 2n\omega_1(\sin\omega_1 t - \frac{\omega_1}{\omega}e^{-nt}\sin\omega t)\}$$

$$+\frac{q\rho - f}{\omega^2 + n^2}[1 - e^{-nt}(\cos\omega t + \frac{n}{\omega}\sin\omega t)]$$

$$+\frac{q\mu}{\omega\tau(\omega^2 + n^2)^2}\{(\omega^2 + n^2)\omega t - 2n\omega$$

$$-e^{-nt}[(\omega^2 - n^2)\sin\omega t - 2n\omega\cos\omega t]\} \tag{16.6.15}$$

b. Case $\omega^2 = 0$

Equating in equation (16.6.14) ω^2 to zero, we obtain:

$$x(l) = \frac{l(v_0 + 2ns_0)}{(l+n)^2} + \frac{l^2 s_0}{(l+n)^2} + \frac{a\omega_1 l}{(l^2 + \omega_1^2)(l+n)^2}\cos\lambda$$

$$+\frac{al^2}{(l^2 + \omega_1^2)(l+n)^2}\sin\lambda + \frac{q\rho - f}{(l+n)^2} + \frac{q\mu}{l\tau(l+n)^2} \tag{16.6.16}$$

Using pairs 1, 37, 38, 46, 47, 36, and 15 from Table 1.1, we invert equation (16.6.16) from the Laplace domain into the time domain and obtain the solution of differential equation of motion (16.6.1) with the initial conditions of motion (16.6.2):

$$x = (v_0 + 2ns_0)te^{-nt} + s_0(1 - nt)e^{-nt}$$

$$+ \frac{a\omega_1 \cos \lambda}{4n^2\omega_1^2 + (\omega_1^2 - n^2)^2} \{2n[e^{-nt}(1+nt) - \cos\omega_1 t]$$

$$-(\omega_1^2 - n^2)(\frac{1}{\omega_1}\sin\omega_1 t - te^{-nt})\}$$

$$+ \frac{a\sin\lambda}{4n^2\omega_1^2 + (\omega_1^2 - n^2)^2} \{2n\omega_1(\sin\omega_1 t - \omega_1 te^{-nt})$$

$$-(\omega_1^2 - n^2)[e^{-nt}(1+nt) - \cos\omega_1 t]\} + \frac{qp - f}{n^2}[1 - e^{-nt}(1+nt)]$$

$$+ \frac{q\mu}{\tau n^2}[t - \frac{2}{n} + e^{-nt}\left(\frac{2}{n} + t\right)] \qquad\qquad (16.6.17)$$

c.   Case $\omega^2 < 0$

In equation (16.6.14), taking ω^2 with the negative sign, we may write:

$$x(l) = \frac{l(v_0 + 2ns_0)}{(l+n)^2 - \omega^2} + \frac{l^2 s_0}{(l+n)^2 - \omega^2} + \frac{a\omega_1 l}{(l^2 + \omega_1^2)[(l+n)^2 - \omega^2]}\cos\lambda$$

$$+ \frac{al^2}{(l^2 + \omega_1^2)[(l+n)^2 - \omega^2]}\sin\lambda + \frac{qp - f}{(l+n)^2 - \omega^2} + \frac{q\mu}{l\tau[(l+n)^2 - \omega^2]}$$

$$(16.6.18)$$

Applying pairs 1, 25, 28, 48, 49, 23, and 40 from Table 1.1, we invert equation (16.6.18) from the Laplace domain into the time domain and obtain the solution of differential equation of motion (16.6.1) with the initial conditions of motion (16.6.2):

$$x = \frac{v_0 + 2ns_0}{\omega} e^{-nt} \sin h\omega t + \frac{s_0}{\omega} e^{-nt}(\omega \cos h\omega t - n \sin h\omega t)$$

$$+ \frac{a\omega_1 \cos \lambda}{4n^2\omega_1^2 + (\omega_1^2 - \omega_0^2)^2} \{e^{-nt}[2n \cos h\omega t + \frac{1}{\omega}(\omega_1^2 + \omega^2)\sin h\omega t]$$

$$-2n \cos \omega_1 t - \frac{1}{\omega_1}(\omega_1^2 - \omega_0^2)\sin \omega_1 t\}$$

$$+ \frac{(\omega_1^2 - \omega_0^2)a \sin \lambda}{4n^2\omega_1^2 + (\omega_1^2 - \omega_0^2)^2} \{\frac{\omega_1^2}{\omega^2}(1 - \cos \omega_1 t)$$

$$- \frac{\omega_1^2}{n^2 - \omega^2}[1 - e^{-nt}(\cos h\omega t - \frac{n}{\omega}\sin h\omega t)]$$

$$+ \frac{2n\omega_1}{\omega_1^2 - \omega_0^2}(\sin \omega_1 t - \frac{\omega_1}{\omega}e^{-nt}\sin h\omega t)\}$$

$$+ \frac{q\rho - f}{n^2 - \omega_2}[1 - e^{-nt}(\cos h\omega t + \frac{n}{\omega}\sin h\omega t)]$$

$$+ \frac{q\mu}{\omega\tau(n^2 - \omega^2)^2}[(n^2 - \omega^2)(\omega t + e^{-nt}\sin h\omega t)$$

$$-2n\omega(1 - \cos h\omega t)] \qquad (16.6.19)$$

16.6.2 Initial Displacement Equals Zero

The initial conditions of motion are:

$$\text{for} \quad t = 0 \quad x = 0; \quad \frac{dx}{dt} = v_0 \qquad (16.6.20)$$

The solutions of differential equation of motion (16.6.1) with the initial conditions of motion (16.6.20) for the following cases are presented below.

a. Case $\omega^2 > 0$

$$x = \frac{1}{\omega} v_0 e^{-nt} \sin \omega t + \frac{a\omega_1 \cos \lambda}{(\omega_1^2 - \omega_0^2)^2 + 4n^2\omega_1^2} \{2n[e^{-nt}(\cos \omega t$$

$$+ \frac{n}{\omega} \sin \omega t) - \cos \omega_1 t] - (\omega_1^2 - \omega_0^2)(\frac{1}{\omega_1} \sin \omega_1 t - \frac{1}{\omega} e^{-nt} \sin \omega t)\}$$

$$+ \frac{a \sin \lambda}{(\omega_1^2 - \omega_0^2)^2 + 4n^2\omega_1^2} \{(\omega_1^2 - \omega_0^2)[e^{-nt}(\cos \omega t + \frac{n}{\omega} \sin \omega t) - \cos \omega_1 t]$$

$$+ 2n\omega_1(\sin \omega_1 t - \frac{\omega_1}{\omega} e^{-nt} \sin \omega t)\}$$

$$+ \frac{q\rho - f}{\omega^2 + n^2}[1 - e^{-nt}(\cos \omega t + \frac{n}{\omega} \sin \omega t)]$$

$$+ \frac{q\mu}{\omega\tau(\omega^2 + n^2)^2} \{(\omega^2 + n^2)\omega t - 2n\omega$$

$$- e^{-nt}[(\omega^2 - n^2)\sin \omega t - 2n\omega \cos \omega t]\} \qquad (16.6.21)$$

b. Case $\omega^2 = 0$

$$x = v_0 t e^{-nt} + \frac{a\omega_1 \cos \lambda}{4n^2\omega_1^2 + (\omega_1^2 - n^2)^2} \{2n[e^{-nt}(1 + nt) - \cos \omega_1 t]$$

$$- (\omega_1^2 - n^2)(\frac{1}{\omega_1} \sin \omega_1 t - te^{-nt})\}$$

$$+ \frac{a \sin \lambda}{4n^2\omega_1^2 + (\omega_1^2 - n^2)^2} \{2n\omega_1(\sin \omega_1 t - \omega_1 te^{-nt})$$

$$- (\omega_1^2 - n^2)[e^{-nt}(1 + nt) - \cos \omega_1 t]\}$$

$$+ \frac{q\rho - f}{n^2}[1 - e^{-nt}(1 + nt)] + \frac{q\mu}{\tau n^2}[t - \frac{2}{n} + e^{-nt}(\frac{2}{n} + t)]$$

$$(16.6.22)$$

c. Case $\omega^2 < 0$

$$x = \frac{v_0}{\omega} e^{-nt} \sin h\omega t + \frac{a\omega_1 \cos \lambda}{4n^2\omega_1^2 + (\omega_1^2 - \omega_0^2)^2} \{e^{-nt}[2n\cosh\omega t$$

$$+\frac{1}{\omega}(\omega_1^2 + \omega^2)\sinh\omega t] - 2n\cos\omega_1 t - \frac{1}{\omega_1}(\omega_1^2 - \omega_0^2)\sin\omega_1 t\}$$

$$+\frac{(\omega_1^2 - \omega_0^2)a\sin\lambda}{4n^2\omega_1^2 + (\omega_1^2 - \omega_0^2)^2}\{\frac{\omega_1^2}{\omega^2}(1 - \cos\omega_1 t)$$

$$-\frac{\omega_1^2}{n^2 - \omega^2}[1 - e^{-nt}(\cos h\omega t - \frac{n}{\omega}\sin h\omega t)]$$

$$+\frac{2n\omega_1}{\omega_1^2 - \omega_0^2}(\sin\omega_1 t - \frac{\omega_1}{\omega}e^{-nt}\sinh\omega t)\}$$

$$+\frac{q\rho - f}{n^2 - \omega_2}[1 - e^{-nt}(\cos h\omega t + \frac{n}{\omega}\sin h\omega t)]$$

$$+\frac{q\mu}{\omega\tau(n^2 - \omega^2)^2}[(n^2 - \omega^2)(\omega t + e^{-nt}\sin h\omega t)$$

$$-2n\omega(1 - \cos h\omega t)] \qquad\qquad (16.6.23)$$

16.6.3 Initial Velocity Equals Zero
The initial conditions of motion are:

$$\text{for} \quad t = 0 \quad x = s_0; \quad \frac{dx}{dt} = 0 \qquad\qquad (16.6.24)$$

The solutions of differential equation of motion (16.6.1) with the initial conditions of motion (16.6.24) for the following cases are presented below.

a. Case $\omega^2 > 0$

$$x = \frac{1}{\omega} 2 n s_0 e^{-nt} \sin \omega t + s_0 e^{-nt} (\cos \omega t - \tfrac{n}{\omega} \sin \omega t)$$

$$+ \frac{a \omega_1 \cos \lambda}{(\omega_1^2 - \omega_0^2)^2 + 4 n^2 \omega_1^2} \{ 2 n [e^{-nt} (\cos \omega t + \frac{n}{\omega} \sin \omega t) - \cos \omega_1 t]$$

$$- (\omega_1^2 - \omega_0^2)(\frac{1}{\omega_1} \sin \omega_1 t - \frac{1}{\omega} e^{-nt} \sin \omega t) \}$$

$$+ \frac{a \sin \lambda}{(\omega_1^2 - \omega_0^2)^2 + 4 n^2 \omega_1^2} \{ (\omega_1^2 - \omega_0^2)[e^{-nt} (\cos \omega t + \frac{n}{\omega} \sin \omega t)$$

$$- \cos \omega_1 t] + 2 n \omega_1 (\sin \omega_1 t - \frac{\omega_1}{\omega} e^{-nt} \sin \omega t) \}$$

$$+ \frac{q \rho - f}{\omega^2 + n^2} [1 - e^{-nt} (\cos \omega t + \frac{n}{\omega} \sin \omega t)]$$

$$+ \frac{q \mu}{\omega \tau (\omega^2 + n^2)^2} \{ (\omega^2 + n^2) \omega t - 2 n \omega$$

$$- e^{-nt} [(\omega^2 - n^2) \sin \omega t - 2 n \omega \cos \omega t] \} \qquad (16.6.25)$$

b. Case $\omega^2 = 0$

$$x = 2 n s_0 t e^{-nt} + s_0 (1 - nt) e^{-nt} + \frac{a \omega_1 \cos \lambda}{4 n^2 \omega_1^2 + (\omega_1^2 - n^2)^2} \{ 2 n [e^{-nt} (1 + nt)$$

$$- \cos \omega_1 t] - (\omega_1^2 - n^2)(\frac{1}{\omega_1} \sin \omega_1 t - t e^{-nt}) \}$$

$$+ \frac{a \sin \lambda}{4 n^2 \omega_1^2 + (\omega_1^2 - n^2)^2} \{ 2 n \omega_1 (\sin \omega_1 t - \omega_1 t e^{-nt})$$

$$- (\omega_1^2 - n^2)[e^{-nt} (1 + nt) - \cos \omega_1 t] \}$$

$$+\frac{q\rho-f}{n^2}[1-e^{-nt}(1+nt)]+\frac{q\mu}{\tau n^2}[t-\frac{2}{n}+e^{-nt}\left(\frac{2}{n}+t\right)]$$

$$(16.6.26)$$

c. Case $\omega^2 < 0$

$$x=\frac{2ns_0}{\omega}e^{-nt}\sin h\omega t+\frac{s_0}{\omega}e^{-nt}(\omega\cos h\omega t-n\sin h\omega t)$$

$$+\frac{a\omega_1\cos\lambda}{4n^2\omega_1^2+(\omega_1^2-\omega_0^2)^2}\{e^{-nt}[2n\cos h\omega t$$

$$+\frac{1}{\omega}(\omega_1^2+\omega^2)\sin h\omega t]-2n\cos\omega_1 t-\frac{1}{\omega_1}(\omega_1^2-\omega_0^2)\sin\omega_1 t\}$$

$$+\frac{(\omega_1^2-\omega_0^2)a\sin\lambda}{4n^2\omega_1^2+(\omega_1^2-\omega_0^2)^2}\{\frac{\omega_1^2}{\omega^2}(1-\cos\omega_1 t)$$

$$-\frac{\omega_1^2}{n^2-\omega^2}[1-e^{-nt}(\cos h\omega t-\frac{n}{\omega}\sin h\omega t)]$$

$$+\frac{2n\omega_1}{\omega_1^2-\omega_0^2}(\sin\omega_1 t-\frac{\omega_1}{\omega}e^{-nt}\sin h\omega t)\}$$

$$+\frac{q\rho-f}{n^2-\omega_2}[1-e^{-nt}(\cos h\omega t+\frac{n}{\omega}\sin h\omega t)]$$

$$+\frac{q\mu}{\omega\tau(n^2-\omega^2)^2}[(n^2-\omega^2)(\omega t+e^{-nt}\sin h\omega t)$$

$$-2n\omega(1-\cos h\omega t)]$$

$$(16.6.27)$$

16.6.4 Both the Initial Displacement and Velocity Equal Zero

The initial conditions of motion are:

$$\text{for}\quad t=0\quad x=0;\quad\frac{dx}{dt}=0$$

$$(16.6.28)$$

The solutions of differential equation of motion (16.6.1) with the initial conditions of motion (16.6.28) for the following cases are presented below.

a. Case $\omega^2 > 0$

$$x = \frac{a\omega_1 \cos\lambda}{(\omega_1^2 - \omega_0^2)^2 + 4n^2\omega_1^2}\{2n[e^{-nt}(\cos\omega t + \frac{n}{\omega}\sin\omega t) - \cos\omega_1 t]$$

$$- (\omega_1^2 - \omega_0^2)(\frac{1}{\omega_1}\sin\omega_1 t - \frac{1}{\omega}e^{-nt}\sin\omega t)\}$$

$$+ \frac{a\sin\lambda}{(\omega_1^2 - \omega_0^2)^2 + 4n^2\omega_1^2}\{(\omega_1^2 - \omega_0^2)[e^{-nt}(\cos\omega t + \frac{n}{\omega}\sin\omega t)$$

$$- \cos\omega_1 t] + 2n\omega_1(\sin\omega_1 t - \frac{\omega_1}{\omega}e^{-nt}\sin\omega t)\}$$

$$+ \frac{qp - f}{\omega^2 + n^2}[1 - e^{-nt}(\cos\omega t + \frac{n}{\omega}\sin\omega t)]$$

$$+ \frac{q\mu}{\omega\tau(\omega^2 + n^2)^2}\{(\omega^2 + n^2)\omega t - 2n\omega$$

$$- e^{-nt}[(\omega^2 - n^2)\sin\omega t - 2n\omega\cos\omega t]\} \qquad \textbf{(16.6.29)}$$

b. Case $\omega^2 = 0$

$$x = \frac{a\omega_1 \cos\lambda}{4n^2\omega_1^2 + (\omega_1^2 - n^2)^2}\{2n[e^{-nt}(1 + nt) - \cos\omega_1 t]$$

$$- (\omega_1^2 - n^2)(\frac{1}{\omega_1}\sin\omega_1 t - te^{-nt})\}$$

$$+ \frac{a\sin\lambda}{4n^2\omega_1^2 + (\omega_1^2 - n^2)^2}\{2n\omega_1(\sin\omega_1 t - \omega_1 te^{-nt})$$

$$-(\omega_1^2 - n^2)[e^{-nt}(1 + nt) - \cos\omega_1 t]\} + \frac{qp - f}{n^2}[1 - e^{-nt}(1 + nt)]$$

$$+ \frac{q\mu}{\tau n^2}[t - \frac{2}{n} + e^{-nt}(\frac{2}{n} + t)] \qquad \textbf{(16.6.30)}$$

c. Case $\omega^2 < 0$

$$x = \frac{a\omega_1 \cos \lambda}{4n^2\omega_1^2 + (\omega_1^2 - \omega_0^2)^2} \{e^{-nt}[2n\cos h\omega t + \frac{1}{\omega}(\omega_1^2 + \omega^2)\sin h\omega t]$$

$$-2n\cos \omega_1 t \frac{1}{\omega_1}(\omega_1^2 - \omega_0^2)\sin \omega_1 t\}$$

$$+\frac{(\omega_1^2 - \omega_0^2)a\sin \lambda}{4n^2\omega_1^2 + (\omega_1^2 - \omega_0^2)^2} \{\frac{\omega_1^2}{\omega^2}(1 - \cos \omega_1 t)$$

$$-\frac{\omega_1^2}{n^2 - \omega^2}[1 - e^{-nt}(\cos h\omega t - \frac{n}{\omega}\sin h\omega t)]$$

$$+\frac{2n\omega_1}{\omega_1^2 - \omega_0^2}(\sin \omega_1 t - \frac{\omega_1}{\omega}e^{-nt}\sin h\omega t)\}$$

$$+\frac{q\rho - f}{n^2 - \omega_2}[1 - e^{-nt}(\cos h\omega t + \frac{n}{\omega}\sin h\omega t)]$$

$$+\frac{q\mu}{\omega\tau(n^2 - \omega^2)^2}[(n^2 - \omega^2)(\omega t + e^{-nt}\sin h\omega t) - 2n\omega(1 - \cos h\omega t)]$$

(16.6.31)

17

DAMPING, STIFFNESS, AND CONSTANT RESISTANCE

This chapter discusses problems associated with engineering systems subjected to the force of inertia, the damping force, the stiffness force, and the constant resisting force as the resisting forces. These forces are marked with the plus sign (+) in Row 17 of Guiding Table 2.1. The problems presented in the sections of this chapter differ from each other by the active forces applied to the systems, as indicated by the intersection of Columns 1 through 6 with Row 17. The left sides of the differential equations of motion presented throughout this chapter are identical. The right sides are different and consist of the active forces applied to the systems.

The engineering problems discussed in this chapter could be associated with the working processes of systems interacting with the viscoelastoplastic media that may exert damping, stiffness, and constant resisting forces as a reaction to their deformation (more related information is presented in section 1.2).

17.1 Active Force Equals Zero

According to Guiding Table 2.1, this section describes engineering problems associated with the force of inertia, the damping force, the stiffness force, and the constant resisting force as the resisting forces (Row 17) in the absence of active forces (Column 1). The considerations related to the motion of a system in the absence of active forces are discussed in section 1.3. The current problem could be related to the deformation of the viscoelastoplastic media that exert damping, stiffness, and constant resisting forces as a reaction to the deformation (see section 1.2).

The system is moving in the horizontal direction. We want to determine the law of motion of the system. Figure 17.1 shows the model of a system subjected to the action of a damping force, a stiffness force, and a constant resisting force.

The considerations above and the model in the Figure 17.1 let us compose the left and right sides of the differential equation of motion. The left side consists of the force of inertia, the damping force, the stiffness force, and the constant resisting force, while the right side equals zero. Therefore, the differential equation of motion reads:

$$m\frac{d^2x}{dt^2} + C\frac{dx}{dt} + Kx + P = 0 \qquad (17.1.1)$$

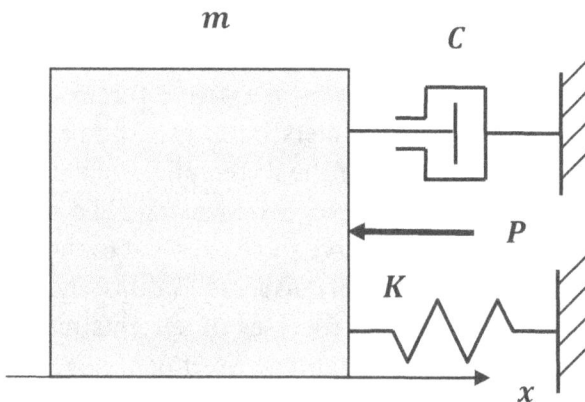

Figure 17.1 Model of a system subjected to a damping force, a stiffness force, and a constant resisting force

Differential equation of motion (17.1.1) has different solutions for various initial conditions of motion. These solutions are presented below.

17.1.1 General Initial Conditions

The general initial conditions of motion are:

$$\text{for} \quad t = 0 \quad x = s_0; \quad \frac{dx}{dt} = v_0 \qquad (17.1.2)$$

where s_0 and v_0 are the initial displacement and initial velocity respectively.

Dividing the equation (17.1.1) by m, we may write:

$$\frac{d^2x}{dt^2} + 2n\frac{dx}{dt} + \omega_0^2 x + p = 0 \qquad (17.1.3)$$

where n is the damping factor and ω_0 is the natural frequency, while:

$$2n = \frac{C}{m} \qquad (17.1.4)$$

$$\omega_0^2 = \frac{K}{m} \qquad (17.1.5)$$

$$p = \frac{P}{m} \qquad (17.1.6)$$

Using Laplace Transform Pairs 3, 4, 5, and 1 from Table 1.1, we convert differential equation of motion (17.1.3) with the initial conditions of motion (17.1.2) from the time domain into the Laplace domain and obtain the corresponding algebraic equation of motion in the Laplace domain:

$$l^2x(l) - lv_0 - l^2s_0 + 2nlx(l) - 2nls_0 + \omega_0^2 x(l) + p = 0 \qquad (17.1.7)$$

Applying some basic algebra to equation (17.1.7), we may write:

$$x(l)(l^2 + 2nl + \omega_0^2) = l(v_0 + 2ns_0) + l^2s_0 - p \qquad (17.1.8)$$

Solving equation (17.1.8) for the displacement $x(l)$ in the Laplace domain, we have:

$$x(l) = \frac{l(v_0 + 2ns_0)}{l^2 + 2nl + \omega_0^2} + \frac{l^2 s_0}{l^2 + 2nl + \omega_0^2} - \frac{p}{l^2 + 2nl + \omega_0^2} \qquad (17.1.9)$$

Based on the appropriate algebraic procedures with the denominators of equation (17.1.9), we may write:

$$l^2 + 2nl + \omega_0^2 + n^2 - n^2 = (l+n)^2 + \omega^2 \qquad (17.1.10)$$

where:

$$\omega^2 = \omega_0^2 - n^2 \qquad (17.1.11)$$

while ω^2 could be positive, equal to zero, or negative.

All these three cases are considered below.

a. Case $\omega^2 > 0$

Adjusting the denominators in equation (17.1.9) according to equation (17.1.10), we have:

$$x(l) = \frac{l(v_0 + 2ns_0)}{(l+n)^2 + \omega^2} + \frac{l^2 s_0}{(l+n)^2 + \omega^2} - \frac{p}{(l+n)^2 + \omega^2} \qquad (17.1.12)$$

Using pairs 1, 24, 27, and 22 from Table 1.1, we invert equation (17.1.12) from the Laplace domain into the time domain and obtain the solution of differential equation of motion (17.1.1) with the initial conditions of motion (17.1.2):

$$x = \frac{v_0 + 2ns_0}{\omega} e^{-nt} \sin \omega t + s_0 e^{-nt} (\cos \omega t - \frac{n}{\omega} \sin \omega t)$$

$$- \frac{p}{\omega^2 + n^2} [1 - e^{-nt} (\cos \omega t + \frac{n}{\omega} \sin \omega t)] \qquad (17.1.13)$$

b. Case $\omega^2 = 0$

Equating in equation (17.1.12) ω^2 to zero, we have:

$$x(l) = \frac{l(v_0 + 2ns_0)}{(l+n)^2} + \frac{l^2 s_0}{(l+n)^2} - \frac{p}{(l+n)^2} \qquad (17.1.14)$$

Based on pairs 1, 37, 38, and 36 from Table 1.1, we invert equation (17.1.14) from the Laplace domain into the time domain and obtain the solution of differential equation of motion (17.1.1) with the initial conditions of motion (17.1.2):

$$x = (v_0 + 2ns_0)te^{-nt} + s_0(1-nt)e^{-nt} - \frac{p}{n^2}[1-e^{-nt}(1+nt)] \qquad (17.1.15)$$

c. Case $\omega^2 < 0$

Taking in equation (17.1.12) ω^2 with the negative sign, we may write:

$$x(l) = \frac{l(v_0 + 2ns_0)}{(l+n)^2 - \omega^2} + \frac{l^2 s_0}{(l+n)^2 - \omega^2} - \frac{p}{(l+n)^2 - \omega^2} \qquad (17.1.16)$$

Using pairs 1, 25, 28, and 23 from Table 1.1, we invert equation (17.1.16) from the Laplace domain into the time domain and obtain the solution of differential equation of motion (17.1.1) with the initial conditions of motion (17.1.2):

$$x = \frac{v_0 + 2ns_0}{\omega}e^{-nt}\sinh\omega t + \frac{s_0}{\omega}e^{-nt}(\omega\cosh\omega t - n\sinh\omega t)$$

$$- \frac{p}{n^2 - \omega^2}[1-e^{-nt}(\cosh\omega t + \frac{n}{\omega}\sinh\omega t)] \qquad (17.1.17)$$

17.1.2 Initial Displacement Equals Zero

The initial conditions of motion are:

$$\text{for} \quad t = 0 \quad x = 0; \quad \frac{dx}{dt} = v_0 \qquad (17.1.18)$$

The solutions of differential equation of motion (17.1.1) with the initial conditions of motion (17.1.18) for the following cases are presented below.

a. Case $\omega^2 > 0$

$$x = \frac{v_0}{\omega}v_0te^{-nt}\sin\omega t - \frac{p}{\omega^2 + n^2}[1-e^{-nt}(\cos\omega t + \frac{n}{\omega}\sin\omega t)] \quad (17.1.19)$$

b. Case $\omega^2 = 0$

$$x = v_0 t e^{-nt} - \frac{p}{n^2}[1 - e^{-nt}(1 + nt)] \tag{17.1.20}$$

c. Case $\omega^2 < 0$

$$x = \frac{1}{\omega} v_0 e^{-nt} \sinh \omega t - \frac{p}{n^2 - \omega^2}[1 - e^{-nt}(\cosh \omega t + \frac{n}{\omega} \sinh \omega t)]$$

$$\tag{17.1.21}$$

17.1.3 Initial Velocity Equals Zero

In this case, the motion of the system into the positive direction, as shown in Figure 17.1, is possible if the initial displacement is taken with the negative sign.

Therefore, the initial conditions of motion are:

$$\text{for} \quad t = 0 \quad x = -s_0; \quad \frac{dx}{dt} = 0 \tag{17.1.22}$$

In addition, it should be emphasized that in the current case the motion is possible if $|Ks_0| > |P|$.

The solutions of differential equation of motion (17.1.1) with the initial conditions of motion (17.1.22) for the following cases are presented below.

a. Case $\omega^2 > 0$

$$x = -\frac{2ns_0}{\omega} e^{-nt} \sin \omega t - s_0 e^{-nt}(\cos \omega t - \frac{n}{\omega} \sin \omega t)$$

$$-\frac{p}{\omega^2 + n^2}[1 - e^{-nt}(\cos \omega t + \frac{n}{\omega} \sin \omega t)] \tag{17.1.23}$$

b. Case $\omega^2 = 0$

$$x = -2ns_0 t e^{-nt} - s_0(1 - nt)e^{-nt} - \frac{p}{n^2}[1 - e^{-nt}(1 + nt)] \tag{17.1.24}$$

c. Case $\omega^2 < 0$

$$x = -\frac{2ns_0}{\omega} e^{-nt} \sinh \omega t - \frac{s_0}{\omega} e^{-nt}(\omega \cosh \omega t - n \sinh \omega t)$$

$$-\frac{p}{n^2 - \omega^2}[1 - e^{-nt}(\cosh \omega t + \frac{n}{\omega} \sinh \omega t)] \tag{17.1.25}$$

17.2 Constant Force R

This section describes the engineering systems subjected to the force of inertia, the damping force, the stiffness force, and the constant resisting force as the resisting forces (Row 17 in Guiding Table 2.1) and the constant active force (Column 2). The current problem could be related to the working processes of engineering systems interacting with viscoelastoplastic media that exert as the reaction to their deformation damping, stiffness, and constant resisting forces. The presence of stiffness forces may cause vibratory motion of the system. Considerations related to the deformation of viscoelastoplastic media as well as to the behavior of a constant resisting force applied to a vibratory system are presented in section 1.2.

The system moves in the horizontal direction. We want to determine the law of motion of this engineering system. Figure 17.2 shows the model of a system that is subjected to the action of a constant active force, a damping force, a stiffness force, and a constant resisting force.

Based on the considerations above and on the model in Figure 17.2, we can assemble the left and right sides of the differential equation of motion of this system. The left side consists of the force of inertia, the damping force, the stiffness force, and the constant resisting force. The right side includes a the constant active force. Therefore, the differential equation of motion reads:

$$m\frac{d^2x}{dt^2} + C\frac{dx}{dt} + Kx + P = R \qquad (17.2.1)$$

Differential equation of motion (17.2.1) has different solutions for various initial conditions of motion. These solutions are presented below.

17.2.1 General Initial Conditions

The general initial conditions of motion are:

$$\text{for} \quad t = 0 \quad x = s_0; \quad \frac{dx}{dt} = v_0 \qquad (17.2.2)$$

where s_0 and v_0 are the initial displacement and initial velocity respectively.

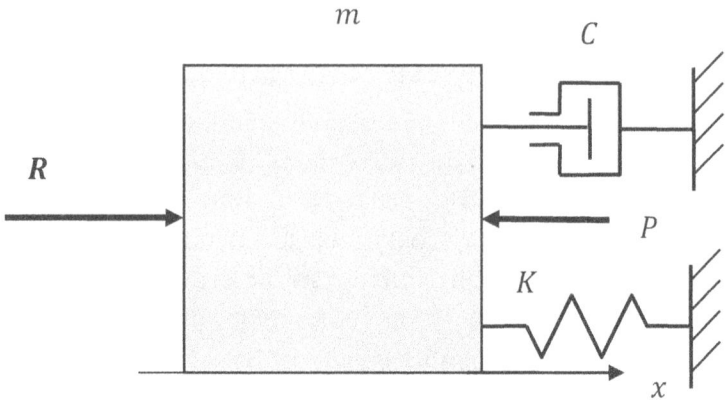

Figure 17.2 Model of a system subjected to a constant active force, a damping force, a stiffness force, and a constant resisting force

Dividing the equation (17.2.1) by m, we may write:

$$\frac{d^2x}{dt^2} + 2n\frac{dx}{dt} + \omega_0^2 x + p = r \qquad (17.2.3)$$

where n is the damping factor and ω_0 is the natural frequency, while:

$$2n = \frac{C}{m} \qquad (17.2.4)$$

$$\omega_0^2 = \frac{K}{m} \qquad (17.2.5)$$

$$p = \frac{P}{m} \qquad (17.2.6)$$

$$r = \frac{R}{m} \qquad (17.2.7)$$

Using Laplace Transform Pairs 3, 4, 5, and 1 from Table 1.1, we convert differential equation of motion (17.2.3) with the initial conditions of motion (17.2.2) from the time domain into the Laplace domain and obtain the corresponding algebraic equation of motion in the Laplace domain:

$$l^2 x(l) - lv_0 - l^2 s_0 + 2nlx(l) - 2nls_0 + \omega_0^2 x(l) + p = r \qquad \textbf{(17.2.8)}$$

Using some basic algebra to equation (17.2.8), we write:

$$x(l)(l^2 + 2nl + \omega_0^2) = r - p + l(v_0 + 2ns_0) + l^2 s_0 \qquad \textbf{(17.2.9)}$$

Solving equation (17.2.9) for the displacement $x(l)$ in the Laplace domain, we have:

$$x(l) = \frac{r - p}{l^2 + 2nl + \omega_0^2} + \frac{l(v_0 + 2ns_0)}{l^2 + 2nl + \omega_0^2} + \frac{l^2 s_0}{l^2 + 2nl + \omega_0^2} \qquad \textbf{(17.2.10)}$$

Applying algebraic procedures to the denominators of equation (17.2.10), we write:

$$l^2 + 2nl + \omega_0^2 + n^2 - n^2 = (l + n)^2 + \omega^2 \qquad \textbf{(17.2.11)}$$

where:

$$\omega^2 = \omega_0^2 - n^2 \qquad \textbf{(17.2.12)}$$

while ω^2 could be positive, equal to zero, or negative.

All these three cases are considered below.

a. Case $\omega^2 > 0$

Transforming the denominators in equation (17.2.10) according to equation (17.2.11), we have:

$$x(l) = \frac{r - p}{(l + n)^2 + \omega^2} + \frac{l(v_0 + 2ns_0)}{(l + n)^2 + \omega^2} + \frac{l^2 s_0}{(l + n)^2 + \omega^2} \qquad \textbf{(17.2.13)}$$

Using pairs 1, 22, 24, and 27 from Table 1.1, we invert equation (17.2.13) from the Laplace domain into the time domain and obtain the solution of differential equation of motion (17.2.1) with the initial conditions of motion (17.2.2):

$$x = \frac{r - p}{\omega^2 + n^2}[1 - e^{-nt}(\cos\omega t + \frac{n}{\omega}\sin\omega t)] + \frac{1}{\omega}(v_0 + 2ns_0)e^{-nt}\sin\omega t$$

$$+ s_0 e^{-nt}(\cos\omega t - \frac{n}{\omega}\sin\omega t) \qquad \textbf{(17.2.14)}$$

b. Case $\omega^2 = 0$

In equation (17.2.13), equating ω^2 to zero, we write:

$$x(l) = \frac{r-p}{(l+n)^2} + \frac{l(v_0 + 2ns_0)}{(l+n)^2} + \frac{l^2 s_0}{(l+n)^2} \qquad (17.2.15)$$

Based on pairs 1, 36, 37, and 38 from Table 1.1, we invert equation (17.2.15) from the Laplace domain into the time domain and obtain the solution of differential equation of motion (17.2.1) with the initial conditions of motion (17.2.2):

$$x = \frac{r-p}{n^2}[1 - e^{-nt}(1+nt)] + (v_0 + 2ns_0)te^{-nt} + s_0 e^{-nt}(1-nt) \quad (17.2.16)$$

c. Case $\omega^2 < 0$

In equation (17.2.13), taking ω^2 with the negative sign, we have:

$$x(l) = \frac{r-p}{(l+n)^2 - \omega^2} + \frac{l(v_0 + 2ns_0)}{(l+n)^2 - \omega^2} + \frac{l^2 s_0}{(l+n)^2 - \omega^2} \qquad (17.2.17)$$

Applying pairs 1, 23, 25, and 28 from Table 1.1, we invert equation (17.2.17) from the Laplace domain into the time domain and obtain the solution of differential equation of motion (17.2.1) with the initial conditions of motion (17.2.2):

$$x = \frac{r-p}{n^2 - \omega^2}\frac{1}{1}[1 - e^{-nt}(\cosh \omega t + \frac{n}{\omega}\sinh \omega t)] + \frac{v_0 + 2ns_0}{\omega}e^{-nt}\sinh \omega t$$

$$+ \frac{s_0}{\omega}e^{-nt}(\omega \cosh \omega t - n \sinh \omega t) \qquad (17.2.18)$$

17.2.2 Initial Displacement Equals Zero

The initial conditions of motion are:

$$\text{for} \quad t = 0 \quad x = 0; \quad \frac{dx}{dt} = v_0 \qquad (17.2.19)$$

The solutions of differential equation of motion (17.2.1) with the initial conditions of motion (17.2.19) for the following cases are presented below.

a. Case $\omega^2 > 0$

$$x = \frac{r-p}{\omega^2 + n^2}[1 - e^{-nt}(\cos\omega t + \frac{n}{\omega}\sin\omega t)] + \frac{1}{\omega}v_0 e^{-nt}\sin\omega t \quad \textbf{(17.2.20)}$$

b. Case $\omega^2 = 0$

$$x = \frac{r-p}{n^2}[1 - e^{-nt}(1 + nt)] + v_0 t e^{-nt} \quad \textbf{(17.2.21)}$$

c. Case $\omega^2 < 0$

$$x = \frac{r-p}{n^2 - \omega^2}[1 - e^{-nt}(\cosh\omega t + \frac{n}{\omega}\sinh\omega t)] + \frac{v_0}{\omega}e^{-nt}\sinh\omega t$$

$$\textbf{(17.2.22)}$$

17.2.3 Initial Velocity Equals Zero

The initial conditions of motion are:

$$\text{for}\quad t = 0 \quad x = s_0;\quad \frac{dx}{dt} = 0 \quad \textbf{(17.2.23)}$$

The solutions of differential equation of motion (17.2.1) with the initial conditions of motion (17.2.23) for the following cases are presented below.

a. Case $\omega^2 > 0$

$$x = \frac{r-p}{\omega^2 + n^2}[1 - e^{-nt}(\cos\omega t + \frac{n}{\omega}\sin\omega t)] + \frac{1}{\omega}2ns_0 e^{-nt}\sin\omega t$$

$$+ s_0 e^{-nt}(\cos\omega t - \frac{n}{\omega}\sin\omega t) \quad \textbf{(17.2.24)}$$

b. Case $\omega^2 = 0$

$$x = \frac{r-p}{n^2}[1 - e^{-nt}(1 + nt)] + 2ns_0 t e^{-nt} + s_0 e^{-nt}(1 - nt) \quad \textbf{(17.2.25)}$$

c. Case $\omega^2 < 0$

$$x = \frac{r-p}{n^2 - \omega^2}[1 - e^{-nt}(\cosh \omega t + \frac{n}{\omega}\sinh \omega t)] + \frac{2ns_0}{\omega}e^{-nt}\sinh \omega t$$

$$+ \frac{s_0}{\omega}e^{-nt}(\omega \cosh \omega t - n\sinh \omega t) \qquad (17.2.26)$$

17.2.4 Both the Initial Displacement and Velocity Equal Zero

The initial conditions of motion are:

$$\text{for} \quad t = 0 \quad x = 0; \quad \frac{dx}{dt} = 0 \qquad (17.2.27)$$

The solutions of differential equation of motion (17.2.1) with the initial conditions of motion (17.2.27) for the following cases are presented below.

a. Case $\omega^2 > 0$

$$x = \frac{r-p}{\omega^2 + n^2}[1 - e^{-nt}(\cos \omega t + \frac{n}{\omega}\sin \omega t)] \qquad (17.2.28)$$

b. Case $\omega^2 = 0$

$$x = \frac{r-p}{n^2}\left[1 - e^{-nt}(1 + nt)\right] \qquad (17.2.29)$$

c. Case $\omega^2 < 0$

$$x = \frac{r-p}{n^2 - \omega^2}[1 - e^{-nt}(\cosh \omega t + \frac{n}{\omega}\sinh \omega t)] \qquad (17.2.30)$$

17.3 Harmonic Force $A \sin(\omega_1 t + \lambda)$

The intersection of Row 17 of Guiding Table 2.1 with Column 3 indicates that this section describes the engineering systems associated with the action of the force of inertia, the damping force, the stiffness force, and the constant resisting force as the resisting forces and the harmonic force as an active force. The current problems could be associated with the working processes of vibratory machines intended for the interaction with the viscoelastoplastic media that exert damping, stiffness, and constant resisting forces as a reaction to their deformation. Additional information regarding the deformation of viscoelastoplastic media as well as regarding the behavior of a constant resisting force applied to a vibratory system is presented in section 1.2.

The system is moving in the horizontal direction. We want to determine the law of motion of the system. Figure 17.3 shows the model of a system subjected to the action of a harmonic force, a damping force, a stiffness force, and a constant resisting force.

Based on the considerations above and on the model in Figure 17.3, we can compose the left and right sides of the differential equation of motion. The left side consists of the force of inertia, the damping force, the stiffness force, and the constant resisting force, while the right side includes the harmonic force. Hence, the differential equation of motion reads:

$$m\frac{d^2x}{dt^2} + C\frac{dx}{dt} + Kx + P = A\sin(\omega_1 t + \lambda) \qquad \textbf{(17.3.1)}$$

Differential equation of motion (17.3.1) has different solutions for various initial conditions of motion. These solutions are presented below.

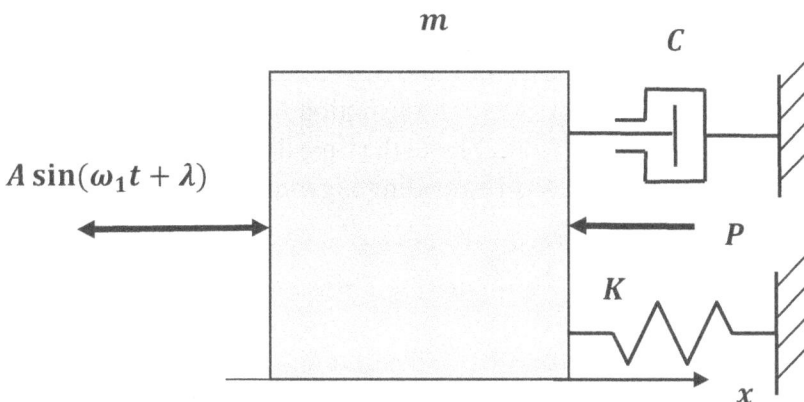

Figure 17.3 Model of a system subjected to a harmonic force, a damping force, a stiffness force, and a constant resisting force

17.3.1 General Initial Conditions

The general initial conditions of motion are:

$$\text{for} \quad t = 0 \quad x = s_0; \quad \frac{dx}{dt} = v_0 \tag{17.3.2}$$

where s_0 and v_0 are the initial displacement and initial velocity respectively.

Transforming the sinusoidal function in equation (17.3.1) and dividing the latter by m, we obtain:

$$\frac{d^2x}{dt^2} + 2n\frac{dx}{dt} + \omega_0^2 x + p = a\sin\omega_1 t\cos\lambda + a\cos\omega_1 t\sin\lambda \tag{17.3.3}$$

where n is the damping factor and ω_0 is the natural frequency, while:

$$2n = \frac{C}{m} \tag{17.3.4}$$

$$\omega_0^2 = \frac{K}{m} \tag{17.3.5}$$

$$p = \frac{P}{m} \tag{17.3.6}$$

$$a = \frac{A}{m} \tag{17.3.7}$$

Using Laplace Transform Pairs 3, 4, 5, 6, and 7 from Table 1.1, we convert differential equation of motion (17.3.3) with the initial conditions of motion (17.3.2) from the time domain into the Laplace domain and obtain the corresponding algebraic equation of motion in the Laplace domain:

$$l^2 x(l) - lv_0 - l^2 s_0 + 2nlx(l) - 2nls_0 + \omega_0^2 x(l) + p$$
$$= \frac{a\omega_1 l}{l^2 + \omega_1^2}\cos\lambda + \frac{al^2}{l^2 + \omega_1^2}\sin\lambda \tag{17.3.8}$$

The solution of equation (17.3.8) for the displacement $x(l)$ in the Laplace domain reads:

$$x(l) = \frac{l(v_0 + 2ns_0)}{l^2 + 2nl + \omega_0^2} + \frac{l^2 s_0}{l^2 + 2nl + \omega_0^2} + \frac{a\omega_1 l}{(l^2 + \omega_1^2)(l^2 + 2nl + \omega_0^2)} \cos\lambda$$

$$+ \frac{al^2}{(l^2 + \omega_1^2)(l^2 + 2nl + \omega_0^2)} \sin\lambda - \frac{p}{l^2 + 2nl + \omega_0^2} \qquad (17.3.9)$$

Applying algebraic procedures to the denominators of equation (17.3.9), we have:

$$l^2 + 2nl + \omega_0^2 + n^2 - n^2 = (l + n)^2 + \omega^2 \qquad (17.3.10)$$

where:

$$\omega^2 = \omega_0^2 - n^2 \qquad (17.3.11)$$

while ω^2 could be positive, equal to zero, or negative.

All these three cases are considered below.

a. Case $\omega^2 > 0$

Transforming the denominators in equation (17.3.9) according to equation (17.3.10), we write:

$$x(l) = \frac{l(v_0 + 2ns_0)}{(l + n)^2 + \omega^2} + \frac{l^2 s_0}{(l + n)^2 + \omega^2} + \frac{a\omega_1 l \cos\lambda}{(l^2 + \omega_1^2)[(l + n)^2 + \omega^2]}$$

$$+ \frac{al^2 \sin\lambda}{(l^2 + \omega_1^2)[(l + n)^2 + \omega^2]} - \frac{p}{(l + n)^2 + \omega^2} \qquad (17.3.12)$$

Based on pairs 1, 24, 27, 42, 43, and 22 from Table 1.1, we invert equation (17.3.12) from the Laplace domain into the time domain and obtain the solution of differential equation of motion (17.3.1) with the initial conditions of motion (17.3.2):

$$x = \frac{1}{\omega}(v_0 + 2ns_0)e^{-nt}\sin\omega t + s_0 e^{-nt}(\cos\omega t - \frac{n}{\omega}\sin\omega t)$$

$$+ \frac{a\omega_1 \cos\lambda}{(\omega_1^2 - \omega_0^2)^2 + 4n^2\omega_1^2}\{2n[e^{-nt}(\cos\omega t + \frac{n}{\omega}\sin\omega t) - \cos\omega_1 t]$$

$$- (\omega_1^2 - \omega_0^2)(\frac{1}{\omega_1}\sin\omega_1 t - \frac{1}{\omega}e^{-nt}\sin\omega t)\}$$

$$+\frac{a\sin\lambda}{(\omega_1^2-\omega_0^2)^2+4n^2\omega_1^2}\{(\omega_1^2-\omega_0^2)[e^{-nt}(\cos\omega t+\frac{n}{\omega}\sin\omega t)$$

$$-\cos\omega_1 t]+2n\omega_1(\sin\omega_1 t-\frac{\omega_1}{\omega}e^{-nt}\sin\omega t)\}$$

$$-\frac{p}{\omega^2+n^2}[1-e^{-nt}(\cos\omega t+\frac{n}{\omega}\sin\omega t)] \tag{17.3.13}$$

b. Case $\omega^2 = 0$

In equation (17.3.12), equating ω^2 to zero, we have:

$$x(l)=\frac{l(v_0+2ns_0)}{(l+n)^2}+\frac{l^2 s_0}{(l+n)^2}+\frac{a\omega_1 l}{(l^2+\omega_1^2)(l+n)^2}\cos\lambda$$

$$+\frac{al^2}{(l^2+\omega_1^2)(l+n)^2}\sin\lambda-\frac{p}{(l+n)^2} \tag{17.3.14}$$

Using pairs 1, 37, 38, 46, 47, and 36 from Table 1.1, we invert equation (17.3.14) from the Laplace domain into the time domain and obtain the solution of differential equation of motion (17.3.1) with the initial conditions of motion (17.3.2):

$$x=(v_0+2ns_0)te^{-nt}+s_0(1-nt)e^{-nt}$$

$$+\frac{a\omega_1\cos\lambda}{4n^2\omega_1^2+(\omega_1^2-n^2)^2}\{2n[e^{-nt}(1+nt)-\cos\omega_1 t]$$

$$-(\omega_1^2-n^2)(\frac{1}{\omega_1}\sin\omega_1 t-te^{-nt})\}$$

$$+\frac{a\sin\lambda}{4n^2\omega_1^2+(\omega_1^2-n^2)^2}\{2n\omega_1(\sin\omega_1 t-\omega_1 te^{-nt})$$

$$-(\omega_1^2-n^2)[e^{-nt}(1+nt)-\cos\omega_1 t]\}$$

$$-\frac{p}{n^2}[1-e^{-nt}(1+nt)] \tag{17.3.15}$$

c. Case $\omega^2 < 0$

In equation (17.3.12), taking ω^2 with the negative sign, we may write:

$$x(l) = \frac{l(v_0 + 2ns_0)}{(l+n)^2 - \omega^2} + \frac{l^2 s_0}{(l+n)^2 - \omega^2} + \frac{a\omega_1 l}{(l^2 + \omega_1^2)[(l+n)^2 - \omega^2]} \cos\lambda$$

$$+ \frac{al^2}{(l^2 + \omega_1^2)[(l+n)^2 - \omega^2]} \sin\lambda - \frac{p}{(l+n)^2 - \omega^2} \qquad (17.3.16)$$

Using pairs 1, 25, 28, 48, 49, and 23 from the Table 1.1, we invert equation (17.3.16) from the Laplace domain into the time domain and obtain the solution of differential equation of motion (17.3.1) with the initial conditions of motion (17.3.2):

$$x = \frac{v_0 + 2ns_0}{\omega} e^{-nt} \sinh\omega t + \frac{s_0}{\omega} e^{-nt} (\omega \cosh\omega t - n\sinh\omega t)$$

$$+ \frac{a\omega_1 \cos\lambda}{4n^2\omega_1^2 + (\omega_1^2 - \omega_0^2)^2} \{e^{-nt}[2n\cosh\omega t + \frac{1}{\omega}(\omega_1^2 + \omega^2 + n^2)\sinh\omega t]$$

$$- 2n\cos\omega_1 t - \frac{1}{\omega_1}(\omega_1^2 + \omega^2 - n^2)\sin\omega_1 t\}$$

$$+ \frac{(\omega_1^2 - \omega_0^2)a\sin\lambda}{4n^2\omega_1^2 + (\omega_1^2 - \omega_0^2)^2} \{\frac{\omega_1^2}{\omega^2}(1 - \cos\omega_1 t)$$

$$- \frac{\omega_1^2}{n^2 - \omega^2}[1 - e^{-nt}(\cosh\omega t - \frac{n}{\omega}\sinh\omega t)]$$

$$+ \frac{2n\omega_1}{\omega_1^2 - \omega_0^2}(\sin\omega_1 t - \frac{\omega_1}{\omega}e^{-nt}\sinh\omega t)\}$$

$$- \frac{p}{n^2 - \omega^2}[1 - e^{-nt}(\cosh\omega t + \frac{n}{\omega}\sinh\omega t)] \qquad (17.3.17)$$

17.3.2 Initial Displacement Equals Zero

The initial conditions of motion are:

$$\text{for} \quad t = 0 \quad x = 0; \quad \frac{dx}{dt} = v_0 \qquad (17.3.18)$$

The solutions of differential equation of motion (17.3.1) with the initial conditions of motion (17.3.18) for the following cases are presented below.

a. Case $\omega^2 > 0$

$$x = \frac{1}{\omega} v_0 e^{-nt} \sin \omega t$$

$$+ \frac{a\omega_1 \cos \lambda}{(\omega_1^2 - \omega_0^2)^2 + 4n^2\omega_1^2} \{ 2n[e^{-nt}(\cos \omega t + \frac{n}{\omega} \sin \omega t) - \cos \omega_1 t]$$

$$- (\omega_1^2 - \omega_0^2)(\frac{1}{\omega_1} \sin \omega_1 t - \frac{1}{\omega} e^{-nt} \sin \omega t) \}$$

$$+ \frac{a \sin \lambda}{(\omega_1^2 - \omega_0^2)^2 + 4n^2\omega_1^2} \{ (\omega_1^2 - \omega_0^2)[e^{-nt}(\cos \omega t + \frac{n}{\omega} \sin \omega t) - \cos \omega_1 t]$$

$$+ 2n\omega_1 (\sin \omega_1 t - \frac{\omega_1}{\omega} e^{-nt} \sin \omega t) \}$$

$$- \frac{p}{\omega^2 + n^2} [1 - e^{-nt}(\cos \omega t + \frac{n}{\omega} \sin \omega t)] \qquad\qquad \textbf{(17.3.19)}$$

b. Case $\omega^2 = 0$

$$x = v_0 t e^{-nt} + \frac{a\omega_1 \cos \lambda}{4n^2\omega_1^2 + (\omega_1^2 - n^2)^2} \{ 2n[e^{-nt}(1 + nt) - \cos \omega_1 t]$$

$$- (\omega_1^2 - n^2)(\frac{1}{\omega_1} \sin \omega_1 t - t e^{-nt}) \}$$

$$+ \frac{a \sin \lambda}{4n^2\omega_1^2 + (\omega_1^2 - n^2)^2} \{ 2n\omega_1 (\sin \omega_1 t - \omega_1 t e^{-nt})$$

$$- (\omega_1^2 - n^2)[e^{-nt}(1 + nt) - \cos \omega_1 t] \} - \frac{p}{n^2}[1 - e^{-nt}(1 + nt)]$$

$$\textbf{(17.3.20)}$$

c. Case $\omega^2 < 0$

$$x = \frac{v_0}{\omega} e^{-nt} \sinh \omega t + \frac{a\omega_1 \cos \lambda}{4n^2\omega_1^2 + (\omega_1^2 - \omega_0^2)^2} \{ e^{-nt}[2n \cosh \omega t$$

$$+ \frac{1}{\omega}(\omega_1^2 + \omega^2 + n^2) \sinh \omega t] - 2n \cos \omega_1 t$$

$$- \frac{1}{\omega_1}(\omega_1^2 + \omega^2 - n^2) \sin \omega_1 t \}$$

$$+\frac{(\omega_1^2-\omega_0^2)a\sin\lambda}{4n^2\omega_1^2+(\omega_1^2-\omega_0^2)^2}\{\frac{\omega_1^2}{\omega^2}(1-\cos\omega_1 t)$$

$$-\frac{\omega_1^2}{n^2-\omega^2}[1-e^{-nt}(\cosh\omega t-\frac{n}{\omega}\sinh\omega t)]$$

$$+\frac{2n\omega_1}{\omega_1^2-\omega_0^2}(\sin\omega_1 t-\frac{\omega_1}{\omega}e^{-nt}\sinh\omega t)\}$$

$$-\frac{p}{n^2-\omega^2}[1-e^{-nt}(\cosh\omega t+\frac{n}{\omega}\sinh\omega t)] \qquad (17.3.21)$$

17.3.3 Initial Velocity Equals Zero

The initial conditions of motion are:

$$\text{for}\quad t=0\quad x=s_0;\quad \frac{dx}{dt}=0 \qquad (17.3.22)$$

The solutions of differential equation of motion (17.3.1) with the initial conditions of motion (17.3.22) for the following cases are presented below.

a. Case $\omega^2>0$

$$x=\frac{1}{\omega}2ns_0e^{-nt}\sin\omega t+s_0e^{-nt}(\cos\omega t-\frac{n}{\omega}\sin\omega t)$$

$$+\frac{a\omega_1\cos\lambda}{(\omega_1^2-\omega_0^2)^2+4n^2\omega_1^2}\{2n[e^{-nt}(\cos\omega t+\frac{n}{\omega}\sin\omega t)-\cos\omega_1 t]$$

$$-(\omega_1^2-\omega_0^2)(\frac{1}{\omega_1}\sin\omega_1 t-\frac{1}{\omega}e^{-nt}\sin\omega t)\}$$

$$+\frac{a\sin\lambda}{(\omega_1^2-\omega_0^2)^2+4n^2\omega_1^2}\{(\omega_1^2-\omega_0^2)[e^{-nt}(\cos\omega t+\frac{n}{\omega}\sin\omega t)-\cos\omega_1 t]$$

$$+2n\omega_1(\sin\omega_1 t-\frac{\omega_1}{\omega}e^{-nt}\sin\omega t)\}$$

$$-\frac{p}{\omega^2+n^2}[1-e^{-nt}(\cos\omega t+\frac{n}{\omega}\sin\omega t)] \qquad (17.3.23)$$

b. Case $\omega^2 = 0$

$$x = 2ns_0 t e^{-nt} + s_0(1 - nt)e^{-nt}$$

$$+ \frac{a\omega_1 \cos \lambda}{4n^2\omega_1^2 + (\omega_1^2 - n^2)^2} \{2n[e^{-nt}(1 + nt) - \cos\omega_1 t]$$

$$- (\omega_1^2 - n^2)(\frac{1}{\omega_1}\sin\omega_1 t - te^{-nt})\}$$

$$+ \frac{a\sin\lambda}{4n^2\omega_1^2 + (\omega_1^2 - n^2)^2} \{2n\omega_1(\sin\omega_1 t - \omega_1 t e^{-nt})$$

$$- (\omega_1^2 - n^2)[e^{-nt}(1 + nt) - \cos\omega_1 t]\} - \frac{p}{n^2}[1 - e^{-nt}(1 + nt)]$$

$$(17.3.24)$$

c. Case $\omega^2 < 0$

$$x = \frac{2ns_0}{\omega}e^{-nt}\sinh\omega t + \frac{s_0}{\omega}e^{-nt}(\omega\cosh\omega t - n\sinh\omega t)$$

$$+ \frac{a\omega_1 \cos\lambda}{4n^2\omega_1^2 + (\omega_1^2 - \omega_0^2)^2} \{e^{-nt}[2n\cosh\omega t$$

$$+ \frac{1}{\omega}(\omega_1^2 + \omega^2 + n^2)\sinh\omega t]$$

$$- 2n\cos\omega_1 t - \frac{1}{\omega_1}(\omega_1^2 + \omega^2 - n^2)\sin\omega_1 t\}$$

$$+ \frac{(\omega_1^2 - \omega_0^2)a\sin\lambda}{4n^2\omega_1^2 + (\omega_1^2 - \omega_0^2)^2} \{\frac{\omega_1^2}{\omega^2}(1 - \cos\omega_1 t)$$

$$- \frac{\omega_1^2}{n^2 - \omega^2}[1 - e^{-nt}(\cosh\omega t - \frac{n}{\omega}\sinh\omega t)]$$

$$+ \frac{2n\omega_1}{\omega_1^2 - \omega_0^2}(\sin\omega_1 t - \frac{\omega_1}{\omega}e^{-nt}\sinh\omega t)\}$$

$$- \frac{p}{n^2 - \omega^2}[1 - e^{-nt}(\cosh\omega t + \frac{n}{\omega}\sinh\omega t)] \qquad (17.3.25)$$

17.3.4 Both the Initial Displacement and Velocity Equal Zero

The initial conditions of motion are:

$$\text{for} \quad t = 0 \quad x = 0; \quad \frac{dx}{dt} = 0 \tag{17.3.26}$$

The solutions of differential equation of motion (17.3.1) with the initial conditions of motion (17.3.26) for the following cases are presented below.

a. Case $\omega^2 > 0$

$$x = \frac{a\omega_1 \cos\lambda}{(\omega_1^2 - \omega_0^2)^2 + 4n^2\omega_1^2} \{2n[e^{-nt}(\cos\omega t + \frac{n}{\omega}\sin\omega t) - \cos\omega_1 t]$$

$$- (\omega_1^2 - \omega_0^2)(\frac{1}{\omega_1}\sin\omega_1 t - \frac{1}{\omega}e^{-nt}\sin\omega t)\}$$

$$+ \frac{a\sin\lambda}{(\omega_1^2 - \omega_0^2)^2 + 4n^2\omega_1^2} \{(\omega_1^2 - \omega_0^2)[e^{-nt}(\cos\omega t + \frac{n}{\omega}\sin\omega t)$$

$$- \cos\omega_1 t] + 2n\omega_1(\sin\omega_1 t - \frac{\omega_1}{\omega}e^{-nt}\sin\omega t)\}$$

$$- \frac{p}{\omega^2 + n^2}[1 - e^{-nt}(\cos\omega t + \frac{n}{\omega}\sin\omega t)] \tag{17.3.27}$$

b. Case $\omega^2 = 0$

$$x = \frac{a\omega_1 \cos\lambda}{4n^2\omega_1^2 + (\omega_1^2 - n^2)^2} \{2n[e^{-nt}(1 + nt) - \cos\omega_1 t]$$

$$- (\omega_1^2 - n^2)(\frac{1}{\omega_1}\sin\omega_1 t - te^{-nt})\}$$

$$+ \frac{a\sin\lambda}{4n^2\omega_1^2 + (\omega_1^2 - n^2)^2} \{2n\omega_1(\sin\omega_1 t - \omega_1 te^{-nt})$$

$$- (\omega_1^2 - n^2)[e^{-nt}(1 + nt) - \cos\omega_1 t]\} - \frac{p}{n^2}[1 - e^{-nt}(1 + nt)]$$

$$\tag{17.3.28}$$

c. Case $\omega^2 < 0$

$$x = \frac{a\omega_1 \cos\lambda}{4n^2\omega_1^2 + (\omega_1^2 - \omega_0^2)^2} \{e^{-nt}[2n\cosh\omega t + \frac{1}{\omega}(\omega_1^2 + \omega^2 + n^2)\sinh\omega t]$$

$$- 2n\cos\omega_1 t - \frac{1}{\omega_1}(\omega_1^2 + \omega^2 - n^2)\sin\omega_1 t\}$$

$$+ \frac{(\omega_1^2 - \omega_0^2)a\sin\lambda}{4n^2\omega_1^2 + (\omega_1^2 - \omega_0^2)^2} \{\frac{\omega_1^2}{\omega^2}(1 - \cos\omega_1 t)$$

$$- \frac{\omega_1^2}{n^2 - \omega^2}[1 - e^{-nt}(\cosh\omega t - \frac{n}{\omega}\sinh\omega t)]$$

$$+ \frac{2n\omega_1}{\omega_1^2 - \omega_0^2}(\sin\omega_1 t - \frac{\omega_1}{\omega}e^{-nt}\sinh\omega t)\}$$

$$- \frac{p}{n^2 - \omega^2}[1 - e^{-nt}(\cosh\omega t + \frac{n}{\omega}\sinh\omega t)] \qquad (17.3.29)$$

17.4 Time-Dependent Force $Q\left(\rho + \frac{\mu t}{\tau}\right)$

The intersection of Row 17 of Guiding Table 2.1 and Column 4 indicates that this section describes engineering systems subjected to the action of the force of inertia, the damping force, the stiffness force, and the constant resisting force as the resisting forces and to the action of the time-dependent force as the active force.

The current problem could be associated with the working process of a system interacting with a viscoelastoplastic medium that exerts the damping force, the stiffness force, and the constant resisting force as a reaction to its deformation (see section 1.2). Sometimes during the initial phase of deformation, the engineering system is subjected to a time-dependent force that acts a predetermined interval of time.

The system is moving in the horizontal direction. We want to determine the law of motion of the system. Figure 17.4 shows the model of a system subjected to the action of a time-dependent force, a damping force, a stiffness force, and a constant resisting force.

The considerations above and the model in Figure 16.4 let us compose the left and right sides of the differential equation of motion. The left side consists of the force of inertia, the damping force, the stiffness force, and the constant resisting force, while the right side includes the time-dependent force. Therefore, the differential equation of motion reads:

$$m\frac{d^2x}{dt^2} + C\frac{dx}{dt} + Kx + P = Q(\rho + \frac{\mu t}{\tau}) \qquad (17.4.1)$$

Differential equation of motion (17.4.1) has different solutions for various initial conditions of motion. These solutions are presented below.

17.4.1 General Initial Conditions

The general initial conditions of motion are:

$$\text{for} \quad t = 0 \quad x = s_0; \quad \frac{dx}{dt} = v_0 \qquad (17.4.2)$$

where s_0 and v_0 are the initial displacement and initial velocity respectively.

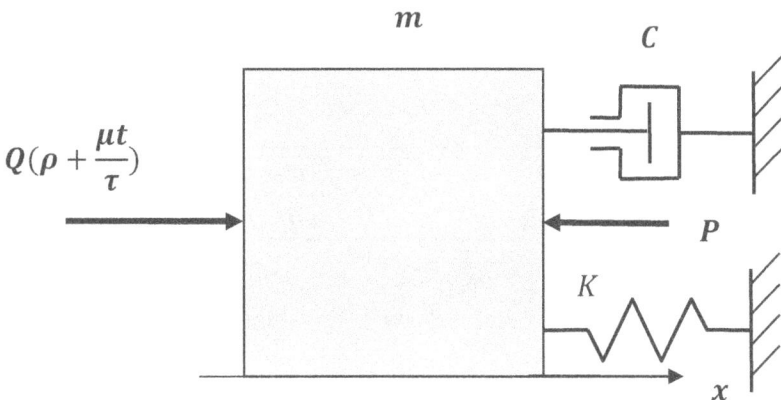

Figure 17.4 Model of a system subjected to a time-dependent force, a damping force, a stiffness force, and a constant resisting force

Dividing equation (17.4.1) by m, we have:

$$\frac{d^2x}{dt^2} + 2n\frac{dx}{dt} + \omega_0^2 x + p = q(\rho + \frac{\mu t}{\tau}) \qquad (17.4.3)$$

where n is the damping factor and ω_0 is the natural frequency, while:

$$2n = \frac{C}{m} \qquad (17.4.4)$$

$$\omega_0^2 = \frac{K}{m} \qquad (17.4.5)$$

$$p = \frac{P}{m} \qquad (17.4.6)$$

$$q = \frac{Q}{m} \qquad (17.4.7)$$

Using Laplace Transform Pairs 3, 4, 5, and 2 from Table 1.1, we convert differential equation of motion (17.4.3) with the initial conditions of motion (17.4.2) from the time domain into the Laplace domain and obtain the corresponding algebraic equation of motion in the Laplace domain:

$$l^2 x(l) - l v_0 - l^2 s_0 + 2nlx(l) - 2nls_0 + \omega_0^2\, x(l) + p = q\rho + \frac{q\mu}{\tau l} \quad (17.4.8)$$

The solution of equation (17.4.8) for the displacement $x(l)$ in the Laplace domain reads:

$$x(l) = \frac{l(v_0 + 2ns_0)}{l^2 + 2nl + \omega_0^2} + \frac{l^2 s_0}{l^2 + 2nl + \omega_0^2}$$

$$+ \frac{q\rho - p}{l^2 + 2nl + \omega_0^2} + \frac{q\mu}{l\tau(l^2 + 2nl + \omega_0^2)} \qquad (17.4.9)$$

Based on algebraic procedures with the denominators of equation (17.4.9), we may write:

$$l^2 + 2nl + \omega_0^2 + n^2 - n^2 = (l+n)^2 + \omega^2 \qquad (17.4.10)$$

where:

$$\omega^2 = \omega_0^2 - n^2 \qquad (17.4.11)$$

while ω^2 could be positive, equal to zero, or negative.

All these three cases are considered below.

a. Case $\omega^2 > 0$

Adjusting the denominators in equation (17.4.9) according to equation (17.4.10), we have:

$$x(l) = \frac{l(v_0 + 2ns_0)}{(l+n)^2 + \omega^2} + \frac{l^2 s_0}{(l+n)^2 + \omega^2} + \frac{q\rho - p}{(l+n)^2 + \omega^2} + \frac{q\mu}{l\tau[(l+n)^2 + \omega^2)]}$$

$$(17.4.12)$$

Based on pairs 1, 24, 27, 22, and 39 from Table 1.1, we invert equation (17.4.12) from the Laplace domain into the time domain and obtain the solution of differential equation of motion (17.4.1) with the initial conditions of motion (17.4.2):

$$x = \frac{v_0 + 2ns_0}{\omega} e^{-nt} \sin \omega t + s_0 (\cos \omega t - \frac{n}{\omega} \sin \omega t) e^{-nt}$$

$$+ \frac{q\rho - p}{\omega^2 + n^2} [1 - e^{-nt} (\cos \omega t + \frac{n}{\omega} \sin \omega t)]$$

$$+ \frac{q\mu}{\omega\tau(\omega^2 + n^2)^2} \{(\omega^2 + n^2)\omega t - 2n\omega$$

$$- e^{-nt} [(\omega^2 - n^2) \sin \omega t - 2n\omega \cos \omega t]\} \qquad (17.4.13)$$

b. Case $\omega^2 = 0$

In equation (17.4.12), equating ω^2 to zero, we write:

$$x(l) = \frac{l(v_0 + 2ns_0)}{(l+n)^2} + \frac{l^2 s_0}{(l+n)^2} + \frac{q\rho - p}{(l+n)^2} + \frac{q\mu}{l\tau(l+n)^2} \qquad (17.4.14)$$

Using pairs 1, 37, 38, 36, 15 from Table 1.1, we invert equation (17.4.14) from the Laplace domain into the time domain and obtain the solution of differential equation of motion (17.4.1) with the initial conditions of motion (17.4.2):

$$x = (v_0 + 2ns_0) te^{-nt} + s_0 (1 - nt) e^{-nt}$$

$$+ \frac{q\rho - p}{n^2} [1 - e^{-nt} (1 + nt)] + \frac{q\mu}{\tau n^2} [t - \frac{2}{n} + e^{-nt} (\frac{2}{n} + t)] \qquad (17.4.15)$$

c. Case $\omega^2 < 0$

In equation (17.4.12), taking ω^2 with the negative sign, we have:

$$x(l) = \frac{l(v_0 + 2ns_0)}{(l+n)^2 - \omega^2} + \frac{l^2 s_0}{(l+n)^2 - \omega^2} + \frac{q\rho - p}{(l+n)^2 - \omega^2} + \frac{q\mu}{l\tau[(l+n)^2 - \omega^2]}$$

$$(17.4.16)$$

Using pairs 1, 25, 28, 23, and 40 from Table 1.1, we invert equation (17.4.16) from the Laplace domain into the time domain and obtain the solution of differential equation (17.4.1) with the initial conditions of motion (17.4.2):

$$x = \frac{v_0 + 2ns_0}{\omega} e^{-nt} \sinh \omega t + \frac{s_0}{\omega} e^{-nt} (\omega \cosh \omega t - n \sinh \omega t)$$

$$+ \frac{q\rho - p}{\omega^2 + n^2} [1 - e^{-nt}(\cosh \omega t + \frac{n}{\omega} \sinh \omega t)]$$

$$+ \frac{q\mu}{\omega\tau(n^2 - \omega^2)^2} [(n^2 - \omega^2)(\omega t + e^{-nt} \sinh \omega t) - 2n\omega(1 - \cosh \omega t)]$$

$$(17.4.17)$$

17.4.2 Initial Displacement Equals Zero

The initial conditions of motion are:

$$\text{for} \quad t = 0 \quad x = 0; \quad \frac{dx}{dt} = v_0 \qquad (17.4.18)$$

The solutions of differential equation of motion (17.4.1) with the initial conditions of motion (17.4.18) for the following cases are presented below.

a. Case $\omega^2 > 0$

$$x = \frac{v_0}{\omega} e^{-nt} \sin \omega t + \frac{q\rho - p}{\omega^2 + n^2} [1 - e^{-nt}(\cos \omega t + \frac{n}{\omega} \sin \omega t)]$$

$$+ \frac{q\mu}{\omega\tau(\omega^2 + n^2)^2} \{(\omega^2 + n^2)\omega t - 2n\omega$$

$$- e^{-nt}[(\omega^2 - n^2)\sin \omega t - 2n\omega \cos \omega t]\} \qquad (17.4.19)$$

b. Case $\omega^2 = 0$

$$x = v_0 t e^{-nt} + \frac{q\rho - p}{n^2}[1 - e^{-nt}(1 + nt)] + \frac{q\mu}{\tau n^2}[t - \frac{2}{n} + e^{-nt}(\frac{2}{n} + t)]$$

$$(17.4.20)$$

c. Case $\omega^2 < 0$

$$x = \frac{v_0}{\omega}e^{-nt}\sinh\omega t + \frac{q\rho - p}{n^2 - \omega^2}[1 - e^{-nt}(\cosh\omega t + \frac{n}{\omega}\sinh\omega t)]$$

$$+ \frac{q\mu}{\omega\tau(n^2 - \omega^2)^2}[(n^2 - \omega^2)(\omega t + e^{-nt}\sinh\omega t) - 2n\omega(1 - \cosh\omega t)]$$

$$(17.4.21)$$

17.4.3 Initial Velocity Equals Zero

The initial conditions of motion are:

$$\text{for} \quad t = 0 \quad x = s_0; \quad \frac{dx}{dt} = 0 \qquad (17.4.22)$$

The solutions of differential equation of motion (17.4.1) with the initial conditions of motion (17.4.22) for the following cases are presented below.

a. Case $\omega^2 > 0$

$$x = \frac{2ns_0}{\omega}e^{-nt}\sin\omega t + s_0(\cos\omega t - \frac{n}{\omega}\sin\omega t)e^{-nt}$$

$$+ \frac{q\rho - p}{\omega^2 + n^2}[1 - e^{-nt}(\cos\omega t + \frac{n}{\omega}\sin\omega t)]$$

$$+ \frac{q\mu}{\omega\tau(\omega^2 + n^2)^2}\{(\omega^2 + n^2)\omega t - 2n\omega$$

$$- e^{-nt}[(\omega^2 - n^2)\sin\omega t - 2n\omega\cos\omega t]\} \qquad (17.4.23)$$

b. Case $\omega^2 = 0$

$$x = 2ns_0 t e^{-nt} + s_0(1 - nt)e^{-nt} + \frac{q\rho - p}{\tau n^2}[1 - e^{-nt}(1 + nt)]$$

$$+ \frac{q\mu}{\tau n^2}[t - \frac{2}{n} + e^{-nt}(\frac{2}{n} + t)] \qquad (17.4.24)$$

c. Case $\omega^2 < 0$

$$x = \frac{2ns_0}{\omega}e^{-nt}\sinh\omega t + \frac{s_0}{\omega}e^{-nt}(\omega\cosh\omega t - n\sinh\omega t)$$

$$+\frac{q\rho - p}{n^2 - \omega^2}[1 - e^{-nt}(\cosh\omega t + \frac{n}{\omega}\sinh\omega t)]$$

$$+\frac{q\mu}{\omega\tau(n^2 - \omega^2)^2}[(n^2 - \omega^2)(\omega t + e^{-nt}\sinh\omega t) - 2n\omega(1 - \cosh\omega t)]$$

$$(17.4.25)$$

17.4.4 Both the Initial Displacement and Velocity Equal Zero

The initial conditions of motion are:

$$\text{for}\quad t = 0\quad x = 0;\quad \frac{dx}{dt} = 0 \qquad (17.4.26)$$

The solutions of differential equation of motion (17.4.1) with the initial conditions of motion (17.4.26) for the following cases are presented below.

a. Case $\omega^2 > 0$

$$x = \frac{q\rho - p}{\omega^2 + n^2}[1 - e^{-nt}(\cos\omega t + \frac{n}{\omega}\sin\omega t)]$$

$$+\frac{q\mu}{\omega\tau(\omega^2 + n^2)^2}\{(\omega^2 + n^2)\omega t$$

$$- 2n\omega - e^{-nt}[(\omega^2 - n^2)\sin\omega t - 2n\omega\cos\omega t]\} \quad (17.4.27)$$

b. Case $\omega^2 = 0$

$$x = \frac{q\rho - p}{n^2}\left[1 - e^{-nt}(1 + nt)\right] + \frac{q\mu}{\tau n^2}[t - \frac{2}{n} + e^{-nt}\left(\frac{2}{n} + t\right)] \qquad (17.4.28)$$

c. Case $\omega^2 < 0$

$$x = \frac{q\rho - p}{n^2 - \omega^2}[1 - e^{-nt}(\cosh\omega t + \frac{n}{\omega}\sinh\omega t)]$$

$$+\frac{q\mu}{\omega\tau(n^2-\omega^2)^2}[(n^2-\omega^2)(\omega t+e^{-nt}\sinh\omega t)-2n\omega(1-\cosh\omega t)]$$

$$(\textbf{17.4.29})$$

17.5 Constant Force R and Harmonic Force $A\sin(\omega_1 t+\lambda)$

This section describes engineering systems subjected to the action of the force of inertia, the damping force, the stiffness force, and the constant resisting force as the resisting forces (Row 17 of Guiding Table 2.1), and the constant active force and the harmonic force as the active forces (Column 5).

The current problem could be associated with the working process of vibratory machines intended for interaction with the viscoelastoplastic media that exert damping, stiffness, and constant resisting forces as the reaction to their deformation. Additional considerations related to the deformation of viscoelastoplastic media as well as to the behavior of a constant resisting force applied to a vibratory system are presented in section 1.2.

The system is moving in the horizontal direction. We want to determine the law of motion of the system. Figure 17.5 shows the model of a system subjected to the action of a constant active force, a harmonic force, a damping force, a stiffness force, and a constant resisting force.

The considerations above and the model shown in Figure 17.5 let us compose the left and the right sides of the differential equation of motion of the system. The left side consists of the force of inertia, the damping force, the stiffness force, and the constant resisting force. The right side consists of the constant active force and the harmonic force. Therefore, the differential equation of motion reads:

$$m\frac{d^2x}{dt^2}+C\frac{dx}{dt}+Kx+P=R+A\sin(\omega_1 t+\lambda) \qquad (\textbf{17.5.1})$$

Differential equation of motion (17.5.1) has different solutions for various initial conditions of motion. These solutions are presented below.

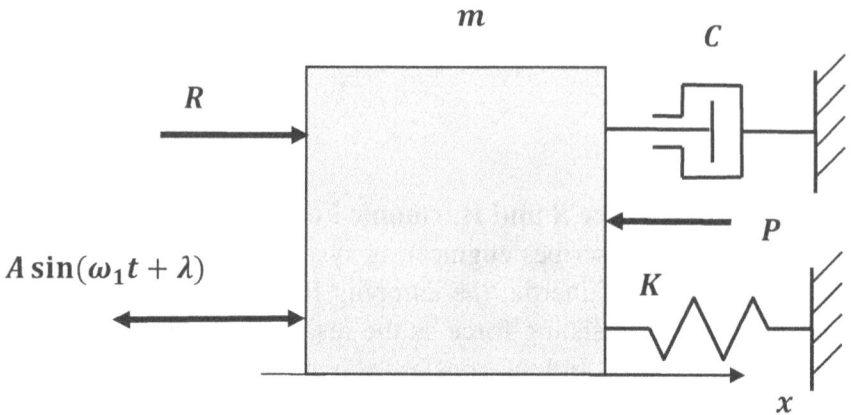

Figure 17.5 Model of a system subjected to a constant active force, a harmonic force, a damping force, a stiffness force, and a constant resisting force

17.5.1 General Initial Conditions

The general initial conditions of motion are:

$$\text{for} \quad t = 0 \quad x = s_0; \quad \frac{dx}{dt} = v_0 \qquad (17.5.2)$$

Transforming the sinusoidal function in equation (17.5.1) and dividing the latter by m, we obtain:

$$\frac{d^2x}{dt^2} + 2n\frac{dx}{dt} + \omega_0^2 x + p = r + a\sin\omega_1 t\cos\lambda + a\cos\omega_1 t\sin\lambda$$

$$(17.5.3)$$

where n is the damping factor and ω_0 in the natural frequency, while:

$$2n = \frac{C}{m} \qquad (17.5.4)$$

$$\omega_0^2 = \frac{K}{m} \qquad (17.5.5)$$

$$p = \frac{P}{m} \qquad (17.5.6)$$

$$r = \frac{R}{m} \qquad (17.5.7)$$

$$a = \frac{A}{m} \qquad (17.5.8)$$

Based on Laplace Transform Pairs 3, 4, 1, 5, 6, and 7 from Table 1.1, we convert the differential equation of motion (17.5.3) with the initial conditions of motion (17.5.2) from the time domain into the Laplace domain, and obtain the corresponding algebraic equation of motion in the Laplace domain:

$$l^2 x(l) - l v_0 - l^2 s_0 + 2nlx(l) - 2nls_0 + \omega_0^2 x(l) + p$$

$$= r + \frac{a\omega_1 l \cos \lambda}{l^2 + \omega_1^2} + \frac{al^2 \sin \lambda}{l^2 + \omega_1^2} \qquad (17.5.9)$$

Applying basic algebra to the equation (17.5.9), we have:

$$x(l)(l^2 + 2nl + \omega_0^2) = r - p + l(v_0 + 2ns_0) + l^2 s_0$$

$$+ \frac{a\omega_1 l \cos \lambda}{l^2 + \omega_1^2} + \frac{al^2 \sin \lambda}{l^2 + \omega_1^2} \qquad (17.5.10)$$

The solution of equation (17.5.10) for the displacement $x(l)$ in the Laplace domain reads:

$$x(l) = \frac{r - p}{l^2 + 2nl + \omega_0^2} + \frac{l(v_0 + 2ns_0)}{l^2 + 2nl + \omega_0^2} + \frac{l^2 s_0}{l^2 + 2nl + \omega_0^2}$$

$$+ \frac{a\omega_1 l \cos \lambda}{(l^2 + \omega_1^2)(l^2 + 2nl + \omega_0^2)} + \frac{al^2 \sin \lambda}{(l^2 + \omega_1^2)(l^2 + 2nl + \omega_0^2)} \qquad (17.5.11)$$

Based on algebraic procedures with the denominators in equation (17.5.11), we may write:

$$l^2 + 2nl + \omega_0^2 + n^2 - n^2 = (l + n)^2 + \omega^2 \qquad (17.5.12)$$

where:

$$\omega^2 = \omega_0^2 - n^2 \qquad (17.5.13)$$

while ω^2 could be positive, equal to zero, or negative.
All these three cases are considered below.

a. Case $\omega^2 > 0$

Adjusting the denominators in equation (17.5.11) according to equation (17.5.12), we have:

$$x(l) = \frac{r-p}{(l+n)^2 + \omega^2} + \frac{l(v_0 + 2ns_0)}{(l+n)^2 + \omega^2} + \frac{l^2 s_0}{(l+n)^2 + \omega^2}$$

$$+ \frac{a\omega_1 l \cos\lambda}{(l^2 + \omega_1^2)[(l+n)^2 + \omega^2]} + \frac{a l^2 \sin\lambda}{(l^2 + \omega_1^2)[(l+n)^2 + \omega^2]} \qquad \textbf{(17.5.14)}$$

Using pairs 1, 22, 24, 27, 42, and 43 from Table 1.1, we invert equation (17.5.14) from the Laplace domain into the time domain and obtain the solution of differential equation of motion (17.5.1) with the initial conditions of motion (17.5.2):

$$x = \frac{r-p}{\omega_0^2}[1 - e^{-nt}(\cos\omega t + \frac{n}{\omega}\sin\omega t)] + \frac{1}{\omega}(v_0 + 2ns_0)e^{-nt}\sin\omega t$$

$$+ s_0 e^{-nt}(\cos\omega t - \frac{n}{\omega}\sin\omega t)$$

$$+ \frac{a\omega_1 \cos\lambda}{(\omega_1^2 - \omega_0^2)^2 + 4n^2\omega_1^2}\{2n[e^{-nt}(\cos\omega t + \frac{n}{\omega}\sin\omega t) - \cos\omega_1 t]$$

$$- (\omega_1^2 - \omega_0^2)(\frac{1}{\omega_1}\sin\omega_1 t - \frac{1}{\omega}e^{-nt}\sin\omega t)\}$$

$$+ \frac{a\sin\lambda}{(\omega_1^2 - \omega_0^2)^2 + 4n^2\omega_1^2}\{(\omega_1^2 - \omega_0^2)[e^{-nt}(\cos\omega t + \frac{n}{\omega}\sin\omega t) - \cos\omega_1 t]$$

$$+ 2n\omega_1(\sin\omega_1 t - \frac{\omega_1}{\omega}e^{-nt}\sin\omega t)\} \qquad \textbf{(17.5.15)}$$

b. Case $\omega^2 = 0$

In equation (17.5.14), equating ω^2 to zero, we have:

$$x(l) = \frac{r-p}{(l+n)^2} + \frac{l(v_0 + 2ns_0)}{(l+n)^2} + \frac{l^2 s_0}{(l+n)^2}$$

$$+ \frac{a\omega_1 l \cos\lambda}{(l^2 + \omega_1^2)(l+n)^2} + \frac{a l^2 \sin\lambda}{(l^2 + \omega_1^2)(l+n)^2} \qquad \textbf{(17.5.16)}$$

Using pairs 1, 36, 37, 38, 46, and 47 from Table 1.1, we invert equation (17.5.16) from the Laplace domain into the time domain and obtain the solution of differential equation of motion (17.5.1) with the initial conditions of motion (17.5.2):

$$x = \frac{r-p}{n^2}[1-e^{-nt}(1+nt)] + (v_0 + 2ns_0)te^{-nt}$$

$$+ s_0(1-nt)e^{-nt} + \frac{a\omega_1\cos\lambda}{4n^2\omega_1^2 + (\omega_1^2-n^2)^2}\{2n[e^{-nt}(1+nt)-\cos\omega_1 t]$$

$$- (\omega_1^2-n^2)(\frac{1}{\omega_1}\sin\omega_1 t - te^{-nt})\}$$

$$+ \frac{a\sin\lambda}{4n^2\omega_1^2 + (\omega_1^2-n^2)^2}\{2n\omega_1(\sin\omega_1 t - \omega_1 te^{-nt})$$

$$- (\omega_1^2-n^2)[e^{-nt}(1+nt)-\cos\omega_1 t]\} \qquad (17.5.17)$$

c. Case $\omega^2 < 0$

In equation (17.5.14), taking ω^2 with the negative sign, we may write:

$$x(l) = \frac{r-p}{(l+n)^2 - \omega^2} + \frac{l(v_0 + 2ns_0)}{(l+n)^2 - \omega^2} + \frac{l^2 s_0}{(l+n)^2 - \omega^2}$$

$$+ \frac{a\omega_1 l\cos\lambda}{(l^2 + \omega_1^2)[(l+n)^2 - \omega^2]} + \frac{al^2\sin\lambda}{(l^2 + \omega_1^2)[(l+n)^2 - \omega^2]} \qquad (17.5.18)$$

Using pairs 1, 23, 25, 28, 48, and 49 from Table 1.1, we invert equation (17.5.18) from the Laplace domain into the time domain and obtain the solution of differential equation of motion (17.5.1) with the initial conditions of motion (17.5.2):

$$x = \frac{r-p}{n^2 - \omega^2}[1 - e^{-nt}(\cosh\omega t + \frac{n}{\omega}\sinh\omega t)]$$

$$+ \frac{v_0 + 2ns_0}{\omega}e^{-nt}\sinh\omega t + \frac{s_0}{\omega}e^{-nt}(\omega\cosh\omega t - n\sinh\omega t)$$

$$+ \frac{a\omega_1\cos\lambda}{4n^2\omega_1^2 + (\omega_1^2-\omega_0^2)^2}\{e^{-nt}[2n\cosh\omega t + \frac{1}{\omega}(\omega_1^2 + \omega^2 + n^2)\sinh\omega t]$$

$$-2n\cos\omega_1 t - \frac{1}{\omega_1}(\omega_1^2 + \omega^2 - n^2)\sin\omega_1 t\}$$

$$+ \frac{(\omega_1^2 - \omega_0^2)a\sin\lambda}{4n^2\omega_1^2 + (\omega_1^2 - \omega_0^2)^2}\{\frac{\omega_1^2}{\omega^2}(1 - \cos\omega_1 t)$$

$$- \frac{\omega_1^2}{n^2 - \omega^2}[1 - e^{-nt}(\cosh\omega t - \frac{n}{\omega}\sinh\omega t)]$$

$$+ \frac{2n\omega_1}{\omega_1^2 - \omega_0^2}(\sin\omega_1 t - \frac{\omega_1}{\omega}e^{-nt}\sinh\omega t)\} \qquad (17.5.19)$$

17.5.2 Initial Displacement Equals Zero

The initial conditions of motion are:

$$\text{for} \quad t = 0 \quad x = 0; \quad \frac{dx}{dt} = v_0 \qquad (17.5.20)$$

The solutions of differential equation of motion (17.5.1) with the initial conditions of motion (17.5.20) for the following cases are presented below.

a. Case $\omega^2 > 0$

$$x = \frac{r-p}{\omega_0^2}[1 - e^{-nt}(\cos\omega t + \frac{n}{\omega}\sin\omega t)] + \frac{1}{\omega}v_0 e^{-nt}\sin\omega t$$

$$+ \frac{a\omega_1\cos\lambda}{(\omega_1^2 - \omega_0^2)^2 + 4n^2\omega_1^2}\{2n[e^{-nt}(\cos\omega t + \frac{n}{\omega}\sin\omega t) - \cos\omega_1 t]$$

$$- (\omega_1^2 - \omega_0^2)(\frac{1}{\omega_1}\sin\omega_1 t - \frac{1}{\omega}e^{-nt}\sin\omega t)\}$$

$$+ \frac{a\sin\lambda}{(\omega_1^2 - \omega_0^2)^2 + 4n^2\omega_1^2}\{(\omega_1^2 - \omega_0^2)[e^{-nt}(\cos\omega t + \frac{n}{\omega}\sin\omega t)$$

$$- \cos\omega_1 t] + 2n\omega_1(\sin\omega_1 t - \frac{\omega_1}{\omega}e^{-nt}\sin\omega t)\} \qquad (17.5.21)$$

b. Case $\omega^2 = 0$

$$x = \frac{r-p}{n^2}\left[1 - e^{-nt}(1 + nt)\right] + v_0 t e^{-nt}$$

$$+\frac{a\omega_1\cos\lambda}{4n^2\omega_1^2+(\omega_1^2-n^2)^2}\{2n[e^{-nt}(1+nt)-\cos\omega_1 t]$$

$$-(\omega_1^2-n^2)(\frac{1}{\omega_1}\sin\omega_1 t-te^{-nt})\}$$

$$+\frac{a\sin\lambda}{4n^2\omega_1^2+(\omega_1^2-n^2)^2}\{2n\omega_1(\sin\omega_1 t-\omega_1 te^{-nt})$$

$$-(\omega_1^2-n^2)[e^{-nt}(1+nt)-\cos\omega_1 t]\} \tag{17.5.22}$$

c. Case $\omega^2 < 0$

$$x=\frac{r-p}{n^2-\omega^2}[1-e^{-nt}(\cosh\omega t+\frac{n}{\omega}\sinh\omega t)]$$

$$+\frac{v_0}{\omega}e^{-nt}\sinh\omega t+\frac{a\omega_1\cos\lambda}{4n^2\omega_1^2+(\omega_1^2-\omega_0^2)^2}\{e^{-nt}[2n\cosh\omega t$$

$$+\frac{1}{\omega}(\omega_1^2+\omega^2+n^2)\sinh\omega t]-2n\cos\omega_1 t$$

$$-\frac{1}{\omega_1}(\omega_1^2+\omega^2-n^2)\sin\omega_1 t\}+\frac{(\omega_1^2-\omega_0^2)a\sin\lambda}{4n^2\omega_1^2+(\omega_1^2-\omega_0^2)^2}\{\frac{\omega_1^2}{\omega^2}(1-\cos\omega_1 t)$$

$$-\frac{\omega_1^2}{n^2-\omega^2}[1-e^{-nt}(\cosh\omega t-\frac{n}{\omega}\sinh\omega t)]$$

$$+\frac{2n\omega_1}{\omega_1^2-\omega_0^2}(\sin\omega_1 t-\frac{\omega_1}{\omega}e^{-nt}\sinh\omega t)\} \tag{17.5.23}$$

17.5.3 Initial Velocity Equals Zero
The initial conditions of motion are:

$$\text{for}\quad t=0\quad x=s_0;\quad \frac{dx}{dt}=0 \tag{17.5.24}$$

The solutions of differential equation of motion (17.5.1) with the initial conditions of motion (17.5.24) for the following cases are presented below.

a. Case $\omega^2 > 0$

$$x = \frac{r-p}{\omega_0^2}[1 - e^{-nt}(\cos\omega t + \frac{n}{\omega}\sin\omega t)] + \frac{1}{\omega}2ns_0 e^{-nt}\sin\omega t$$

$$+ s_0 e^{-nt}(\cos\omega t - \frac{n}{\omega}\sin\omega t)$$

$$+ \frac{a\omega_1\cos\lambda}{(\omega_1^2 - \omega_0^2)^2 + 4n^2\omega_1^2}\{2n[e^{-nt}(\cos\omega t + \frac{n}{\omega}\sin\omega t) - \cos\omega_1 t]$$

$$- (\omega_1^2 - \omega_0^2)(\frac{1}{\omega_1}\sin\omega_1 t - \frac{1}{\omega}e^{-nt}\sin\omega t)\}$$

$$+ \frac{a\sin\lambda}{(\omega_1^2 - \omega_0^2)^2 + 4n^2\omega_1^2}\{(\omega_1^2 - \omega_0^2)[e^{-nt}(\cos\omega t + \frac{n}{\omega}\sin\omega t) - \cos\omega_1 t]$$

$$+ 2n\omega_1(\sin\omega_1 t - \frac{\omega_1}{\omega}e^{-nt}\sin\omega t)\} \tag{17.5.25}$$

b. Case $\omega^2 = 0$

$$x = \frac{r-p}{n^2}[1 - e^{-nt}(1 + nt)] + 2ns_0 t e^{-nt}$$

$$+ s_0(1 - nt)e^{-nt} + \frac{a\omega_1\cos\lambda}{4n^2\omega_1^2 + (\omega_1^2 - n^2)^2}\{2n[e^{-nt}(1 + nt) - \cos\omega_1 t]$$

$$- (\omega_1^2 - n^2)(\frac{1}{\omega_1}\sin\omega_1 t - te^{-nt})\}$$

$$+ \frac{a\sin\lambda}{4n^2\omega_1^2 + (\omega_1^2 - n^2)^2}\{2n\omega_1(\sin\omega_1 t - \omega_1 te^{-nt})$$

$$- (\omega_1^2 - n^2)[e^{-nt}(1 + nt) - \cos\omega_1 t]\} \tag{17.5.26}$$

c. Case $\omega^2 < 0$

$$x = \frac{r-p}{n^2 - \omega^2}[1 - e^{-nt}(\cosh\omega t + \frac{n}{\omega}\sinh\omega t)] + \frac{2ns_0}{\omega}e^{-nt}\sinh\omega t$$

$$+ \frac{s_0}{\omega}e^{-nt}(\omega\cosh\omega t - n\sinh\omega t)$$

$$+ \frac{a\omega_1 \cos \lambda}{4n^2\omega_1^2 + (\omega_1^2 - \omega_0^2)^2} \{e^{-nt}[2n\cosh \omega t$$

$$+ \frac{1}{\omega}(\omega_1^2 + \omega^2 + n^2)\sinh \omega t]$$

$$- 2n\cos \omega_1 t - \frac{1}{\omega_1}(\omega_1^2 + \omega^2 - n^2)\sin \omega_1 t\}$$

$$+ \frac{(\omega_1^2 - \omega_0^2)a \sin \lambda}{4n^2\omega_1^2 + (\omega_1^2 - \omega_0^2)^2} \{\frac{\omega_1^2}{\omega^2}(1 - \cos \omega_1 t)$$

$$- \frac{\omega_1^2}{n^2 - \omega^2}[1 - e^{-nt}(\cosh \omega t - \frac{n}{\omega}\sinh \omega t)]$$

$$+ \frac{2n\omega_1}{\omega_1^2 - \omega_0^2}(\sin \omega_1 t - \frac{\omega_1}{\omega}e^{-nt}\sinh \omega t)\} \qquad (17.5.27)$$

17.5.4 Both the Initial Displacement and Velocity Equal Zero

The initial conditions of motion are:

$$\text{for} \quad t = 0 \quad x = 0; \quad \frac{dx}{dt} = 0 \qquad (17.5.28)$$

The solutions of differential equation of motion (17.5.1) with the initial conditions of motion (17.5.28) for the following cases are presented below.

a. Case $\omega^2 > 0$

$$x = \frac{r - p}{\omega_0^2}[1 - e^{-nt}(\cos \omega t + \frac{n}{\omega}\sin \omega t)]$$

$$+ \frac{a\omega_1 \cos \lambda}{(\omega_1^2 - \omega_0^2)^2 + 4n^2\omega_1^2} \{2n[e^{-nt}(\cos \omega t + \frac{n}{\omega}\sin \omega t) - \cos \omega_1 t]$$

$$- (\omega_1^2 - \omega_0^2)(\frac{1}{\omega_1}\sin \omega_1 t - \frac{1}{\omega}e^{-nt}\sin \omega t)\}$$

$$+ \frac{a \sin \lambda}{(\omega_1^2 - \omega_0^2)^2 + 4n^2\omega_1^2} \{(\omega_1^2 - \omega_0^2)[e^{-nt}(\cos \omega t + \frac{n}{\omega}\sin \omega t) - \cos \omega_1 t]$$

$$+ 2n\omega_1(\sin \omega_1 t - \frac{\omega_1}{\omega}e^{-nt}\sin \omega t)\} \qquad (17.5.29)$$

b. Case $\omega^2 = 0$

$$x = \frac{r-p}{n^2}[1-e^{-nt}(1+nt)]$$

$$+\frac{a\omega_1\cos\lambda}{4n^2\omega_1^2+(\omega_1^2-n^2)^2}\{2n[e^{-nt}(1+nt)-\cos\omega_1 t]$$

$$-(\omega_1^2-n^2)(\frac{1}{\omega_1}\sin\omega_1 t-te^{-nt})\}$$

$$+\frac{a\sin\lambda}{4n^2\omega_1^2+(\omega_1^2-n^2)^2}\{2n\omega_1(\sin\omega_1 t-\omega_1 te^{-nt})$$

$$-(\omega_1^2-n^2)[e^{-nt}(1+nt)-\cos\omega_1 t]\} \tag{17.5.30}$$

c. Case $\omega^2 < 0$

$$x = \frac{r-p}{n^2-\omega^2}[1-e^{-nt}(\cosh\omega t+\frac{n}{\omega}\sinh\omega t)]$$

$$+\frac{a\omega_1\cos\lambda}{4n^2\omega_1^2+(\omega_1^2-\omega_0^2)^2}\{e^{-nt}[2n\cosh\omega t+\frac{1}{\omega}(\omega_1^2+\omega^2+n^2)\sinh\omega t]$$

$$-2n\cos\omega_1 t-\frac{1}{\omega_1}(\omega_1^2+\omega^2-n^2)\sin\omega_1 t\}$$

$$+\frac{(\omega_1^2-\omega_0^2)a\sin\lambda}{4n^2\omega_1^2+(\omega_1^2-\omega_0^2)^2}\{\frac{\omega_1^2}{\omega^2}(1-\cos\omega_1 t)$$

$$-\frac{\omega_1^2}{n^2-\omega^2}[1-e^{-nt}(\cosh\omega t-\frac{n}{\omega}\sinh\omega t)]$$

$$+\frac{2n\omega_1}{\omega_1^2-\omega_0^2}(\sin\omega_1 t-\frac{\omega_1}{\omega}e^{-nt}\sinh\omega t)\} \tag{17.5.31}$$

17.6 Harmonic Force $A\sin(\omega_1 t+\lambda)$ and Time-Dependent Force $Q(\rho+\frac{\mu t}{\tau})$

This section, which is related to the intersection of Row 17 and Column 6 of Guiding Table 2.1, describes the engineering systems subjected to the force of inertia, the damping force, the stiffness force, and the constant resisting force as the resisting forces, and the harmonic force and time-dependent force as the active forces. The current problem could be related to the working process of a vibratory system that interacts with a viscoelastoplastic medium. The time-dependent force along with the harmonic force is acting for a limited interval of time during the initial phase of the working process. Considerations related to the deformation of the viscoelastoplastic media as well as to the behavior of a constant resisting force applied to a vibratory system are presented in section 1.2.

The system is moving in the horizontal direction. We want to determine the law of motion of the system. Figure 17.6 shows the model of a system subjected to the action of a harmonic force, a time-dependent force, a damping force, a stiffness force, and a constant resisting force.

Based on the considerations above and on the model in Figure 17.6, we can assemble the left and right sides of the differential equation of motion. The left side consists of the force of inertia, the damping force, the stiffness force, and the constant resisting force. The right side includes the harmonic force and the time-dependent force. Hence, the differential equation of motion reads:

$$m\frac{d^2x}{dt^2} + C\frac{dx}{dt} + Kx + P = A\sin(\omega_1 t + \lambda) + Q(\rho + \frac{\mu t}{\tau}) \quad \textbf{(17.6.1)}$$

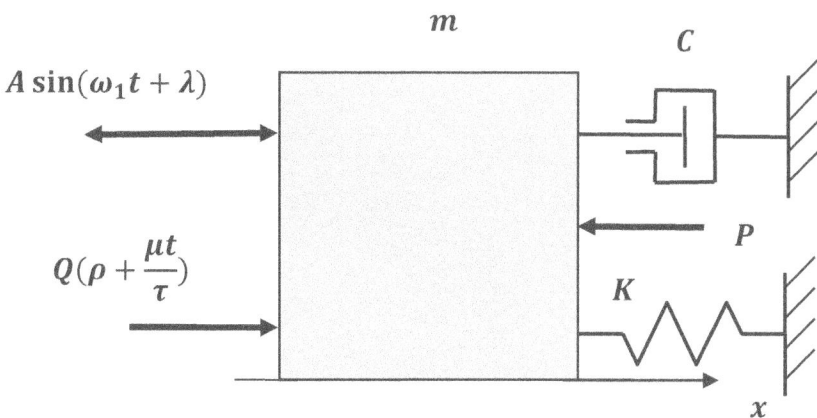

Figure 17.6 Model of a system subjected to a harmonic force, a time-dependent force, a damping force, a stiffness force, and a constant resisting force

Differential equation of motion (17.6.1) has different solutions for various initial conditions of motion. These solutions are presented below.

17.6.1 General Initial Conditions

The general initial conditions of motion are:

$$\text{for} \quad t = 0 \quad x = s_0; \quad \frac{dx}{dt} = v_0 \tag{17.6.2}$$

where s_0 and v_0 are the initial displacement and initial velocity respectively.

Transforming equation (17.6.1), we write:

$$m\frac{d^2x}{dt^2} + C\frac{dx}{dt} + Kx + P = A\sin\omega_1 t \cos\lambda$$

$$+ A\cos\omega_1 t \sin\lambda + Q\rho + Q\frac{\mu t}{\tau} \tag{17.6.3}$$

Dividing equation (17.6.3) by m, we have:

$$\frac{d^2x}{dt^2} + 2n\frac{dx}{dt} + \omega_0^2 x + p = a\sin\omega_1 t \cos\lambda + a\cos\omega_1 t \sin\lambda + q\rho + \frac{q\mu}{\tau l}$$

$$\tag{17.6.4}$$

where n is the damping factor and ω_0 is the natural frequency, while:

$$2n = \frac{C}{m} \tag{17.6.5}$$

$$\omega_0^2 = \frac{K}{m} \tag{17.6.6}$$

$$p = \frac{P}{m} \tag{17.6.7}$$

$$a = \frac{A}{m} \tag{17.6.8}$$

$$q = \frac{Q}{m} \tag{17.6.9}$$

Using Laplace Transform Pairs 3, 4, 5, 6, 7, and 2 from Table 1.1, we convert differential equation of motion (17.6.4) with the initial conditions of motion (17.6.2) from the time domain into the Laplace domain and obtain the resulting algebraic equation of motion in the Laplace domain:

$$l^2 x(l) - l v_0 - l^2 s_0 + 2nlx(l) - 2nls_0 + \omega_0^2 x(l) + p$$

$$= \frac{a\omega_1 l \cos \lambda}{l^2 + \omega_1^2} + \frac{al^2 \sin \lambda}{l^2 + \omega_1^2} + q\rho + \frac{q\mu}{l\tau} \qquad (17.6.10)$$

The solution of equation (17.6.10) for the displacement $x(l)$ in the Laplace domain reads:

$$x(l) = \frac{l(v_0 + 2ns_0)}{l^2 + 2nl + \omega_0^2} + \frac{l^2 s_0}{l^2 + 2nl + \omega_0^2}$$

$$+ \frac{a\omega_1 l \cos \lambda}{(l^2 + \omega_1^2)(l^2 + 2nl + \omega_0^2)} + \frac{al^2 \sin \lambda}{(l^2 + \omega_1^2)(l^2 + 2nl + \omega_0^2)}$$

$$+ \frac{q\rho - p}{l^2 + 2nl + \omega_0^2} + \frac{q\mu}{\tau l(l^2 + 2nl + \omega_0^2)} \qquad (17.6.11)$$

Applying basic algebra to the denominators in equation (17.6.11), we may write:

$$l^2 + 2nl + \omega_0^2 + n^2 - n^2 = (l+n)^2 + \omega^2 \qquad (17.6.12)$$

where:

$$\omega^2 = \omega_0^2 - n^2 \qquad (17.6.13)$$

while ω^2 could be positive, equal to zero, or negative.

All these three cases are considered below.

a. Case $\omega^2 > 0$

Transforming the denominators in equation (17.6.11) according to equation (17.6.17), we have:

$$x(l) = \frac{l(v_0 + 2ns_0)}{(l+n)^2 + \omega^2} + \frac{l^2 s_0}{(l+n)^2 + \omega^2}$$

$$+ \frac{a\omega_1 l \cos \lambda}{(l^2 + \omega_1^2)[(l+n)^2 + \omega^2]} + \frac{al^2 \sin \lambda}{(l^2 + \omega_1^2)[(l+n)^2 + \omega^2]}$$

$$+\frac{q\rho-p}{(l+n)^2+\omega^2}+\frac{q\mu}{l\tau[(l+n)^2+\omega^2]} \tag{17.6.14}$$

Based on pairs $1, 24, 27, 42, 43, 22,$ and 39 from Table 1.1, we invert equation (17.6.14) from the Laplace domain into the time domain and obtain the solution of differential equation of motion (17.6.1) with the initial conditions of motion (17.6.2):

$$x=\frac{1}{\omega}(v_0+2ns_0)e^{-nt}\sin\omega t+s_0e^{-nt}(\cos\omega t-\frac{n}{\omega}\sin\omega t)$$

$$+\frac{a\omega_1\cos\lambda}{(\omega_1^2-\omega_0^2)^2+4n^2\omega_1^2}\{2n[e^{-nt}(\cos\omega t+\frac{n}{\omega}\sin\omega t)-\cos\omega_1 t]$$

$$-(\omega_1^2-\omega_0^2)(\frac{1}{\omega_1}\sin\omega_1 t-\frac{1}{\omega}e^{-nt}\sin\omega t)\}$$

$$+\frac{a\sin\lambda}{(\omega_1^2-\omega_0^2)^2+4n^2\omega_1^2}\{(\omega_1^2-\omega_0^2)[e^{-nt}(\cos\omega t+\frac{n}{\omega}\sin\omega t)-\cos\omega_1 t]$$

$$+2n\omega_1(\sin\omega_1 t-\frac{\omega_1}{\omega}e^{-nt}\sin\omega t)\}$$

$$+\frac{q\rho-p}{\omega^2+n^2}[1-e^{-nt}(\cos\omega t+\frac{n}{\omega}\sin\omega t)]$$

$$+\frac{q\mu}{\omega\tau(\omega^2+n^2)^2}\{(\omega^2+n^2)\omega t-2n\omega$$

$$-e^{-nt}[(\omega^2-n^2)\sin\omega t-2n\omega\cos\omega t]\} \tag{17.6.15}$$

b. Case $\omega^2=0$

In equation (17.6.14), equating ω^2 to zero, we obtain:

$$x(l)=\frac{l(v_0+2ns_0)}{(l+n)^2}+\frac{l^2s_0}{(l+n)^2}+\frac{a\omega_1 l}{(l^2+\omega_1^2)(l+n)^2}\cos\lambda$$

$$+\frac{al^2}{(l^2+\omega_1^2)(l+n)^2}\sin\lambda+\frac{q\rho-p}{(l+n)^2}+\frac{q\mu}{l\tau(l+n)^2} \tag{17.6.16}$$

Using pairs $1, 37, 38, 46, 47, 36,$ and 15 from Table 1.1, we invert equation (17.6.16) from the Laplace domain into the time domain

and obtain the solution of differential equation of motion (17.6.1) with the initial conditions of motion (17.6.2):

$$x = (v_0 + 2ns_0)te^{-nt} + s_0(1-nt)e^{-nt}$$

$$+ \frac{a\omega_1 \cos\lambda}{4n^2\omega_1^2 + (\omega_1^2 - n^2)^2}\{2n[e^{-nt}(1+nt) - \cos\omega_1 t]$$

$$- (\omega_1^2 - n^2)(\frac{1}{\omega_1}\sin\omega_1 t - te^{-nt})\}$$

$$+ \frac{a\sin\lambda}{4n^2\omega_1^2 + (\omega_1^2 - n^2)^2}\{2n\omega_1(\sin\omega_1 t - \omega_1 te^{-nt})$$

$$- (\omega_1^2 - n^2)[e^{-nt}(1+nt) - \cos\omega_1 t]\}$$

$$+ \frac{q\rho - p}{n^2}[1 - e^{-nt}(1+nt)] + \frac{q\mu}{\tau n^2}[t - \frac{2}{n} + e^{-nt}(\frac{2}{n} + t)] \quad \textbf{(17.6.17)}$$

c. Case $\omega^2 < 0$

In equation (17.6.14), taking ω^2 with the negative sign, we may write:

$$x(l) = \frac{l(v_0 + 2ns_0)}{(l+n)^2 - \omega^2} + \frac{l^2 s_0}{(l+n)^2 - \omega^2}$$

$$+ \frac{a\omega_1 l}{(l^2 + \omega_1^2)[(l+n)^2 - \omega^2]}\cos\lambda + \frac{al^2}{(l^2 + \omega_1^2)[(l+n)^2 - \omega^2]}\sin\lambda$$

$$+ \frac{q\rho - p}{(l+n)^2 - \omega^2} + \frac{q\mu}{l\tau[(l+n)^2 - \omega^2]} \quad \textbf{(17.6.18)}$$

Using pairs 1, 25, 28, 48, 49, 23, and 40 from Table 1.1, we invert equation (17.6.18) from the Laplace domain into the time domain and obtain the solution of differential equation of motion (17.6.1) with the initial conditions of motion (17.6.2):

$$x = \frac{v_0 + 2ns_0}{\omega}e^{-nt}\sinh\omega t + \frac{s_0}{\omega}e^{-nt}(\omega\cosh\omega t - n\sinh\omega t)$$

$$+\frac{a\omega_1\cos\lambda}{4n^2\omega_1^2+(\omega_1^2-\omega_0^2)^2}\{e^{-nt}[2n\cosh\omega t+\frac{1}{\omega}(\omega_1^2+\omega^2+n^2)\sinh\omega t]$$

$$-2n\cos\omega_1 t-\frac{1}{\omega_1}(\omega_1^2+\omega^2-n^2)\sin\omega_1 t\}$$

$$+\frac{(\omega_1^2-\omega_0^2)a\sin\lambda}{4n^2\omega_1^2+(\omega_1^2-\omega_0^2)^2}\{\frac{\omega_1^2}{\omega^2}(1-\cos\omega_1 t)$$

$$-\frac{\omega_1^2}{n^2-\omega^2}[1-e^{-nt}(\cosh\omega t-\frac{n}{\omega}\sinh\omega t)]$$

$$+\frac{2n\omega_1}{\omega_1^2-\omega_0^2}(\sin\omega_1 t-\frac{\omega_1}{\omega}e^{-nt}\sinh\omega t)\}$$

$$+\frac{q\rho-p}{n^2-\omega^2}[1-e^{-nt}(\cosh\omega t+\frac{n}{\omega}\sinh\omega t)]$$

$$+\frac{q\mu}{\omega\tau(n^2-\omega^2)^2}[(n^2-\omega^2)(\omega t+e^{-nt}\sinh\omega t)-2n\omega(1-\cosh\omega t)]$$

$$(17.6.19)$$

17.6.2 Initial Displacement Equals Zero

The initial conditions of motion are:

$$\text{for}\quad t=0\quad x=0;\quad \frac{dx}{dt}=v_0 \qquad (17.6.20)$$

The solutions of differential equation of motion (17.6.1) with the initial conditions of motion (17.6.20) for the following cases are presented below.

a. Case $\omega^2>0$

$$x=\frac{1}{\omega}v_0 e^{-nt}\sin\omega t$$

$$+\frac{a\omega_1\cos\lambda}{(\omega_1^2-\omega_0^2)^2+4n^2\omega_1^2}\{2n[e^{-nt}(\cos\omega t+\frac{n}{\omega}\sin\omega t)-\cos\omega_1 t]$$

$$-(\omega_1^2-\omega_0^2)(\frac{1}{\omega_1}\sin\omega_1 t-\frac{1}{\omega}e^{-nt}\sin\omega t)\}$$

$$+\frac{a\sin\lambda}{(\omega_1^2-\omega_0^2)^2+4n^2\omega_1^2}\{(\omega_1^2-\omega_0^2)[e^{-nt}(\cos\omega t+\frac{n}{\omega}\sin\omega t)$$

$$-\cos\omega_1 t]+2n\omega_1(\sin\omega_1 t-\frac{\omega_1}{\omega}e^{-nt}\sin\omega t)\}$$

$$+\frac{q\rho-p}{\omega^2+n^2}[1-e^{-nt}(\cos\omega t+\frac{n}{\omega}\sin\omega t)]$$

$$+\frac{q\mu}{\omega\tau(\omega^2+n^2)^2}\{(\omega^2+n^2)\omega t-2n\omega$$

$$-e^{-nt}[(\omega^2-n^2)\sin\omega t-2n\omega\cos\omega t]\}\qquad(17.6.21)$$

b. Case $\omega^2=0$

$$x=v_0 te^{-nt}+\frac{a\omega_1\cos\lambda}{4n^2\omega_1^2+(\omega_1^2-n^2)^2}\{2n[e^{-nt}(1+nt)-\cos\omega_1 t]$$

$$-(\omega_1^2-n^2)(\frac{1}{\omega_1}\sin\omega_1 t-te^{-nt})\}$$

$$+\frac{a\sin\lambda}{4n^2\omega_1^2+(\omega_1^2-n^2)^2}\{2n\omega_1(\sin\omega_1 t-\omega_1 te^{-nt})$$

$$-(\omega_1^2-n^2)[e^{-nt}(1+nt)-\cos\omega_1 t]\}$$

$$+\frac{q\rho-p}{n^2}[1-e^{-nt}(1+nt)]+\frac{q\mu}{\tau n^2}[t-\frac{2}{n}+e^{-nt}(\frac{2}{n}+t)]\quad(17.6.22)$$

c. Case $\omega^2<0$

$$x=\frac{v_0}{\omega}e^{-nt}\sinh\omega t+\frac{a\omega_1\cos\lambda}{4n^2\omega_1^2+(\omega_1^2-\omega_0^2)^2}\{e^{-nt}[2n\cosh\omega t$$

$$+\frac{1}{\omega}(\omega_1^2+\omega^2+n^2)\sinh\omega t]-2n\cos\omega_1 t$$

$$-\frac{1}{\omega_1}(\omega_1^2+\omega^2-n^2)\sin\omega_1 t\}+\frac{(\omega_1^2-\omega_0^2)a\sin\lambda}{4n^2\omega_1^2+(\omega_1^2-\omega_0^2)^2}\{\frac{\omega_1^2}{\omega^2}(1-\cos\omega_1 t)$$

$$-\frac{\omega_1^2}{n^2-\omega^2}[1-e^{-nt}(\cosh\omega t-\frac{n}{\omega}\sinh\omega t)]$$

$$+\frac{2n\omega_1}{\omega_1^2-\omega_0^2}(\sin\omega_1 t-\frac{\omega_1}{\omega}e^{-nt}\sinh\omega t)\}$$

$$+\frac{q\rho-p}{n^2-\omega^2}[1-e^{-nt}(\cosh\omega t+\frac{n}{\omega}\sinh\omega t)]$$

$$+\frac{q\mu}{\omega\tau(n^2-\omega^2)^2}[(n^2-\omega^2)(\omega t+e^{-nt}\sinh\omega t)-2n\omega(1-\cosh\omega t)]$$

$$\text{(17.6.23)}$$

17.6.3 Initial Velocity Equals Zero
The initial conditions of motion are:

$$\text{for}\quad t=0\quad x=s_0;\quad \frac{dx}{dt}=0 \qquad \text{(17.6.24)}$$

The solutions of differential equation of motion (17.6.1) with the initial conditions of motion (17.6.24) for the following cases are presented below.

a. Case $\omega^2>0$

$$x=\frac{1}{\omega}2ns_0e^{-nt}\sin\omega t+s_0e^{-nt}(\cos\omega t-\frac{n}{\omega}\sin\omega t)$$

$$+\frac{a\omega_1\cos\lambda}{(\omega_1^2-\omega_0^2)^2+4n^2\omega_1^2}\{2n[e^{-nt}(\cos\omega t+\frac{n}{\omega}\sin\omega t)-\cos\omega_1 t]$$

$$-(\omega_1^2-\omega_0^2)(\frac{1}{\omega_1}\sin\omega_1 t-\frac{1}{\omega}e^{-nt}\sin\omega t)\}$$

$$+\frac{a\sin\lambda}{(\omega_1^2-\omega_0^2)^2+4n^2\omega_1^2}\{(\omega_1^2-\omega_0^2)[e^{-nt}(\cos\omega t+\frac{n}{\omega}\sin\omega t)-\cos\omega_1 t]$$

$$+2n\omega_1(\sin\omega_1 t-\frac{\omega_1}{\omega}e^{-nt}\sin\omega t)\}$$

$$+\frac{q\rho-p}{\omega^2+n^2}[1-e^{-nt}(\cos\omega t+\frac{n}{\omega}\sin\omega t)]$$

$$+\frac{q\mu}{\omega\tau(\omega^2+n^2)^2}\{(\omega^2+n^2)\omega t-2n\omega$$

$$-e^{-nt}[(\omega^2-n^2)\sin\omega t-2n\omega\cos\omega t]\} \qquad \text{(17.6.25)}$$

b. Case $\omega^2 = 0$

$$x = 2ns_0te^{-nt} + s_0(1-nt)e^{-nt}$$

$$+ \frac{a\omega_1 \cos\lambda}{4n^2\omega_1^2 + (\omega_1^2 - n^2)^2} \{2n[e^{-nt}(1+nt) - \cos\omega_1 t]$$

$$- (\omega_1^2 - n^2)(\frac{1}{\omega_1}\sin\omega_1 t - te^{-nt})\}$$

$$+ \frac{a\sin\lambda}{4n^2\omega_1^2 + (\omega_1^2 - n^2)^2} \{2n\omega_1(\sin\omega_1 t - \omega_1 te^{-nt})$$

$$- (\omega_1^2 - n^2)[e^{-nt}(1+nt) - \cos\omega_1 t]\}$$

$$+ \frac{q\rho - p}{n^2}[1 - e^{-nt}(1+nt)] + \frac{q\mu}{\tau n^2}[t - \frac{2}{n} + e^{-nt}\left(\frac{2}{n} + t\right)] \quad \textbf{(17.6.26)}$$

c. Case $\omega^2 < 0$

$$x = \frac{2ns_0}{\omega}e^{-nt}\sinh\omega t + \frac{s_0}{\omega}e^{-nt}(\omega\cosh\omega t - n\sinh\omega t$$

$$+ \frac{a\omega_1 \cos\lambda}{4n^2\omega_1^2 + (\omega_1^2 - \omega_0^2)^2} \{e^{-nt}[2n\cosh\omega t + \frac{1}{\omega}(\omega_1^2 + \omega^2 + n^2)\sinh\omega t]$$

$$- 2n\cos\omega_1 t - \frac{1}{\omega_1}(\omega_1^2 + \omega^2 - n^2)\sin\omega_1 t\}$$

$$+ \frac{(\omega_1^2 - \omega_0^2)a\sin\lambda}{4n^2\omega_1^2 + (\omega_1^2 - \omega_0^2)^2} \{\frac{\omega_1^2}{\omega^2}(1 - \cos\omega_1 t)$$

$$- \frac{\omega_1^2}{n^2 - \omega^2}[1 - e^{-nt}(\cosh\omega t - \frac{n}{\omega}\sinh\omega t)]$$

$$+ \frac{2n\omega_1}{\omega_1^2 - \omega_0^2}(\sin\omega_1 t - \frac{\omega_1}{\omega}e^{-nt}\sinh\omega t)\}$$

$$+ \frac{q\rho - p}{n^2 - \omega^2}[1 - e^{-nt}(\cosh\omega t + \frac{n}{\omega}\sinh\omega t)]$$

$$+ \frac{q\mu}{\omega\tau(n^2 - \omega^2)^2}[(n^2 - \omega^2)(\omega t + e^{-nt}\sinh\omega t) - 2n\omega(1 - \cosh\omega t)]$$

$$\textbf{(17.6.27)}$$

17.6.4 Both the Initial Displacement and Velocity Equal Zero

The initial conditions of motion are:

$$\text{for} \quad t = 0 \quad x = 0; \quad \frac{dx}{dt} = 0 \qquad (17.6.28)$$

The solutions of differential equation of motion (17.6.1) with the initial conditions of motion (17.6.28) for the following cases are presented below.

a. Case $\omega^2 > 0$

$$x = \frac{a\omega_1 \cos \lambda}{(\omega_1^2 - \omega_0^2)^2 + 4n^2\omega_1^2} \{2n[e^{-nt}(\cos\omega t + \frac{n}{\omega}\sin\omega t) - \cos\omega_1 t]$$

$$- (\omega_1^2 - \omega_0^2)(\frac{1}{\omega_1}\sin\omega_1 t - \frac{1}{\omega}e^{-nt}\sin\omega t)\}$$

$$+ \frac{a\sin\lambda}{(\omega_1^2 - \omega_0^2)^2 + 4n^2\omega_1^2} \{(\omega_1^2 - \omega_0^2)[e^{-nt}(\cos\omega t + \frac{n}{\omega}\sin\omega t) - \cos\omega_1 t]$$

$$+ 2n\omega_1(\sin\omega_1 t - \frac{\omega_1}{\omega}e^{-nt}\sin\omega t)\}$$

$$+ \frac{q\rho - p}{\omega^2 + n^2}[1 - e^{-nt}(\cos\omega t + \frac{n}{\omega}\sin\omega t)]$$

$$+ \frac{q\mu}{\omega\tau(\omega^2 + n^2)^2} \{(\omega^2 + n^2)\omega t - 2n\omega$$

$$- e^{-nt}[(\omega^2 - n^2)\sin\omega t - 2n\omega\cos\omega t]\} \qquad (17.6.29)$$

b. Case $\omega^2 = 0$

$$x = \frac{a\omega_1 \cos \lambda}{4n^2\omega_1^2 + (\omega_1^2 - n^2)^2} \{2n[e^{-nt}(1 + nt) - \cos\omega_1 t]$$

$$- (\omega_1^2 - n^2)(\frac{1}{\omega_1}\sin\omega_1 t - te^{-nt})\}$$

$$+ \frac{a\sin\lambda}{4n^2\omega_1^2 + (\omega_1^2 - n^2)^2} \{2n\omega_1(\sin\omega_1 t - \omega_1 te^{-nt})$$

$$- (\omega_1^2 - n^2)[e^{-nt}(1 + nt) - \cos\omega_1 t]\}$$

$$+\frac{qp-p}{n^2}[1-e^{-nt}(1+nt)]+\frac{q\mu}{\tau n^2}[t-\frac{2}{n}+e^{-nt}\left(\frac{2}{n}+t\right)] \qquad \textbf{(17.6.30)}$$

c. Case $\omega^2 < 0$

$$x = \frac{a\omega_1 \cos\lambda}{4n^2\omega_1^2+(\omega_1^2-\omega_0^2)^2}\{e^{-nt}[2n\cosh\omega t+\frac{1}{\omega}(\omega_1^2+\omega^2+n^2)\sinh\omega t]$$

$$-2n\cos\omega_1 t-\frac{1}{\omega_1}(\omega_1^2+\omega^2-n^2)\sin\omega_1 t\}$$

$$+\frac{(\omega_1^2-\omega_0^2)a\sin\lambda}{4n^2\omega_1^2+(\omega_1^2-\omega_0^2)^2}\{\frac{\omega_1^2}{\omega^2}(1-\cos\omega_1 t)$$

$$-\frac{\omega_1^2}{n^2-\omega^2}[1-e^{-nt}(\cosh\omega t-\frac{n}{\omega}\sinh\omega t)]$$

$$+\frac{2n\omega_1}{\omega_1^2-\omega_0^2}(\sin\omega_1 t-\frac{\omega_1}{\omega}e^{-nt}\sinh\omega t)\}$$

$$+\frac{qp-p}{n^2-\omega^2}[1-e^{-nt}(\cosh\omega t+\frac{n}{\omega}\sinh\omega t)]$$

$$+\frac{q\mu}{\omega\tau(n^2-\omega^2)^2}[(n^2-\omega^2)(\omega t+e^{-nt}\sinh\omega t)-2n\omega(1-\cosh\omega t)]$$

$$\textbf{(17.6.31)}$$

18

DAMPING, STIFFNESS, CONSTANT
RESISTANCE, AND FRICTION

This chapter describes engineering systems subjected to the force of inertia, the damping force, the stiffness force, the constant resisting force, and the friction force as the resisting forces. On Row 18 in Guiding Table 2.1 these forces are marked with the plus sign (+). This chapter discusses systems that are subjected to the action of all possible resisting forces for common engineering problems. The numbers of the sections describing the systems subjected to the action of various active forces are shown on the intersection of Row 18 with Columns 1 through 6. Throughout the chapter, the left sides of the differential equations of motion for the engineering systems are identical whereas the right sides of these equations are different, depending on the active forces applied to the systems.

The problems discussed in this chapter could be related to the interaction of engineering systems with viscoelastoplastic media that may exert the resisting forces as a reaction to their deformation (see section 1.2).

18.1 Active Force Equals Zero

The intersection of Row 18 and Column 1 of Guiding Table 2.1 indicates that this section describes engineering systems subjected to the action of the force of inertia, the damping force, the stiffness force, the constant resisting force, and the friction force as the resisting forces in the absence of an active force. The considerations related to the motion of a system in the absence of active forces are discussed in section 1.3. The presence of the stiffness force may cause vibratory motion of the system.

The current problem could be associated with the deformation of the viscoelastoplastic media that in some cases exert damping, stiffness, constant resisting, and friction forces as a reaction to their deformation. Additional information regarding the deformation of a viscoelastoplastic medium as well as the considerations related to the behavior of a constant resisting force and a friction force applied to a vibratory system are presented in the section 1.2.

The system is moving in the horizontal direction. We want to determine the law of motion of the system. Figure 18.1 shows the model of a system subjected to the action of a damping force, a stiffness force, a constant resisting force, and a friction force.

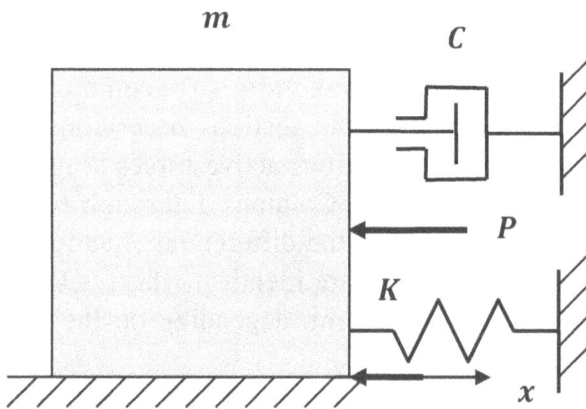

Figure 18.1 Model of a system subjected to a damping force, a stiffness force, a constant resisting force, and a friction force

The considerations above and the model in Figure 18.1 let us compose the left and right sides of the differential equation of motion. The left side consists of the force of inertia, the damping force, the stiffness force, the constant resisting force, and the friction force, while the right side equals zero. Therefore, the differential equation of motion reads:

$$m\frac{d^2x}{dt^2} + C\frac{dx}{dt} + Kx + P + F = 0 \qquad (18.1.1)$$

Differential equation of motion (18.1.1) has different solutions for various initial conditions of motion. These solutions are presented below.

18.1.1 General Initial Conditions

The general initial conditions of motion are:

$$\text{for} \quad t = 0 \quad x = s_0; \quad \frac{dx}{dt} = v_0 \qquad (18.1.2)$$

where s_0 and v_0 are the initial displacement and initial velocity respectively.

Dividing equation (18.1.1) by m, we may write:

$$\frac{d^2x}{dt^2} + 2n\frac{dx}{dt} + \omega_0^2 x + p + f = 0 \qquad (18.1.3)$$

where n is the damping factor and ω_0 is the natural frequency, while:

$$2n = \frac{C}{m} \qquad (18.1.4)$$

$$\omega_0^2 = \frac{K}{m} \qquad (18.1.5)$$

$$p = \frac{P}{m} \qquad (18.1.6)$$

$$f = \frac{F}{m} \qquad (18.1.7)$$

Based on Laplace Transform Pairs 3, 4, 5, and 1 from Table 1.1, we convert differential equation of motion (18.1.3) with the initial conditions of motion (18.1.2) from the time domain into the Laplace domain and obtain the corresponding algebraic equation of motion in the Laplace domain:

$$l^2 x(l) - lv_0 - l^2 s_0 + 2nlx(l) - 2nls_0 + \omega_0^2 x(l) + p + f = 0 \quad \textbf{(18.1.8)}$$

Applying basic algebra to equation (18.1.8), we have:

$$x(l)(l^2 + 2nl + \omega_0^2) = l(v_0 + 2ns_0) + l^2 s_0 - p - f \quad \textbf{(18.1.9)}$$

The solution of equation (18.1.9) for the displacement $x(l)$ in the Laplace domain reads:

$$x(l) = \frac{l(v_0 + 2ns_0)}{l^2 + 2nl + \omega_0^2} + \frac{l^2 s_0}{l^2 + 2nl + \omega_0^2} - \frac{p + f}{l^2 + 2nl + \omega_0^2} \quad \textbf{(18.1.10)}$$

Applying some algebraic procedures to the denominators in equation (18.1.10), we may write:

$$l^2 + 2nl + \omega_0^2 + n^2 - n^2 = (l + n)^2 + \omega^2 \quad \textbf{(18.1.11)}$$

where:

$$\omega^2 = \omega_0^2 - n^2 \quad \textbf{(18.1.12)}$$

while ω^2 could be positive, equal to zero, or negative.

All these three cases are considered below.

a. Case $\omega^2 > 0$

Adjusting the denominators in equation (18.1.10) according to equation (18.1.11), we have:

$$x(l) = \frac{l(v_0 + 2ns_0)}{(l + n)^2 + \omega^2} + \frac{l^2 s_0}{(l + n)^2 + \omega^2} - \frac{p + f}{(l + n)^2 + \omega^2} \quad \textbf{(18.1.13)}$$

Using pairs 1, 24, 27, and 22 from Table 1.1, we invert equation (18.1.13) from the Laplace domain into the time domain and obtain the solution of differential equation of motion (18.1.1) with the initial conditions of motion (18.1.2):

$$x = \frac{v_0 + 2ns_0}{\omega} e^{-nt} \sin \omega t + s_0 e^{-nt} (\cos \omega t - \frac{n}{\omega} \sin \omega t)$$

$$-\frac{p+f}{\omega^2 + n^2} [1 - e^{-nt} (\cos \omega t + \frac{n}{\omega} \sin \omega t)] \qquad \textbf{(18.1.14)}$$

b. Case $\omega^2 = 0$

In equation (18.1.13), equating ω^2 to zero, we have:

$$x(l) = \frac{l(v_0 + 2ns_0)}{(l+n)^2} + \frac{l^2 s_0}{(l+n)^2} - \frac{p+f}{(l+n)^2} \qquad \textbf{(18.1.15)}$$

Using pairs 1, 37, 38, and 36 from Table 1.1, we invert equation (18.1.15) from the Laplace domain into the time domain and obtain the solution of differential equation (18.1.1) with the initial conditions of motion (18.1.2):

$$x = (v_0 + 2ns_0)te^{-nt} + s_0(1 - nt)e^{-nt} - \frac{p+f}{n^2}[1 - e^{-nt}(1+nt)]$$

$$\textbf{(18.1.16)}$$

c. Case $\omega^2 < 0$

In equation (18.1.13), taking ω^2 with the negative sign, we may write:

$$x(l) = \frac{l(v_0 + 2ns_0)}{(l+n)^2 - \omega^2} + \frac{l^2 s_0}{(l+n)^2 - \omega^2} - \frac{p+f}{(l+n)^2 - \omega^2} \qquad \textbf{(18.1.17a)}$$

Applying pairs 1, 25, 28, and 23 from Table 1.1, we invert equation (18.1.17a) from the Laplace domain into the time domain and obtain the solution of differential equation of motion (18.1.1) with the initial conditions of motion (18.1.2):

$$x = \frac{v_0 + 2ns_0}{\omega} e^{-nt} \sinh \omega t + \frac{s_0}{\omega} e^{-nt} (\omega \cosh \omega t - n \sinh \omega t)$$

$$-\frac{p+f}{n^2 - \omega^2} [1 - e^{-nt} (\cosh \omega t + \frac{n}{\omega} \sinh \omega t)] \qquad \textbf{(18.1.17b)}$$

18.1.2 Initial Displacement Equals Zero

The initial conditions of motion are:

$$\text{for} \quad t = 0 \quad x = 0; \quad \frac{dx}{dt} = v_0 \qquad \textbf{(18.1.18)}$$

The solutions of differential equation of motion (18.1.1) with the initial conditions of motion (18.1.18) for the following cases are presented below.

a. Case $\omega^2 > 0$

$$x = \frac{v_0}{\omega} e^{-nt} \sin \omega t - \frac{p+f}{\omega^2 + n^2} [1 - e^{-nt} (\cos \omega t + \frac{n}{\omega} \sin \omega t)] \qquad \textbf{(18.1.19)}$$

b. Case $\omega^2 = 0$

$$x = v_0 t e^{-nt} - \frac{p+f}{n^2} [1 - e^{-nt} (1 + nt)] \qquad \textbf{(18.1.20)}$$

c. Case $\omega^2 < 0$

$$x = \frac{1}{\omega} v_0 e^{-nt} \sinh \omega t - \frac{p+f}{n^2 - \omega^2} [1 - e^{-nt} (\cosh \omega t + \frac{n}{\omega} \sinh \omega t)]$$

$$\textbf{(18.1.21)}$$

18.1.3 Initial Velocity Equals Zero

As shown in Figure 18.1, the system moves in the positive direction. This can occur if the initial displacement is negative. According to this note, the initial conditions of motion in this case are:

$$\text{for} \quad t = 0 \quad x = -s_0; \quad \frac{dx}{dt} = 0 \qquad \textbf{(18.1.22)}$$

It should be also mentioned that in this case the motion is possible if $|Ks_0| > |P + F|$.

The solutions of differential equation of motion (18.1.1) with the initial conditions of motion (18.1.22) for the following cases are presented below.

a. Case $\omega^2 > 0$

$$x = -\frac{2ns_0}{\omega}e^{-nt}\sin\omega t - s_0 e^{-nt}(\cos\omega t - \frac{n}{\omega}\sin\omega t)$$

$$-\frac{p+f}{\omega^2+n^2}[1-e^{-nt}(\cos\omega t + \frac{n}{\omega}\sin\omega t)] \qquad (18.1.23)$$

b. Case $\omega^2 = 0$

$$x = -2ns_0 te^{-nt} - s_0(1-nt)e^{-nt} - \frac{p+f}{n^2}[1-e^{-nt}(1+nt)] \qquad (18.1.24)$$

c. Case $\omega^2 < 0$

$$x = -\frac{2ns_0}{\omega}e^{-nt}\sinh\omega t - \frac{s_0}{\omega}e^{-nt}(\omega\cosh\omega t - n\sinh\omega t)$$

$$-\frac{p+f}{n^2-\omega^2}[1-e^{-nt}(\cosh\omega t + \frac{n}{\omega}\sinh\omega t)] \qquad (18.1.25)$$

18.2 Constant Force R

Guiding Table 2.1 indicates that this section describes engineering systems subjected to the force of inertia, the damping force, the stiffness force, the constant resisting force, and the friction force as the resisting forces (Row 18) and the constant active force (Column 2). The current problem could be related to the working processes of engineering systems interacting with viscoelastoplastic media that may exert the damping, stiffness, constant resisting, and friction forces. These forces represent the reaction of the media to their deformation. Additional information regarding the deformation of the viscoelastoplastic media is presented in section 1.2.

The system is moving in the horizontal direction. We want to determine the law of motion of the system. Figure 18.2 shows the model of a system subjected to the action of a constant active force, a damping force, a stiffness force, a constant resisting force, and a friction force.

Based on the considerations above and on the model in Figure 18.2, we can assemble the left and right sides of the differential equation of motion of this system. The left side consists of the force of inertia, the damping force, the stiffness force, the constant resisting force, and the friction force. The right side includes the constant active force. Therefore, the differential equation of motion reads:

$$m\frac{d^2x}{dt^2} + C\frac{dx}{dt} + Kx + P + F = R \qquad (18.2.1)$$

Due to the presence of the stiffness force, the system may perform vibratory motion. The considerations related to the behavior of a constant resisting force as well as a friction force applied to a vibratory system are discussed in section 1.2.

Differential equation of motion (18.2.1) has different solutions for various initial conditions of motion. These solutions are presented below.

18.2.1 General Initial Conditions
The general initial conditions of motion are:

$$\text{for} \quad t = 0 \quad x = s_0; \quad \frac{dx}{dt} = v_0 \qquad (18.2.2)$$

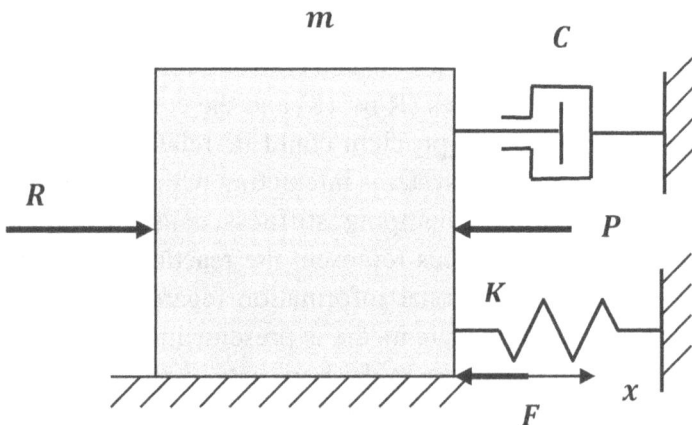

Figure 18.2 Model of a system subjected to a constant active force, a damping force, a stiffness force, a constant resisting force, and a friction force

where s_0 and v_0 are the initial displacement and initial velocity respectively.

Dividing equation (18.2.1) by m, we may write:

$$\frac{d^2x}{dt^2} + 2n\frac{dx}{dt} + \omega_0^2 x + p + f = r \qquad (18.2.3)$$

where n is the damping factor and ω_0 is the natural frequency, while:

$$2n = \frac{C}{m} \qquad (18.2.4)$$

$$\omega_0^2 = \frac{K}{m} \qquad (18.2.5)$$

$$p = \frac{P}{m} \qquad (18.2.6)$$

$$f = \frac{P}{m} \qquad (18.2.7)$$

$$r = \frac{R}{m} \qquad (18.2.8)$$

Using Laplace Transform Pairs 3, 4, 5, and 1 from Table 1.1, we convert differential equation of motion (18.2.3) with the initial conditions of motion (18.2.2) from the time domain into the Laplace domain and obtain the resulting algebraic equation of motion in the Laplace domain:

$$l^2x(l) - lv_0 - l^2s_0 + 2nlx(l) - 2nls_0 + \omega_0^2 x(l) + p + f = r \qquad (18.2.9)$$

Using some basic algebra to equation (18.2.9), we write:

$$x(l)(l^2 + 2nl + \omega_0^2) = r - p - f + l(v_0 + 2ns_0) + l^2s_0 \qquad (18.2.10)$$

The solution of equation (18.2.10) for the displacement $x(l)$ in the Laplace domain reads:

$$x(l) = \frac{r - p - f}{l^2 + 2nl + \omega_0^2} + \frac{l(v_0 + 2ns_0)}{l^2 + 2nl + \omega_0^2} + \frac{l^2s_0}{l^2 + 2nl + \omega_0^2} \qquad (18.2.11)$$

Based on corresponding algebraic procedures with the denominators in equation (18.2.11), we may write:

$$l^2 + 2nl + \omega_0^2 + n^2 - n^2 = (l+n)^2 + \omega^2 \qquad \textbf{(18.2.12)}$$

where:

$$\omega^2 = \omega_0^2 - n^2 \qquad \textbf{(18.2.13)}$$

while ω^2 could be positive, equal to zero, or negative.

All these three cases are considered below.

a. Case $\omega^2 > 0$

Adjusting the denominators in equation (18.2.11) according to equation (18.2.13), we may write:

$$x(l) = \frac{r-p-f}{(l+n)^2 + \omega^2} + \frac{l(v_0 + 2ns_0)}{(l+n)^2 + \omega^2} + \frac{l^2 s_0}{(l+n)^2 + \omega^2} \qquad \textbf{(18.2.14)}$$

Based on pairs 1, 22, 24, and 27 from Table 1.1, we invert equation (18.2.14) from the Laplace domain into the time domain and obtain the solution of differential equation of motion (18.2.1) with the initial conditions of motion (18.2.2):

$$x = \frac{r-p-f}{\omega^2 + n^2}[1 - e^{-nt}(\cos\omega t + \frac{n}{\omega}\sin\omega t)] + \frac{1}{\omega}(v_0 + 2ns_0)e^{-nt}\sin\omega t$$

$$+ s_0 e^{-nt}(\cos\omega t - \frac{n}{\omega}\sin\omega t) \qquad \textbf{(18.2.15)}$$

b. Case $\omega^2 = 0$

In equation (18.2.14), equating ω^2 to zero, we have:

$$x(l) = \frac{r-p-f}{(l+n)^2} + \frac{l(v_0 + 2ns_0)}{(l+n)^2} + \frac{l^2 s_0}{(l+n)^2} \qquad \textbf{(18.2.16)}$$

Using pairs 1, 36, 37, and 38 from Table 1.1, we invert equation (18.2.16) from the Laplace domain into the time domain and obtain the solution of differential equation of motion (18.2.1) with the initial conditions of motion (18.2.2):

$$x = \frac{r-p-f}{n^2}\left[1 - e^{-nt}(1+nt)\right] + (v_0 + 2ns_0)te^{-nt} + s_0 e^{-nt}(1-nt)$$

$$\textbf{(18.2.17)}$$

c. Case $\omega^2 < 0$

In equation (18.2.14), taking ω^2 with the negative sign, we obtain:

$$x(l) = \frac{r-p-f}{(l+n)^2 - \omega^2} + \frac{l(v_0 + 2ns_0)}{(l+n)^2 - \omega^2} + \frac{l^2 s_0}{(l+n)^2 - \omega^2} \qquad (18.2.18)$$

Using pairs 1, 23, 25, and 28 from Table 1.1, we invert equation (18.2.18) from the Laplace domain into the time domain and obtain the solution of differential equation of motion (18.2.1) with the initial conditions of motion (18.2.2):

$$x = \frac{r-p-f}{n^2 - \omega^2}[1 - e^{-nt}(\cosh \omega t + \frac{n}{\omega}\sinh \omega t)] + \frac{v_0 + 2ns_0}{\omega}e^{-nt}\sinh \omega t$$

$$+ \frac{s_0}{\omega}e^{-nt}(\omega \cosh \omega t - n \sinh \omega t) \qquad (18.2.19)$$

18.2.2 Initial Displacement Equals Zero

The initial conditions of motion are:

$$\text{for} \quad t = 0 \quad x = 0; \quad \frac{dx}{dt} = v_0 \qquad (18.2.20)$$

The solutions of differential equation of motion (18.2.1) with the initial conditions of motion (18.2.20) for the following cases are presented below.

a. Case $\omega^2 > 0$

$$x = \frac{r-p-f}{\omega^2 + n^2}[1 - e^{-nt}(\cos \omega t + \frac{n}{\omega}\sin \omega t)] + \frac{1}{\omega}v_0 e^{-nt}\sin \omega t \quad (18.2.21)$$

b. Case $\omega^2 = 0$

$$x = \frac{r-p-f}{n^2}\left[1 - e^{-nt}(1 + nt)\right] + v_0 t e^{-nt} \qquad (18.2.22)$$

c. Case $\omega^2 < 0$

$$x = \frac{r-p-f}{n^2 - \omega^2}[1 - e^{-nt}(\cosh \omega t + \frac{n}{\omega}\sinh \omega t)] + \frac{v_0}{\omega}e^{-nt}\sinh \omega t$$

$$(18.2.23)$$

18.2.3 Initial Velocity Equals Zero

The initial conditions of motion are:

$$\text{for} \quad t = 0 \quad x = s_0; \quad \frac{dx}{dt} = 0 \tag{18.2.24}$$

The solutions of differential equation of motion (18.2.1) with the initial conditions of motion (18.2.24) for the following cases are presented below.

a. Case $\omega^2 > 0$

$$x = \frac{r - p - f}{\omega^2 + n^2}[1 - e^{-nt}(\cos \omega t + \frac{n}{\omega}\sin \omega t)] + \frac{2ns_0}{\omega}e^{-nt}\sin \omega t$$

$$+ s_0 e^{-nt}(\cos \omega t - \frac{n}{\omega}\sin \omega t) \tag{18.2.25}$$

b. Case $\omega^2 = 0$

$$x = \frac{r - p - f}{n^2}[1 - e^{-nt}(1 + nt)] + 2ns_0 te^{-nt} + s_0 e^{-nt}(1 - nt) \tag{18.2.26}$$

c. Case $\omega^2 < 0$

$$x = \frac{r - p - f}{n^2 - \omega^2}[1 - e^{-nt}(\cosh \omega t + \frac{n}{\omega}\sinh \omega t)] + \frac{2ns_0}{\omega}e^{-nt}\sinh \omega t$$

$$+ \frac{s_0}{\omega}e^{-nt}(\omega \cosh \omega t - n \sinh \omega t) \tag{18.2.27}$$

18.2.4 Both the Initial Displacement and Velocity Equal Zero

The initial conditions of motion are:

$$\text{for} \quad t = 0 \quad x = 0; \quad \frac{dx}{dt} = 0 \tag{18.2.28}$$

The solutions of differential equation of motion (18.2.1) with the initial conditions of motion (18.2.28) for the following cases are presented below.

a. Case $\omega^2 > 0$

$$x = \frac{r - p - f}{\omega^2 + n^2}[1 - e^{-nt}(\cos\omega t + \frac{n}{\omega}\sin\omega t)] \qquad (18.2.29)$$

b. Case $\omega^2 = 0$

$$x = \frac{r - p - f}{n^2}[1 - e^{-nt}(1 + nt)] \qquad (18.2.30)$$

c. Case $\omega^2 < 0$

$$x = \frac{r - p - f}{n^2 - \omega^2}[1 - e^{-nt}(\cosh\omega t + \frac{n}{\omega}\sinh\omega t)] \qquad (18.2.31)$$

18.3 Harmonic Force $A\sin(\omega_1 t + \lambda)$

The intersection of Row 18 with Column 3 in Guiding Table 2.1 indicates that this section describing engineering systems experiencing the force of inertia, the damping force, the stiffness force, the constant resisting force, and the friction force as the resisting forces and the harmonic force as the active force. The current problem could be related to the working processes of vibratory machines intended for the deformation of viscoelasto-plastic media. Considerations related to the deformation of these media as well as to the behavior of a constant resisting force and a friction force applied to a vibratory system are presented in section 1.2.

The system is moving in the horizontal direction. We want to determine the law of motion of the system. Figure 18.3 shows the model of a system subjected to the action of a harmonic force, a damping force, a stiffness force, a constant resisting force, and a friction force.

Based on the considerations above and on the model in Figure 18.3, we can compose the left and right sides of the differential equation of motion. The left side consists of the force of inertia, the damping force, the stiffness force, the constant resisting force,

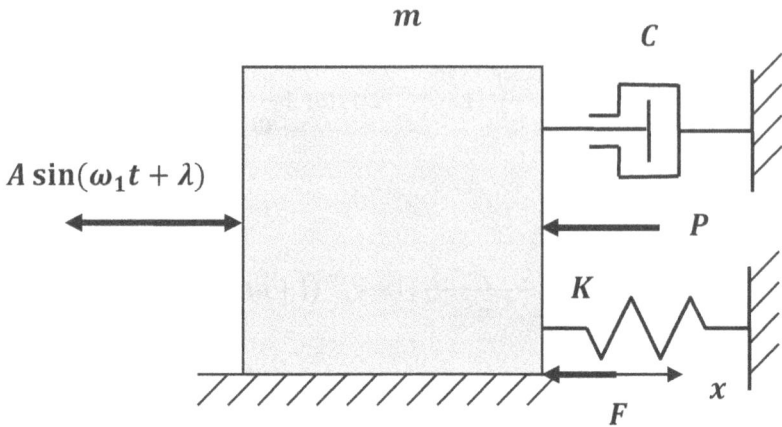

Figure 18.3 Model of a system subjected to a harmonic force, a damping force, a stiffness force, a constant resisting force, and a friction force

and the friction force. The right side includes the harmonic force. Hence, the differential equation of motion reads:

$$m\frac{d^2x}{dt^2} + C\frac{dx}{dt} + Kx + P + F = A\sin(\omega_1 t + \lambda) \qquad (18.3.1)$$

Differential equation of motion (18.3.1) has different solutions for various initial conditions of motion. These solutions are presented below.

18.3.1 General Initial Conditions

The general initial conditions of motion are:

$$\text{for} \quad t = 0 \quad x = s_0; \quad \frac{dx}{dt} = v_0 \qquad (18.3.2)$$

where s_0 and v_0 are the initial displacement and initial velocity respectively.

Transforming the sinusoidal function in equation (18.3.1) and dividing the latter by m, we obtain:

$$\frac{d^2x}{dt^2} + 2n\frac{dx}{dt} + \omega_0^2 x + p + f = a\sin\omega_1 t\cos\lambda + a\cos\omega_1 t\sin\lambda \quad (18.3.3)$$

where n is the damping factor and ω_0 is the natural frequency, while:

$$2n = \frac{C}{m} \tag{18.3.4}$$

$$\omega_0^2 = \frac{K}{m} \tag{18.3.5}$$

$$p = \frac{P}{m} \tag{18.3.6}$$

$$f = \frac{F}{m} \tag{18.3.7}$$

$$a = \frac{A}{m} \tag{18.3.8}$$

Based on Laplace Transform Pairs 3, 4, 5, 6, and 7 from Table 1.1, we convert differential equation of motion (18.3.3) with the initial conditions of motion (18.3.2) from the time domain into the Laplace domain and obtain the resulting algebraic equation of motion in the Laplace domain:

$$l^2 x(l) - l v_0 - l^2 s_0 + 2nlx(l) - 2nls_0 + \omega_0^2 x(l) + p + f$$

$$= \frac{a\omega_1 l}{l^2 + \omega_1^2} \cos \lambda + \frac{al^2}{l^2 + \omega_1^2} \sin \lambda \tag{18.3.9}$$

Solving equation (18.3.9) for the displacement $x(l)$ in the Laplace domain, we may write:

$$x(l) = \frac{l(v_0 + 2ns_0)}{l^2 + 2nl + \omega_0^2} + \frac{l^2 s_0}{l^2 + 2nl + \omega_0^2} + \frac{a\omega_1 l}{(l^2 + \omega_1^2)(l^2 + 2nl + \omega_0^2)}$$

$$\cos \lambda + \frac{al^2}{(l^2 + \omega_1^2)(l^2 + 2nl + \omega_0^2)} \sin \lambda - \frac{p + f}{l^2 + 2nl + \omega_0^2}$$

$$\tag{18.3.10}$$

Applying basic algebra to the denominators of equation (18.3.10), we have:

$$l^2 + 2nl + \omega_0^2 + n^2 - n^2 = (l + n)^2 + \omega^2 \tag{18.3.11}$$

where:

$$\omega^2 = \omega_0^2 - n^2 \qquad (18.3.12)$$

while ω^2 could be positive, equal to zero, or negative.

All these three cases are considered below.

a. Case $\omega^2 > 0$

Adjusting the denominators in equation (18.3.10) according to equation (18.3.11), we obtain:

$$x(l) = \frac{l(v_0 + 2ns_0)}{(l+n)^2 + \omega^2} + \frac{l^2 s_0}{(l+n)^2 + \omega^2} + \frac{a\omega_1 l \cos\lambda}{(l^2 + \omega_1^2)[(l+n)^2 + \omega^2]}$$

$$+ \frac{al^2 \sin\lambda}{(l^2 + \omega_1^2)[(l+n)^2 + \omega^2]} - \frac{p+f}{(l+n)^2 + \omega^2} \qquad (18.3.13)$$

Using Laplace Transform Pairs 1, 24, 27, 42, 43, and 22 from Table 1.1, we invert equation (18.3.13) from the Laplace domain into the time domain and obtain the solution of differential equation of motion (18.3.1) with the initial conditions of motion (18.3.2):

$$x = \frac{1}{\omega}(v_0 + 2ns_0)e^{-nt}\sin\omega t + s_0 e^{-nt}(\cos\omega t - \frac{n}{\omega}\sin\omega t)$$

$$+ \frac{a\omega_1 \cos\lambda}{(\omega_1^2 - \omega_0^2)^2 + 4n^2\omega_1^2}\{2n[e^{-nt}(\cos\omega t + \frac{n}{\omega}\sin\omega t) - \cos\omega_1 t]$$

$$-(\omega_1^2 - \omega_0^2)(\frac{1}{\omega_1}\sin\omega_1 t - \frac{1}{\omega}e^{-nt}\sin\omega t)\}$$

$$+ \frac{a\sin\lambda}{(\omega_1^2 - \omega_0^2)^2 + 4n^2\omega_1^2}\{(\omega_1^2 - \omega_0^2)[e^{-nt}(\cos\omega t + \frac{n}{\omega}\sin\omega t) - \cos\omega_1 t]$$

$$+ 2n\omega_1(\sin\omega_1 t - \frac{\omega_1}{\omega}e^{-nt}\sin\omega t)\}$$

$$- \frac{p+f}{\omega^2 + n^2}[1 - e^{-nt}(\cos\omega t + \frac{n}{\omega}\sin\omega t)] \qquad (18.3.14)$$

b. Case $\omega^2 = 0$

In equation (18.3.13), equating ω^2 to zero, we have:

$$x(l) = \frac{l(v_0 + 2ns_0)}{(l+n)^2} + \frac{l^2 s_0}{(l+n)^2} + \frac{a\omega_1 l}{(l^2 + \omega_1^2)(l+n)^2} \cos \lambda$$

$$+ \frac{al^2}{(l^2 + \omega_1^2)(l+n)^2} \sin \lambda - \frac{p+f}{(l+n)^2} \qquad (18.3.15)$$

Applying pairs 1, 37, 38, 46, 47, and 36 from Table 1.1, we invert equation (18.3.15) from the Laplace domain into the time domain and obtain the solution of differential equation of motion (18.3.1) with the initial conditions of motion (18.3.2):

$$x = (v_0 + 2ns_0)te^{-nt} + s_0(1 - nt)e^{-nt}$$

$$+ \frac{a\omega_1 \cos \lambda}{4n^2\omega_1^2 + (\omega_1^2 - n^2)^2} \{2n[e^{-nt}(1 + nt) - \cos \omega_1 t]$$

$$- (\omega_1^2 - n^2)(\frac{1}{\omega_1} \sin \omega_1 t - te^{-nt})\}$$

$$+ \frac{a \sin \lambda}{4n^2\omega_1^2 + (\omega_1^2 - n^2)^2} \{2n\omega_1(\sin \omega_1 t - \omega_1 te^{-nt})$$

$$- (\omega_1^2 - n^2)[e^{-nt}(1 + nt) - \cos \omega_1 t]\}$$

$$- \frac{p+f}{n^2}[1 - e^{-nt}(1 + nt)] \qquad (18.3.16)$$

c. Case $\omega^2 < 0$

In equation (18.3.13), taking ω^2 with the negative sign, we may write:

$$x(l) = \frac{l(v_0 + 2ns_0)}{(l+n)^2 - \omega^2} + \frac{l^2 s_0}{(l+n)^2 - \omega^2} + \frac{a\omega_1 l}{(l^2 + \omega_1^2)[(l+n)^2 - \omega^2]} \cos \lambda$$

$$+ \frac{al^2}{(l^2 + \omega_1^2)[(l+n)^2 - \omega^2]} \sin \lambda - \frac{p+f}{(l+n)^2 - \omega^2} \qquad (18.3.17)$$

Using pairs $1, 25, 28, 48, 49$, and 23 from Table 1.1, we invert equation (18.3.17) from the Laplace domain into the time domain and obtain the solution of differential equation of motion (18.3.1) with the initial conditions of motion (18.3.2):

$$x = \frac{v_0 + 2ns_0}{\omega} e^{-nt} \sinh \omega t + \frac{s_0}{\omega} e^{-nt} (\omega \cosh \omega t - n \sinh \omega t)$$

$$+ \frac{a\omega_1 \cos \lambda}{4n^2 \omega_1^2 + (\omega_1^2 - \omega_0^2)^2} \{ e^{-nt} [2n \cosh \omega t$$

$$+ \frac{1}{\omega} (\omega_1^2 + \omega^2 + n^2) \sinh \omega t] - 2n \cos \omega_1 t - \frac{1}{\omega_1} (\omega_1^2 + \omega^2 - n^2) \sin \omega_1 t \}$$

$$+ \frac{(\omega_1^2 - \omega_0^2) a \sin \lambda}{4n^2 \omega_1^2 + (\omega_1^2 - \omega_0^2)^2} \{ \frac{\omega_1^2}{\omega^2} (1 - \cos \omega_1 t)$$

$$- \frac{\omega_1^2}{n^2 - \omega^2} [1 - e^{-nt} (\cosh \omega t - \frac{n}{\omega} \sinh \omega t)]$$

$$+ \frac{2n\omega_1}{\omega_1^2 - \omega_0^2} (\sin \omega_1 t - \frac{\omega_1}{\omega} e^{-nt} \sinh \omega t) \}$$

$$- \frac{p+f}{n^2 - \omega^2} [1 - e^{-nt} (\cosh \omega t + \frac{n}{\omega} \sinh \omega t)] \tag{18.3.18}$$

18.3.2 Initial Displacement Equals Zero

The initial conditions of motion are:

$$\text{for} \quad t = 0 \quad x = 0; \quad \frac{dx}{dt} = v_0 \tag{18.3.19}$$

The solutions of differential equation of motion (18.3.1) with the initial conditions of motion (18.3.19) for the following cases are presented below.

a. Case $\omega^2 > 0$

$$x = \frac{1}{\omega} v_0 e^{-nt} \sin \omega t$$

$$+\frac{a\omega_1\cos\lambda}{(\omega_1^2-\omega_0^2)^2+4n^2\omega_1^2}\{2n[e^{-nt}(\cos\omega t+\frac{n}{\omega}\sin\omega t)-\cos\omega_1 t]$$

$$-(\omega_1^2-\omega_0^2)(\frac{1}{\omega_1}\sin\omega_1 t-\frac{1}{\omega}e^{-nt}\sin\omega t)\}$$

$$+\frac{a\sin\lambda}{(\omega_1^2-\omega_0^2)^2+4n^2\omega_1^2}\{(\omega_1^2-\omega_0^2)[e^{-nt}(\cos\omega t+\frac{n}{\omega}\sin\omega t)$$

$$-\cos\omega_1 t]+2n\omega_1(\sin\omega_1 t-\frac{\omega_1}{\omega}e^{-nt}\sin\omega t)\}$$

$$-\frac{p+f}{\omega^2+n^2}[1-e^{-nt}(\cos\omega t+\frac{n}{\omega}\sin\omega t)] \qquad\qquad (18.3.20)$$

b. Case $\omega^2 = 0$

$$x = v_0 t e^{-nt}+\frac{a\omega_1\cos\lambda}{4n^2\omega_1^2+(\omega_1^2-n^2)^2}\{2n[e^{-nt}(1+nt)-\cos\omega_1 t]$$

$$-(\omega_1^2-n^2)(\frac{1}{\omega_1}\sin\omega_1 t-te^{-nt})\}$$

$$+\frac{a\sin\lambda}{4n^2\omega_1^2+(\omega_1^2-n^2)^2}\{2n\omega_1(\sin\omega_1 t-\omega_1 te^{-nt})$$

$$-(\omega_1^2-n^2)[e^{-nt}(1+nt)-\cos\omega_1 t]\}$$

$$-\frac{p+f}{n^2}[1-e^{-nt}(1+nt)] \qquad\qquad (18.3.21)$$

c. Case $\omega^2 < 0$

$$x = \frac{v_0}{\omega}e^{-nt}\sinh\omega t+\frac{a\omega_1\cos\lambda}{4n^2\omega_1^2+(\omega_1^2-\omega_0^2)^2}\{e^{-nt}[2n\cosh\omega t$$

$$+\frac{1}{\omega}(\omega_1^2+\omega^2+n^2)\sinh\omega t]$$

$$-2n\cos\omega_1 t-\frac{1}{\omega_1}(\omega_1^2+\omega^2-n^2)\sin\omega_1 t\}$$

$$+\frac{(\omega_1^2-\omega_0^2)a\sin\lambda}{4n^2\omega_1^2+(\omega_1^2-\omega_0^2)^2}\{\frac{\omega_1^2}{\omega^2}(1-\cos\omega_1 t)$$

$$-\frac{\omega_1^2}{n^2-\omega^2}[1-e^{-nt}(\cosh\omega t - \frac{n}{\omega}\sinh\omega t)]$$

$$+\frac{2n\omega_1}{\omega_1^2-\omega_0^2}(\sin\omega_1 t - \frac{\omega_1}{\omega}e^{-nt}\sinh\omega t)\}$$

$$-\frac{p+f}{n^2-\omega^2}[1-e^{-nt}(\cosh\omega t + \frac{n}{\omega}\sinh\omega t)] \qquad (18.3.22)$$

18.3.3 Initial Velocity Equals Zero

The initial conditions of motion are:

$$\text{for} \quad t=0 \quad x=s_0; \quad \frac{dx}{dt}=0 \qquad (18.3.23)$$

The solutions of differential equation of motion (18.3.1) with the initial conditions of motion (18.3.23) for the following cases are presented below.

a. **Case $\omega^2 > 0$**

$$x = \frac{1}{\omega}2ns_0e^{-nt}\sin\omega t + s_0e^{-nt}(\cos\omega t - \frac{n}{\omega}\sin\omega t)$$

$$+\frac{a\omega_1\cos\lambda}{(\omega_1^2-\omega_0^2)^2+4n^2\omega_1^2}\{2n[e^{-nt}(\cos\omega t + \frac{n}{\omega}\sin\omega t)-\cos\omega_1 t]$$

$$-(\omega_1^2-\omega_0^2)(\frac{1}{\omega_1}\sin\omega_1 t - \frac{1}{\omega}e^{-nt}\sin\omega t)\}$$

$$+\frac{a\sin\lambda}{(\omega_1^2-\omega_0^2)^2+4n^2\omega_1^2}\{(\omega_1^2-\omega_0^2)[e^{-nt}(\cos\omega t + \frac{n}{\omega}\sin\omega t)$$

$$-\cos\omega_1 t]+2n\omega_1(\sin\omega_1 t - \frac{\omega_1}{\omega}e^{-nt}\sin\omega t)\}$$

$$-\frac{p+f}{\omega^2+n^2}[1-e^{-nt}(\cos\omega t + \frac{n}{\omega}\sin\omega t)] \qquad (18.3.24)$$

b. **Case $\omega^2 = 0$**

$$x = 2ns_0te^{-nt} + s_0(1-nt)e^{-nt}$$

$$+\frac{a\omega_1\cos\lambda}{4n^2\omega_1^2+(\omega_1^2-n^2)^2}\{2n[e^{-nt}(1+nt)-\cos\omega_1t]$$

$$-(\omega_1^2-n^2)(\frac{1}{\omega_1}\sin\omega_1t-te^{-nt})\}$$

$$+\frac{a\sin\lambda}{4n^2\omega_1^2+(\omega_1^2-n^2)^2}\{2n\omega_1(\sin\omega_1t-\omega_1te^{-nt})$$

$$-(\omega_1^2-n^2)[e^{-nt}(1+nt)-\cos\omega_1t]\}$$

$$-\frac{p+f}{n^2}[1-e^{-nt}(1+nt)] \tag{18.3.25}$$

c. **Case $\omega^2<0$**

$$x=\frac{2ns_0}{\omega}e^{-nt}\sinh\omega t+\frac{s_0}{\omega}e^{-nt}(\omega\cosh\omega t-n\sinh\omega t)$$

$$+\frac{a\omega_1\cos\lambda}{4n^2\omega_1^2+(\omega_1^2-\omega_0^2)^2}\{e^{-nt}[2n\cosh\omega t$$

$$+\frac{1}{\omega}(\omega_1^2+\omega^2+n^2)\sinh\omega t]$$

$$-2n\cos\omega_1t-\frac{1}{\omega_1}(\omega_1^2+\omega^2-n^2)\sin\omega_1t\}$$

$$+\frac{(\omega_1^2-\omega_0^2)a\sin\lambda}{4n^2\omega_1^2+(\omega_1^2-\omega_0^2)^2}\{\frac{\omega_1^2}{\omega^2}(1-\cos\omega_1t)$$

$$-\frac{\omega_1^2}{n^2-\omega^2}[1-e^{-nt}(\cosh\omega t-\frac{n}{\omega}\sinh\omega t)]$$

$$+\frac{2n\omega_1}{\omega_1^2-\omega_0^2}(\sin\omega_1t-\frac{\omega_1}{\omega}e^{-nt}\sinh\omega t)\}$$

$$-\frac{p+f}{n^2-\omega^2}[1-e^{-nt}(\cosh\omega t+\frac{n}{\omega}\sinh\omega t)] \tag{18.3.26}$$

18.3.4 Both the Initial Displacement and Velocity Equal Zero
The initial conditions of motion are:

$$\text{for}\quad t=0\quad x=0;\quad \frac{dx}{dt}=0 \tag{18.3.27}$$

The solutions of differential equation of motion (18.3.1) with the initial conditions of motion (18.3.27) for the following cases are presented below.

a. **Case $\omega^2 > 0$**

$$x = \frac{a\omega_1 \cos \lambda}{(\omega_1^2 - \omega_0^2)^2 + 4n^2\omega_1^2} \{2n[e^{-nt}(\cos \omega t + \frac{n}{\omega}\sin \omega t) - \cos \omega_1 t]$$

$$-(\omega_1^2 - \omega_0^2)(\frac{1}{\omega_1}\sin \omega_1 t - \frac{1}{\omega}e^{-nt}\sin \omega t)\}$$

$$+\frac{a \sin \lambda}{(\omega_1^2 - \omega_0^2)^2 + 4n^2\omega_1^2}\{(\omega_1^2 - \omega_0^2)[e^{-nt}(\cos \omega t + \frac{n}{\omega}\sin \omega t)$$

$$-\cos \omega_1 t] + 2n\omega_1(\sin \omega_1 t - \frac{\omega_1}{\omega}e^{-nt}\sin \omega t)\}$$

$$-\frac{p+f}{\omega^2 + n^2}[1 - e^{-nt}(\cos \omega t + \frac{n}{\omega}\sin \omega t)] \qquad (18.3.28)$$

b. **Case $\omega^2 = 0$**

$$x = \frac{a\omega_1 \cos \lambda}{4n^2\omega_1^2 + (\omega_1^2 - n^2)^2}\{2n[e^{-nt}(1+nt) - \cos \omega_1 t]$$

$$-(\omega_1^2 - n^2)(\frac{1}{\omega_1}\sin \omega_1 t - te^{-nt})\}$$

$$+\frac{a \sin \lambda}{4n^2\omega_1^2 + (\omega_1^2 - n^2)^2}\{2n\omega_1(\sin \omega_1 t - \omega_1 te^{-nt})$$

$$-(\omega_1^2 - n^2)[e^{-nt}(1+nt) - \cos \omega_1 t]\}$$

$$-\frac{p+f}{n^2}[1 - e^{-nt}(1+nt)] \qquad (18.3.29)$$

c. **Case $\omega^2 < 0$**

$$x = \frac{a\omega_1 \cos \lambda}{4n^2\omega_1^2 + (\omega_1^2 - \omega_0^2)^2}\{e^{-nt}[2n \cosh \omega t$$

$$+\frac{1}{\omega}(\omega_1^2 + \omega^2 + n^2)\sinh \omega t]$$

$$-2n \cos \omega_1 t - \frac{1}{\omega_1}(\omega_1^2 + \omega^2 - n^2)\sin \omega_1 t\}$$

$$+\frac{(\omega_1^2-\omega_0^2)a\sin\lambda}{4n^2\omega_1^2+(\omega_1^2-\omega_0^2)^2}\{\frac{\omega_1^2}{\omega^2}(1-\cos\omega_1 t)$$

$$-\frac{\omega_1^2}{n^2-\omega^2}[1-e^{-nt}(\cosh\omega t-\frac{n}{\omega}\sinh\omega t)]$$

$$+\frac{2n\omega_1}{\omega_1^2-\omega_0^2}(\sin\omega_1 t-\frac{\omega_1}{\omega}e^{-nt}\sinh\omega t)\}$$

$$-\frac{p+f}{n^2-\omega^2}[1-e^{-nt}(\cosh\omega t+\frac{n}{\omega}\sinh\omega t)] \qquad (18.3.30)$$

18.4   Time-Dependent Force $Q\left(\rho+\frac{\mu t}{\tau}\right)$

This section discusses engineering systems subjected to the force of inertia, the damping force, the stiffness force, the constant resisting force, and the friction force as the resisting forces (Row 18 in Guiding Table 2.1) and to the time-dependent force as the active force (Column 4). The current problem could be related to the working process of a system intended for the deformation of viscoelastoplastic media. These media in certain conditions can exert the damping force, the stiffness force, the constant resisting force, and the friction force as the reaction to their deformation (more considerations related to the deformation of viscoelasto-plastic media are presented in section 1.2). Sometimes during the initial phase of deformation, the engineering system could be subjected to a time-dependent force that acts a predetermined interval of time.

The system is moving in the horizontal direction. We want to determine the law of motion of the system. Figure 18.4 shows the model of a system subjected to the action of a time-dependent force, a damping force, a stiffness force, a constant resisting force, and a friction force.

The considerations above and the model in Figure 18.4 let us compose the left and right sides of the differential equation of motion. The left side comprises the force of inertia, the damping force, the stiffness force, the constant resisting force, and the friction force.

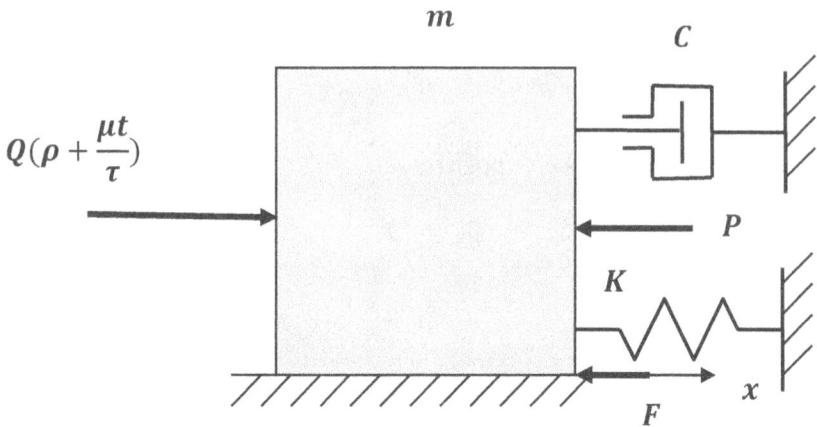

Figure 18.4 Model of a system subjected to a time-dependent force, a damping force, a stiffness force, a constant resisting force, and a friction force

The right side consists of the time-dependent force. Therefore the differential equation of motion reads:

$$m\frac{d^2x}{dt^2}+C\frac{dx}{dt}+Kx+P+F=Q(\rho+\frac{\mu t}{\tau}) \qquad \textbf{(18.4.1)}$$

Differential equation of motion (18.4.1) has different solutions for various initial conditions of motion. These solutions are presented below.

18.4.1 General Initial Conditions
The general initial conditions of motion are:

$$\text{for} \quad t=0 \quad x=s_0; \quad \frac{dx}{dt}=v_0 \qquad \textbf{(18.4.2)}$$

where s_0 and v_0 are the initial displacement and initial velocity respectively.

Dividing equation (18.4.1) by m, we have:

$$\frac{d^2x}{dt^2}+2n\frac{dx}{dt}+\omega_0^2x+p+f=q(\rho+\frac{\mu t}{\tau}) \qquad \textbf{(18.4.3)}$$

where n is the damping factor and ω_0 is the natural frequency, while:

$$2n = \frac{C}{m} \qquad (18.4.4)$$

$$\omega_0^2 = \frac{K}{m} \qquad (18.4.5)$$

$$p = \frac{P}{m} \qquad (18.4.6)$$

$$f = \frac{F}{m} \qquad (18.4.7)$$

$$q = \frac{Q}{m} \qquad (18.4.8)$$

Based on Laplace Transform Pairs 3, 4, 5, and 2 from Table 1.1, we convert differential equation of motion (18.4.3) with the initial conditions of motion (18.4.2) from the time domain into the Laplace domain and obtain the resulting algebraic equation of motion in the Laplace domain:

$$l^2 x(l) - l v_0 - l^2 s_0 + 2nlx(l) - 2nls_0 + \omega_0^2 x(l) + p + f = qp + \frac{q\mu}{\tau l}$$

$$(18.4.9)$$

Solving equation (18.4.9) for the displacement $x(l)$ in the Laplace domain, we have:

$$x(l) = \frac{l(v_0 + 2ns_0)}{l^2 + 2nl + \omega_0^2} + \frac{l^2 s_0}{l^2 + 2nl + \omega_0^2} + \frac{qp - p - f}{l^2 + 2nl + \omega_0^2}$$

$$+ \frac{q\mu}{l\tau(l^2 + 2nl + \omega_0^2)} \qquad (18.4.10)$$

Applying basic algebra to the denominators of equation (18.4.10), we write:

$$l^2 + 2nl + \omega_0^2 + n^2 - n^2 = (l + n)^2 + \omega^2 \qquad (18.4.11)$$

where:

$$\omega^2 = \omega_0^2 - n^2 \qquad (18.4.12)$$

while ω^2 could be positive, equal to zero, or negative.
All these three cases are considered below.

a. Case $\omega^2 > 0$

Transforming the denominators of equation (18.4.10) according to equation (18.4.11), we have:

$$x(l) = \frac{l(v_0 + 2ns_0)}{(l+n)^2 + \omega^2} + \frac{l^2 s_0}{(l+n)^2 + \omega^2} + \frac{q\rho - p - f}{(l+n)^2 + \omega^2}$$

$$+ \frac{q\mu}{l\tau[(l+n)^2 + \omega^2)]} \qquad (18.4.13)$$

Using pairs 1, 24, 27, 22 and 39 from Table 1.1, we invert equation (18.4.13) from the Laplace domain into the time domain and obtain the solution of differential equation of motion (18.4.1) with the initial conditions of motion (18.4.2):

$$x = \frac{v_0 + 2ns_0}{\omega} e^{-nt} \sin \omega t + s_0 (\cos \omega t - \frac{n}{\omega} \sin \omega t) e^{-nt}$$

$$+ \frac{q\rho - p - f}{\omega^2 + n^2} [1 - e^{-nt} (\cos \omega t + \frac{n}{\omega} \sin \omega t)]$$

$$+ \frac{q\mu}{\omega\tau(\omega^2 + n^2)^2} \{ (\omega^2 + n^2)\omega t$$

$$- 2n\omega - e^{-nt} [(\omega^2 - n^2) \sin \omega t - 2n\omega \cos \omega t] \} \qquad (18.4.14)$$

b. Case $\omega^2 = 0$

In equation (18.4.13), equating ω^2 to zero, we write:

$$x(l) = \frac{l(v_0 + 2ns_0)}{(l+n)^2} + \frac{l^2 s_0}{(l+n)^2} + \frac{q\rho - p - f}{(l+n)^2}$$

$$+ \frac{q\mu}{l\tau(l+n)^2} \qquad (18.4.15)$$

Using pairs 1, 37, 38, 36, 15 from Table 1.1, we invert equation (18.4.15) from the Laplace domain into the time domain and obtain the solution of differential equation of motion (18.4.1) with the initial conditions of motion (18.4.2):

$$x = (v_0 + 2ns_0)te^{-nt} + s_0(1-nt)e^{-nt} + \frac{q\rho - p - f}{n^2}[1 - e^{-nt}(1+nt)]$$

$$+ \frac{q\mu}{\tau n^2}[t - \frac{2}{n} + e^{-nt}\left(\frac{2}{n} + t\right)] \qquad (18.4.16)$$

c. Case $\omega^2 < 0$

In equation (18.4.13), taking ω^2 with the negative sign, we may write:

$$x(l) = \frac{l(v_0 + 2ns_0)}{(l+n)^2 - \omega^2} + \frac{l^2 s_0}{(l+n)^2 - \omega^2} + \frac{q\rho - p - f}{(l+n)^2 - \omega^2}$$

$$+ \frac{q\mu}{l\tau[(l+n)^2 - \omega^2]} \qquad (18.4.17)$$

Applying pairs 1, 25, 28, 23, and 40 from Table 1.1, we invert equation (18.4.17) from the Laplace domain into the time domain and obtain the solution of differential equation of motion (18.4.1) with the initial conditions of motion (18.4.2):

$$x = \frac{v_0 + 2ns_0}{\omega}e^{-nt}\sinh\omega t + \frac{s_0}{\omega}e^{-nt}(\omega\cosh\omega t - n\sinh\omega t)$$

$$+ \frac{q\rho - p - f}{n^2 - \omega^2}[1 - e^{-nt}(\cosh\omega t + \frac{n}{\omega}\sinh\omega t)]$$

$$+ \frac{q\mu}{\omega\tau(n^2 - \omega^2)^2}[(n^2 - \omega^2)(\omega t + e^{-nt}\sinh\omega t)$$

$$-2n\omega(1 - \cosh\omega t)] \qquad (18.4.18)$$

18.4.2 Initial Displacement Equals Zero

The initial conditions of motion are:

$$\text{for} \quad t = 0 \quad x = 0; \quad \frac{dx}{dt} = v_0 \qquad (18.4.19)$$

The solutions of differential equation of motion (18.4.1) with the initial conditions of motion (18.4.19) for the following cases are presented below.

a. **Case $\omega^2 > 0$**

$$x = \frac{v_0}{\omega} e^{-nt} \sin \omega t + \frac{q\rho - p - f}{\omega^2 + n^2} [1 - e^{-nt} (\cos \omega t + \frac{n}{\omega} \sin \omega t)]$$

$$+ \frac{q\mu}{\omega\tau(\omega^2 + n^2)^2} \{(\omega^2 + n^2)\omega t - 2n\omega$$

$$- e^{-nt} [(\omega^2 - n^2)\sin \omega t - 2n\omega \cos \omega t]\} \qquad (18.4.20)$$

b. **Case $\omega^2 = 0$**

$$x = v_0 t e^{-nt} + \frac{q\rho - p - f}{n^2} [1 - e^{-nt} (1 + nt)]$$

$$+ \frac{q\mu}{\tau n^2} [t - \frac{2}{n} + e^{-nt} (\frac{2}{n} + t)] \qquad (18.4.21)$$

c. **Case $\omega^2 < 0$**

$$x = \frac{v_0}{\omega} e^{-nt} \sinh \omega t + \frac{q\rho - p - f}{n^2 - \omega^2} [1 - e^{-nt} (\cosh \omega t + \frac{n}{\omega} \sinh \omega t]$$

$$+ \frac{q\mu}{\omega\tau(n^2 - \omega^2)^2} [(n^2 - \omega^2)(\omega t + e^{-nt} \sinh \omega t)$$

$$- 2n\omega(1 - \cosh \omega t)] \qquad (18.4.22)$$

18.4.3 Initial Velocity Equals Zero

The initial conditions of motion are:

$$\text{for} \quad t = 0 \quad x = s_0; \quad \frac{dx}{dt} = 0 \qquad (18.4.23)$$

The solutions of differential equation of motion (18.4.1) with the initial conditions of motion (18.4.23) for the following cases are presented below.

a. Case $\omega^2 > 0$

$$x = \frac{2ns_0}{\omega}e^{-nt}\sin\omega t + s_0(\cos\omega t - \frac{n}{\omega}\sin\omega t)e^{-nt}$$

$$+\frac{q\rho - p - f}{\omega^2 + n^2}[1 - e^{-nt}(\cos\omega t + \frac{n}{\omega}\sin\omega t)]$$

$$+\frac{q\mu}{\omega\tau(\omega^2 + n^2)^2}\{(\omega^2 + n^2)\omega t - 2n\omega$$

$$-e^{-nt}[(\omega^2 - n^2)\sin\omega t - 2n\omega\cos\omega t]\} \qquad (18.4.24)$$

b. Case $\omega^2 = 0$

$$x = 2ns_0te^{-nt} + s_0(1 - nt)e^{-nt} + \frac{q\rho - p - f}{\tau n^2}[1 - e^{-nt}(1 + nt)]$$

$$+\frac{q\mu}{\tau n^2}[t - \frac{2}{n} + e^{-nt}\left(\frac{2}{n} + t\right)] \qquad (18.4.25)$$

c. Case $\omega^2 < 0$

$$x = \frac{2ns_0}{\omega}e^{-nt}\sinh\omega t + \frac{s_0}{\omega}e^{-nt}(\omega\cosh\omega t - n\sinh\omega t)$$

$$+\frac{q\rho - p - f}{n^2 - \omega^2}[1 - e^{-nt}(\cosh\omega t + \frac{n}{\omega}\sinh\omega t)]$$

$$+\frac{q\mu}{\omega\tau(n^2 - \omega^2)^2}[(n^2 - \omega^2)(\omega t + e^{-nt}\sinh\omega t)$$

$$-2n\omega(1 - \cosh\omega t)] \qquad (18.4.26)$$

18.4.4 Both the Initial Displacement and Velocity Equal Zero
The initial conditions of motion are:

$$\text{for}\quad t = 0\quad x = 0;\quad \frac{dx}{dt} = 0 \qquad (18.4.27)$$

The solutions of differential equation of motion (18.4.1) with the initial conditions of motion (18.4.27) for the following cases are presented below.

a. **Case $\omega^2 > 0$**

$$x = \frac{q\rho - p - f}{\omega^2 + n^2}[1 - e^{-nt}(\cos\omega t + \frac{n}{\omega}\sin\omega t)]$$

$$+ \frac{q\mu}{\omega\tau(\omega^2 + n^2)^2}\{(\omega^2 + n^2)\omega t - 2n\omega$$

$$- e^{-nt}[(\omega^2 - n^2)\sin\omega t - 2n\omega\cos\omega t]\} \qquad (18.4.28)$$

b. **Case $\omega^2 = 0$**

$$x = \frac{q\rho - p - f}{n^2}[1 - e^{-nt}(1 + nt)]$$

$$+ \frac{q\mu}{\tau n^2}[t - \frac{2}{n} + e^{-nt}(\frac{2}{n} + t)] \qquad (18.4.29)$$

c. **Case $\omega^2 < 0$**

$$x = \frac{q\rho - p - f}{n^2 - \omega^2}[1 - e^{-nt}(\cosh\omega t + \frac{n}{\omega}\sinh\omega t)]$$

$$+ \frac{q\mu}{\omega\tau(n^2 - \omega^2)^2}[(n^2 - \omega^2)(\omega t + e^{-nt}\sinh\omega t)$$

$$- 2n\omega(1 - \cosh\omega t)] \qquad (18.4.30)$$

18.5 Constant Force R and Harmonic Force $A\sin(\omega_1 t + \lambda)$

According to Guiding Table 2.1, this section describes the engineering systems subjected to the force of inertia, the damping force, the stiffness force, the constant resisting force, and the friction force as the resisting forces (Row 18) and to the constant active force and the harmonic force as the active forces (Column 5).

The current problem could be related to the working process of a vibratory system that interacts with a viscoelastoplastic

medium. As discussed in section 1.2, in certain cases the viscoelastoplastic media can exert the damping force, the stiffness force, the
constant resisting force, and the friction force as the reaction to their
deformation. The same section presents the considerations related to
the behavior of a constant resisting force as well as a friction force
applied to a vibratory system.

The system is moving in the horizontal direction. We want to
determine the law of motion of the system. Figure 18.5 shows the
model of a system subjected to the action of a constant active force,
a harmonic force, a damping force, a stiffness force, a constant resisting force, and a friction force.

The considerations above and the model shown in Figure 18.5
let us compose the left and the right sides of the differential equation
of motion of the system. The left side consists of the force of inertia,
the damping force, the stiffness force, the constant resisting force,
and the friction force. The right side includes the constant active
force and the harmonic force. Therefore, the differential equation of
motion reads:

$$m\frac{d^2x}{dt^2} + C\frac{dx}{dt} + Kx + P + F = R + A\sin(\omega_1 t + \lambda) \quad (18.5.1)$$

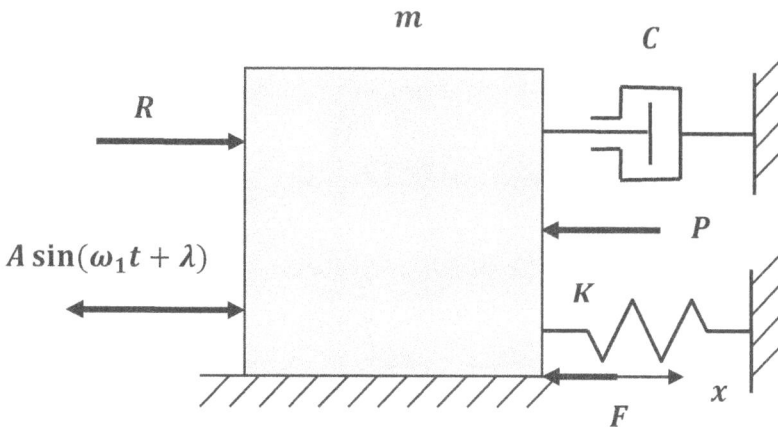

**Figure 18.5 Model of a system subjected to a constant active
force, a harmonic force, a damping force, a stiffness force, a
constant resisting force, and a friction force**

Differential equation of motion (18.5.1) has different solutions for various initial conditions of motion. These solutions are presented below.

18.5.1 General Initial conditions

The general initial conditions of motion are:

$$\text{for} \quad t = 0 \quad x = s_0; \quad \frac{dx}{dt} = v_0 \qquad (18.5.2)$$

Transforming the sinusoidal function in equation (18.5.1) and dividing the latter by m, we have:

$$\frac{d^2 x}{dt^2} + 2n \frac{dx}{dt} + \omega_0^2 x + p + f = r + a \sin \omega_1 t \cos \lambda$$

$$+ a \cos \omega_1 t \sin \lambda \qquad (18.5.3)$$

where n is the damping factor and ω_0 in the natural frequency, while:

$$2n = \frac{C}{m} \qquad (18.5.4)$$

$$\omega_0^2 = \frac{K}{m} \qquad (18.5.5)$$

$$p = \frac{P}{m} \qquad (18.5.6)$$

$$f = \frac{F}{m} \qquad (18.5.7)$$

$$r = \frac{R}{m} \qquad (18.5.8)$$

$$a = \frac{A}{m} \qquad (18.5.9)$$

Using Laplace Transform Pairs 3, 4, 1, 5, 6, and 7 from Table 1.1, we convert differential equation of motion (18.5.3) with the initial conditions of motion (18.5.2) from the time domain into the Laplace domain, and obtain the resulting algebraic equation of motion in the Laplace domain:

$$l^2 x(l) - l v_0 - l^2 s_0 + 2nlx(l) - 2nls_0 + \omega_0^2 x(l) + p + f$$

$$= r + \frac{a\omega_1 l \cos \lambda}{l^2 + \omega_1^2} + \frac{al^2 \sin \lambda}{l^2 + \omega_1^2} + \frac{al^2 \sin \lambda}{l^2 + \omega_1^2} \qquad (18.5.10)$$

Based on some algebraic procedures with equation (18.5.10), we may write:

$$x(l)(l^2 + 2nl + \omega_0^2) = r - p - f + l(v_0 + 2ns_0) + l^2 s_0$$

$$+ \frac{a\omega_1 l \cos \lambda}{l^2 + \omega_1^2} + \frac{al^2 \sin \lambda}{l^2 + \omega_1^2} \qquad (18.5.11)$$

The solution of equation (18.5.11) for the displacement $x(l)$ in the Laplace domain reads:

$$x(l) = \frac{r - p - f}{l^2 + 2nl + \omega_0^2} + \frac{l(v_0 + 2ns_0)}{l^2 + 2nl + \omega_0^2} + \frac{l^2 s_0}{l^2 + 2nl + \omega_0^2}$$

$$+ \frac{a\omega_1 l \cos \lambda}{(l^2 + \omega_1^2)(l^2 + 2nl + \omega_0^2)}$$

$$+ \frac{al^2 \sin \lambda}{(l^2 + \omega_1^2)(l^2 + 2nl + \omega_0^2)} \qquad (18.5.12)$$

Applying basic algebra to the denominators in equation (18.5.12), we have:

$$l^2 + 2nl + \omega_0^2 + n^2 - n^2 = (l+n)^2 + \omega^2 \qquad (18.5.13)$$

where:

$$\omega^2 = \omega_0^2 - n^2 \qquad (18.5.14)$$

while ω^2 could be positive, equal to zero, or negative.
All these three cases are considered below.

a. Case $\omega^2 > 0$

Adjusting the denominators in equation (18.5.12) according to equation (18.5.13), we have:

$$x(l) = \frac{r - p - f}{(l+n)^2 + \omega^2} + \frac{l(v_0 + 2ns_0)}{(l+n)^2 + \omega^2} + \frac{l^2 s_0}{(l+n)^2 + \omega^2}$$

$$+ \frac{a\omega_1 l \cos \lambda}{(l^2 + \omega_1^2)[(l+n)^2 + \omega^2]} + \frac{al^2 \sin \lambda}{(l^2 + \omega_1^2)[(l+n)^2 + \omega^2]} \quad \textbf{(18.5.15)}$$

Based on pairs 1, 22, 24, 27, 42, and 43 from Table 1.1, we invert equation (18.5.15) from the Laplace domain into the time domain and obtain the solution of differential equation of motion (18.5.1) with the initial conditions of motion (18.2.2):

$$x = \frac{r - p - f}{\omega^2 + n^2}[1 - e^{-nt}(\cos \omega t + \frac{n}{\omega}\sin \omega t)]$$

$$+ \frac{1}{\omega}(v_0 + 2ns_0)e^{-nt}\sin \omega t + s_0 e^{-nt}(\cos \omega t - \frac{n}{\omega}\sin \omega t)$$

$$+ \frac{a\omega_1 \cos \lambda}{(\omega_1^2 - \omega_0^2)^2 + 4n^2 \omega_1^2}\{2n[e^{-nt}(\cos \omega t + \frac{n}{\omega}\sin \omega t) - \cos \omega_1 t]$$

$$- (\omega_1^2 - \omega_0^2)(\frac{1}{\omega_1}\sin \omega_1 t - \frac{1}{\omega}e^{-nt}\sin \omega t)\}$$

$$+ \frac{a \sin \lambda}{(\omega_1^2 - \omega_0^2)^2 + 4n^2 \omega_1^2}\{(\omega_1^2 - \omega_0^2)[e^{-nt}(\cos \omega t + \frac{n}{\omega}\sin \omega t)$$

$$- \cos \omega_1 t] + 2n\omega_1(\sin \omega_1 t - \frac{\omega_1}{\omega}e^{-nt}\sin \omega t)\} \quad \textbf{(18.5.16)}$$

b. Case $\omega^2 = 0$

Equating ω^2 to zero in equation (18.5.15), we have:

$$x(l) = \frac{r - p - f}{(l+n)^2} + \frac{l(v_0 + 2ns_0)}{(l+n)^2} + \frac{l^2 s_0}{(l+n)^2}$$

$$+ \frac{a\omega_1 l \cos \lambda}{(l^2 + \omega_1^2)(l+n)^2} + \frac{al^2 \sin \lambda}{(l^2 + \omega_1^2)(l+n)^2} \quad \textbf{(18.5.17)}$$

Using pairs 1, 36, 37, 38, 46, and 47 from Table 1.1, we invert equation (18.5.17) from the Laplace domain into the time domain and obtain the solution of differential equation of motion (18.5.1) with the initial conditions of motion (18.5.2):

$$x = \frac{r-p-f}{n^2}[1-e^{-nt}(1+nt)]+(v_0+2ns_0)te^{-nt}+s_0(1-nt)e^{-nt}$$

$$+\frac{a\omega_1\cos\lambda}{4n^2\omega_1^2+(\omega_1^2-n^2)^2}\{2n[e^{-nt}(1+nt)-\cos\omega_1 t]$$

$$-(\omega_1^2-n^2)(\frac{1}{\omega_1}\sin\omega_1 t-te^{-nt})\}$$

$$+\frac{a\sin\lambda}{4n^2\omega_1^2+(\omega_1^2-n^2)^2}\{2n\omega_1(\sin\omega_1 t-\omega_1 te^{-nt})$$

$$-(\omega_1^2-n^2)[e^{-nt}(1+nt)-\cos\omega_1 t]\} \qquad (18.5.18)$$

c. Case $\omega^2 < 0$

In equation (18.5.15), taking ω^2 with the negative sign, we may write:

$$x(l) = \frac{r-p-f}{(l+n)^2-\omega^2}+\frac{l(v_0+2ns_0)}{(l+n)^2-\omega^2}+\frac{l^2 s_0}{(l+n)^2-\omega^2}$$

$$+\frac{a\omega_1 l\cos\lambda}{(l^2+\omega_1^2)[(l+n)^2-\omega^2]}+\frac{al^2\sin\lambda}{(l^2+\omega_1^2)[(l+n)^2-\omega^2]} \qquad (18.5.19)$$

Based on pairs 1, 23, 25, 28, 48, and 49 from Table 1.1, we invert equation (18.5.19) from the Laplace domain into the time domain and obtain the solution of differential equation of motion (18.5.1) with the initial conditions of motion (18.2.2):

$$x = \frac{r-p-f}{n^2-\omega^2}[1-e^{-nt}(\cosh\omega t+\frac{n}{\omega}\sinh\omega t)]$$

$$+\frac{v_0+2ns_0}{\omega}e^{-nt}\sinh\omega t+\frac{s_0}{\omega}e^{-nt}(\omega\cosh\omega t-n\sinh\omega t)$$

$$+\frac{a\omega_1\cos\lambda}{4n^2\omega_1^2+(\omega_1^2-\omega_0^2)^2}\{e^{-nt}[2n\cosh\omega t$$

$$+\frac{1}{\omega}(\omega_1^2+\omega^2+n^2)\sinh\omega t]$$

$$-2n\cos\omega_1 t-\frac{1}{\omega_1}(\omega_1^2+\omega^2-n^2)\sin\omega_1 t\}$$

$$+\frac{(\omega_1^2-\omega_0^2)a\sin\lambda}{4n^2\omega_1^2+(\omega_1^2-\omega_0^2)^2}\{\frac{\omega_1^2}{\omega^2}(1-\cos\omega_1 t)$$

$$-\frac{\omega_1^2}{n^2-\omega^2}[1-e^{-nt}(\cosh\omega t-\frac{n}{\omega}\sinh\omega t)]$$

$$+\frac{2n\omega_1}{\omega_1^2-\omega_0^2}(\sin\omega_1 t-\frac{\omega_1}{\omega}e^{-nt}\sinh\omega t)\}\qquad\textbf{(18.5.20)}$$

18.5.2 Initial Displacement Equals Zero

The initial conditions of motion are:

$$\text{for}\quad t=0\quad x=0;\quad \frac{dx}{dt}=v_0\qquad\textbf{(18.5.21)}$$

The solutions of differential equation of motion (18.5.1) with the initial conditions of motion (18.5.21) for the following cases are presented below.

a. Case $\omega^2>0$

$$x=\frac{r-p-f}{\omega^2+n^2}[1-e^{-nt}(\cos\omega t+\frac{n}{\omega}\sin\omega t)]+\frac{1}{\omega}v_0 e^{-nt}\sin\omega t$$

$$+\frac{a\omega_1\cos\lambda}{(\omega_1^2-\omega_0^2)^2+4n^2\omega_1^2}\{2n[e^{-nt}(\cos\omega t+\frac{n}{\omega}\sin\omega t)-\cos\omega_1 t]$$

$$-(\omega_1^2-\omega_0^2)(\frac{1}{\omega_1}\sin\omega_1 t-\frac{1}{\omega}e^{-nt}\sin\omega t)\}$$

$$+\frac{a\sin\lambda}{(\omega_1^2-\omega_0^2)^2+4n^2\omega_1^2}\{(\omega_1^2-\omega_0^2)[e^{-nt}(\cos\omega t+\frac{n}{\omega}\sin\omega t)$$

$$-\cos\omega_1 t]+2n\omega_1(\sin\omega_1 t-\frac{\omega_1}{\omega}e^{-nt}\sin\omega t)\}\qquad\textbf{(18.5.22)}$$

b. Case $\omega^2=0$

$$x=\frac{r-p-f}{n^2}\left[1-e^{-nt}(1+nt)\right]+v_0 te^{-nt}$$

$$+\frac{a\omega_1\cos\lambda}{4n^2\omega_1^2+(\omega_1^2-n^2)^2}\{2n[e^{-nt}(1+nt)-\cos\omega_1 t]$$

$$-(\omega_1^2-n^2)(\frac{1}{\omega_1}\sin\omega_1 t-te^{-nt})\}$$

$$+\frac{a\sin\lambda}{4n^2\omega_1^2+(\omega_1^2-n^2)^2}\{2n\omega_1(\sin\omega_1 t-\omega_1 te^{-nt})$$

$$-(\omega_1^2-n^2)[e^{-nt}(1+nt)-\cos\omega_1 t]\} \tag{18.5.23}$$

c. Case $\omega^2 < 0$

$$x=\frac{r-p-f}{n^2-\omega^2}[1-e^{-nt}(\cosh\omega t+\frac{n}{\omega}\sinh\omega t)]+\frac{v_0}{\omega}e^{-nt}\sinh\omega t$$

$$+\frac{a\omega_1\cos\lambda}{4n^2\omega_1^2+(\omega_1^2-\omega_0^2)^2}\{e^{-nt}[2n\cosh\omega t$$

$$+\frac{1}{\omega}(\omega_1^2+\omega^2+n^2)\sinh\omega t]-2n\cos\omega_1 t$$

$$-\frac{1}{\omega_1}(\omega_1^2+\omega^2-n^2)\sin\omega_1 t\}$$

$$+\frac{(\omega_1^2-\omega_0^2)a\sin\lambda}{4n^2\omega_1^2+(\omega_1^2-\omega_0^2)^2}\{\frac{\omega_1^2}{\omega^2}(1-\cos\omega_1 t)$$

$$-\frac{\omega_1^2}{n^2-\omega^2}[1-e^{-nt}(\cosh\omega t-\frac{n}{\omega}\sinh\omega t)]$$

$$+\frac{2n\omega_1}{\omega_1^2-\omega_0^2}(\sin\omega_1 t-\frac{\omega_1}{\omega}e^{-nt}\sinh\omega t)\} \tag{18.5.24}$$

18.5.3 Initial Velocity Equals Zero

The initial conditions of motion are:

$$\text{for}\quad t=0\quad x=s_0;\quad \frac{dx}{dt}=0 \tag{18.5.25}$$

The solutions of differential equation of motion (18.5.1) with the initial conditions of motion (18.5.25) for the following cases are presented below.

a. Case $\omega^2 > 0$

$$x = \frac{r-p-f}{\omega^2+n^2}[1-e^{-nt}(\cos\omega t + \frac{n}{\omega}\sin\omega t)] + \frac{1}{\omega}2ns_0e^{-nt}\sin\omega t$$

$$+ s_0e^{-nt}(\cos\omega t - \frac{n}{\omega}\sin\omega t)$$

$$+ \frac{a\omega_1\cos\lambda}{(\omega_1^2-\omega_0^2)^2+4n^2\omega_1^2}\{2n[e^{-nt}(\cos\omega t + \frac{n}{\omega}\sin\omega t) - \cos\omega_1 t]$$

$$-(\omega_1^2-\omega_0^2)(\frac{1}{\omega_1}\sin\omega_1 t - \frac{1}{\omega}e^{-nt}\sin\omega t)\}$$

$$+ \frac{a\sin\lambda}{(\omega_1^2-\omega_0^2)^2+4n^2\omega_1^2}\{(\omega_1^2-\omega_0^2)[e^{-nt}(\cos\omega t + \frac{n}{\omega}\sin\omega t)$$

$$- \cos\omega_1 t] + 2n\omega_1(\sin\omega_1 t - \frac{\omega_1}{\omega}e^{-nt}\sin\omega t)\} \qquad \textbf{(18.5.26)}$$

b. Case $\omega^2 = 0$

$$x = \frac{r-p-f}{n^2}[1-e^{-nt}(1+nt)] + 2ns_0te^{-nt}$$

$$+ s_0(1-nt)e^{-nt} + \frac{a\omega_1\cos\lambda}{4n^2\omega_1^2+(\omega_1^2-n^2)^2}\{2n[e^{-nt}(1+nt) - \cos\omega_1 t]$$

$$-(\omega_1^2-n^2)(\frac{1}{\omega_1}\sin\omega_1 t - te^{-nt})\}$$

$$+ \frac{a\sin\lambda}{4n^2\omega_1^2+(\omega_1^2-n^2)^2}\{2n\omega_1(\sin\omega_1 t - \omega_1 te^{-nt})$$

$$-(\omega_1^2-n^2)[e^{-nt}(1+nt) - \cos\omega_1 t]\} \qquad \textbf{(18.5.27)}$$

c. Case $\omega^2 < 0$

$$x = \frac{r-p-f}{n^2-\omega^2}[1-e^{-nt}(\cosh\omega t + \frac{n}{\omega}\sinh\omega t)] + \frac{2ns_0}{\omega}e^{-nt}\sinh\omega t$$

$$+ \frac{s_0}{\omega}e^{-nt}(\omega\cosh\omega t - n\sinh\omega t) + \frac{a\omega_1\cos\lambda}{4n^2\omega_1^2+(\omega_1^2-\omega_0^2)^2}$$

$$\times \{e^{-nt}[2n\cosh\omega t + \frac{1}{\omega}(\omega_1^2 + \omega^2 + n^2)\sinh\omega t] - 2n\cos\omega_1 t$$

$$-\frac{1}{\omega_1}(\omega_1^2 + \omega^2 - n^2)\sin\omega_1 t\}$$

$$+\frac{(\omega_1^2 - \omega_0^2)a\sin\lambda}{4n^2\omega_1^2 + (\omega_1^2 - \omega_0^2)^2}\{\frac{\omega_1^2}{\omega^2}(1 - \cos\omega_1 t)$$

$$-\frac{\omega_1^2}{n^2 - \omega^2}[1 - e^{-nt}(\cosh\omega t - \frac{n}{\omega}\sinh\omega t)]$$

$$+\frac{2n\omega_1}{\omega_1^2 - \omega_0^2}(\sin\omega_1 t - \frac{\omega_1}{\omega}e^{-nt}\sinh\omega t)\} \qquad (18.5.28)$$

18.5.4 Both the Initial Displacement and Velocity Equal Zero

The initial conditions of motion are:

$$\text{for} \quad t = 0 \quad x = 0; \quad \frac{dx}{dt} = 0 \qquad (18.5.29)$$

The solutions of differential equation of motion (18.5.1) with the initial conditions of motion (18.5.29) for the following cases are presented below.

a. Case $\omega^2 > 0$

$$x = \frac{r - p - f}{\omega^2 + n^2}[1 - e^{-nt}(\cos\omega t + \frac{n}{\omega}\sin\omega t)]$$

$$+\frac{a\omega_1\cos\lambda}{(\omega_1^2 - \omega_0^2)^2 + 4n^2\omega_1^2}\{2n[e^{-nt}(\cos\omega t + \frac{n}{\omega}\sin\omega t) - \cos\omega_1 t]$$

$$-(\omega_1^2 - \omega_0^2)(\frac{1}{\omega_1}\sin\omega_1 t - \frac{1}{\omega}e^{-nt}\sin\omega t)\}$$

$$+\frac{a\sin\lambda}{(\omega_1^2 - \omega_0^2)^2 + 4n^2\omega_1^2}\{(\omega_1^2 - \omega_0^2)[e^{-nt}(\cos\omega t + \frac{n}{\omega}\sin\omega t)$$

$$-\cos\omega_1 t] + 2n\omega_1(\sin\omega_1 t - \frac{\omega_1}{\omega}e^{-nt}\sin\omega t)\} \qquad (18.5.30)$$

b. Case $\omega^2 = 0$

$$x = \frac{r-p-f}{n^2}[1 - e^{-nt}(1+nt)]$$

$$+ \frac{a\omega_1 \cos\lambda}{4n^2\omega_1^2 + (\omega_1^2 - n^2)^2} \{2n[e^{-nt}(1+nt) - \cos\omega_1 t]$$

$$-(\omega_1^2 - n^2)(\frac{1}{\omega_1}\sin\omega_1 t - te^{-nt})\}$$

$$+ \frac{a\sin\lambda}{4n^2\omega_1^2 + (\omega_1^2 - n^2)^2} \{2n\omega_1(\sin\omega_1 t - \omega_1 te^{-nt})$$

$$-(\omega_1^2 - n^2)[e^{-nt}(1+nt) - \cos\omega_1 t]\} \tag{18.5.31}$$

c. Case $\omega^2 < 0$

$$x = \frac{r-p-f}{n^2 - \omega^2}[1 - e^{-nt}(\cosh\omega t + \frac{n}{\omega}\sinh\omega t)]$$

$$+ \frac{a\omega_1 \cos\lambda}{4n^2\omega_1^2 + (\omega_1^2 - \omega_0^2)^2} \{e^{-nt}[2n\cosh\omega t$$

$$+ \frac{1}{\omega}(\omega_1^2 + \omega^2 + n^2)\sinh\omega t] - 2n\cos\omega_1 t$$

$$- \frac{1}{\omega_1}(\omega_1^2 + \omega^2 - n^2)\sin\omega_1 t\}$$

$$+ \frac{(\omega_1^2 - \omega_0^2)a\sin\lambda}{4n^2\omega_1^2 + (\omega_1^2 - \omega_0^2)^2} \{\frac{\omega_1^2}{\omega^2}(1 - \cos\omega_1 t)$$

$$- \frac{\omega_1^2}{n^2 - \omega^2}[1 - e^{-nt}(\cosh\omega t - \frac{n}{\omega}\sinh\omega t)]$$

$$+ \frac{2n\omega_1}{\omega_1^2 - \omega_0^2}(\sin\omega_1 t - \frac{\omega_1}{\omega}e^{-nt}\sinh\omega t)\} \tag{18.5.32}$$

18.6 Harmonic Force $A\sin(\omega_1 t + \lambda)$ and Time-Dependent Force $Q\left(\rho + \frac{\mu t}{\tau}\right)$

The intersection of Row 18 and Column 6 in Guiding Table 2.1 indicates that this section describes engineering systems subjected to the force of inertia, the damping force, the stiffness force, the

constant resisting force, and the friction force as the resisting forces and the harmonic force and time-dependent force as the active
forces.

The current problem could be related to the working process
of a vibratory system that interacts with a viscoelastoplastic media.
Along with the harmonic force, the time-dependent force is acting a
limited interval of time during the initial phase of the working process. Additional considerations related to the deformation of viscoelastoplastic media as well as to the behavior of a constant resisting
force and a friction force applied to a vibratory system are presented
in section 1.2.

The system is moving in the horizontal direction. We want to
determine the law of motion of the system. Figure 18.6 shows the
model of a system subjected to the action of a harmonic force, a
time-dependent force, a damping force, a stiffness force, a constant
resisting force, and a friction force.

Based on the considerations above and on the model in
Figure 18.6, we can assemble the left and right sides of the differential equation of motion. The left side consists of the force of inertia,
the damping force, the stiffness force, the constant resisting force,

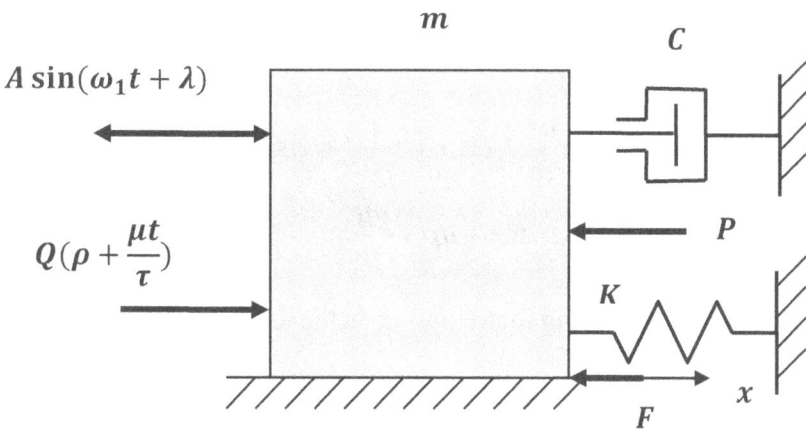

**Figure 18.6 Model of a system subjected to a harmonic force,
a time-dependent force, a damping force, a stiffness force, a
constant resisting force, and a friction force**

and the friction force. The right side includes the harmonic force and the time-dependent force. Hence, the differential equation of motion reads:

$$m\frac{d^2x}{dt^2}+C\frac{dx}{dt}+Kx+P+F=A\sin(\omega_1 t+\lambda)+Q(\rho+\frac{\mu t}{\tau}) \qquad \textbf{(18.6.1)}$$

Differential equation of motion (18.6.1) has different solutions for various initial conditions of motion. These solutions are presented below.

18.6.1 General Initial Conditions

The general initial conditions of motion are:

$$\text{for} \quad t=0 \quad x=s_0; \quad \frac{dx}{dt}=v_0 \qquad \textbf{(18.6.2)}$$

where s_0 and v_0 are the initial displacement and initial velocity respectively.

Transforming equation (18.6.1), we have:

$$m\frac{d^2x}{dt^2}+C\frac{dx}{dt}+Kx+P+F=A\sin\omega_1 t\cos\lambda$$

$$+A\cos\omega_1 t\sin\lambda+Q\rho+Q\frac{\mu t}{\tau} \qquad \textbf{(18.6.3)}$$

Dividing equation (18.6.3) by m, we may write:

$$\frac{d^2x}{dt^2}+2n\frac{dx}{dt}+\omega_0^2 x+p+f=a\sin\omega_1 t\cos\lambda$$

$$+a\cos\omega_1 t\sin\lambda+q\rho+\frac{q\mu}{\tau l} \qquad \textbf{(18.6.4)}$$

where n is the damping factor and ω_0 is the natural frequency, while:

$$2n=\frac{C}{m} \qquad \textbf{(18.6.5)}$$

$$\omega_0^2=\frac{K}{m} \qquad \textbf{(18.6.6)}$$

$$p = \frac{P}{m} \qquad\qquad (18.6.7)$$

$$f = \frac{F}{m} \qquad\qquad (18.6.8)$$

$$a = \frac{A}{m} \qquad\qquad (18.6.9)$$

$$q = \frac{Q}{m} \qquad\qquad (18.6.10)$$

Based on Laplace Transform Pairs 3, 4, 5, 6, 7, and 2 from Table 1.1, we convert differential equation of motion (18.6.4) with the initial conditions of motion (18.6.2) from the time domain into the Laplace domain and obtain the resulting algebraic equation of motion in the Laplace domain:

$$l^2 x(l) - l v_0 - l^2 s_0 + 2nlx(l) - 2nls_0 + \omega_0^2\, x(l) + p + f$$

$$= \frac{a\omega_1 l \cos \lambda}{l^2 + \omega_1^2} + \frac{al^2 \sin \lambda}{l^2 + \omega_1^2} + qp + \frac{q\mu}{l\tau} \qquad (18.6.11)$$

Solving equation (18.6.11) for the displacement $x(l)$ in the Laplace domain, we may write:

$$x(l) = \frac{l(v_0 + 2ns_0)}{l^2 + 2nl + \omega_0^2} + \frac{l^2 s_0}{l^2 + 2nl + \omega_0^2} + \frac{a\omega_1 l \cos \lambda}{(l^2 + \omega_1^2)(l^2 + 2nl + \omega_0^2)}$$

$$+ \frac{al^2 \sin \lambda}{(l^2 + \omega_1^2)(l^2 + 2nl + \omega_0^2)} + \frac{qp - p - f}{l^2 + 2nl + \omega_0^2}$$

$$+ \frac{q\mu}{\tau l(l^2 + 2nl + \omega_0^2)} \qquad\qquad (18.6.12)$$

Using algebraic procedures with the denominators in equation (18.6.12), we have:

$$l^2 + 2nl + \omega_0^2 + n^2 - n^2 = (l + n)^2 + \omega^2 \qquad (18.6.13)$$

where:

$$\omega^2 = \omega_0^2 - n^2 \qquad (18.6.14)$$

while ω^2 could be positive, equal to zero, or negative.

All these three cases are considered below.

a. Case $\omega^2 > 0$

Transforming the denominators in equation (18.6.12) according to equation (18.6.13), we write:

$$x(l) = \frac{l(v_0 + 2ns_0)}{(l+n)^2 + \omega^2} + \frac{l^2 s_0}{(l+n)^2 + \omega^2} + \frac{a\omega_1 l \cos\lambda}{(l^2 + \omega_1^2)[(l+n)^2 + \omega^2]}$$

$$+ \frac{al^2 \sin\lambda}{(l^2 + \omega_1^2)[(l+n)^2 + \omega^2]} + \frac{q\rho - p - f}{(l+n)^2 + \omega^2}$$

$$+ \frac{q\mu}{l\tau[(l+n)^2 + \omega^2]} \qquad (18.6.15)$$

Using pairs 1, 24, 27, 42, 43, 22, and 39 from Table 1.1, we invert equation (18.6.15) from the Laplace domain into the time domain and obtain the solution of differential equation of motion (18.6.1) with the initial conditions of motion (18.6.2):

$$x = \frac{1}{\omega}(v_0 + 2ns_0)e^{-nt}\sin\omega t + s_0 e^{-nt}(\cos\omega t - \frac{n}{\omega}\sin\omega t)$$

$$+ \frac{a\omega_1\cos\lambda}{(\omega_1^2 - \omega_0^2)^2 + 4n^2\omega_1^2}\{2n[e^{-nt}(\cos\omega t + \frac{n}{\omega}\sin\omega t) - \cos\omega_1 t]$$

$$-(\omega_1^2 - \omega_0^2)(\frac{1}{\omega_1}\sin\omega_1 t - \frac{1}{\omega}e^{-nt}\sin\omega t)\}$$

$$+ \frac{a\sin\lambda}{(\omega_1^2 - \omega_0^2)^2 + 4n^2\omega_1^2}\{(\omega_1^2 - \omega_0^2)[e^{-nt}(\cos\omega t + \frac{n}{\omega}\sin\omega t)$$

$$- \cos\omega_1 t] + 2n\omega_1(\sin\omega_1 t - \frac{\omega_1}{\omega}e^{-nt}\sin\omega t)\}$$

$$+\frac{q\rho-p-f}{\omega^2+n^2}[1-e^{-nt}(\cos\omega t+\frac{n}{\omega}\sin\omega t)]$$

$$+\frac{q\mu}{\omega\tau(\omega^2+n^2)^2}\{(\omega^2+n^2)\omega t-2n\omega$$

$$-e^{-nt}[(\omega^2-n^2)\sin\omega t-2n\omega\cos\omega t]\}\qquad\text{(18.6.16)}$$

b. Case $\omega^2=0$

Equating ω^2 to zero in equation (18.6.15), we obtain:

$$x(l)=\frac{l(v_0+2ns_0)}{(1+n)^2}+\frac{l^2s_0}{(1+n)^2}+\frac{a\omega_1 l}{(l^2+\omega_1^2)(1+n)^2}\cos\lambda$$

$$+\frac{al^2}{(l^2+\omega_1^2)(1+n)^2}\sin\lambda+\frac{q\rho-p}{(1+n)^2}+\frac{q\mu}{l\tau(1+n)^2}\qquad\text{(18.6.17)}$$

Based on pairs $1, 37, 38, 46, 47, 36$, and 15 from Table 1.1, we invert equation (18.6.17) from the Laplace domain into the time domain and obtain the solution of differential equation of motion (18.6.1) with the initial conditions of motion (18.6.2):

$$x=(v_0+2ns_0)te^{-nt}+s_0(1-nt)e^{-nt}$$

$$+\frac{a\omega_1\cos\lambda}{4n^2\omega_1^2+(\omega_1^2-n^2)^2}\{2n[e^{-nt}(1+nt)-\cos\omega_1 t]$$

$$-(\omega_1^2-n^2)(\frac{1}{\omega_1}\sin\omega_1 t-te^{-nt})\}$$

$$+\frac{a\sin\lambda}{4n^2\omega_1^2+(\omega_1^2-n^2)^2}\{2n\omega_1(\sin\omega_1 t-\omega_1 te^{-nt})$$

$$-(\omega_1^2-n^2)[e^{-nt}(1+nt)-\cos\omega_1 t]\}+\frac{q\rho-p-f}{n^2}[1-e^{-nt}(1+nt)]$$

$$+\frac{q\mu}{\tau n^2}[t-\frac{2}{n}+e^{-nt}\left(\frac{2}{n}+t\right)]\qquad\text{(18.6.18)}$$

c. Case $\omega^2 < 0$

In equation (18.3.13), taking ω^2 with the negative sign, we may write:

$$x(l) = \frac{l(v_0 + 2ns_0)}{(l+n)^2 - \omega^2} + \frac{l^2 s_0}{(l+n)^2 - \omega^2} + \frac{a\omega_1 l}{(l^2 + \omega_1^2)[(l+n)^2 - \omega^2]} \cos \lambda$$

$$+ \frac{al^2}{(l^2 + \omega_1^2)[(l+n)^2 - \omega^2]} \sin \lambda + \frac{q\rho - p - f}{(l+n)^2 - \omega^2}$$

$$+ \frac{q\mu}{l\tau[(l+n)^2 - \omega^2]} \qquad\qquad \textbf{(18.6.19)}$$

Using pairs 1, 25, 28, 48, 49, 23, and 40 from Table 1.1, we invert equation (18.6.19) from the Laplace domain into the time domain and obtain the solution of differential equation of motion (18.6.1) with the initial conditions of motion (18.6.2):

$$x = \frac{v_0 + 2ns_0}{\omega} e^{-nt} \sinh \omega t + \frac{s_0}{\omega} e^{-nt} (\omega \cosh \omega t - n \sinh \omega t)$$

$$+ \frac{a\omega_1 \cos \lambda}{4n^2 \omega_1^2 + (\omega_1^2 - \omega_0^2)^2} \{ e^{-nt} [2n \cosh \omega t$$

$$+ \frac{1}{\omega} (\omega_1^2 + \omega^2 + n^2) \sinh \omega t]$$

$$-2n \cos \omega_1 t - \frac{1}{\omega_1} (\omega_1^2 + \omega^2 - n^2) \sin \omega_1 t \}$$

$$+ \frac{(\omega_1^2 - \omega_0^2) a \sin \lambda}{4n^2 \omega_1^2 + (\omega_1^2 - \omega_0^2)^2} \{ \frac{\omega_1^2}{\omega^2} (1 - \cos \omega_1 t)$$

$$- \frac{\omega_1^2}{n^2 - \omega^2} [1 - e^{-nt} (\cosh \omega t - \frac{n}{\omega} \sinh \omega t)]$$

$$+ \frac{2n\omega_1}{\omega_1^2 - \omega_0^2} (\sin \omega_1 t - \frac{\omega_1}{\omega} e^{-nt} \sinh \omega t) \}$$

$$+ \frac{q\rho - p - f}{n^2 - \omega^2} [1 - e^{-nt} (\cosh \omega t + \frac{n}{\omega} \sinh \omega t)]$$

$$+ \frac{q\mu}{\omega\tau(n^2 - \omega^2)^2} [(n^2 - \omega^2)(\omega t + e^{-nt} \sinh \omega t)$$

$$-2n\omega(1 - \cosh \omega t)] \qquad\qquad \textbf{(18.6.20)}$$

18.6.2 Initial Displacement Equals Zero

The initial conditions of motion are:

$$\text{for} \quad t = 0 \quad x = 0; \quad \frac{dx}{dt} = v_0 \qquad (18.6.21)$$

The solutions of differential equation of motion (18.6.1) with the initial conditions of motion (18.6.21) for the following cases are presented below.

a. Case $\omega^2 > 0$

$$x = \frac{1}{\omega} v_0 e^{-nt} \sin \omega t$$

$$+ \frac{a\omega_1 \cos \lambda}{(\omega_1^2 - \omega_0^2)^2 + 4n^2 \omega_1^2} \{2n[e^{-nt}(\cos \omega t + \frac{n}{\omega} \sin \omega t) - \cos \omega_1 t]$$

$$-(\omega_1^2 - \omega_0^2)(\frac{1}{\omega_1} \sin \omega_1 t - \frac{1}{\omega} e^{-nt} \sin \omega t)\}$$

$$+ \frac{a \sin \lambda}{(\omega_1^2 - \omega_0^2)^2 + 4n^2 \omega_1^2} \{(\omega_1^2 - \omega_0^2)[e^{-nt}(\cos \omega t + \frac{n}{\omega} \sin \omega t)$$

$$- \cos \omega_1 t] + 2n\omega_1 (\sin \omega_1 t - \frac{\omega_1}{\omega} e^{-nt} \sin \omega t)\}$$

$$+ \frac{q\rho - p - f}{\omega^2 + n^2}[1 - e^{-nt}(\cos \omega t + \frac{n}{\omega} \sin \omega t)]$$

$$+ \frac{q\mu}{\omega \tau (\omega^2 + n^2)^2} \{(\omega^2 + n^2)\omega t - 2n\omega$$

$$- e^{-nt}[(\omega^2 - n^2)\sin \omega t - 2n\omega \cos \omega t]\} \qquad (18.6.22)$$

b. Case $\omega^2 = 0$

$$x = v_0 t e^{-nt} + \frac{a\omega_1 \cos \lambda}{4n^2 \omega_1^2 + (\omega_1^2 - n^2)^2} \{2n[e^{-nt}(1 + nt) - \cos \omega_1 t]$$

$$-(\omega_1^2 - n^2)(\frac{1}{\omega_1} \sin \omega_1 t - t e^{-nt})\}$$

$$+\frac{a\sin\lambda}{4n^2\omega_1^2+(\omega_1^2-n^2)^2}\{2n\omega_1(\sin\omega_1 t-\omega_1 te^{-nt})$$

$$-(\omega_1^2-n^2)[e^{-nt}(1+nt)-\cos\omega_1 t]\}$$

$$+\frac{q\rho-p-f}{n^2}[1-e^{-nt}(1+nt)]$$

$$+\frac{q\mu}{\tau n^2}[t-\frac{2}{n}+e^{-nt}\left(\frac{2}{n}+t\right)] \tag{18.6.23}$$

c. Case $\omega^2 < 0$

$$x=\frac{v_0}{\omega}e^{-nt}\sinh\omega t+\frac{a\omega_1\cos\lambda}{4n^2\omega_1^2+(\omega_1^2-\omega_0^2)^2}\{e^{-nt}[2n\cosh\omega t$$

$$+\frac{1}{\omega}(\omega_1^2+\omega^2+n^2)\sinh\omega t]$$

$$-2n\cos\omega_1 t-\frac{1}{\omega_1}(\omega_1^2+\omega^2-n^2)\sin\omega_1 t\}$$

$$+\frac{(\omega_1^2-\omega_0^2)a\sin\lambda}{4n^2\omega_1^2+(\omega_1^2-\omega_0^2)^2}\{\frac{\omega_1^2}{\omega^2}(1-\cos\omega_1 t)$$

$$-\frac{\omega_1^2}{n^2-\omega^2}[1-e^{-nt}(\cosh\omega t-\frac{n}{\omega}\sinh\omega t)]$$

$$+\frac{2n\omega_1}{\omega_1^2-\omega_0^2}(\sin\omega_1 t-\frac{\omega_1}{\omega}e^{-nt}\sinh\omega t)\}$$

$$+\frac{q\rho-p-f}{n^2-\omega^2}[1-e^{-nt}(\cosh\omega t+\frac{n}{\omega}\sinh\omega t)]$$

$$+\frac{q\mu}{\omega\tau(n^2-\omega^2)^2}[(n^2-\omega^2)(\omega t+e^{-nt}\sinh\omega t)$$

$$-2n\omega(1-\cosh\omega t)] \tag{18.6.24}$$

18.6.3 Initial Velocity Equals Zero

The initial conditions of motion are:

$$\text{for}\quad t=0\quad x=s_0;\quad \frac{dx}{dt}=0 \tag{18.6.25}$$

The solutions of differential equation of motion (18.6.1) with the initial conditions of motion (18.6.25) for the following cases are presented below.

a. **Case** $\omega^2 > 0$

$$x = \frac{1}{\omega} 2ns_0 e^{-nt} \sin \omega t + s_0 e^{-nt} (\cos \omega t - \frac{n}{\omega} \sin \omega t)$$

$$+ \frac{a\omega_1 \cos \lambda}{(\omega_1^2 - \omega_0^2)^2 + 4n^2 \omega_1^2} \{2n[e^{-nt}(\cos \omega t + \frac{n}{\omega} \sin \omega t) - \cos \omega_1 t]$$

$$- (\omega_1^2 - \omega_0^2)(\frac{1}{\omega_1} \sin \omega_1 t - \frac{1}{\omega} e^{-nt} \sin \omega t)\}$$

$$+ \frac{a \sin \lambda}{(\omega_1^2 - \omega_0^2)^2 + 4n^2 \omega_1^2} \{(\omega_1^2 - \omega_0^2)[e^{-nt}(\cos \omega t + \frac{n}{\omega} \sin \omega t)$$

$$- \cos \omega_1 t] + 2n\omega_1 (\sin \omega_1 t - \frac{\omega_1}{\omega} e^{-nt} \sin \omega t)\}$$

$$+ \frac{q\rho - p - f}{\omega^2 + n^2} [1 - e^{-nt}(\cos \omega t + \frac{n}{\omega} \sin \omega t)]$$

$$+ \frac{q\mu}{\omega\tau(\omega^2 + n^2)^2} \{(\omega^2 + n^2)\omega t - 2n\omega$$

$$- e^{-nt}[(\omega^2 - n^2)\sin \omega t - 2n\omega \cos \omega t]\} \qquad (18.6.26)$$

b. **Case** $\omega^2 = 0$

$$x = 2ns_0 te^{-nt} + s_0(1 - nt)e^{-nt}$$

$$+ \frac{a\omega_1 \cos \lambda}{4n^2 \omega_1^2 + (\omega_1^2 - n^2)^2} \{2n[e^{-nt}(1 + nt) - \cos \omega_1 t]$$

$$- (\omega_1^2 - n^2)(\frac{1}{\omega_1} \sin \omega_1 t - te^{-nt})\}$$

$$+ \frac{a \sin \lambda}{4n^2 \omega_1^2 + (\omega_1^2 - n^2)^2} \{2n\omega_1 (\sin \omega_1 t - \omega_1 te^{-nt})$$

$$-(\omega_1^2 - n^2)[e^{-nt}(1+nt) - \cos\omega_1 t]\}$$

$$+\frac{q\rho - p - f}{n^2}[1 - e^{-nt}(1+nt)]$$

$$+\frac{q\mu}{\tau n^2}[t - \frac{2}{n} + e^{-nt}\left(\frac{2}{n} + t\right)] \qquad (18.6.27)$$

c. **Case $\omega^2 < 0$**

$$x = \frac{2ns_0}{\omega}e^{-nt}\sinh\omega t + \frac{s_0}{\omega}e^{-nt}(\omega\cosh\omega t - n\sinh\omega t)$$

$$+\frac{a\omega_1\cos\lambda}{4n^2\omega_1^2 + (\omega_1^2 - \omega_0^2)^2}\{e^{-nt}[2n\cosh\omega t$$

$$+\frac{1}{\omega}(\omega_1^2 + \omega^2 + n^2)\sinh\omega t]$$

$$-2n\cos\omega_1 t - \frac{1}{\omega_1}(\omega_1^2 + \omega^2 - n^2)\sin\omega_1 t\}$$

$$+\frac{(\omega_1^2 - \omega_0^2)a\sin\lambda}{4n^2\omega_1^2 + (\omega_1^2 - \omega_0^2)^2}\{\frac{\omega_1^2}{\omega^2}(1 - \cos\omega_1 t)$$

$$-\frac{\omega_1^2}{n^2 - \omega^2}[1 - e^{-nt}(\cosh\omega t - \frac{n}{\omega}\sinh\omega t)]$$

$$+\frac{2n\omega_1}{\omega_1^2 - \omega_0^2}(\sin\omega_1 t - \frac{\omega_1}{\omega}e^{-nt}\sinh\omega t)\}$$

$$+\frac{q\rho - p - f}{n^2 - \omega^2}[1 - e^{-nt}(\cosh\omega t + \frac{n}{\omega}\sinh\omega t)]$$

$$+\frac{q\mu}{\omega\tau(n^2 - \omega^2)^2}[(n^2 - \omega^2)(\omega t + e^{-nt}\sinh\omega t)$$

$$-2n\omega(1 - \cosh\omega t)] \qquad (18.6.28)$$

18.6.4 Both the Initial Displacement and Velocity Equal Zero

The initial conditions of motion are:

$$\text{for}\quad t = 0\quad x = 0;\quad \frac{dx}{dt} = 0 \qquad (18.6.29)$$

The solutions of differential equation of motion (18.6.1) with the initial conditions of motion (18.6.29) for the following cases are presented below.

a. Case $\omega^2 > 0$

$$x = \frac{a\omega_1 \cos\lambda}{(\omega_1^2 - \omega_0^2)^2 + 4n^2\omega_1^2}\{2n[e^{-nt}(\cos\omega t + \frac{n}{\omega}\sin\omega t) - \cos\omega_1 t]$$

$$-(\omega_1^2 - \omega_0^2)(\frac{1}{\omega_1}\sin\omega_1 t - \frac{1}{\omega}e^{-nt}\sin\omega t)\}$$

$$+\frac{a\sin\lambda}{(\omega_1^2 - \omega_0^2)^2 + 4n^2\omega_1^2}\{(\omega_1^2 - \omega_0^2)[e^{-nt}(\cos\omega t + \frac{n}{\omega}\sin\omega t)$$

$$-\cos\omega_1 t] + 2n\omega_1(\sin\omega_1 t - \frac{\omega_1}{\omega}e^{-nt}\sin\omega t)\}$$

$$+\frac{q\rho - p - f}{\omega^2 + n^2}[1 - e^{-nt}(\cos\omega t + \frac{n}{\omega}\sin\omega t)]$$

$$+\frac{q\mu}{\omega\tau(\omega^2 + n^2)^2}\{(\omega^2 + n^2)\omega t - 2n\omega$$

$$-e^{-nt}[(\omega^2 - n^2)\sin\omega t - 2n\omega\cos\omega t]\} \qquad (18.6.30)$$

b. Case $\omega^2 = 0$

$$x = \frac{a\omega_1 \cos\lambda}{4n^2\omega_1^2 + (\omega_1^2 - n^2)^2}\{2n\left[e^{-nt}(1 + nt) - \cos\omega_1 t\right]$$

$$-(\omega_1^2 - n^2)(\frac{1}{\omega_1}\sin\omega_1 t - te^{-nt})\}$$

$$+\frac{a\sin\lambda}{4n^2\omega_1^2 + (\omega_1^2 - n^2)^2}\{2n\omega_1(\sin\omega_1 t - \omega_1 te^{-nt})$$

$$-(\omega_1^2 - n^2)[e^{-nt}(1 + nt) - \cos\omega_1 t]\}$$

$$+\frac{q\rho - p - f}{n^2}[1 - e^{-nt}(1 + nt)] + \frac{q\mu}{\tau n^2}[t - \frac{2}{n} + e^{-nt}\left(\frac{2}{n} + t\right)]$$

$$ \qquad (18.6.31)$$

c. Case $\omega^2 < 0$

$$x = \frac{a\omega_1 \cos \lambda}{4n^2\omega_1^2 + (\omega_1^2 - \omega_0^2)^2} \{e^{-nt}[2n \cosh \omega t$$

$$+ \frac{1}{\omega}(\omega_1^2 + \omega^2 + n^2)\sinh \omega t]$$

$$-2n\cos\omega_1 t - \frac{1}{\omega_1}(\omega_1^2 + \omega^2 - n^2)\sin\omega_1 t\}$$

$$+ \frac{(\omega_1^2 - \omega_0^2)a\sin\lambda}{4n^2\omega_1^2 + (\omega_1^2 - \omega_0^2)^2} \{\frac{\omega_1^2}{\omega^2}(1 - \cos\omega_1 t)$$

$$- \frac{\omega_1^2}{n^2 - \omega^2}[1 - e^{-nt}(\cosh\omega t - \frac{n}{\omega}\sinh\omega t)]$$

$$+ \frac{2n\omega_1}{\omega_1^2 - \omega_0^2}(\sin\omega_1 t - \frac{\omega_1}{\omega}e^{-nt}\sinh\omega t)\}$$

$$+ \frac{q\rho - p - f}{n^2 - \omega^2}[1 - e^{-nt}(\cosh\omega t + \frac{n}{\omega}\sinh\omega t)]$$

$$+ \frac{q\mu}{\omega\tau(n^2 - \omega^2)^2}[(n^2 - \omega^2)(\omega t + e^{-nt}\sinh\omega t)$$

$$-2n\omega(1 - \cosh\omega t)] \tag{18.6.32}$$

19

EXAMPLES OF TWO-DIMENSIONAL MOTION

The trajectory of a body's two-dimensional motion represents a flat curved line having a variable radius of curvature. The analytical description of this motion is based on a system of two simultaneous differential equations of motion. These equations describe the rectilinear motion of a body in two mutually perpendicular directions on a coordinate plane. If the curved line has a constant radius of curvature, we have a case of rotational motion that is completely determined by one equation of motion, the solution of which represents the angular displacement as a function of time.

Thus, in order to investigate an engineering problem associated with a two-dimensional motion, we must compose a system of two simultaneous differential equations of motion. One of these equations describes the rectilinear motion in the horizontal direction (x), while the second describes the motion in the vertical direction (y). As it turns out, these equations and their solutions can be found in Chapters 3 through 18. Therefore, for a problem associated with the two-dimensional motion, we must identify in Guiding Table 2.1 the two corresponding sections that contain the desired solutions.

The purpose of this chapter is to demonstrate a few examples, using Guiding Table 2.1, to solve engineering problems related to two-dimensional motion.

19.1 Projectile

At the moment when the projectile leaves the weapon's barrel, it possesses a certain initial velocity directed at a certain angle to the horizon. During its flight, the projectile is subjected to the action of the force of inertia, the air resistance, and the force of gravity. We assume that the damping coefficient C of the air resistance has a constant value and does not depend on the direction of the motion of the projectile or of its altitude. We want to determine the basic parameters of the projectile's motion.

In Figure 19.1, the curved line OA represents the initial part of the projectile's trajectory, where the vector v_0 represents the projectile's initial velocity. This vector is tangent to the curve OA at the origin O and shows the magnitude and the direction of the projectile's initial velocity. In this figure, α represents the angle between the velocity vector and the horizontal axis x. The velocity vector is resolved into its horizontal component v_{0x} and vertical component v_{0y}.

According to the diagram in Figure 19.1, we determine the values of the initial velocity in the horizontal and vertical directions respectively:

$$v_{0x} = v_0 \cos \alpha \qquad (19.1.1)$$

$$v_{0y} = v_0 \sin \alpha \qquad (19.1.2)$$

The movement of the projectile could be described by a system of two simultaneous differential equations of motion, one of which describes the displacement in the horizontal (x) direction, and the other in the vertical (y) direction.

During the horizontal motion, the projectile is subjected to the force of inertia and damping force as the resisting forces, while the active force equals zero. Referring to Guiding Table 2.1, we find

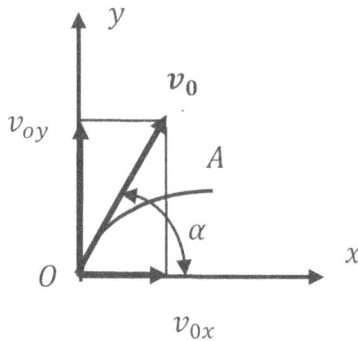

Figure 19.1 Diagram of the initial velocity vector and its components

that section 11.1 contains the description of an identical problem. Figure 11.1 represents the model of the system moving in the horizontal direction, while differential equation of motion (11.1.1) describes the motion of the system. The initial conditions of motion are similar to the initial conditions of motion (11.1.15) in which the initial velocity v_0 should be replaced by v_{0x}. The same replacement should be made in the solution of differential equation of motion (11.1.16). Thus, we obtain:

$$x = \frac{v_{0x}}{2n}(1 - e^{-2nt})$$ (19.1.3)

The same replacements should be performed in equations (11.1.9) and (11.1.10), and we determine the velocity and acceleration of the projectile respectively:

$$\frac{dx}{dt} = v_{0x}e^{-2nt}$$ (19.1.4)

$$\frac{d^2x}{dt^2} = -2nv_0xe^{-2nt}$$ (19.1.5)

Now let us consider the movement of the projectile in the vertical direction. In the upward motion, the projectile is subjected to the force of inertia, the damping force, and the force of gravity as the resisting forces, while the active force equals zero. Obviously, the force of gravity has a constant value. Actually, as seen in Guiding Table 2.1, the solution of a similar problem is presented in section 13.1. In Figure 13.1, replacing the x-coordinate by the y-coordinate, and the constant resisting force P by the force of gravity W, and also revolving Figure 13.1 by 90 degrees, we obtain the model of the system for the current case. Performing similar replacements in differential equation of motion (13.1.1), we obtain the differential equation of motion of the projectile in the vertical direction:

$$m\frac{d^2y}{dt^2} + C\frac{dy}{dt} + W = 0 \qquad (19.1.6)$$

Similar replacements should be made in the initial conditions of motion (13.1.21) and in equations (13.1.22), (13.1.10), and (13.1.11) that represent the displacement, the velocity, and the acceleration respectively. Thus, we obtain:

$$y = \frac{1}{2n}[v_{0y} + \frac{g}{2n} - gt - \left(v_{0y} + \frac{g}{2n} \right)e^{-2nt}] \qquad (19.1.7)$$

$$\frac{dy}{dt} = (v_{0y} + \frac{g}{2n})e^{-2nt} - \frac{g}{2n} \qquad (19.1.8)$$

$$\frac{d^2y}{dt^2} = -2n(v_{0y} + \frac{g}{2n})e^{-2nt} \qquad (19.1.9)$$

where g is the acceleration of gravity and:

$$g = \frac{W}{m} \qquad (19.1.10)$$

19.2 Ground-to-Air Missile

In this example, we want to determine the basic parameters of motion of a ground-to-air missile that is powered by a rocket engine, developing a thrust force of a constant value. The missile is launched in the direction that forms an angle β with the horizon. We

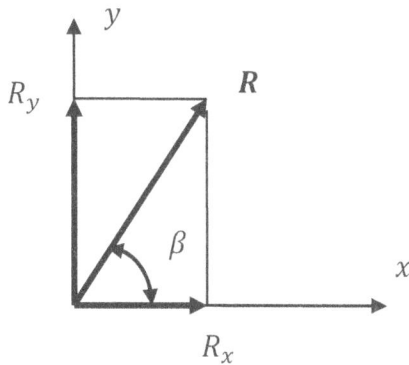

Figure 19.2 Diagram of the thrust force vector and its components

assume that the air resistance does not depend on the direction of motion of the missile or on its altitude.

The movement of the missile represents a two-dimensional motion the description of which involves a system of two simultaneous differential equations of motion. One of these equations describes the motion in the horizontal (x) direction, while the other is in the vertical (y) direction.

Figure 19.2 shows the vector diagram of the thrust force R as well as its horizontal R_x and vertical R_y components. According to the diagram these components are:

$$R_x = R\cos\beta \qquad \textbf{(19.2.1)}$$

$$R_y = R\sin\beta \qquad \textbf{(19.2.2)}$$

During the horizontal motion, the missile experiences the force of inertia and the damping force as the resisting forces, and the action of the horizontal component of the thrust force as the active force. Guiding Table 2.1 shows that a similar problem is described in section 11.2. In this section, Figure 11.2 represents the model of the system, while differential equation of motion (11.2.1) describes the motion of the missile in the horizontal direction. In this equation, the constant active force R should be replaced by the force R_x. The solution of this differential equation with the initial conditions of motion (11.2.23) is presented by equation (11.2.24) in which we replace r by r_x. Thus, we may write:

$$x = \frac{1}{2n}(r_x t + \frac{r_x}{2n}e^{-2nt} - \frac{r_x}{2n}) \tag{19.2.3}$$

where:

$$r_x = \frac{R_x}{m} \tag{19.2.4}$$

The same replacements should be performed in equations (11.2.21) and (11.2.24) that represent the velocity and the acceleration respectively. Therefore, the equations for the velocity and acceleration in the horizontal direction of the missile's motion respectively read:

$$\frac{dx}{dt} = \frac{r_x}{2n}(1 - e^{-2nt}) \tag{19.2.5}$$

$$\frac{d^2 x}{dt^2} = r_x e^{-2nt} \tag{19.2.6}$$

The motion of the missile in the vertical direction is characterized by the force of inertia, the damping force, and the force of gravity as the resisting forces, and by the vertical thrust component as the active force. Referring to Guiding Table 2.1, we can see that a similar problem is described in section 13.2. In Figure 13.2, replacing the x-coordinate by the y-coordinate, and the constant resisting force P by the force of gravity W, and also the constant active force R by the vertical component of the thrust force R_y, and also revolving Figure 13.2 by 90 degrees, we obtain the model of the system for the current case.

Substituting the appropriate replacements in equation (13.2.1), we obtain the differential equation of motion of the missile in the vertical direction:

$$m\frac{d^2 y}{dt^2} + C\frac{dy}{dt} + W = R_y \tag{19.2.7}$$

Performing similar substitutions in the initial conditions of motion (13.2.27), we write:

$$\text{for} \quad t = 0 \quad y = 0; \quad \frac{dy}{dt} = 0 \qquad \textbf{(19.2.8)}$$

We obtain the solution of differential equation of motion (19.2.7) with the initial conditions of motion (19.2.9) by making the appropriate replacements in equation (13.2.28). Thus, we have:

$$y = \frac{1}{2n}[(r_y - g)t + \frac{r_y - g}{2n}e^{-2nt} - \frac{r_y - g}{2n}] \qquad \textbf{(19.2.9)}$$

where:

$$r_y = \frac{R_y}{m} \qquad \textbf{(19.2.10)}$$

$$g = \frac{W}{m} \qquad \textbf{(19.2.11)}$$

while g is the acceleration of gravity.

Performing the appropriate replacements in equations (13.2.23) and (13.2.24), we determine the velocity and acceleration of the missile respectively:

$$\frac{dy}{dt} = \frac{r_y - g}{2n}(1 - e^{-2nt}) \qquad \textbf{(19.2.12)}$$

$$\frac{d^2y}{dt^2} = (r_y - p)e^{-2nt} \qquad \textbf{(19.2.13)}$$

INDEX

Note: *f* in italic page numbers indicates a figure; *t* indicates a table

693

and constant active force and harmonic
force, 166–171, *166f*
and harmonic force, 156–160, *156f*
and harmonic force and time-
dependent force, 171–176, *172f*
and time-dependent force, 160–166, *161f*
constant value, 3, 4, 35
constant velocity, 2

D
damping coefficient
braking systems, 35, *35f*
and damping, 11
in general differential equation of
motion, 45
introduction, 9–10
notation, 53
projectiles, 686
water vessel acceleration, 39, 40
damping force, 343–373
active force equals zero, 344–347, *344f*
braking systems, 35, *35*, 36
and constant active force, 347–352, *347f*
and constant active force and harmonic
force, 362–367, *362f*
definition, 9
in differential equation of motion, 9,
35, 36, 45
and harmonic force, 352–357, *352f*
and harmonic force and time-
dependent force, 367–373, *368f*
introduction, 9–11, *9f*
as loading factor characteristic, 5
missiles, 689, 690
multiple, 10–11
notation, 53
projectiles, 687
in schematic images, 10
and time-dependent force, 357–362, *357f*
and velocity, 9, *9f*, 10
viscoelastic media, 13–14
viscoelastoplastic media, 13–14, 18
water vessel acceleration, 39, 40, *40f*

damping force and constant resistance,
409–442
active force equals zero, 410–414, *411f*
and constant active force, 414–420,
415f
and constant active force and harmonic
force, 431–437, *432f*
and harmonic force, 420–426, *421f*
and harmonic force and time-
dependent force, 437–442, *438f*
and time-dependent force, 426–431,
427f
damping force, constant resistance, and
friction, 443–475
active force equals zero, 444–447, *445f*
constant force, 448–453, *449f*
and constant force and harmonic force,
464–469, *465f*
and harmonic force, 453–459, *454f*
and harmonic force and time-
dependent force, 470–475, *470f*
and time-dependent force, 459–464,
460f
damping force and friction, 375–408
active force equals zero, 375–380, *376f*
and constant active force and harmonic
force, 396–402, *397f*
and constant force, 380–386, *381f*
and harmonic force, 386–391, *386f*
and harmonic force and time-
dependent force, 402–408, *403f*
and time-dependent force, 392–396,
392f
damping force and stiffness, 477–528
active force equals zero, 478–485, *479f*
and constant active force, 485–491, *486f*
constant force and harmonic force,
507–517, *508f*
and harmonic force, 491–500, *491f*
and harmonic force and time-
dependent force, 517–528, *518f*
and time-dependent force, 500–507,
501f

www.ingramcontent.com/pod-product-compliance
Lightning Source LLC
Chambersburg PA
CBHW060514220326
41598CB00025B/3661